W9-CHJ-728

TOPOLOGY

by

JOHN G. HOCKING

Michigan State University

and

GAIL S. YOUNG

Tulane University

ADDISON-WESLEY PUBLISHING COMPANY, INC.

READING, MASSACHUSETTS, U.S.A.

LONDON, ENGLAND

This book is in the
ADDISON-WESLEY SERIES IN MATHEMATICS

Printed in the United States of America

Library of Congress Catalog Card No. 61–5023

PREFACE

We have designed this book as a text for a one-year first course in topology. As we have used it ourselves, the first four chapters cover the material presented in the first semester, and the second semester is taken from the last four chapters. The core of the first half-year has been the following: Sections 1–1 through 1–12, Sections 2–1 through 2–10, Sections 3–1 through 3–7, and Sections 4–1 through 4–6. The second half-year usually continues with Sections 4–7 through 4–10, Chapter 5, Chapter 6, Sections 8–1 through 8–3, and as much as possible of the remaining material.

Many important topics in topology are passed over with no more than a brief mention. It is hoped that such mention will stimulate the reader to follow the indicated paths to new ideas. Also, these digressions are planned so as to give a broad, if ill-defined, framework upon which the reader may build as he progresses. This is part of our deliberate intent to leave the book "open" at the far end. That is, while we present the beginnings of topology, we point out that there is much beyond the confines of this book. And in many instances we attempt to show the direction in which further material may be found.

A few words about prerequisites for a successful study of this book are in order. It has been our experience that the student should have had the elements of set theory and should have had, or be studying concurrently, courses in group theory and the theory of functions. A knowledge of higher geometry is also desirable. It is not that this much "mathematical maturity" is necessary, although it is decidedly advantageous. But we do draw examples and motivation from analysis and geometry and depend heavily upon group-theoretic results in the later chapters.

We use the conventional square brackets in referring to the numbered bibliography. Thus, for instance, [14] refers to item number fourteen in the bibliography. For purposes of internal reference, theorems, lemmas, and corollaries are numbered consecutively within each chapter. A reference such as Theorem 4–11 refers to the eleventh theorem of Chapter 4. The exercises are also numbered consecutively within each chapter. Sections are numbered, too; Section 2–11 refers to Section 11 of Chapter 2, for example.

We have adopted two conventions that should be mentioned. An equivalence class is always denoted by special heavy brackets; $[f]$ denotes the equivalence class of which f is a representative. The other convention is a modification of a space-saving device due to Halmos. We use a hollow square, $\square$, to indicate the end of the proof of a theorem.

It would be difficult to properly acknowledge the assistance and cooperation we have had from colleagues and students. We have profited by discussions with Professors Raoul Bott, E. E. Moise, Hans Samelson, and R. L. Wilder of the University of Michigan and with Professor L. M. Kelly of Michigan State University. We owe much to Professors Kelly and P. H. Doyle of Michigan State University, who read and criticized parts of the manuscript. Careful study and criticism of parts of the manuscript was also undertaken by Mr. James Kiesler of Louisiana State University and Dr. Delia Koo of Michigan State University. Mr. T. S. Wu and Mr. Y. F. Lin of Tulane University gave valuable help with the proofs.

J. G. H.

G. S. Y.

A NOTE ON SET-THEORETIC CONCEPTS

Throughout this book, and all of mathematics, certain ideas from the theory of sets are fundamental. We review these basic concepts here and introduce notation that will be used in our subsequent developments. The reader who is unfamiliar with set theory is advised to consult a standard work on the foundations of mathematics, for example Wilder [43].

In set theory, there are two undefined or primitive concepts. The first of these is *set* itself. This word is used more or less as a synonym for such words as *collection, class, family, system,* or *aggregate.* The second primitive concept is the phrase *is an element of* as used in the statement "x is an element of the set S." We also use such synonyms as "x is in S," "x belongs to S," "x lies in S," etc.

A set U is a *subset* of a set S if every element of U is also an element of S. Then U is a *proper subset* of S if there are elements of S that are not elements of U but not conversely. Given a subset U of the set S, the *complement* of U in S, denoted by $S - U$, is the set of all elements of S which are not elements of U. If U is a subset but not a proper subset of S, then $S - U$ contains no elements. The set containing no elements is called the *empty* set and is denoted by $\emptyset$. The definition of a subset implies that $\emptyset$ is a subset of every set.

Two sets S and T are identical, $S = T$, if every element of S is an element of T and conversely. This is equivalent to saying that S is a subset of T and that T is a subset of S.

Given two sets S and T, two new sets may be formed. The *union* $S \cup T$ of S and T is the set composed of all elements that are in S *or* are in T. The *intersection* $S \cap T$ is the set of all elements that are in S *and* are in T. The two sets S and T are said to be *disjoint* if $S \cap T = \emptyset$. These definitions provide two binary operations $\cup$ and $\cap$ on sets, and these operations satisfy certain basic logical equations. Some of these are as follows: Let A, B, and C be any sets, then the commutative laws are

$$A \cup B = B \cup A \quad \text{and} \quad A \cap B = B \cap A, \tag{1}$$

the associative laws are

$$A \cup (B \cup C) = (A \cup B) \cup C \quad \text{and} \quad A \cap (B \cap C) = (A \cap B) \cap C, \tag{2}$$

and the distributive laws are

$$A \cup (B \cap C) = (A \cup B) \cap (A \cup C)$$

and

$$A \cap (B \cup C) = (A \cap B) \cup (A \cap C). \tag{3}$$

If U and V are subsets of a set S, then

$$S - (S - U) = U, \tag{4}$$

$$U - (U - V) = U \cap V, \tag{5}$$

and, de Morgan's laws,

$$S - (A \cup B) = (S - A) \cap (S - B)$$

and

$$S - (A \cap B) = (S - A) \cup (S - B). \tag{6}$$

These and other properties will be used more or less implicitly as the need arises. The reader may prove these laws himself or may consult Wilder and the references given there. In Section 1–9 we also mention another important concept in set theory, the axiom of choice and its logical equivalents.

CONTENTS

(*Starred sections may be omitted without loss of continuity*)

CHAPTER 1

TOPOLOGICAL SPACES AND FUNCTIONS

1–1 Introduction. Topology may be considered as an abstract study of the limit-point concept. As such, it stems in part from a recognition of the fact that many important mathematical topics depend entirely upon the properties of limit points. The very definition of a continuous function is an example of this dependence. Another example is the precise meaning of the connectedness of a geometric figure. To exaggerate, one might view topology as the complement of modern algebra in that together they cover the two fundamental types of operations found in mathematics.

In applying the unifying principle of abstraction, we study concrete examples and try to isolate the basic properties upon which the interesting phenomena depend. In the final analysis, of course, the determination of the "correct" properties to be abstracted is largely an experimental process. For instance, although the limit of a sequence of real numbers is a widely used idea, experience has shown that a more basic concept is that of a limit point of a *set* of real numbers.

DEFINITION 1–1. The real number p is a *limit point of a set X of real numbers* provided that for every positive number ϵ, there is an element x of the set X such that $0 < |p - x| < \epsilon$.

As an example, let X consist of all real numbers of the two forms $1/n$ and $(n - 1)/n$, where n is an integer greater than 2. Then 0 and 1 are the only limit points of X. Thus a limit point of a set need not belong to that set. On the other hand, every real number is a limit point of the set of all rational numbers, indicating that a set may have limit points belonging to itself.

Some terminology is needed before we pursue this abstraction further. Let S be any set of elements. These may be such mathematical entities as points in the Euclidean plane, curves in a given class, infinite sequences of real numbers, elements of an algebraic group, etc., but in general we take S to be an abstract undefined set. To reflect the geometric content of topology, we refer to the elements of S by the generic name *point*. We may now name our fundamental structure.

DEFINITION 1–2. The set S *has a topology* (or *is topologized*) provided that, for every point p in S and every subset X of S, the question "Is p a limit point of X?" can be answered.

This definition is so extremely general as to be almost useless in practice. There is nothing in it to impose certain desirable properties upon the limit-point relation (more on this point shortly), and also nothing in it indicates the means whereby the pertinent question can be answered. An economical method of accomplishing the latter is to adopt some rule or test whose application will answer the question in every case. For the set of real numbers, Definition 1–1 serves this purpose and hence defines a topology for the real numbers. [The use of the word *topology* here differs from its use as the name of a subject. Loosely speaking, topology (the subject) is the study of topologies (as in Definition 1–2).]

A set S may be assigned many different topologies, but there are two extremes. For the first, we always answer the question in Definition 1–2 in the affirmative; that is, every point is a limit point of every subset. This yields a worthless topology: there are simply too many limit points! For the other extreme, we assume that the answer is always "no," that is, no point is a limit point of any set. The resulting topology is called *the discrete topology* for S. The very fact that it is dignified with a name would indicate that this extreme is not quite so useless as the first.

Those factors that dictate the choice of a topology for a given set S should become more apparent as we progress. In many cases, a "natural" topology exists, a topology agreeing with our intuitive idea of what a limit point should be. Definition 1–1 furnishes such a topology for the real numbers, for instance. In general, however, we require only a structure within the set S which will define *limit point* in a simple manner and in such a way that certain basic relations concerning limit points are maintained. To illustrate this latter requirement, it is intuitively evident that if p is a limit point of a subset X and X is contained in another subset Y, then we would want p to be also a limit point of Y. There are many such structures one may impose upon a set and we will develop the more commonly used topologies in this chapter. Before doing this, however, we continue our preliminary discussion with a few general remarks upon the aims and tools of topology.

The study of topologized sets (or any other abstract system) involves two broad and interrelated questions. The first of these concerns the investigation and classification of the various concrete realizations, or models, which we may encounter. This entails the recognition of equivalent models, as is done for isomorphic groups or congruent geometric figures, for example. In turn, this equivalence of models is usually defined in terms of a one-to-one reversible transformation of one model onto another. This equivalence transformation is so chosen as to leave invariant the fundamental properties of the models. As examples, we have the rigid motions in geometry, the isomorphisms in group theory, etc.

One of the first to perceive the importance of these underlying transformations was Felix Klein. In his famous Erlanger Program (1870), he characterized the various geometries in terms of these basic transformations. For instance, we may define Euclidean geometry as the study of those properties of geometric figures that are invariant under the group of rigid motions.

Insofar as topology is an abstract form of geometry and fits into the Klein Erlanger Program, its basic transformations are the homeomorphisms (which we will define shortly).

The second broad question in studying an abstract system such as our topologized sets involves consideration of transformations more general than the one-to-one equivalence transformation. The requirement that the transformation be one-to-one and reversible is dropped and we retain only the requirement that the basic structure is to be preserved. The homomorphisms in group theory illustrate this situation. In topology, the corresponding transformations are those that preserve limit points. Such a transformation is said to be *continuous* and is a true generalization of the continuous functions used in analysis. It follows that second aspect of topology finds many applications in function theory.

Since we are to be dealing with very general sets, we must give precise meaning to the word *transformation.*

DEFINITION 1–3. Given two sets X and Y, a *transformation* (also called a *function* or a *mapping*) $f:X \rightarrow Y$ of X into Y is a triple (X, Y, G), where G itself is a collection of ordered pairs (x, y), the first element of each pair being an element of X, and the second an element of Y, with the condition that each element of X appears as the first element of exactly one pair in G.

If each element of Y appears as the second element of some pair in G, then the transformation f is said to be *onto.*

If each element of Y which appears at all, appears as the second element of exactly one pair in G, then f is said to be *one-to-one*. Note that a transformation can be onto without being one-to-one and conversely.

As an aid in understanding Definition 1–3, consider the equation $y = x^2$, x a real number. We may take X to be the set of all real numbers and then the collection G is the set of pairs (x, x^2). From this alone, we cannot determine the set Y, however. Certainly Y must contain all nonnegative real numbers since each such number appears as the second element of at least one pair (x, x^2). Taking Y to be just the set of nonnegative reals will cause f to be onto. But if Y is all real numbers, or all reals greater than -7, or any other set containing the nonnegative reals as a proper subset, the transformation is not onto. With each new choice of Y, we change the triple and hence the transformation.

Continuing with the same example, we could assume that X is the set of nonnegative reals also. Then the transformation is one-to-one, as is easily seen. Depending upon the choice of Y, the transformation may or may not be onto, of course. Thus we see that we have stated explicitly the conditions usually left implicit in defining a function in elementary analysis. The reader will find that the seemingly pedantic distinctions made here are really quite necessary.

If $f:X \to Y$ is a transformation of X into Y and x is an element of the set X, then we let $f(x)$ denote the second element of the pair in G whose first element is x. That is, $f(x)$ is the "functional value" in Y of the point x. Similarly, if Z is a subset of X, then $f(Z)$ denotes that subset of Y composed of all points $f(z)$, where z is a point in Z. If y is a point of Y, then by $f^{-1}(y)$ is meant the set of all points x in X for which $f(x) = y$; and if W is a subset of Y, then $f^{-1}(W)$ is the set-theoretic union of the sets $f^{-1}(w)$, w in W. Note that f^{-1} can be used as a symbol to denote the triple (Y, X, G'), where G' consists of all pairs (y, x) that are reversals of pairs in G. But *f^{-1} is a transformation only if f is both one-to-one and onto.* If A is a subset of X and if $f:X \to Y$, then f may be restricted to A to yield a transformation denoted by $f|A: A \to Y$, and called the *restriction* of f to A.

We can now define the transformations that underlie the study of topology. Let S and T be topologized sets. A *homeomorphism* of S onto T is a one-to-one transformation $f:S \to T$ which is onto, and such that a point p is a limit point of a subset X of S if and only if $f(p)$ is a limit point of $f(X)$. This last condition means that a homeomorphism preserves limit points, a condition that is certainly natural enough if we expect to study limit points. Note that since a homeomorphism f is both one-to-one and onto, its inverse f^{-1} is also a transformation. Furthermore the "if and only if" condition implies that f^{-1} is also a homeomorphism $f^{-1}:T \to S$.

One might consider the homeomorphism as the analogue of an isomorphism in algebra, or a conformal mapping in analysis, or a rigid motion in geometry. The less restricted class of continuous transformations mentioned earlier are then analogous to the homomorphisms in algebra, or analytic functions in analysis, or projections onto a lower-dimensional subspace in geometry. A transformation $f:S \to T$ is *continuous* provided that if p is a limit point of a subset X of S, then $f(p)$ is a limit point or a point of $f(X)$.

By introducing a new symbol, we can express continuity more concisely. If X is a subset of the topologized set S, we let $\overline{X}$ denote the set-theoretic union of X and all its limit points and call $\overline{X}$ the *closure* of X. The continuity requirement on f then may be expressed by assuming that if p is a point of $\overline{X}$, then $f(p)$ is a point of $\overline{f(X)}$.

EXERCISE 1–1. Show that if S is a set with the discrete topology and $f:S \to T$ is any transformation of S into a topologized set T, then f is continuous.

EXERCISE 1–2. A real-valued function $y = f(x)$ defined on an interval $[a, b]$ is continuous provided that if $a \leq x_0 \leq b$ and $\epsilon > 0$, then there is a number

$\delta > 0$ such that if $|x - x_0| < \delta$, x in $[a, b]$, then $|f(x) - f(x_0)| < \epsilon$. Show that this is equivalent to our definition, using Definition 1–1.

1–2 Topological spaces. In attempting to formulate a rule to use in answering the pertinent question in Definition 1–2, we should be guided by the properties of limit points and their relationships as found in analysis, where this abstraction began. For instance, we would not welcome a situation in which a point p is a limit point of the set of limit points of a set X and yet p is not a limit point of X itself. The structure we present first to accomplish our aims is widely adopted.

Consider a set S. Let $\{O_\alpha\}$ be a collection of subsets of S, called *open sets*, satisfying the following axioms:

O_1. The union of any number of open sets is an open set.
O_2. The intersection of a finite number of open sets is an open set.
O_3. Both S and the empty set $\emptyset$ are open.

With such a collection $\{O_\alpha\}$ we now determine the limit points of a subset as follows. A point p is a *limit point of a subset* X of S provided that every open set containing p also contains a point of X distinct from p. This definition yields a topology for S and, with such a topology, S is called a *topological space.*

Note that not every set with a topology is a topological space. If S is a topologized set, then for S to be a topological space, it must be possible to obtain the given topology by selecting certain subsets of S as open sets satisfying O_1, O_2, and O_3 and to recover the given limit-point relations, using these open sets.

We now suppose that we have a topological space S with open sets $\{O_\alpha\}$. We define a subset X of S to be *closed* if $S - X$ is open.

THEOREM 1–1. If X is any subset of S, then X is closed if and only if $X = \overline{X}$.

Proof: Suppose $X = \overline{X}$. Then no point of $S - X$ is a point or a limit point of X. About each point p in $S - X$, then, there is an open set O_p containing no point of X. By Axiom O_1, the union of all the sets O_p, p in $S - X$, is an open set. Clearly this union is $S - X$.

Conversely, if X is closed, then $S - X$ is open. If p is any point of $S - X$, then $S - X$ itself is an open set containing p but no point of X. Hence no point of $S - X$ can be a limit point of X. □

THEOREM 1–2. The closed subsets $\{C_\alpha\}$ of a topological space S satisfy the following properties:

C_1. The intersection of any number of closed sets is closed.
C_2. The union of a finite number of closed sets is closed.
C_3. Both S and the empty set $\emptyset$ are closed.

Proof: Let $\{C'_\beta\}$ be any subcollection of $\{C_\alpha\}$. For each set C'_β, $S - C'_\beta$ is open. Hence by O_1, $\cup_\beta(S - C'_\beta)$ is open and therefore $S - \cup_\beta(S - C'_\beta)$ is closed. But $S - \cup_\beta(S - C'_\beta) = \cap_\beta C'_\beta$ by de Morgan's law. This proves C_1.

Let $C'_1, \ldots, C'_n$ be any finite subcollection of $\{C_\alpha\}$. Then each $S - C'_i$ is open and by O_2, $\cap_{i=1}^n (S - C'_i)$ is open. Again applying de Morgan's law, we see that $\cap_{i=1}^n (S - C'_i) = S - \cup_{i=1}^n C'_i$. Thus $\cup_{i=1}^n C_i$ is closed, proving C_2. Property C_3 follows immediately from O_3. □

This result is actually a theorem in pure set theory, not in topology. It depends only upon de Morgan's law, which asserts that if S is any set and $\{X_\alpha\}$ is any collection of subsets of S, then $\cap_\alpha X_\alpha = S - \cup_\alpha(S - X_\alpha)$. A proof of this property is rather easy and is available in any treatise on the theory of sets. For example, see Fraenkel [8].

We might point out the obvious formal duality between Properties C_1, C_2, C_3 and Axioms O_1, O_2, O_3. One may always pass from true statements about open sets to true statements about closed sets by interchanging "open set" with "closed set" and "union" with "intersection" throughout. This would be much too formal an approach, however, and defining a topological space via its closed sets lacks certain advantages which we will bring out in the next section.

1–3 Basis and subbasis of a topology. One justification for considering open sets is a desire to reduce the number of subsets that one must study in order to define a topology. If c is the cardinal number of the set of real numbers, for example, then the set of all subsets of the real numbers has cardinal number 2^c, a "larger" infinity than c. To decide set by set and point by point which points are to be limit points of which sets would require $c \cdot 2^c = 2^c$ decisions. But the collection of open sets in the topology determined by Definition 1–1 has only cardinal number c. A proof of this is presented later.

It is natural to ask if we can select a still smaller collection of subsets and use these to define the open sets. The answer is often affirmative, and the following definition provides such a collection.

A collection of subsets $\{B_\alpha\}$ of a given set S is a *basis for a topology* in S provided that

(1) $\cup B_\alpha = S$ and that

(2) if p is a point of $B_\alpha \cap B_\beta$, then there is an element B_γ of $\{B_\alpha\}$ which contains p and which itself is contained in $B_\alpha \cap B_\beta$.

We note that the collection of open sets satisfying Axioms O_1, O_2, and O_3 is a basis according to this definition.

Suppose that $\mathcal{B} = \{B_\alpha\}$ is such a basis in a set S. We define "open set," and hence a topology, by agreeing that a subset of S is open if it is a union of elements of $\mathcal{B}$. We may either agree that the empty set is a union of no elements of $\mathcal{B}$, or explicitly include the empty set in $\mathcal{B}$. The resulting collection of open sets satisfies Axiom O_3, for by (1) S is open, and by agreement $\emptyset$ is open. Also, the satisfaction of Axiom O_1 is obvious, for a union of unions of basis elements is a union of basis elements. To establish Axiom O_2, we first point out that condition (2) can be formulated as follows:

(2′) If p is a point of $B_{\alpha_1} \cap B_{\alpha_2} \cap \ldots \cap B_{\alpha_n}$, then there is an element of $\mathcal{B}$ that contains p and is contained in $\bigcap_{i=1}^{n} B_{\alpha_i}$. (The proof is by induction, of course.) Now if $\{B_{\alpha_1}, \ldots, B_{\alpha_n}\}$ is any finite collection of basis elements, then for each point p in $\bigcap_{i=1}^{n} B_{\alpha_i}$ there is a basis element $B_{\alpha(p)}$ containing p and lying in the intersection. It follows that $\bigcup_p B_{\alpha(p)}$, the union of all such basis elements, must be contained in $\bigcap_{i=1}^{n} B_{\alpha_i}$. But since $B_{\alpha(p)}$ contains p for each point p, we also know that $\bigcup_p B_{\alpha(p)}$ contains $\bigcap_{i=1}^{n} B_{\alpha_i}$. Thus this intersection is a union of basis elements and is open. The same kind of argument will also show that the intersection of a finite number of open sets is a union of basis elements and hence is open. We may therefore state the following result.

THEOREM 1–3. If $\mathcal{B}$ is a basis for a topology in S, then the collection of open sets defined by $\mathcal{B}$ satisfies the axioms for a topological space.

Given a set S and some intuitive idea of what its topology should be, it is usually much easier to find a basis that agrees with the intuition than it is to describe the open sets in general. However, there may be many choices for a basis, all giving the same topology. For example, in the Euclidean plane we can take as a basis the collection of all interiors of circles or the set of all interiors of squares. Since any union of interiors of circles is a union of interiors of squares and conversely, it is obvious that both collections define the same open sets in the plane. Either of these collections defines the *Euclidean topology* for the plane. We also could have used as a basis the collection of all interiors of ellipses, or all interiors of triangles, or all interiors of crescents, and achieved the same topology. This is an example of the equivalence of different bases.

Two *bases are equivalent* if they determine identical collections of open sets.

THEOREM 1–4. A necessary and sufficient condition that two bases $\mathcal{B}$ and $\mathcal{B}'$ for topologies in a set S be equivalent is that if p is a point of an element B of $\mathcal{B}$, then there is an element B' of $\mathcal{B}'$ containing the point p and contained in B and conversely.

Proof: If $\mathcal{B}$ and $\mathcal{B}'$ are equivalent, then the condition is obviously satisfied. Suppose that the condition holds, and let O be a union of elements of $\mathcal{B}$. Then each point of O lies in an element of $\mathcal{B}'$, and this element is contained in O. Thus O is also a union of elements of $\mathcal{B}'$. A converse argument shows that the open sets defined by $\mathcal{B}'$ are the same as those defined by $\mathcal{B}$. □

Of course, it is also possible to choose nonequivalent bases, but this will lead to different topologies. For instance, the set of all half-planes $x > x_0$ for all real numbers x_0 satisfies the two conditions for a basis for a topology of the plane. It is easy to see that the only nonempty open sets of this topology are the plane itself and the elements of the basis. It is true that each such "open set" is open in the Euclidean topology, but the Euclidean topology has many open sets that are not open in this new topology. Thus the two topologies are not equivalent, although in a sense to be discussed shortly, they are comparable.

Another example of a different basis for the plane is the set of all horizontal open line segments. It is left as an easy exercise to show that every Euclidean open set is open in this new topology but not conversely. That is, there are more open sets in this new topology than in the Euclidean topology.

It is often the case that we have a topological space S but still find it convenient to select a basis for S. That is, we choose a particular subcollection of the open subsets of S as a basis, in such a way that the new basis is equivalent to the basis of all open sets of S. A subcollection $\mathcal{B}$ of open sets of a topological space S is a basis for S if and only if every open set in S is a union of elements of $\mathcal{B}$. (This is a slightly different use of the word "basis" than that given by the previous definition, but we will not discriminate between them.) The concept of a countable basis illustrates this situation. A *countable basis* for a space S is a basis that contains only countably many sets. This term is used almost always in the sense of a basis for a topology already given in S.

EXERCISE 1–3. The collection of all circles in the plane with rational radii and with centers having rational coordinates is a countable collection. Show that the interiors of such circles form a basis for the Euclidean topology of the plane.

Now suppose we have a set S and any collection $\{X_\alpha\}$ of subsets of S such that $\cup_\alpha X_\alpha = S$. Can we define a topology for S in which each X_α is an open set? The answer is "yes" because we may always assign to S the discrete topology in which there are no limit points. In the discrete topology, every set is closed and hence every set is open. It appears that our question should have been, "Is there a topology for S in which each X_α is open, and in which there are no 'extraneous' open sets?" By this we mean that no proper subcollection of the open sets contains all of the sets X_α and satisfies Axioms O_1, O_2, and O_3. The answer is still "yes." Any collection of sets satisfying Axiom O_2 and containing all the sets X_α must also contain all finite intersections of sets in $\{X_\alpha\}$. Then if the same collection satisfies Axiom O_1, it contains all unions of such finite intersections. Thus the collection $\mathcal{B}$ of all finite intersections of sets in $\{X_\alpha\}$ (each X_α is such an intersection) satisfies the conditions for a basis and hence determines a collection $\mathcal{O}$ of open sets for a topology in S. The topology so determined answers our question affirmatively, and this situation motivates our next definition.

A subcollection $\mathcal{B}$ of all open sets of a topological space S is a *subbasis* of S provided that the collection of all finite intersections of elements of $\mathcal{B}$ is a basis for S.

EXERCISE 1–4. Show that the collection of all open half-planes is a subbasis for the Euclidean topology of the plane.

EXERCISE 1–5. Let S be any infinite set. Show that requiring every infinite subset of S to be open imposes the discrete topology on S.

Let $\{O_\alpha\}$ and $\{R_\beta\}$ be two collections of subsets of a set S, both satisfying Axioms O_1, O_2, and O_3. That is, S has two topologies. We will say that the topology $\mathfrak{I}_1$ determined by $\{O_\alpha\}$ is a *finer topology* than the topology $\mathfrak{I}_2$ determined by $\{R_\beta\}$ if every set R_β is a union of sets O_α, that is, each R_β is open in the $\mathfrak{I}_1$ topology. We will denote this situation with the symbol $\mathfrak{I}_1 \geqq \mathfrak{I}_2$. We easily see that the two topologies are equivalent if we have both $\mathfrak{I}_1 \geqq \mathfrak{I}_2$ and $\mathfrak{I}_2 \geqq \mathfrak{I}_1$. We now consider the collection of all possible topologies on a given set S. As an exercise, the reader may prove the following result due to Birkhoff [63]: the collection of all topologies on a given set S constitutes a lattice under the partial ordering defined above.

1–4 Metric spaces and metric topologies. In this section, we give the most direct generalization of the topology used in real numbers in analysis.

Let M be a set of points, and assume that there exists a real-valued function $d(x, y)$ on pairs of elements of M satisfying the following conditions:

1. $d(x, y) \geqq 0$.
2. $d(x, y) = 0$ if and only if $x = y$.
3. $d(x, y) = d(y, x)$.
4. $d(x, y) + d(y, z) \geqq d(x, z)$ (the triangle inequality).

We say that M is a *metric space* with *metric d*, or with *distance function d*.

The spaces that are most familiar to the reader are metric spaces. For example, if we define the distance between two real numbers x and y by setting $d(x, y) = |x - y|$, we have converted the real numbers into a metric space.

A metric provides an easy way to define a topology in a metric space. For let x be any point of a metric space M with metric d, and let r be a positive number. The *spherical neighborhood* $S(x, r)$ of the point x is the set of all points y in M such that $d(x, y) < r$, the number r being the *radius* of the neighborhood.

The set of all spherical neighborhoods in M satisfies the conditions for a basis. The first condition is satisfied trivially, of course. To prove that the second condition holds, let p be any point in an intersection $S(x_1, r_1) \cap S(x_2, r_2)$. Let r be the smaller of the two numbers $r_1 - d(p, x_1)$ and $r_2 - d(p, x_2)$. Since p is in both spherical neighborhoods, it follows that r is positive. Now suppose q is a point in $S(p, r)$. Then for $i = 1$ or 2, we have

$$\begin{aligned} d(q, x_i) &\leqq d(p, q) + d(p, x_i) < r + d(p, x_i) \\ &\leqq (r_i - d(p, x_i)) + d(p, x_i) = r_i. \end{aligned}$$

Thus q lies in $S(x_i, r_i)$, $i = 1, 2$, and hence $S(p, r)$ is contained in the intersection $S(x_1, r_1) \cap S(x_2, r_2)$.

The topology defined in a metric space M by the basis of all spherical neighborhoods in M is the *metric topology* of M.

As an important example we define *Euclidean n-dimensional space* E^n. The points of E^n are all ordered n-tuples $(x_1, x_2, \ldots, x_n)$ of real numbers. If $x = (x_1, \ldots, x_n)$ and $y = (y_1, \ldots, y_n)$, then we define

$$d(x, y) = \left[\sum_{i=1}^{n} (x_i - y_i)^2\right]^{1/2}.$$

It is left as an exercise for the reader to prove that this is indeed a metric. (It is evident that we are using nothing more than the usual formula for the distance between two points as we find it in analytic geometry.)

One may consider a metric space from two standpoints. To the topologist, the particular metric used on a space is merely a convenient way to define open sets. For instance, we may use the metric

$$d(x, y) = \sum_{i=1}^{n} |x_i - y_i|$$

for E^n and obtain exactly the Euclidean topology. The metric is often a convenience in proving theorems and, to the topologist, the choice between equivalent metrics is merely a question of expediency.

On the other hand, to a metric geometer the metric is important in itself. A change in the metric changes the metric space. As we pointed out earlier, the natural equivalence relation between topological spaces is the homeomorphism. For the geometer, the corresponding transformation on metric spaces is the *isometry*, a one-to-one *distance-preserving* transformation of one metric space onto another. A question that might interest a metric geometer is this: does a certain type of metric space M with metric d have the midpoint property, i.e., for each two points x and y in M is there a point z such that $d(x, z) = d(y, z) = \frac{1}{2} d(x, y)$? This is not a topological question at all. To see this, we note that the closed interval $[0, 1]$ in E^1 and the closed semicircle $\rho = 1, 0 \leq \theta \leq \pi$ (in polar coordinates) are homeomorphic under the homeomorphism $f(x) = (1, \pi x)$. Then in the Euclidean metrics the first example has the midpoint property and the second does not. A topologist might be interested in knowing whether a certain type of space has at least one metric with the midpoint property or if the space were such that *every* metric has the midpoint property.

We will use the term "metric space" to mean a topological space that has a metric such that the basis of spherical neighborhoods yields the original topology. Of course, any set may be assigned a distance function. We simply let the distance between distinct points equal unity in every case, and the axioms for a metric will be satisfied. This metric will impose the discrete topology on the set, however. The crux of the matter here is the requirement that the metric topology be the original topology. In this sense, a metric space is often called a *metrizable space.*

As an example of the topological power of a metric, we give the following result. First, a set X of points in a space S is said to be *dense* in S if every point of S is a point or a limit point of X, that is, if $S = \overline{X}$. A space is *separable* if it has a countable dense subset. For instance, E^n is separable since the set of all points whose coordinates are all rational is countable and dense.

Theorem 1–5. Every separable metric space has a countable basis.

Proof: Let M be a metric space with metric $d(x, y)$ and having a countable dense subset $X = \{x_i\}$. For each rational number $r > 0$ and each integer $i > 0$, there is a spherical neighborhood $S(x_i, r)$, and the set $\mathcal{B}$ of all these is countable. We will show that $\mathcal{B}$ is a basis. Let p be any point of M and let O be an open set containing p. Then there is a positive number ϵ such that $S(p, \epsilon)$ is contained in O, by definition. There is a point x_i of X such that $d(x_i, p) < \epsilon/3$ since X is dense. Let r be a rational number satisfying $\epsilon/3 < r < 2\epsilon/3$, and consider $S(x_i, r)$. Certainly $S(x_i, r)$ contains p, and if y is any point of $S(x_i, r)$, then

$$d(y, p) \leq d(y, x_i) + d(x_i, p) < \frac{2\epsilon}{3} + \frac{\epsilon}{3} = \epsilon.$$

Thus y is in $S(p, \epsilon)$ and so $S(x_i, r)$ is an element of $\mathcal{B}$ that contains p and lies in O. It follows that O is a union of elements of $\mathcal{B}$ and that $\mathcal{B}$ is a basis for the topology of M. □

Without the assumption of metricity, Theorem 1–5 is not true. In E^2, consider the set P of all points (x, y) with $y \geqq 0$. Let a basis for P consist of (1) all interiors of circles in P but not touching the x-axis, and (2) the union of a point on the x-axis and the interior of a circle tangent from above to the x-axis at that point. The set of points in P both of whose coordinates are rational is both countable and dense in P. But no element of the basis just defined contains two points on the x-axis. If there were a countable basis $\mathcal{B}$ for P, then each basis element above would be a union of elements of $\mathcal{B}$. This would imply that there is a subcollection of $\mathcal{B}$ such that each point of the x-axis lies in one and only one element of that subcollection. This contradicts the fact the real numbers are uncountable.

EXERCISE 1–6. In E^n, let $x = (x_1, \ldots, x_n)$ and $y = (y_1, \ldots, y_n)$, and define $d'(x, y) = \sum_{i=1}^{n} |x_i - y_i|$ and $d''(x, y) = \max_i |x_i - y_i|$. Show that both d' and d'' give the same topology as the Euclidean metric. What do the basis elements look like?

1–5 Continuous mappings.

The definition of a continuous transformation given in Section 1–1 is not easy to apply. A more useful criterion for continuity is contained in Theorem 1–6. In fact, this condition is usually given as the definition of continuity.

THEOREM 1–6. Let $f:S \to T$ be a transformation of the space S into the space T. A necessary and sufficient condition that f be continuous is that if O is any open subset of T, then its inverse image $f^{-1}(O)$ is open in S.

[Note that to speak of $f^{-1}(O)$, it is not necessary that each point of O be the image of a point of S. Indeed, $f^{-1}(O)$ may very well be empty.]

Proof: Suppose first that f is continuous, and let O be open in T. If $f^{-1}(O)$ is not open, then $S - f^{-1}(O)$ is not closed. Hence there is some point p in $f^{-1}(O)$ that is a limit point of $S - f^{-1}(O)$. By the definition of continuity, $f(p)$ is a limit point or a point of $f[S - f^{-1}(O)]$. It is certainly possible that disjoint sets have intersecting images in general, but not if one of these is an inverse set. That is, we can assert that $f[f^{-1}(O)] \subset O$ and $f[S - f^{-1}(O)]$ are disjoint. This implies that $f(p)$ cannot be a point of $f[S - f^{-1}(O)]$, so it must be a limit point of this set. But O is an open set of T that contains $f(p)$ but no point of $f[S - f^{-1}(O)]$. This contradicts the definition of limit point, and hence $f^{-1}(O)$ must be open.

The argument in the other direction is even easier. Suppose p is a limit point of a subset X of S. If $f(p)$ is not in $\overline{f(X)}$, then $T - \overline{f(X)}$ is an open set containing $f(p)$. Hence $f^{-1}[T - \overline{f(X)}]$ is an open set containing p but not intersecting X, another contradiction. $\square$

THEOREM 1–7. A necessary and sufficient condition that the transformation $f:S \to T$ of the space S into the space T be continuous is that if x is a point of S, and V is an open subset of T containing $f(x)$, then there is an open set U in S containing x and such that $f(U)$ lies in V.

Proof: To establish the sufficiency, we show that if O is an open set in T, then $f^{-1}(O)$ is open in S. To do so, let x be a point of $f^{-1}(O)$. Then O

is an open set containing $f(x)$ so that there is an open set U_x containing x and such that $f(U_x)$ lies in O. It follows that U_x is in $f^{-1}(O)$ and that $f^{-1}(O) = \cup_x U_x$. Hence $f^{-1}(O)$ is open. For the necessity, take $U = f^{-1}(V)$. □

Exercise 1–7. Show that a one-to-one transformation $f:S \to T$ of a space S onto a space T is a homeomorphism if and only if both f and f^{-1} are continuous.

A rewording of Theorem 1–7 for metric spaces strongly resembles the classic definition of continuity in analysis.

Theorem 1–8. *Let $f:M \to N$ be a transformation of the metric space M with metric d into the metric space N with metric ρ. A necessary and sufficient condition that f be continuous is that if ϵ is any positive number and x is a point of M, then there is a number $\delta > 0$ such that if $d(x, y) < \delta$, then $\rho[f(x), f(y)] < \epsilon$.*

Proof: The sufficiency of the condition is easily established. Let V be an open set in N and y be a point of V. There is a spherical neighborhood $S(y, \epsilon)$ lying in V. The given condition implies that the neighborhood $S(x, \delta)$, x in $f^{-1}(y)$, in M is such that $f[S(x, \delta)]$ is contained in $S(y, \epsilon)$ and hence lies in V. Thus the condition of Theorem 1–7 is satisfied. Again a proof of necessity is easy and is left as an exercise. □

At an early stage in his study of topology, the student may not recall whether Theorem 1–6 says that the inverse of an open set is open or that the image of an open set is open. Both conditions seem equally sensible. It may help to give a name to the second possibility, which is, moreover, an important type of transformation.

A transformation $f:S \to T$ of the space S into the space T is said to be *interior* if f is continuous and if the image of every open subset of S is open in T.

Some writers discuss transformations that carry open sets into open sets but that are not necessarily continuous. Such transformations are usually called *open*.

We will refer to a continuous transformation as a *mapping* from now on.

Theorem 1–9. *A necessary and sufficient condition that the one-to-one mapping $f:S \to T$ of the space S onto the space T be a homeomorphism is that f be interior.*

Proof: According to Exercise 1–7, we need only show that f^{-1} is continuous. But this follows from Theorem 1–6, for if O is open in S, then $(f^{-1})^{-1}(O) = f(O)$ is open in T. Thus f is a homeomorphism. The necessity also follows immediately from Theorem 1–6. □

It is now easy to give examples of one-to-one mappings that are not homeomorphisms. Let S be the set of all nonnegative real numbers with their metric topology, and let T be the unit circle in its metric topology. For each x in S, let $f(x) = (1, 2\pi x^2/(1 + x^2))$, a point in polar coordinates on T. It is easily shown that f is continuous and one-to-one. But the set of all x in S such that $x < 1$ is open in S while its image is not open in T. Hence f is not interior and is not a homeomorphism.

1–6 Connectedness. Subspace topologies. Perhaps the reader feels that some examples of *useful* topological results are overdue. One important example of the usefulness of our development is embodied in this section.

A topological space is *separated* if it is the union of two disjoint, nonempty open sets. A space is *connected* if it is not separated. It should be obvious that either property is invariant under a homeomorphism.

We may leave the proofs of the following lemmas as exercises:

LEMMA 1–10. A space is separated if and only if it is the union of two disjoint, nonempty closed sets.

LEMMA 1–11. A space S is connected if and only if the only sets in S that are both open and closed are S and the empty set.

THEOREM 1–12. The real line E^1 is connected.

To prove such a theorem, we must use some properties of the real number system. We have assumed implicitly that the reader already knows a good deal about the real numbers, and we do not intend to make a detailed study here. We do state one important property, however, and take it to be an axiom.

DEDEKIND CUT AXIOM. Let L and R be two subsets of E^1 with the three properties that (1) neither L nor R is empty, (2) $R \cup L = E^1$, and (3) every number in L is less than any number of R. Then there is either a largest number in L or else a smallest number in R, but not both.

Proof of Theorem 1–12. Suppose E^1 is not connected. Then it is the union of two disjoint nonempty open sets, U and V. Let u be some point in U and v be some point in V. It is, at most, a renaming of the sets to assume that $u < v$. Let L consist of (a) all numbers, whether in U or V, that are less than u, together with (b) all numbers x such that every point in the closed interval $[u, x]$ belongs to U. Let R be all other numbers. Certainly L is nonempty, and since v must lie in R, R is also nonempty. By definition, every number is in L or in R. Also, by construction, every number in L is less than every number in R. Thus L and R form a Dedekind cut, and there is a number m that is either the largest in L or the smallest in R. The number m must lie in U or in V; suppose first that m is in U. Then there is an open interval (a, b) containing m and lying in the open set U. We may assume that a and b are also in U. If m is in L, then $m \geq u$, and we have $u \leq m < b$. But then $[u, b]$ lies in U, so b is in L, although $m < b$. Hence m cannot be in L. If m is the smallest number in R, there must be a point y of V between u and m. But then y must be in R, although y is less than m, another contradiction. Thus m cannot be in U. If m belongs to V, we can choose $[a, b]$ to lie in V with $a < m < b$. If m is in R, then a is in R and is less than m. If m lies in L, then b is also in L and is greater than m. Hence m cannot be in V. This means we have a contradiction in any case, so E^1 must be connected. □

We have defined a connected space, but it should be obvious that there are separated spaces that contain connected sets. For instance, consider the union of two parallel lines. There is a general principle for changing a definition so that it applies to a subset of a space.

Let S be a topological space and X be a subset of S. The *subspace topology* of X is that obtained by defining a subset U of X to be *open in* X if it is the intersection of X with some open subset of S. That is, we take for open sets of X all sets of the form $X \cap O$, where O is open in S. It is easy to prove that, with this topology, X is a topological space, a *subspace* of S. This implies that we have here a general method for constructing many topological spaces.

Furthermore, we can now say that a property defined for spaces is a property of a subset X if X has the property as a subspace. Thus X is a *connected subset* of a space S if X is a connected subspace of S. Expressed without using subspace topology, this says that a subset X of S is connected if there do not exist two open sets U and V in S such that $U \cap X$ and $V \cap X$ are disjoint and nonempty, and such that $U \cup V = X$.

The subspace topology is also called the *relative topology*. We speak of a subset A of a subset X of a space S as being *open relative to* X or as being closed relative to X, etc., if A is open, closed, etc., in X in the subspace topology.

A subset X of a space S is separated, we have implied, if there exist two open sets U and V of S such that $U \cap X$ and $V \cap X$ are disjoint and nonempty, and such that $U \cup V \supset X$. We cannot assume that U and V are disjoint in S, however. Consider a space S consisting of three points a, b, and c, with the open sets being S, $\emptyset$, $a \cup c$, and $a \cup b$. Then $b \cup c$ is not a connected subset of S, but there are no disjoint open sets in S, one containing b and the other containing c.

THEOREM 1–13. A subset X of a space S is connected if and only if there do not exist two nonempty subsets A and B of X such that $X = A \cup B$ and such that $(\overline{A} \cap B) \cup (A \cap \overline{B})$ is empty.

Proof: If two such subsets exist, then $S - \overline{A}$ is an open set containing B, and $S - \overline{B}$ is an open set containing A. Thus we have that $(S - \overline{A}) \cap$

X and $(S - \overline{B}) \cap X$ form a separation of X. Therefore if X is connected, the two sets A and B cannot exist, and if these sets do not exist, X cannot be separated. □

THEOREM 1–14. Suppose that C is a connected subset of a space S and that $\{C_\alpha\}$ is a collection of connected subsets of S, each of which intersects C. Then $S' = C \cup (\cup_\alpha C_\alpha)$ is connected.

Proof: There is no loss in supposing that $S' = S$, since we may take S' as a subspace. Suppose to the contrary that $S = U \cup V$, where U and V are disjoint, open, nonempty sets. Then for each α, C_α must lie entirely in U or entirely in V. For if C_α meets both U and V, we would have $C_\alpha = (C_\alpha \cap U) \cup (C_\alpha \cap V)$, which gives a separation of C_α, although C_α is connected. Similarly, C lies entirely in U or in V. But if C is in V, say, then each C_α meets V, and hence, for each α, C_α lies in V. Then $C \cup (\cup_\alpha C_\alpha)$ is in V, and U is empty, which is a contradiction of the assumption that U and V were nonempty. □

COROLLARY 1–15. For each n, E^n is connected.

Proof: Let $x = (x_1, \ldots, x_n)$ be a point of E^n. For each real number t, let $tx = (tx_1, tx_2, \ldots, tx_n)$, the ordinary scalar product of the vector x by a scalar t. Let l_x denote the set of all such points tx. The mapping $f{:}l_x \to E^1$ defined by $f(tx) = t$ is a homeomorphism of the subset l_x onto E^1. Hence by Theorem 1–12, l_x is connected. Each set l_x contains the origin, and $E^n = \cup_x l_x$, so by Theorem 1–14, E^n is connected. □

Connectedness is not only a *topological invariant*, preserved by homeomorphisms, but it is also preserved by continuous mappings.

THEOREM 1–16. Every continuous image of a connected space is connected.

Proof: Suppose that S is any space, that $f{:}S \to T$ is continuous and onto, and that T is separated. If $T = U \cup V$, where U and V are disjoint, nonempty, open sets, then $f^{-1}(U)$ and $f^{-1}(V)$ are open (Theorem 1–6) and are clearly disjoint and nonempty. Hence S is separated. If S were connected, this would provide a contradiction. □

LEMMA 1–17. For $n > 1$, the complement of the origin in E^n is connected.

Proof: The hyperplane P whose equation is $x_n = 1$ in E^n is homeomorphic (indeed isometric) to E^{n-1}. Let $x = (x_1, \ldots, x_n)$. If x is not the origin, at least one coordinate x_j is not zero. If $j \neq n$, the line consisting of all points $(x_1, \ldots, x_j, \ldots, x_{n-1}, t)$, t a real number, contains the point x, intersects the plane P, and does not pass through the origin. (This line is normal to P.) If x_n is the only nonzero coordinate of the point x, the line of points joining x and the point $(1, 0, \ldots, 0, 1)$, that is, points $[t, 0, \ldots, t + (1 - t)x_n]$, does not pass through the origin, does intersect P, and does contain x. The union of P and all these lines is connected, by Theorem 1–14, and obviously fills up $E^n - 0$. □

The *n-dimensional sphere* S^n is defined to be the set of all points $x = (x_1, x_2, \ldots, x_{n+1})$ in E^{n+1} satisfying $x_1^2 + \cdots + x_{n+1}^2 = 1$. S^0 consists of the two points ± 1 in E^1.

THEOREM 1–18. For each $n > 0$, S^n is connected.

Proof: The mapping f carrying each point $(x_1, x_2, \ldots, x_{n+1})$ of $E^{n+1} - 0$ onto the point $(x_1/|x|^2, \ldots, x_{n+1}/|x|^2)$ is easily shown to be continuous (see Exercise 1–8). By Lemma 1–17, $E^{n+1} - 0$ is connected, and hence Theorem 1–16 proves this theorem. □

It is clear that in Theorem 1–16 and Lemma 1–17 we are assuming some properties of the "analytic geometry" of E^n that we have not proved. In Chapter 5 we give a more detailed treatment of this topic.

As an example of the use of the fact that connectedness is preserved by a continuous function, consider the following familiar situation. Let $y = f(x)$ be a real-valued function continuous on the closed interval $[a, b]$ in E^1. Assume that $f(a) \cdot f(b) < 0$. Then there is a point x_0 in (a, b) such that $f(x_0) = 0$. To prove this, we note that $[a, b]$ is connected (see Exercise 1–9) and that, by Theorem 1–16, $f([a, b])$ is connected. Since $f(a) \cdot f(b) < 0$, it follows that $f([a, b])$ contains the point 0. Hence $f^{-1}(0)$ is not empty.

Another situation familiar from analysis may be generalized in our present context. Suppose $f:S \to T$ and $g:T \to X$ are transformations. Then the transformation $h:S \to X$ defined by $h(x) = g(f(x))$ for each point x in S is called the *composition* of f and g, denoted by $h = gf$.

THEOREM 1–19. If both $f:S \to T$ and $g:T \to X$ are continuous, then the composition gf is also continuous.

Proof: Let O be an open set in X. By Theorem 1–6, $g^{-1}(O)$ is open in T, and then $f^{-1}(g^{-1}(O))$ is open in S. Hence $(gf)^{-1}(O)$ is open in S, implying that gf is continuous. □

EXERCISE 1–8. Show that the mapping f used in Theorem 1–18 is continuous.

EXERCISE 1–9. Given any closed interval $[a, b]$ in E^1, find a continuous mapping of E^1 onto $[a, b]$, thereby proving that $[a, b]$ is connected.

EXERCISE 1–10. Show that $E^{n+1} - S^n$ is the union of two disjoint open connected sets.

EXERCISE 1–11. Let P be a hyperplane in E^n, given by

$$A_1x_1 + \cdots + A_nx_n = B.$$

Show that $E^n - P$ is the union of two disjoint open sets.

EXERCISE 1–12. Show that the torus (the surface of a doughnut) is the continuous image of E^2 and is therefore connected.

1–7 Compactness. As we progress, we shall see that the concept of a covering becomes increasingly important. A collection of sets $\{X_\alpha\}$ is said to *cover* a set X, or is said to be a *covering* of X, if the union $\cup_\alpha X_\alpha$ contains X. Thus the collection of all vertical lines in E^2 covers E^2 and, indeed, covers any subset of E^2. Most often, we will be concerned with coverings whose individual elements are open sets or, as we shall say, *open coverings*. An important instance of the use of open coverings is to be found in the following definition.

Let S be a topological space. Then S is *compact* provided that, if $\{O_\alpha\}$ is any open covering of S, then some finite subcollection $\{O_{\alpha_1}, \ldots, O_{\alpha_n}\}$ of $\{O_\alpha\}$ covers S. The reader will perhaps recognize this as being related to the Heine-Borel theorem for closed intervals in E^1 (see below).

A subset X of a space S is a *compact subset* of S if X is a compact subspace. We note that this says that compactness of a subset is defined in terms of relatively open sets. Our first result indicated that we could have used open subsets of S instead.

LEMMA 1–20. A subset X of a space S is compact if and only if every covering of X by open sets in S contains a finite covering of X.

Proof: If X is compact, and $\{O_\alpha\}$ is a collection of open sets of S covering X, then $\{O_\alpha \cap X\}$ is a collection of relatively open sets covering X. A finite subcollection $\{O_{\alpha_1} \cap X, \ldots, O_{\alpha_n} \cap X\}$ covers X by definition, and hence the collection $\{O_{\alpha_1}, \ldots, O_{\alpha_n}\}$ covers X. Conversely, if $\{U_\alpha\}$ is any collection of relatively open sets covering X, for each α there is an open set O_α such that $O_\alpha \cap X = U_\alpha$. The collection $\{O_\alpha\}$ covers X, so some finite number $O_{\alpha_1}, \ldots, O_{\alpha_n}$ covers X. Then $U_{\alpha_1} = O_{\alpha_1} \cap X, \ldots, U_{\alpha_n} = O_{\alpha_n} \cap X$ covers X. □

It may clarify matters if we prove the Heine-Borel theorem.

THEOREM 1–21. A closed interval $[a, b]$ in E^1 is compact.

Proof: Let $\{O_\alpha\}$ be a collection of open sets in E^1 covering $[a, b]$. We construct a Dedekind cut (L, R) of E^1 as follows. A point p is put into L if (1) $x < a$, or if (2) $a \leqq x \leqq b$ and a finite number of open sets O_α covers the closed interval $[a, x]$. A point is in R otherwise. It is easy to see that this defines a cut. Hence there is a point m that is either the largest in L or the smallest in R. In either case, m is in $[a, b]$, so some open set $O_{\alpha'}$ contains m. Because all open intervals constitute a basis for E^1, there is an interval $[u, v]$ in $O_{\alpha'}$ (we may assume $[u, v]$ to be closed) such that $a < u < m < v$. Regardless of whether m is in L or in R, u is in L, so a finite number $O_{\alpha_1}, \ldots, O_{\alpha_n}$ of open sets in $\{O_\alpha\}$ covers $[a, u]$. The sets $O_{\alpha_1}, \ldots, O_{\alpha_n}, O_{\alpha'}$ therefore cover $[a, v]$, so v is also in L. But $v > m$, contradicting either of the two possibilities for m. □

We mentioned earlier the duality between open and closed sets. One place where this duality is put to use is in describing a condition dual to compactness. A space S has the *finite intersection property* provided that if $\{C_\alpha\}$ is any collection of closed sets such that any finite number of them has a nonempty intersection, then the total intersection $\bigcap_\alpha C_\alpha$ is nonempty. A family of closed sets, in any space, such that any finite number of them has a nonempty intersection, will be said to satisfy the *finite intersection hypothesis.*

THEOREM 1–22. Compactness is equivalent to the finite intersection property.

Proof: Suppose that S is compact and that $\{C_\alpha\}$ is a family of closed sets with an empty intersection. Then each point of S is in the complement of at least one set C_α. Thus the open sets $\{S - C_\alpha\}$ cover S, and some finite number of these, $S - C_{\alpha_1}, \ldots, S - C_{\alpha_n}$, covers S. It follows that $\bigcap_{i=1}^{n} C_{\alpha_i}$ is empty. Hence if $\{C_\alpha\}$ satisfied the finite intersection hypothesis, we would have a contradiction.

On the other hand, suppose that S has the finite intersection property, and let $\{O_\alpha\}$ be an open covering of S. If no finite subcollection of $\{O_\alpha\}$ covers S, then given any such subcollection $O_{\alpha_1}, \ldots, O_{\alpha_n}$, there is some point of S not in any of these. In other words, $\bigcap_{i=1}^{n} (S - O_{\alpha_i})$ is not empty. Hence the sets $\{S - O_\alpha\}$ satisfy the finite intersection hypothesis, and the intersection $\bigcap_\alpha(S - O_\alpha)$ is not empty. But this implies then that $\bigcup_\alpha O_\alpha$ is not S, contradicting the assumption that $\{O_\alpha\}$ covers S. □

Closely related to compactness is the notion of countable compactness. A space S is *countably compact* if every infinite subset of S has a limit point in S. (The reason for using the word "countable" here is that if every countably infinite set has a limit point, then every infinite set does, because every infinite set contains a countably infinite subset.) In general topological spaces, this property is not equivalent to compactness, although we do have this equivalence in metric space. (See Section 2–8.) The following implication holds in general, however.

THEOREM 1–23. A compact space is countably compact.

Proof: Suppose that S is a compact space and that X is any subset. If X has no limit point, then each point x of S lies in an open set O_x containing at most one point of X. The sets $\{O_x\}$ cover S, and hence some finite number $O_{x_1}, \ldots, O_{x_n}$ covers S. But then there are at most n points in X. It follows that any infinite subset must have a limit point. □

Suppose that S is a compact space and that $f{:}S \to T$ is a continuous mapping of S *onto* a space T. If $\{O_\alpha\}$ is a covering of T by open sets, then the sets $\{f^{-1}(O_\alpha)\}$ are nonempty and cover S. Hence some finite subcollection $\{f^{-1}(O_{\alpha_i})\}$, $i = 1, \ldots, n$, covers S. Then $f[f^{-1}(O_{\alpha_i})] = O_{\alpha_i}$ covers T, and T is also compact.

If S is countably compact, and $f:S \to T$ is continuous and onto, consider an infinite subset X of T. For each point x in X, select a point y in S such that $f(y) = x$. The set of all such points y is an infinite set Y in S and hence has a limit point p. Continuity, in the form given by Theorem 1–7, then implies that every open set in T containing $f(p)$ also contains infinitely many points of X. Thus X has a limit point in T. Both of these situations may be summed up as follows:

THEOREM 1–24. Compactness and countable compactness are both invariant under continuous transformations.

The above result is a generalization of the following well-known situation in the calculus.

Let $y = f(x)$ be a continuous real-valued function on a closed interval $[a, b]$. Then $f([a, b])$ is connected, by Theorem 1–16, and compact, by Theorem 1–24. It follows that $f([a, b])$ is also a closed interval, and hence we find: *a real-valued function continuous on a closed interval attains both a maximum and a minimum value.*

The reader may be familiar with the strong form of the Heine-Borel theorem, which states that every open covering of a closed and bounded subset of E^1 contains a finite subcovering. This situation generalizes completely.

THEOREM 1–25. A closed subset of a compact space is compact.

Proof: Let C be a closed subset of a compact space S. Recalling Lemma 1–20, let $\{O_\alpha\}$ be a collection of open sets of S covering C. Then $S - C$ and the sets $\{O_\alpha\}$ constitute an open covering of S, so some finite number of them forms a covering of S. Those elements of this finite covering that contain points of C are all in $\{O_\alpha\}$, and these form a finite subcovering of C. □

We point out that, in the very general spaces now under consideration, a compact subset of a compact space need *not* be closed. For example, let S consist of two points a and b, with open sets S and $\emptyset$ only. Then as a subset, the point a is compact, but it is not closed.

The word "compact" has been defined in so many (related) ways that one must be quite careful in reading the literature. For a long time, a *space* was said to be compact if it were what we have called countably compact. And a *subset* X of a space S was said to be compact if every infinite subset of X had a limit point in S. In metric spaces, which were then the most widely studied, our compactness for spaces was proved as a theorem, and our compactness for subsets was shown to be equivalent to the old compactness for subsets *plus* closure. In more general spaces, however, it was found that countable compactness does not give the "right" theorems, whereas covering compactness does. A new term *bicompact* was introduced and used for a while to mean our covering compactness, but the prefix was later dropped. The terms *countably compact* and *sequentially compact* were coined to replace the older notion of compactness for spaces; for subsets, the terms *conditionally compact* and *precompact* are sometimes used to mean that a subset is compact in the older sense. All these terms are in current use, but the terminology we have adopted is the most common.

EXERCISE 1-13. Show that countable compactness is equivalent to the following condition. If $\{C_n\}$ is a countable collection of closed sets in S satisfying the finite intersection hypothesis, then $\bigcap_{i=1}^{\infty} C_i$ is nonempty.

EXERCISE 1-14. Prove that a compact subset of a metric space is closed.

EXERCISE 1-15. Find a space in which every uncountable subset has a limit point but no countable subset has a limit point.

EXERCISE 1-16. Is the open interval (a, b) compact?

1-8 Product spaces. Euclidean n-space is defined as n-tuples of real numbers or, in other words, by taking n-tuples of points of E^1. This is one instance of another general process for constructing new spaces from old. (The first such process was that of taking subspaces.) For another example (see Fig. 1-1), we may consider the torus obtained by rotating about the z-axis the circle C_1 in the xz-plane whose equation is $(x - 2)^2 + z^2 = 1$. The circle $x^2 + y^2 = 1$ in the xy-plane lies on this torus. We call this circle C_2. We may now assign to each point on the torus a pair of "coordinates" consisting of a point p_1 on C_1 and a point p_2 on C_2. The point (p_1, p_2) on the torus is found by rotating C_1 about the z-axis until it lies in a vertical plane containing p_2 and letting (p_1, p_2) be the image of p_1 under this rotation.

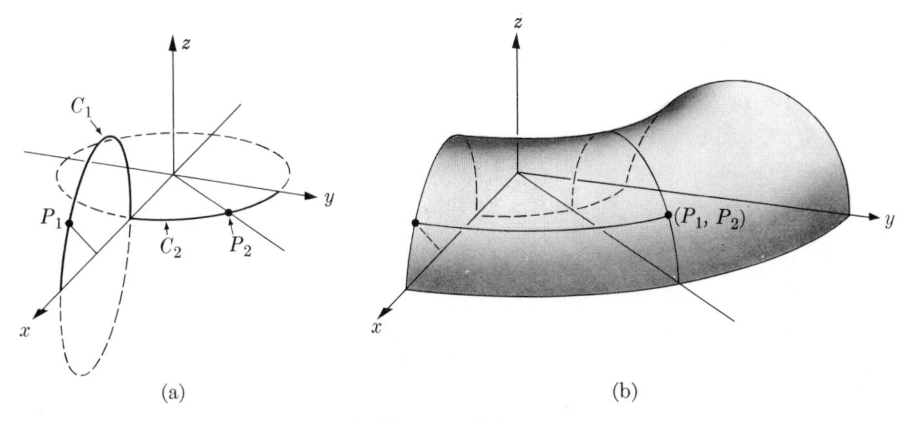

FIGURE 1-1

EXERCISE 1–17. Show that the plane set consisting of all points (x, y) satisfying $1 \leqq x^2 + y^2 \leqq 4$ can be given coordinates consisting of a point on the circle $x^2 + y^2 = 1$ and a point on the interval $[0, 1]$ in E^1.

These considerations lead to the concept of a product space, which we will formulate initially for the case of a finite number of factors. Let $S_1, S_2, \ldots, S_n$ be n spaces, not necessarily distinct. We first form the *product set* $\mathsf{P}_{i=1}^n S_i$, the collection of all ordered n-tuples $(x_1, x_2, \ldots, x_n)$, each x_i being a point of S_i. We topologize this collection so as to obtain the *product space* of $S_1, \ldots, S_n$, for which we use either the same symbol $\mathsf{P}_{i=1}^n S_i$ or the symbol $S_1 \times S_2 \times \cdots \times S_n$. A basis for the topology in $\mathsf{P}_{i=1}^n S_i$ is the collection of all sets of the form $\mathsf{P}_{i=1}^n U_i$, where each U_i is open in S_i. It should be evident that we obtain an equivalent basis if, for each i, we restrict the set U_i to be in a basis for S_i. For instance, considering E^2 as $E^1 \times E^1$, we might take the basis elements for E^2 to be all open rectangle regions, the products of pairs of open intervals.

An alternative phrasing of this topology may be made as follows. For each j, there is a "natural" mapping $\pi_j:\mathsf{P}_{i=1}^n S_i \to S_j$, called a *projection*, defined by $\pi_j(x_1, \ldots, x_j, \ldots, x_n) = x_j$. For π_j to be continuous, it is necessary that if U is an open set in S_j, then $\pi_j^{-1}(U)$ be open in $\mathsf{P}_{i=1}^n S_i$. Suppose that we wish to have no more open sets in $\mathsf{P}_{i=1}^n S_i$ than are required to make each π_j continuous. Then we must take all sets $\pi_j^{-1}(U)$ as a subbasis for a topology. This requires that we take all finite intersections of sets of the form $\pi_j^{-1}(U)$ as basis elements. Thus the requirement that each projection be continuous imposes precisely the same topology as we introduced.

We wish now to extend the definitions above to the case of an infinite number of factors. To prepare for this, let us analyze the concepts used. In the first place, the integers $1, 2, \ldots, n$ used as subscripts for the spaces S_i were functioning not as numbers but as labels. We had several spaces, and to enable us to distinguish them we appended subscripts. Looking at this properly, to each integer we assigned a space, thus forming an *indexing function* from the numbers $1, 2, \ldots, n$ to the collection of spaces. Next we formed n-tuples $(x_1, x_2, \ldots, x_n)$, x_i in S_i. But these n-tuples are again functions, functions f from the integers $1, 2, \ldots, n$ into the union of the sets S_i. We imposed the restriction that $f(i)$ must be a point of S_i, so as to have "ordered" n-tuples.

In the general case, we utilize this new approach. We start with a collection G of spaces and a set $A = \{\alpha\}$, an *indexing set*. We select some definite indexing function $\varphi:A \to G$ and designate $\varphi(\alpha)$ by S_α. As points of the product space $\mathsf{P}_A S_\alpha$ we take all functions $f:A \to \bigcup_A S_\alpha$ with the restriction that $f(\alpha)$ be a point in $\varphi(\alpha) = S_\alpha$. We now wish to topologize $\mathsf{P}_A S_\alpha$. We have two natural choices, which correspond to the two methods used for the finite case. Unfortunately, the two are not equivalent in the infinite case, and the first one does not yield the "correct" results. That is, we do *not* say that sets of the form $\mathbb{P}_A U_\alpha$, U_α open in S_α, constitute a basis. The proper definition, yielding the so-called *Tychonoff topology* [126], is to take a set $\mathbb{P}_A U_\alpha$ to be a basis element if each U_α is open in S_α and, for all but a finite number of values of α, $U_\alpha = S_\alpha$.

As justification for the Tychonoff topology, we note that it is precisely the one giving a topological space and defined by requiring projections to be continuous. If we define the projection $\pi_\beta : \mathbb{P}_A S_\alpha \to S_\beta$ by setting $\pi_\beta(f) = f(\beta)$, and require each π_β to be continuous, we immediately see that for every finite number of elements in A, $\alpha_1, \ldots, \alpha_n$, the intersection $\bigcap_{i=1}^{n} \pi_{\alpha_i}^{-1}(U_{\alpha_i})$, where U_{α_i} is open in S_{α_i}, must be open. But this is precisely the same as the set $\mathbb{P}_A V_\alpha$, where for $\alpha = \alpha_i$ we have $V_{\alpha_i} = U_{\alpha_i}$ and for $\alpha \neq \alpha_i$ we have $V_\alpha = S_\alpha$. It follows that the Tychonoff topology yields the "smallest" number of open sets in terms of which the projections are all continuous. Further discussion of this point will be found in Section 1–10.

1–9 Some theorems in logic. In dealing with product spaces that have infinitely many factors, we are led inevitably to certain results in pure logic. These results form the content of this section, which is a digression from the main interests of the book.

It was Zermelo who first recognized that, without explicit formulation or proof, mathematicians were constantly making use of the following proposition.

AXIOM OF CHOICE. Let $\mathcal{G}$ be a collection of disjoint, nonempty sets. Then there exists a set A such that for each element G of $\mathcal{G}$, $A \cap G$ is precisely one point.

There are cases, of course, in which we can define such a set A explicitly. For example, let $\mathcal{G}$ be the collection of all vertical lines in the plane. Then the x-axis is a set A. On each line we can tell exactly which one point belongs to A. Again, consider a collection $\mathcal{G}$ of disjoint closed intervals in E^2. We can form A in several ways. One way is to note that on each interval in $\mathcal{G}$ there is a point nearest the origin, and to take A to be the set of all such points. Another method would be to note that if an interval in $\mathcal{G}$ is not horizontal, it has a point with largest ordinate, but if the interval is horizontal, it has a point with largest abscissa. We could then take A to be the union of the points thus singled out. On the other hand, to give a famous example, we may form disjoint sets of real numbers by the following criterion. Two real numbers x and y are in the same class if $x - y$ is rational. This is actually an equivalence relation on E^1, and the resulting equivalence classes are disjoint and nonempty sets. (In fact, these sets are all congruent.) We may assert, using the axiom of choice, that there is a set A consisting of one point from each of these sets. Unlike the previous two cases, however, no one has ever *constructed* such a set A. By this we mean that, given one of the sets, no way of telling which point of this set is in A is known. This last example, incidentally, is of a set that is not Lebesgue-measurable. No nonmeasurable set is known which does not depend upon the axiom of choice for its definition.

From the axiom of choice we can establish the Zermelo proposition.

THEOREM 1–26. Let $\mathcal{G}$ be a collection of nonempty sets, G. Then there is a set B of pairs (G, x), where G is an element of $\mathcal{G}$ and x is a point of G, and where each element of $\mathcal{G}$ is the first element of exactly one pair in B.

Proof: Even though the elements of $\mathcal{G}$ may intersect, the collection $\mathcal{G}^* = \{S^*\}$ of sets S^*, where S^* is the set of all pairs (S, y) for all choices of y in S, is a collection of disjoint nonempty sets. We then apply the axiom of choice to $\mathcal{G}^*$. □

We have already made implicit use of the Zermelo proposition in one of its formulations. Suppose that we have a collection $\mathcal{G} = \{S_\alpha\}$ of sets, indexed by means of an indexing function φ from the index set $A = \{\alpha\}$. The collection $B = \{(S_\alpha, x_\alpha)\}$, where x_α is in S_α, given by Theorem 1–26, determines a function $c: A \to \cup_A S_\alpha$, defined by $c(\alpha) = x_\alpha$. This certainly satisfies the condition that $c(\alpha)$ is a point of S_α. Such a function is called a *choice function*. It is evident that the existence of such choice functions is equivalent to the Zermelo proposition. Thus the points in the infinite product $\mathbb{P}_A S_\alpha$ are precisely the choice functions for the collection $\{S_\alpha\}$, and the existence of such "points" depends in general upon the Zermelo proposition.

Several other forms of the axiom of choice can be given, but first we need some definitions. Suppose that we have a set A and a binary relation $<$, defined between elements of A. The relation $<$ is a *simple-order relation*, and A is *simply-ordered* by $<$, provided that

(1) for each two elements x, y of A, either $x < y$ or $y < x$,
(2) if $x < y$, then $y < x$ is false, and
(3) if $x < y$ and $y < z$, then $x < z$.

If (1) is not satisfied for each pair x, y, then $<$ is called a *partial-order relation*, and A is *partially-ordered* by $<$. A simply-ordered set A is *well-ordered* if every nonempty subset A' of A has a first element; that is, there is an element a of A' such that if a' is any other element of A', then $a < a'$. For instance, the positive integers, ordered according to size, are well-ordered.

The axiom of choice is equivalent to the well-ordering theorem of Zermelo.

THEOREM 1-27. Every set can be well-ordered.

We do not prove this. We merely observe that this theorem means that into every set a simple-order relation can be introduced, using the axiom of choice, in such a way that the set is well-ordered under this order relation. The real numbers are not well-ordered by size; there is no smallest positive number, for example. But by Theorem 1-27, there exists an order relation in which the reals are well-ordered although no such relation has ever been explicitly defined.

The last proposition that is equivalent to the axiom of choice and that we will need is the following.

MAXIMAL PRINCIPLE. Let A be a set partially-ordered by a relation $<$. Let B be a subset of A, and assume that B is simply-ordered by $<$. Then there is a subset M of A that is simply-ordered by $<$, contains B, and is not a proper subset of any other subset of A with these properties.

For instance, we may partially-order the set A of all points on lines $x = n$, n a positive integer, by saying that $(n, y_1) < (m, y_2)$ if $m - n$ is positive. The points $(p, 5)$, p a prime number, form a simply-ordered subset. One maximal subset M is the set of all points $(n, 5)$, $n > 0$. Another is the set (n, y), where $y = 5$ if n is a prime and $y = 3$ otherwise.

For a more complete discussion of the various forms of the axiom of choice, the reader is referred to Wilder [43], or to Fraenkel-Bar Hillel [8(a)].

1-10 The Tychonoff theorem.

A principal justification for adopting the Tychonoff topology in product spaces is the following result.

THEOREM 1-28. If $\{S_\alpha\}$ is any collection of compact spaces, indexed by an index set A, then the product space $\mathsf{P}_A S_\alpha$ is compact.

Before proving this result, let us examine the chief difficulty. Suppose we want to prove that $I^1 \times I^1$, where I^1 is the unit interval $[0, 1]$ in E^1, has the finite intersection property. Let $\{C_\beta\}$ be a family of closed subsets of $I^1 \times I^1$ satisfying the finite intersection hypothesis. It would seem natural to proceed as follows. Let π_1 denote the projection of $I^1 \times I^1$ onto its first factor, and let π_2 denote the same for the second factor. The closed sets C_β are compact, hence the images $\pi_i(C_\beta)$ are compact and, by Exercise 1-14, are closed. The closed sets $\pi_i(C_\beta)$ satisfy the finite intersection hypothesis and, by the compactness of I^1, there is a point x_i in $\bigcap_\beta \pi_i(C_\beta)$. Then, one might hope, the point (x_1, x_2) should be in $\bigcap_\beta C_\beta$.

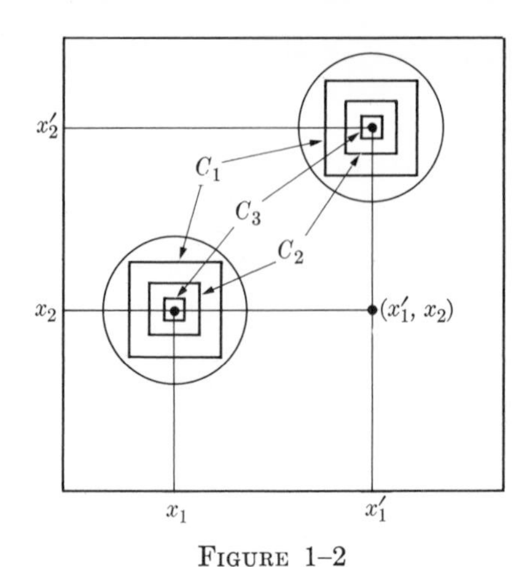

FIGURE 1–2

But this need not be the case. In Fig. 1–2, ignore the circles temporarily, and consider the pairs of closed square regions $C_1, C_2, C_3, \ldots$. We see that x_1' is in $\cap\pi_1(C_j)$, and x_2 is in $\cap\pi_2(C_j)$. But (x_1', x_2) is not in $\cap C_j$. If we could know *a priori*, we might have picked x_1 and x_2, or x_1' and x_2', but in any case some careful choosing is obviously needed.

Given a closed subset C of $\mathbb{P}_A S_\alpha$, it may be possible to add it to the collection $\{C_\alpha\}$ in such a way that the new collection still satisfies the finite intersection hypothesis. If so, then it may be possible to add another closed set C', etc. We could hope to continue, finally enlarging the collection of closed sets until it could not be further enlarged. The point of such a procedure can be seen in Fig. 1–2. If the collection $\{C_j\}$ of pairs of closed square regions were enlarged to include either one of the circular regions, we would have a new family satisfying the finite intersection hypothesis. Whichever disk we add, we eliminate the possibility of selecting "extraneous" pairs of points. It is true that the construction of a maximal family of closed sets in $\mathbb{P}_A S_\alpha$ will require the use of the maximal principle, for there are very many closed sets in $\mathbb{P}_A S_\alpha$, and a constructive proof would, in general, be impossible.

LEMMA 1–29. Let S be a space and $\mathfrak{F} = \{C_\alpha\}$ be a collection of closed subsets of S satisfying the finite intersection hypothesis. Then there exists a collection $\mathfrak{F}'$ of subsets of S such that $\mathfrak{F}'$ contains $\mathfrak{F}$, $\mathfrak{F}'$ satisfies the finite intersection hypothesis, and $\mathfrak{F}'$ is not a proper subcollection of any other collection of sets having the first two properties.

Proof: Let $\Omega = \{\mathfrak{F}_\beta\}$ be the family of all collections of subsets, not necessarily closed, of S, such that $\mathfrak{F}_\beta$ satisfies the finite intersection hypothesis. Partially order Ω by defining $\mathfrak{F}_\alpha < \mathfrak{F}_\beta$ to mean that every set in $\mathfrak{F}_\alpha$ is a set in $\mathfrak{F}_\beta$ but not conversely. The single collection $\mathfrak{F}$ is a trivial simply-ordered subfamily of Ω. Hence the maximal principle states that there is a maximal simply-ordered subfamily Ω' of Ω containing $\mathfrak{F}$. The desired collection $\mathfrak{F}'$ will turn out to be the largest element of Ω', which we must prove exists. A standard technique is used here for the first time. Let $\mathcal{G}$ be the union of all elements in Ω'. We wish to show that $\mathcal{G}$ is in Ω'. If this is done, then $\mathcal{G}$ is certainly the largest element of Ω', for every other element of Ω' is contained in $\mathcal{G}$. Also if $\mathcal{G}$ is in Ω', then $\mathcal{G}$ is not a proper subset of any other element of Ω, for any element containing $\mathcal{G}$ would be in Ω' since such an element would be comparable to every element of Ω'.

Suppose that $C_1, \ldots, C_n$ are sets in $\mathcal{G}$. For each j, $1 \leqq j \leqq n$, C_j is a set in some collection $\mathfrak{F}_j$ in Ω'. Since Ω' is simply-ordered, some one of these, say $\mathfrak{F}_k$, contains all the others and hence contains all the sets $C_1, \ldots, C_n$. Then, since $\mathfrak{F}_k$ satisfies the finite intersection hypothesis, we have $\bigcap_{i=1}^n C_i$ as a nonempty set. It follows that $\mathcal{G}$ satisfies the finite intersection hypothesis and is in the family Ω. Also, $\mathcal{G}$ is comparable with every element of Ω', so it is in Ω'. Thus $\mathcal{G}$ is the collection $\mathfrak{F}'$ claimed by the lemma. □

Corollary 1–30. If $\mathfrak{F} = \{C_\alpha\}$ is a collection of subsets of a space S, and $\mathfrak{F}$ is maximal with respect to satisfying the finite intersection hypothesis, then (a) each intersection of a finite number of elements of $\mathfrak{F}$ is in $\mathfrak{F}$, and (b) every set that intersects each C_α is in $\mathfrak{F}$.

Proof: Let $C_1, \ldots, C_n$ be in $\mathfrak{F}$, and let $C = C_1 \cap \cdots \cap C_n$. If $C'_1, \ldots, C'_k$ are in $\mathfrak{F}$, then $C_1 \cap C_2 \cap \cdots \cap C_n \cap C'_1 \cap \cdots \cap C'_k$ is nonempty. But this is precisely the set $C \cap C'_1 \cap \cdots \cap C'_k$, which proves part (a). Next, if K is a set that intersects every element of $\mathfrak{F}$, and $C_1, \ldots, C_n$ are elements of $\mathfrak{F}$, then $K \cap C_1 \cap \cdots \cap C_n = K \cap (\bigcap_{i=1}^n C_j)$, and this is nonempty because $\bigcap C_j$ is in $\mathfrak{F}$, by (a). □

The reader may be surprised that we did not prove the existence of a maximal collection of *closed* subsets of S containing $\mathfrak{F}$. We could have done so earlier, but a difficulty would have arisen in the proof of Theorem 1–28. That theorem may now be proved.

Proof of Theorem 1–28: Suppose that $\mathfrak{F} = \{C_\alpha\}$ is a collection of closed subsets of $\mathbf{P}_A S_\alpha$ satisfying the finite intersection hypothesis. By Lemma 1–29, there is a collection $\mathcal{G} = \{D_\beta\}$ of subsets of $\mathbf{P}_A S_\alpha$, such that $\mathcal{G}$ contains $\mathfrak{F}$ and is maximal with respect to the finite intersection hypothesis. For each α, the collection $\{\pi_\alpha(D_\beta)\}$ satisfies the finite intersection hypothesis. However, these sets are not necessarily closed. (Incidentally, this is conceivable even if we had required the D_β to be closed. For instance, the graph of $y = 1/x$ is closed in E^2, but its projection onto the x-axis is not closed.) The collection $\overline{\pi_\alpha(D_\beta)}$ still satisfies the finite intersection hypothesis, so there is a point x_α common to each $\overline{\pi_\alpha(D_\beta)}$, by the compactness of S_α. Now let U_α be an open set in S_α containing x_α. Since $U_\alpha \cap$

$\pi_\alpha(D_\beta)$ is nonempty for each β, it follows that $\pi_\alpha^{-1}(U_\alpha)$ meets each D_β. Hence, by Corollary 1–30, $\pi_\alpha^{-1}(U_\alpha)$ is an element of $\mathcal{G}$. (This is where the fact that the elements of $\mathcal{G}$ need not be closed first comes in.) Let p be the point of $\mathbb{P}_A S_\alpha$ whose coordinates are the points x_α, and let $\mathbb{P}_A V_\alpha = V$ be a basis element containing this point. Only a finite number $V_{\alpha_1}, \ldots,$ V_{α_n} of the sets V_α are proper subsets of their corresponding spaces, and for those we have that $\bigcap_{j=1}^{n} \pi_{\alpha_j}^{-1}(V_{\alpha_j}) = V$. Each $\pi_{\alpha_j}^{-1}(V_{\alpha_j})$ is in $\mathcal{G}$, and hence their intersection is in $\mathcal{G}$, by Corollary 1–30(b). But this means that V is in $\mathcal{G}$, so that V meets each C_α. Since V is arbitrary, p must be a limit point of each C_α. Each C_α is closed, hence p is in each C_α; that is, $\bigcap_\alpha C_\alpha$ is nonempty. □

This proof is a fine example of one of Pólya's principles (see his book [25]), which we paraphrase here: "The greater the generality, the easier the proof." In the proof above, it would seem that adding closed sets to the collection $\mathfrak{F}$ should make things more difficult, whereas adding sets that are not even closed should be pure insanity. But this unlikely procedure has just the right amount of generality.

The extensive use of the logical propositions of Section 1–9 in the proofs above is unavoidable. Indeed, it has been proved by J. L. Kelley [89] that the Tychonoff theorem is actually equivalent to the axiom of choice. Not only is the Tychonoff theorem implied by the axiom of choice (through the maximal principle), but it in turn implies the axiom of choice.

For each α in some index set A, let S_α denote a space consisting of two points x_α and y_α with the discrete topology.

Exercise 1–18. Show that if A is infinite, then $\mathbb{P}_A S_\alpha$ has limit points although none of its factor spaces has limit points.

Exercise 1–19. Show that $\mathbb{P}_A S_\alpha$ has no nondegenerate connected subsets.

Exercise 1–20. The (middle-third) Cantor set is composed of all points in the closed interval [0, 1] whose triadic expansion (base 3) contains no units. Show that if A denotes the positive integers, then $\mathbb{P}_A S_\alpha$ is homeomorphic to the Cantor set.

1–11 Function spaces. We have seen two methods of constructing new spaces from known ones, i.e., by taking subspaces and by making product spaces. A third method, one of particular importance in analysis, is introduced here. This method makes a space out of a collection of functions of one known space into another. We begin the discussion with several well-known examples that present some standard procedures.

Example 1. Let I denote a closed interval $[a, b]$ in E^1, and let $C(I)$ be the collection of all real-valued continuous functions defined on I. We topologize $C(I)$ by means of the following metric. For two functions f and g in $C(I)$, define $d(f, g) = \max_I |f(x) - g(x)|$. It is left as an easy exercise for the reader to verify that this is indeed a metric.

EXAMPLE 2. Let $I = [a, b]$ again, and let $R(I)$ be the collection of all bounded real-valued functions on I, continuous or not. We use the metric $d(f, g) = \sup_I |f(x) - g(x)|$. Note that $C(I)$ is a subspace of $R(I)$.

EXAMPLE 3. Let S be any space, and let $(E^1)^S$ denote the set of all bounded real-valued continuous functions defined on S with the metric $d(f, g) = \sup_S |f(x) - g(x)|$. Observe that this yields a metric space, no matter what the space S may be. Observe also that there always are such functions, even if only constant functions.

EXAMPLE 4. Let S be any space, and let M be a metric space with a bounded metric $\delta(x, y)$. Let M^S be the collection of all continuous mappings $f:S \to M$, which we metrize by setting $d(f, g) = \sup_S \delta[f(x), g(x)]$. This is a general procedure and has many applications.

EXAMPLE 5. Let f and g be two real-valued functions on $I^1 = [0, 1]$, such that f, g, f^2, and g^2 are all integrable in the sense of Lebesgue. If two functions f and f' differ only on a set of measure zero, then the integral $\int_0^1 [f(x) - f'(x)]^2 \, dx$ is zero, and we say that f and f' are equivalent. We may then form the corresponding equivalence classes $[f]$. (We use here for the first time our generic symbol $[\;]$ for an equivalence class.) Let L^2 denote the space of all such equivalence classes using the metric

$$d([f], [g]) = \left[\int_0^1 [f'(x) - g'(x)]^2 \, dx\right]^{1/2},$$

where f' and g' are any representatives of the classes $[f]$ and $[g]$, respectively. A proof that this is indeed a metric is somewhat harder than the previous examples; the triangle inequality presents the chief difficulty. (For the reader who has not yet studied Lebesgue integration, it will be quite permissible to think of ordinary Riemann integration here, although the resulting space will be a proper subspace of L^2, not L^2 itself.)

We may abstract the situations given in these examples as follows. Let $F = \{f\}$ be a family of transformations of a space S into a space T. Then F is certainly contained in the product set $\mathsf{P}_S T_x$, where for each x in S, $T_x = T$. The topology introduced into F by means of the metrics used in the first four examples is essentially that subspace topology induced by the topology of the product space $\mathsf{P}_S T_x$.

There is another method of introducing a topology into the family F of transformations. This method stems from the following consideration. Given a mapping $f:S \to T$, we know that the functional value $f(x)$ is a continuous function of x. Can we so choose a topology for F that $f(x)$ is a continuous function of f? The answer is "yes," and the development which follows is largely due to R. H. Fox [78]. For each compact subset C of S and each open subset U of T, let $F(C, U)$ denote the collection of all mappings f in F, such that $f(C)$ is contained in U. The family of all such collections $F(C, U)$ is taken as a subbasis for the *compact-open topology* of F.

Thus each member of the basis of the compact-open topology is of the form $\bigcap_{i=1}^{n} F(C_i, U_i)$, where each C_i is compact in S and each U_i is open in T.

We will not derive properties of function spaces at this point. Some isolated results will appear as we encounter further topological concepts that make such results meaningful. For a detailed treatment of function spaces, the reader is referred to Chapter 7 of Kelley [17] where, of course, further references will be found.

1–12 Uniform continuity and uniform spaces. A well-known result in analysis is the theorem that a continuous real-valued function on a closed interval is uniformly continuous. We generalize this result and then briefly describe an abstraction of the situation, which leads to the concept of a uniform space.

Let M and N be metric spaces with metrics d and ρ, respectively. A transformation $f:M \to N$ is continuous provided that for each point x in M and each positive real number ϵ, there exists a positive real number $\delta(x, \epsilon)$, in general depending on both x and ϵ, such that $\rho[f(x), f(x')] < \epsilon$ whenever $d(x, x') < \delta(x, \epsilon)$. Then if the number $\delta(x, \epsilon)$ can be chosen to be independent of the point x, we say that the transformation f is *uniformly continuous.*

THEOREM 1–31. Let M be a compact metric space, and let f be a continuous mapping of M into a metric space N. Then f is uniformly continuous.

Before proving Theorem 1–31, we present an auxiliary result that has many applications elsewhere.

THEOREM 1–32. Let M be a compact metric space, and let $\mathfrak{U} = \{U_i\}$ be a finite open covering of M. Then there exists a positive number $d(\mathfrak{U})$, such that each subset of M of diameter less than $d(\mathfrak{U})$ is contained in at least one element of $\mathfrak{U}$. [The number $d(\mathfrak{U})$ is called the *Lebesgue number* of the covering $\mathfrak{U}$.]

Proof: Suppose that the theorem is false. Then for each positive integer n, there exists a subset A_n of diameter $< 1/n$ in M, such that A_n does not lie entirely in any element of $\mathfrak{U}$. Let x_n be a point of A_n for each n. Since M is compact, there is a point x in M such that each open set containing x also contains infinitely many points x_n. (This is true even if the x_n are not all distinct.) Let U_j be an element of $\mathfrak{U}$ containing x, and take $d =$ glb $d(x, z)$, z in $M - U_j$. Choosing an integer n such that $n > 2/d$ and such that $d(x, x_n) < d/2$, we have for each point y in A_n,

$$d(x, y) \leqq d(y, x_n) + d(x_n, x) < \frac{1}{n} + \frac{d}{2} < d.$$

Hence y is also in U_j. This then implies that A_n lies in U_j, which is a contradiction. □

Proof of Theorem 1-31: Given $\epsilon > 0$, consider the spherical neighborhoods $S(y, \epsilon/2)$ for each point y in N. Since f is continuous, each inverse set $f^{-1}[S(y, \epsilon/2)]$ is open. The collection of all such sets is an open covering of M, and by the compactness of M, a finite subcollection $\mathfrak{U} = \{f^{-1}[S(y_i, \epsilon/2)]\}$ covers M. Take $\delta(\epsilon) = d(\mathfrak{U})$, where $d(\mathfrak{U})$ is the Lebesgue number of $\mathfrak{U}$. For any two points x and x' in M, for which $d(x, x') < \delta(\epsilon)$ the set $x \cup x'$ lies in some open set $f^{-1}[S(y_i, \epsilon/2)]$, by Theorem 1-32, hence $f[x \cup x']$ lies in $S(y_j, \epsilon/2)$. It follows that

$$\rho[f(x), f(x')] < \epsilon.$$

Therefore f is uniformly continuous. □

The concept of uniform continuity is not strictly of a topological nature. By this we mean that if f is a uniformly continuous mapping from M to N, and if h is a homeomorphism of N onto N', it need not follow that the composite mapping hf of M into N' is uniformly continuous. Nor, if h' is a homeomorphism of M' onto M, does it follow that fh' is uniformly continuous from M' to N. The reader may easily construct counterexamples by observing that if h (or h') is not a uniformly continuous homeomorphism, then the composite mapping need not be uniformly continuous.

Theorem 1-31 actually proves that the topological property of compactness has nontopological consequences. A similar but converse situation occurs frequently, too. For instance, the concept of a Cauchy sequence is not topological, and yet it has topological significance (see Section 2-13). A recent generalization of a metric space has proved to be valuable in studying such properties. This is the concept of a *uniform space* and the resulting *uniform topology*. We describe this briefly.

Let S be a set, and let $S \times S$ be the cartesian product set of S with itself, that is, $S \times S$ is the set of all ordered pairs (x, y), x and y in S. If U is a subset of $S \times S$, then by U^{-1} we mean the set of all pairs (y, x) where (x, y) is in U. If U and V are subsets of $S \times S$, then by $U \circ V$ is meant the set of all pairs (x, z), such that for some point y we have that (x, y) is in V and (y, z) is in U. The set of all pairs (x, x), x in S, is the *diagonal set* Δ.

A *uniformity* for a set S is a nonempty collection $\mathfrak{U}$ of subsets of $S \times S$ satisfying the properties

(1) each member of $\mathfrak{U}$ contains the diagonal Δ,
(2) if U is in $\mathfrak{U}$, then U^{-1} is in $\mathfrak{U}$,
(3) if U is in $\mathfrak{U}$, there is a V in $\mathfrak{U}$ such that $V \circ V$ is contained in U,
(4) if U and V are in $\mathfrak{U}$, then $U \cap V$ is in $\mathfrak{U}$, and
(5) if U is in $\mathfrak{U}$ and the subset V contains U, then V is in $\mathfrak{U}$.

The pair $(S, \mathfrak{U})$ is called a *uniform space.*

We might observe that (1) is a condition derived from the property of a metric d that $d(x, x) = 0$ for each x. Condition (2) reflects the symmetry of the metric d. Condition (3) is a very primitive form of the triangle inequality and says that there are "small" sets in $\mathcal{U}$. Conditions (4) and (5) are the analogues of properties of spherical neighborhoods in a metric space and play a similar role in proving theorems about the uniform topology.

If $(S, \mathcal{U})$ is a uniform space, C is any subset of S, and U is any element of $\mathcal{U}$, we denote by $U[C]$ the set of all points y in S such that (x, y) is in U for some point x in C. Then the *uniform topology* of S for the uniformity $\mathcal{U}$ is obtained by defining a subset O of S to be open if for each point x in O, there is an element U in $\mathcal{U}$ such that $U[x]$ lies in O. We must verify that this does give us a topological space. But Axiom O_1 is satisfied by the very definition of an open set here. If O and O' are open by this definition, then let x be a point in $O \cap O'$. By definition, there are elements U, V of $\mathcal{U}$, such that $U[x]$ lies in O and $V[x]$ lies in O'. By condition (4), $U \cap V$ is in $\mathcal{U}$ and, clearly, $(U \cap V)[x]$ lies in $O \cap O'$. Hence Axiom O_2 is satisfied. Finally, Axiom O_3 is satisfied trivially, which proves that the uniform topology does indeed yield a topological space.

We will not use the uniform topology in this book and therefore will not develop any of its properties. The interested reader, and this should include every student of analysis, is referred to Chapter 6 of Kelley [17].

1–13 Kuratowski's closure operation. Another method of introducing a topology into a set has been given by Kuratowski [18]. The resulting space is not quite so general as was defined in Section 1–2 (see Section 2–2).

Consider a set S, and suppose that there is an operation which assigns to each subset X of S another subset $\overline{X}$, called the *closure* of X. With Kuratowski, we assume three axioms concerning this *closure operation:*

K_1. For any two subsets X and Y, $\overline{X \cup Y} = \overline{X} \cup \overline{Y}$.

K_2. If X is either empty or consists of only a single point, then $\overline{X} = X$.

K_3. For any set X, $\overline{\overline{X}} = \overline{X}$.

The reader may follow the scheme given below and write out a complete proof of the following basic rules of calculation.

1. If X lies in Y, then $\overline{X}$ lies in $\overline{Y}$.
2. $\overline{X \cap Y}$ lies in $\overline{X} \cap \overline{Y}$.
3. $\overline{X} - \overline{Y}$ lies in $\overline{X - Y}$.
4. $\overline{\cap_\alpha X_\alpha}$ lies in $\cap_\alpha \overline{X}$.
5. $\cup_\alpha \overline{X}_\alpha$ lies in $\overline{\cup_\alpha X_\alpha}$.
6. If X is finite, then $\overline{X} = X$.
7. X lies in $\overline{X}$.
8. $\overline{S} = S$.

These rules may be proved in the following order. Rule 1 is implied by Axiom K_1 applied to the fact that if X lies in Y, then $X \cup Y = Y$. Rule 2 follows from Rule 1 and the fact that $X \cap Y$ is a subset of both X and Y. Rule 3 may be proved by using the identity $X \cup Y = (X - Y) \cup Y$ and taking the intersection of $\overline{X} \cup \overline{Y}$ with $S - \overline{Y}$. Rule 4 follows from the fact that $\cap_\alpha X_\alpha$ lies in X_α for each α and hence, by Rule 1, $\overline{\cap_\alpha X_\alpha}$ lies in $\overline{X}_\alpha$ for each α. Rule 5 comes from the fact that for each α, X_α lies in $\cup_\alpha X_\alpha$, and hence, by Rule 1, $\overline{X}_\alpha$ lies in $\overline{\cup_\alpha X_\alpha}$ for each α. Rule 6 is an immediate consequence of Axioms K_1 and K_2. Rule 7 follows from Rule 1 and Axiom K_2. For if x is any point of X, then x is a subset consisting of a single point. By Axiom K_2, $\overline{x} = x$ and by Rule 1, $\overline{x}$ lies in $\overline{X}$. Thus each point of X lies in $\overline{X}$. Finally, Rule 8 is immediately obvious from Rule 7.

THEOREM 1–33. Let S be a set with a closure operation satisfying Axioms K_1, K_2 and K_3. Then S is a topological space.

Proof: Define a subset X of S to be *closed* if $\overline{X} = X$. Axioms C_1, C_2 and C_3 for closed sets (see Theorem 1–2) may be verified as follows. Axiom C_1 follows immediately from Axiom K_1. Axiom C_3 is explicitly given by Axiom K_2 and Rule 8. Axiom C_2 is implied by Rules 4 and 7. For if $\{X_\alpha\}$ is any collection of closed sets, then $\cap_\alpha X_\alpha = \cap_\alpha \overline{X}_\alpha$ contains $\overline{\cap_\alpha X_\alpha} = \cap_\alpha X_\alpha$ and is a closed set. □

The reader will observe that Axiom K_2 says explicitly that "a point is a closed set" of the space S. This property is *not* necessarily shared by the general topological spaces of our Section 1–2. We postpone further discussion of this point until it arises again in Section 2–2.

1–14 Topological groups. A *topological group* G is a collection of elements on which there are two interrelated structures. First, G is a group under an operation which we will designate by "$\circ$." Then G is a topological space having a collection of distinguished subsets satisfying Axioms O_1, O_2 and O_3. Finally the two structures are interrelated by assuming that the function $\pi: G \times G \to G$ given by $\pi(g_1, g_2) = g_1 \circ g_2^{-1}$ is continuous.

We observe that by setting $g_1 = e$ (the identity element of the group G), $\pi(e, g_2) = g_2^{-1}$ is continuous and one-to-one and, indeed, is a homeomorphism of G onto itself. Similarly, the mapping carrying (g_1, g_2) onto $g_1 \circ g_2$ is continuous on $G \times G$ to G. In fact, these two conditions are equivalent to the one condition given in the definition above.

With such a composite concept as a topological group, we may apply both group-theoretical and topological ideas. Thus a topological group may be abelian (an algebraic concept) or may be connected, compact, etc. (topological concepts), or any mixture of these.

Examples come to mind readily. The real numbers constitute both a group under addition and a space (metric), and the function sending the pair (x, y) into $x - y$ is continuous. The additive group of real numbers modulo 2π constitutes a group and a space (unit circle S^1) with the desired continuity of the group operations. Many of the classical groups such as the general linear group (all nonsingular $n \times n$ matrices with complex elements) are also topological groups. Finally, any group whatsoever may be assigned the discrete topology and be considered as a topological group.

If H is a subgroup of the topological group G, then we may apply the subspace topology to H and obtain a *topological subgroup*. This permits us to speak of open or closed (or compact, etc.) subgroups of a topological group. The following result will be of value in Section 7–16.

THEOREM 1–34. Let $\mathcal{R}$ denote the additive group of real numbers modulo 1, and let H be a closed proper subgroup of $\mathcal{R}$. Then H is a finite cyclic group.

Proof: For any point r of $\mathcal{R}$, there is an interval U containing r and containing at most one point of H. If this were not true, then H would contain elements in the intervals $(0, 1/n)$ for all n. This implies that H has arbitrarily small elements. If h_1 is an element of H such that $0 < h_1 < 1/n$, then the set of all multiples of h_1 has an element in each interval of length $1/n$ in $\mathcal{R}$. Hence H would be dense in $\mathcal{R}$ and, being closed, $H = \mathcal{R}$.

Since $\mathcal{R}$ is compact, a finite number of intervals U, each containing just one point of H, covers $\mathcal{R}$. Therefore H is finite. Assuming that H contains an element other than 0, let h_1 be the element of H such that $d(0, h_1)$ is a minimum. Then H consists of all multiples of h_1. For if there were an element h_2 in H that was not a multiple of h_1, then for some integer n, $nh_1 - h_2$ would be closer to zero than h_1. Therefore H is cyclic. □

A product of topological groups is again a topological group. We utilize the usual direct product for the group operation and the Tychonoff topology. Precisely, if $\{G_\alpha\}$ is a collection of topological groups indexed by a set $A = \{\alpha\}$, then the product $\mathbb{P}_A G_\alpha$ has the group operation $g \circ g' = \{g_\alpha\} \circ \{g'_\alpha\} = \{g_\alpha \circ g'_\alpha\}$. Since the mapping of the product $\mathbb{P}_A G_\alpha \times \mathbb{P}_A G_\alpha$ into $\mathbb{P}_A G_\alpha$ given by $(g, h) = g \circ h^{-1}$ has continuous coordinates $g_\alpha \circ h_\alpha^{-1}$, we satisfy the conditions for a topological group.

We have as an example the product of the real numbers with themselves n times, which yields a topological group whose underlying space is Euclidean n-space, E^n. Also, since the topological group $\mathcal{R}_{2\pi}$ of real numbers modulo 2π has S^1 as its underlying space, it follows that $\mathcal{R}_{2\pi} \times \mathcal{R}_{2\pi}$ is a topological group whose underlying space is $S^1 \times S^1$, the two-dimensional torus. Similarly, the product $\mathcal{R}_{2\pi} \times \cdots \times \mathcal{R}_{2\pi}$, n factors, has the n-dimensional torus T^n as its underlying space.

Given a fixed element a of a topological group G, each of the mappings $g \to a \circ g$ and $g \to g \circ a$ is a homeomorphism of G (as a space) onto itself. These homeomorphisms are, respectively, the *left translation* of G by a, and the *right translation* of G by a. If U is a subset of G, then we define the left and right translates of U by the element a as $a \circ U$ and $U \circ a$, respectively.

If U is an open set containing the identity element e of G, then for any element g of G, the translates $g \circ U$ and $U \circ g$ are both open sets containing the element g. Conversely, if U is an open set containing an element g, then both $g^{-1} \circ U$ and $U \circ g^{-1}$ are open sets containing the identity element e. This implies that knowing only the open sets containing the identity e, we know the entire topology of G and conversely. It also follows that all the local properties (see Section 2–10, for example) may be determined by studying the identity element alone. We will make use of this fact later.

For references that develop the theory of topological groups in greater detail than we can do in this work, the reader may consult Chevalley [6], Pontrjagin [26], and Chapter II of Lefschetz [20].

Exercise 1–21. Let E_ω consist of all sequences $x = \{x_n\}$ of real numbers with the metric

$$d(x, y) = \sum_{n=1}^{\infty} \frac{1}{n!} \frac{|x_n - y_n|}{1 + |x_n - y_n|}.$$

First prove that this is a metric. Then show that a necessary and sufficient condition for the sequence $p_i = \{p_{ni}\}$ to converge to $p = \{p_n\}$ is that p_{ni} converge to p_n for each n.

Exercise 1–22. Let M and N be metric spaces, and let $f:M \to N$ be a transformation. Show that f is continuous if and only if the convergence of a sequence $\{x_n\}$ to a point x in M implies the convergence of $\{f(x_n)\}$ to $f(x)$ in N.

Exercise 1–23. Suppose that M is a separable metric space and $\{a_n\}$ is a countable dense subset of M. If a is any point of M, let $f(a) = (x_1, x_2, \ldots, x_n, \ldots)$, where

$$\begin{aligned} x_1 &= d(a_2, a) - d(a_2, a_1) \\ x_2 &= d(a_3, a) - d(a_3, a_1) \\ &\vdots \\ x_n &= d(a_{n+1}, a) - d(a_{n+1}, a_1) \\ &\vdots \end{aligned}$$

Prove that f is a homeomorphism of M into the space E_ω of Exercise 1–21.

Exercise 1–24. Let X be a compact metric space, and let Y have a countable basis. Let the function space Y^X be assigned the compact-open topology, and prove that Y^X also has a countable basis.

Exercise 1–25. Prove that $(I^1)^{I^1}$ is not compact.

Exercise 1–26. If X consists of n points with the discrete topology, prove that Y^X is homeomorphic to $Y^n = Y \times \cdots \times Y$, n factors.

EXERCISE 1–27. If $d(x, y)$ is a metric on a set M, show that

$$\rho(x, y) = \frac{d(x, y)}{1 + d(x, y)}$$

is a metric on M and that the two metric topologies are equivalent.

EXERCISE 1–28. Let M and N be metric spaces with metrics d and ρ, respectively. Show that the product set $M \times N$ is metrized by

$$\delta[(x_1, y_1), (x_2, y_2)] = [d^2(x_1, x_2) + \rho^2(y_1, y_2)]^{1/2}.$$

(This is called the *product metric.*)

EXERCISE 1–29. Prove the following results.

THEOREM 1–35. If S is a separated space and is the union of the two disjoint sets A and B, then any connected subset C of S must lie in either A or B.

COROLLARY 1–36. If X is a connected subset of a space S, and Y is a subset containing X and contained in $\overline{X}$, then Y is connected.

EXERCISE 1–30. Prove the following theorem as a consequence of Theorem 1–7.

THEOREM 1–37. Let $\mathbb{P}_A X_\alpha$ and $\mathbb{P}_A Y_\alpha$ be two product spaces over the same index set A, and let $f_\alpha: X_\alpha \to Y_\alpha$ be continuous for each α in A. Then the mapping $f(x) = y$, $x = \{x_\alpha\}$, $y = \{r_\alpha(x_\alpha)\}$ is continuous.

EXERCISE 1–31. Let M be a metric space, and let A be a closed subset of M. If U is an open set in $M \times I^1$ containing $A \times I^1$, prove that there exists an open set V in M such that $V \times I^1$ contains $A \times I^1$ and is itself contained in U.

CHAPTER 2

THE ELEMENTS OF POINT-SET TOPOLOGY

2–1 Introduction. Suppose that a mathematician is confronted by some concrete collection of objects into which he wishes to introduce a topology so that, for example, he may define continuous functions on the collection. There may be many ways to do this, but it is usually convenient to introduce a topology that is as "strong" as possible in the sense that much is known about the particular topology. For instance, if it were possible to topologize the collection as a compact metric space, our mathematician would probably do so. One reason for this choice is that compact metric spaces have been studied very extensively, and another reason is that it is relatively easy to prove new theorems in such spaces.

A similar problem is to find a new and equivalent topology for a space that is already given. Again, the ideal perhaps is a compact metric topology. Unfortunately, not every space can be assigned a metric, equivalent to its original topology. Still, there are many stages between the most general topological space and a compact metric space and, failing to achieve the ideal, there is yet a chance to choose some well-studied topology for the given collection. Some of these topologies are given in this chapter.

2–2 Separation axioms. A widely used set of successively stronger conditions to be placed upon a topological space are the "trennungsaxioms" of Alexandroff and Hopf [2], the so-called T_i-axioms. The first three of these are

Axiom T_0. Given two points of a topological space S, at least one of them is contained in an open set not containing the other.

Axiom T_1. Given two points of S, each of them lies in an open set not containing the other.

Axiom T_2 (Hausdorff axiom). Given two points of S, there are *disjoint* open sets, each containing just one of the two points.

These axioms are obviously in increasing order of strength in the sense that T_2 implies T_1 and T_1 implies T_0. If we add Axiom T_i, $i = 0$, 1, or 2, to Axioms O_1, O_2, and O_3 for a topological space, we obtain a *T_i-space*. There are T_0-spaces that are not T_1-spaces and T_1-spaces that are not T_2-spaces, so these axioms are indeed successively stronger.

The first axiom, T_0, may be rephrased to say that in a set consisting of two points, at least one of the points is not a limit point of the other. It does not follow that a set consisting of a single point is closed. For, let S consist of all integers, with the open sets containing a given integer n, defined to be the set of all integers k, with $k \geqq n$. It is easy to see that S is a T_0-space and that each integer n is a limit point of each integer larger than n. Thus a single integer k cannot be a closed set, since each integer less than k is a limit point of k. The following exercise is about all we can say for T_0-spaces in this direction.

EXERCISE 2–1. Show that no T_0-space contains a finite set of points $p_1, p_2, \ldots, p_n$ such that for each $k < n$, p_{k+1} is a limit point of p_k and p_1 is a limit point of p_n.

Axiom T_1 readily implies that each point is a closed subset. For each point in the complement of any particular point p lies in an open set that does not contain the point p. Hence the complement of p is a union of open sets and, by Axiom O_1, is open. It follows by induction and Axiom O_2 that *in a T_1-space every finite set is closed.* This is sometimes taken as a definition of the T_1-spaces. Since the requirement that points be closed seems very natural, spaces that are more general than the T_1-spaces are rarely studied.

EXERCISE 2–2. Find a finite space that is a T_0-space but not a T_1-space.

In many ways the Hausdorff axiom is the most interesting of the three. The term *Hausdorff space* is usually used for T_2-spaces in the literature, and we follow this usage here. Hausdorff spaces will be discussed at some length.

THEOREM 2–1. In a Hausdorff space S, let p be a point, and let C be a compact subset not containing p. Then there exist disjoint open sets, one containing p and the other containing C.

Proof: For each point x in C, by Axiom T_2 there are two disjoint open sets U_x and V_x, such that U_x contains p, and V_x contains x. Since C is covered by the collection of open sets $\{V_x\}$, there exists a finite subcollection $\{V_{x_1}, \ldots, V_{x_n}\}$ of $\{V_x\}$, which covers C. Let V be the union $\bigcup_{i=1}^{n} V_{x_i}$, and let U be the intersection $\bigcap_{i=1}^{n} U_{x_i}$. Clearly, V contains C, and U contains p, and these two sets are disjoint by construction. Then U is open by Axiom O_2, and V is open by Axiom O_1. □

COROLLARY 2–2. In a Hausdorff space, compact sets are closed.

Proof: If C is a compact set in a Hausdorff space S, then each point p in the complement $S - C$ lies in an open set not meeting C, by Theorem 2–1. Hence $S - C$ is a union of open sets and so is open. □

This corollary states one of the important properties enjoyed by the Hausdorff spaces and not by more general spaces (see Exercise 2–4 below). In fact, this property is so useful as to make the Hausdorff spaces the most general topological spaces usually studied.

Our Theorem 2–1 is a special case of the following result.

THEOREM 2–3. In a Hausdorff space S, let H and K be two disjoint compact sets. Then there exist two disjoint open sets, one containing H and the other containing K.

Proof: In view of Theorem 2–1, for each point x in H, there are disjoint open sets U_x and V_x, with x in U_x, and K contained in V_x. The collection of all sets $\{U_x\}$ covers the compact set H, and hence a finite subcollection $U_{x_1}, \ldots, U_{x_n}$ covers H. Let $U_1 = \bigcup_{i=1}^n U_{x_i}$ and $V_1 = \bigcap_{i=1}^n V_{x_i}$. Clearly, U_1 contains H, and V_1 contains K, and also $U_1 \cap V_1$ is empty. □

An interesting property of Hausdorff spaces, a property which is sometimes taken as a definition, is stated in the following exercise. First, though, let S be any space, and let $S \times S$ be the product of S with itself. The *diagonal set* Δ in $S \times S$ is the collection $\{(x, x)\}$ of all points of $S \times S$ with equal coordinates.

EXERCISE 2–3. Show that a necessary and sufficient condition that a topological space S be a Hausdorff space is that the diagonal set Δ in $S \times S$ be closed in the Tychonoff topology.

The Hausdorff axiom is a *separation axiom* in this sense: given distinct points x and y in a Hausdorff space S, there are disjoint open sets U and V in S, with x in U and y in V. The set $S - (U \cup V)$ is closed. Now a subset X of a space S is said to *separate the nonempty sets H and K* if $S - X$ is the union of two disjoint sets A and B, where A contains H, and B contains K and $\overline{A} \cap B \cup \overline{B} \cap A = \emptyset$ (see Section 1–6). We see that the Hausdorff axiom may be rephrased as: *each two points of the space can be separated by a closed set.* Similarly, Theorem 2–3 says that each two disjoint compact sets in a Hausdorff space can be separated by a closed set.

The following example shows that Theorem 2–3 cannot be strengthened to yield a separation of noncompact disjoint closed sets by a closed set in an arbitrary Hausdorff space. In this example, "most" such pairs cannot be separated.

EXAMPLE. Let S consist of the real numbers, with the topology given by a basis of all sets consisting of a number x together with all the rational numbers in an open interval (open in the usual topology) containing x. Since in the usual topology of E^1, each two points lie in disjoint open intervals, the new topology is readily shown to be Hausdorff. But no set of irrational numbers now has a limit point, because no basis element contains more than one irrational number. Hence if X is a subset of the irrational numbers J, then X and $J - X$ are disjoint closed subsets of S. Since there are c irrational numbers (c is the cardinal number of the reals), there are 2^c subsets of J and hence 2^c pairs of disjoint closed subsets of S. Suppose that each such pair can be separated by a closed set. Then for each subset X of J, there exist disjoint open sets of S, $U(X)$, and $V(X)$, with X in $U(X)$ and $J - X$ in $V(X)$. If we let R denote the set of rational numbers, the sets $U(X) \cap R$ and $[E^1 - U(X)] \cap R$ give us a partition of the rationals. Furthermore, if X and Y are subsets of J, and Y is neither X nor $J - X$, then the sets $U(X) \cap R$ and $U(Y) \cap R$ are distinct. For if $X - Y$ contains a point x, then the set $U(X) \cap R$ will contain a sequence of rationals converging to the point x, and all but a finite number of points of this sequence will not be in $U(Y)$. Thus as to each subset X of irrationals, this assumption that X and $J - X$ can be separated allows us to assign a subset $U(X) \cap R$ of the rationals in such a way that if $X \neq Y$, then $U(X) \cap R$ and $U(Y) \cap R$ are distinct. But there are 2^c subsets of the irrationals and only c subsets of the rationals. Since $c < 2^c$, we have a contradiction that shows that the separation of disjoint closed subsets in S is not always possible.

EXERCISE 2-4. Construct an example of a T_1-space in which not all compact sets are closed.

EXERCISE 2-5. A property of a space S is *hereditary* if every subspace of S also has the property. Show that for $i = 0$, 1, and 2, the T_i property is hereditary.

EXERCISE 2-6. Is there a T_0-space S such that $S \times S$ is not a T_0-space?

EXERCISE 2-7. For $i = 1$ or $i = 2$, prove that the product of any number of T_i-spaces is a T_i-space.

EXERCISE 2-8. If $f:X \to Y$ is continuous and one-to-one, and if Y is Hausdorff, then prove that X is also Hausdorff.

2-3 T_3- and T_4-spaces. The next two trennungsaxioms of Alexandroff-Hopf are merely the conclusions of Theorems 2-1 and 2-3 stated as axioms for the case of closed sets (rather than compact sets).

AXIOM T_3. If C is a closed set in the space S, and if p is a point not in C, then there are disjoint open sets in S, one containing C and the other containing p.

This axiom could be satisfied vacuously if there were no proper closed subsets in the space S. Therefore, in order that there be a large number of closed sets and that we obtain a condition stronger than the Hausdorff, a space is defined to be a T_3-space if it satisfies both Axiom T_1 and Axiom T_3. A T_3-space is usually called a *regular space.* The following theorem states a condition that is often used as the definition of a regular space.

THEOREM 2–4. A T_1-space S is regular if and only if for each point p in S and each open set U containing p, there is an open set V containing p whose closure $\overline{V}$ is contained in U.

Proof: If S is regular and the point p lies in an open set U, then Axiom T_3 states that there exist disjoint open sets V and W, with p in V, and W containing the closed set $S - U$. Since $\overline{V} \cap (S - U)$ is empty, $\overline{V}$ lies in U.

On the other hand, if p is a point of S, and C is any closed set not containing p, then $S - C$ is an open set containing p. By assumption, there is an open set V containing p, with $\overline{V}$ contained in $S - C$. Thus V is an open set containing p, and $S - \overline{V}$ is an open set containing C. The two open sets V and $S - \overline{V}$ are disjoint. Therefore S is regular. □

THEOREM 2–5. If S is a regular space, p is a point of S, and C is a closed set not containing p, then there exist open sets with disjoint closures, one containing p and the other containing C.

Proof: By Theorem 2–4, there is an open set V containing p, such that $\overline{V}$ lies in $S - C$. By the same theorem, there is an open set V' containing p, with the closure $\overline{V'}$ contained in V. Then V' and $S - \overline{V}$ are the desired open sets. □

AXIOM T_4. If H and K are disjoint closed sets in the space S, then there exist disjoint open sets, one containing H and the other containing K.

Again, a T_4-space, or a *normal space*, is one that satisfies both Axiom T_1 and Axiom T_4. Returning to the example at the end of Section 1–4, we may easily see by a cardinality argument like that at the end of Section 2–2 that this is not only a Hausdorff space that fails to be normal, it is actually a regular space that is not normal.

Since every closed subset of a compact space is compact, Theorem 2–3 can now be reworded to say that *every compact Hausdorff space is normal.* Indeed, it seems to say more in that it states that the open sets found there have disjoint closures—a property not assumed in Axiom T_4. The next result shows that this greater strength is only apparent, not actual.

THEOREM 2–6. If H and K are disjoint closed subsets of a normal space S, then there exist open sets with disjoint closures, one containing H and the other containing K.

Proof: Let U and V be disjoint open sets, U containing H and V containing K. The set $S - U$ is closed, and does not meet H, so that there exist by normality two disjoint open sets U^* and V^*, U^* containing H, and V^* containing $S - U$. Then the sets U^* and V are the desired open sets. □

Let us see whether these separation axioms hold in a metric space. First we note that *every metric space is regular;* for if x is a point and U is an open set containing x in a metric space M, then by definition there is a spherical neighborhood of x with a positive radius, say r, which is contained in U. Then the spherical neighborhood of x with radius $r/2$ is certainly closure-contained in U, and Theorem 2–4 applies to show that M is regular.

To show that every metric space is normal requires more effort, but this will be simplified by introduction of the notion of *distance between sets.* If H and K are subsets of a metric space M with metric d, we define the *distance between* H *and* K, $d(H, K)$, as the greatest lower bound of the numbers $d(x, y)$ for all x in H and y in K. This is *not* a metric on the subsets of M. For example, if H and K are distinct but not disjoint, then $d(H, K) = 0$.

The topologist often finds use for the *Hausdorff metric* defined on the continua in a metric space. If C_1 and C_2 are continua in a metric space M with metric d, then we may define

$$\rho(C_1, C_2) = \max d(x, y),$$

where the maximum is taken over all pairs x in C_1 and y in C_2. It is an interesting exercise to prove that this is a metric on the set of continua in M.

Our proof will actually show that a metric space is completely normal. A space S is *completely normal* provided that it is T_1 and if H and K are any two separated subsets of S (that is, $(H \cap \overline{K}) \cup (\overline{H} \cap K) = \emptyset$), then there are disjoint open sets, one containing H and the other containing K. This property is the T_5 axiom of Alexandroff-Hopf. Obviously, complete normality implies normality.

THEOREM 2–7. Every metric space is completely normal.

Proof: Let M be a metric space with metric d, and let H and K be any two separated subsets of M. Let U be the set of all points x for which $d(x, H) < d(x, K)$, and let V be the set of all points y for which $d(y, H) > d(y, K)$. Since $d(h, K) > 0$ for all points h in H [while $d(h, H) = 0$], we see that U contains H; similarly, V contains K, so neither U nor V is empty. The two sets are disjoint because $d(z, H) > d(z, K)$ and $d(z, H) < d(z, K)$ cannot hold simultaneously. It remains to show that U and V are open. Let x be any point of U. Let $d(x, K) - d(x, H) = \delta$, and let y be an arbitrary point in the spherical neighborhood $S(x, \delta/2)$. Then from the triangle inequality, $d(y, H) < d(x, H) + \delta/2$. Also, $d(y, K) + \delta/2 > d(x, K)$, or $d(y, K) + \delta/2 > d(x, H) + \delta$, or $d(y, K) > d(x, H) + \delta/2$. Therefore the point y lies in U and, since y was taken to be an arbitrary point of $S(x, \delta/2)$, all of this spherical neighborhood lies in U. Finally, U is the union of all such spherical neighborhoods $S(x, \delta/2)$ as x ranges over U, so by Axiom O_1, U is open. The proof that V is open is identical. □

In view of Theorem 2–7, it is apparent that a T_i-space, $i = 0, 1, 2, 3, 4$, or 5, is a valid generalization of a metric space. The reader might well ask how far we must go in this direction before obtaining a metric space. More precisely, he might ask for conditions to be placed upon a topological space S which permit us to introduce a metric on S in such a way that the resulting metric topology is equivalent to the original topology of S. This is the *metrization problem*, which we discuss again in Sections 2–9 and 2–13.

Exercise 2–9. Construct a normal space that is not completely normal.

Exercise 2–10. Show by means of an example that complete normality does not always permit the inclusion of two separated sets in open sets with disjoint closures.

Exercise 2–10(a). Show that in a regular space the conclusion of Theorem 2–3 can be strengthened to require the two open sets to have disjoint closures.

2–4 Continua in Hausdorff spaces. A compact connected set is called a *continuum*. Many important spaces such as $I^n = I^1 \times \cdots \times I^1$, n factors, and S^n, $n > 0$, are themselves continua. Also, problems concerning the structure of a space often find their natural expression in terms of the continua in the space. We note that, since both compactness and connectedness are continuous invariants, *any continuous image of a continuum is a continuum.*

Lemma 2–8. Let a and b be distinct points of a compact Hausdorff space S, and let $\{H_\alpha\}$ be a collection of closed set with index set $\mathfrak{A}$, and suppose that $\{H_\alpha\}$ is simply-ordered by inclusion. If each H_α contains both a and b but is not the union of two separated sets, one containing a and the other containing b, then the intersection $\bigcap_{\mathfrak{A}} H_\alpha$ also has this property.

Proof: Let $H = \bigcap_{\mathfrak{A}} H_\alpha$, and suppose that H is the union of two separated sets A and B, with a in A and b in B. Since H is closed (Axiom C_1), and A and B are closed in H, it follows that A and B are closed in the space S and hence are compact. By Theorem 2–3, there are disjoint open sets U and V in S, with A lying in U and B in V. For each α in $\mathfrak{A}$, $H_\alpha \cap U$ and $H_\alpha \cap V$ are nonempty sets. If the set $K_\alpha = H_\alpha \cap (S - (U \cup V))$ were empty, then $H_\alpha = (H_\alpha \cap U) \cup (H_\alpha \cap V)$ would be a separation of H_α of the prohibited type. Hence K_α is not empty. Also the sets K_α are simply-ordered by inclusion; for, given any subset X, if H_α is contained in H_β, then $H_\alpha \cap X$ lies in $H_\beta \cap X$. The subsets K_α therefore satisfy the finite intersection hypothesis and, since S is compact, the intersection $\cap_\alpha K_\alpha$ is not empty. But this intersection lies in $\cap_\alpha H_\alpha$, which implies that H meets $S - (U \cup V)$, a contradiction. □

THEOREM 2–9. *If a and b are two points of a compact Hausdorff space S, and if S is not the union of two disjoint open sets, one containing a and the other containing b, then S contains a continuum containing both a and b.*

Proof: Let $\{H_\alpha\}$ be the collection of all closed subsets of S, each of which contains $a \cup b$ but in none of which are a and b separated. The collection $\{H_\alpha\}$ is not empty, because the entire space S is one such closed set. Let $\{H_\alpha\}$ be partially-ordered by inclusion. Using the maximal principle, we extract a maximal simply-ordered subcollection $\{K_{\alpha'}\}$ of $\{H_\alpha\}$. In view of Lemma 2–8, the set $K = \cap K_{\alpha'}$ also is in the collection $\{H_\alpha\}$. Suppose that K is not connected, that is, K is the union of two separated sets K_1 and K_2. One of these, say K_1, must contain $a \cup b$. Hence K_1 is also an element of $\{H_\alpha\}$ and is a proper subset of K. This contradicts the maximality of $\{K_{\alpha'}\}$ and proves that K is connected. Then, as an intersection of closed sets, K is closed and hence compact. Therefore K is a continuum containing $a \cup b$. □

A continuum C is said to be *irreducible between two disjoint sets* if C intersects each set but no proper subcontinuum of C intersects both sets.

THEOREM 2–10. *If a continuum is a Hausdorff space, then each two of its points lie in a subcontinuum irreducible between the two points.*

Proof: The set K of the previous proof is such an irreducible continuum. For if it contained a proper subcontinuum K' containing $a \cup b$, then K' would be in the collection $\{H_\alpha\}$ and this would contradict the maximality of $\{K_{\alpha'}\}$. □

A related idea is that of a continuum C's being *irreducible about a set* A, which means that C contains A, but no proper subcontinuum of C contains A. Note that if $A = a \cup b$, the two concepts "irreducible about $a \cup b$" and "irreducible between a and b" coincide. In Hausdorff spaces, there is no loss of generality in assuming the set A to be closed. For in any space, if a closed set H contains a set A, then H also contains $\overline{A}$. In a Hausdorff space, a continuum is closed and so if the continuum C is irreducible about A, then C is *a fortiori* irreducible about $\overline{A}$.

THEOREM 2–11. *If A is any subset of a Hausdorff continuum S, then S contains a subcontinuum irreducible about A.*

Proof: Let $\{H_\alpha\}$ be the collection of all continua in S that contain the set A. This collection is not empty, because S itself is such a continuum. Partially-order $\{H_\alpha\}$ by inclusion and, using the maximal principle, extract a maximal simply-ordered subcollection $\{K_{\alpha'}\}$. We let $K = \cap K_{\alpha'}$. As in the proof of Theorem 2–9 above, we show that K is a continuum containing A. If K were not irreducible about A, the same proper subcontinuum K' of K would contain A, thus contradicting the maximality of $\{K_{\alpha'}\}$. □

The complete reliance upon the maximal principle in these proofs is apparently unavoidable. No proof is known as yet, but it may be true that these theorems imply the axiom of choice, as does the Tychonoff theorem. We remark that in a compact *metric* space, constructive arguments can be given for these results. The essential distinction, and this is a hint to aid the reader in Exercise 2–14 below, is that a compact metric space has a countable basis, whereas a compact Hausdorff space need not.

Even in a metric space, however, the conclusions of Theorems 2–9 and 2–11 may be false if the space is not assumed to be compact. Figure 2–1 shows a plane set K consisting of disjoint closed intervals converging upon a limit interval $[a, b]$ whose midpoint c is deleted. Consider K as a (metric) subspace of E^2. There is no separation of K which separates the points a and b, but still no *connected* subset of K contains $a \cup b$.

We notice that the set K in Fig. 2–1 is neither compact nor connected. For an example that fails only to be compact, see the set M, shown in Fig. 2–2. This set M is a connected subset of E^2 for which Theorem 2–9 fails. For any closed connected subset of M that contains $a \cup b$ must contain all of M except perhaps for a half-open arc beginning at the point c. But *any* such half-open arc can be deleted, and the result is a closed connected subset containing $a \cup b$. Hence there is no minimal

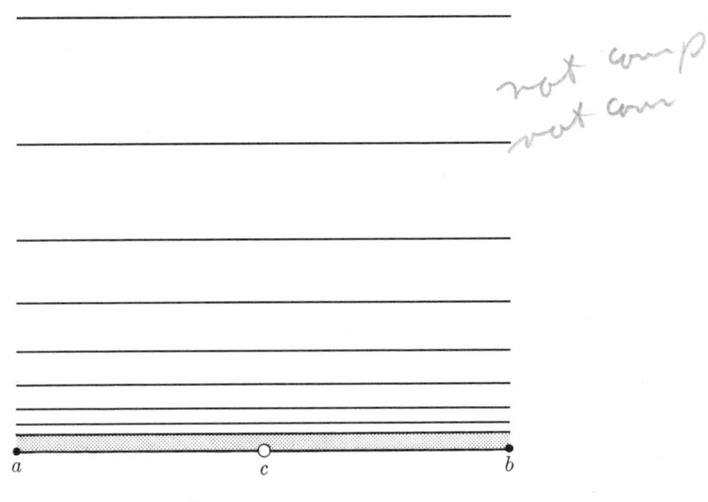

FIGURE 2–1

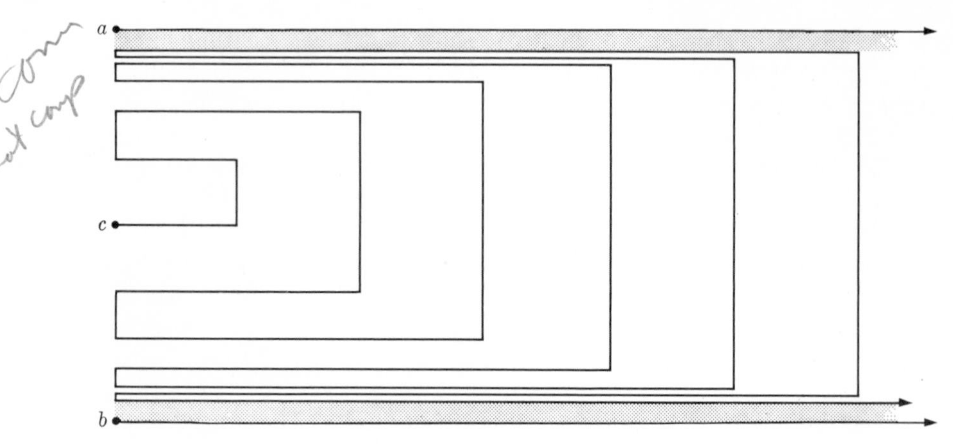

FIGURE 2–2

subset with these properties. The reader should apply the proofs of Theorems 2–9 and 2–11 to these examples to see where the arguments break down.

A study of Theorem 2–9 leads to a useful concept. A subset C of a space S is a *component* of S, provided that C is connected, but is not a proper subset of another connected set in S. In Fig. 2–1, for example, each of the closed intervals above the limit interval is a component of K, as is each half-open interval $[a, c)$ and $(c, b]$. It is true that no separation of K separates the points a and b, but still a and b do not lie together in a connected set. Such examples as this have led to another definition. A subset Q of a space S is a *quasicomponent* of S provided that, for any separation $S = A \cup B$ with A and B separated, Q lies in either A or B but Q is not a proper subset of another set in S with the same property. In the space K of Fig. 2–1 again, the union $[a, c) \cup (c, b]$ is a quasicomponent that is not a component.

Our next result is an existence theorem.

THEOREM 2–12. Every point of a space S lies in a component and in a quasicomponent of S.

Proof: Given any point x in S, consider the set of all points y in S such that x and y lie together in some connected subset of S. This set is connected by virtue of Theorem 1–14 and is maximal by construction. Hence x lies in a component of S.

Next consider the set of all points y in S such that no separation $S = A \cup B$, with A and B separated, has x in A and y in B. This set is easily seen to be a quasicomponent. □

THEOREM 2–13. Every component and every quasicomponent of a space is closed. Each component is a subset of some quasicomponent.

Proof: Let C be a component of a space S. By Corollary 1-36, $\overline{C}$ is connected because C is. Thus if C does not equal $\overline{C}$, we contradict the maximality of C. Therefore every component is closed.

Next, given any separation $S = A \cup B$, both A and B are open and closed subsets in S. A quasicomponent Q of S is the intersection of all such sets A that contain Q in each such separation. As an intersection of closed sets, Q is closed.

Since no separation of a space S can separate a connected subset of S, it follows that each component lies in some quasicomponent. □

In this new terminology we may rephrase Theorem 2-9 as follows.

THEOREM 2-14. In a compact Hausdorff space, every quasicomponent is a component.

And this result can be reworded again to read: *given a component C of a compact Hausdorff space S and a point p in $S - C$, C and p are separated by the empty set.* The technique expressed in Theorem 2-3 permits us to replace the point p by any closed set in $S - C$ and still have a separation by the empty set.

Another definition will carry us a step further in this direction. The *boundary of a set* X, $\beta(X)$, in a space S is the collection of all points p of S such that every open set containing p intersects both X and $S - X$. It is easy to see that the following formula holds:

$$\beta(X) = [\overline{X} \cap (S - X)] \cup [X \cap (\overline{S - X})] = \overline{X} \cap (\overline{S - X}).$$

The proof of the following theorem is left as an exercise.

THEOREM 2-15. Let C be a component of a compact Hausdorff space, and let U be any open set containing C. Then U contains an open set V containing C, such that V has an empty boundary.

This last result permits us to prove one of the most useful results in the theory of connected spaces.

THEOREM 2-16. Let S be a Hausdorff continuum, let U be an open subset of S, and let C be a component of U. Then $\overline{U} - U$ contains a limit point of C.

Proof: Suppose that $\overline{U} - U$ does not contain a limit point of C. Then C is closed. The subspace $\overline{U}$ of S is also compact and Hausdorff. Hence by Theorem 1-13, there are two disjoint relatively open sets, D and E, of $\overline{U}$, D containing C and E containing $\overline{U} - U$. Then D lies entirely in U, and C is a component of $\overline{D}$. Applying Theorem 2-15 to $\overline{D}$, we see that there is an open set D' in D that contains C and that has an empty boundary. Then D' is both open and closed in S, so S is not connected. □

It is interesting to note that there are connected (but not compact) subsets of the plane E^2 that do not have the property given in Theorem 2–16. F. B. Jones [87a] has given an example of a solution of the linear functional equation

$$f(a+b) = f(a) + f(b) \tag{1}$$

whose graph is connected but which is not of the form

$$f(x) = mx. \tag{2}$$

[It is known that if a solution of (1) is *bounded* over an interval, then the solution must be of the form (2)]. Now let U be a bounded open set in E^2, and consider the points of the graph G of Jones' function that lie in U. If (x_1, y_1) and (x_2, y_2) are points of $G \cap U$, there is some point $(\bar{x}, \bar{y})$ in G such that $x_1 < \bar{x} < x_2$ and such that $(\bar{x}, \bar{y})$ is not in U. Otherwise, G is bounded over $[x_1, x_2]$. Hence there is a separation of $U \cap G$ into those points for which $x < \bar{x}$ and those points for which $x > \bar{x}$. It follows that the components of $U \cap G$ are single points. In other words, every bounded open subset of G is totally disconnected, although G is connected!

We now go on to investigate certain structural properties of continua. These properties lead to the topological characterizations of the arc and the simple closed curve to be found in the next section. Let S be a connected space. A point p of S is called a *cut point* of S provided that $S - p = A \cup B$, where A and B are separated; otherwise p is a *non-cut point* of S. As examples, we may point out that every point of E^1 is a cut point, while the end points of the interval I^1 are its only non-cut points. On the other hand, E^n for $n > 1$ and S^n for $n \geqq 1$ have no cut points at all. Note that *the property of being a cut point is a topological invariant* but is not a continuous invariant. To prove the last of this statement, consider the continuous real-valued function $f(x) = 1 - x^2$. This carries the interval $[-1, 1]$ onto I^1 with the cut point 0 of $[-1, 1]$ mapped onto the end point 1 of I^1. Similarly, the property of being a non-cut point *is a topological invariant* but not a continuous invariant. For every point of the square I^2 is a non-cut point, while a projection of I^2 onto I^1 yields a continuous image with cut points. There is a sort of a converse result here, however.

THEOREM 2–17. Let $f:S \to T$ be a continuous mapping of a space S onto a connected space T. If p is a cut point of T, then the inverse set $f^{-1}(p)$ separates S.

Proof: By definition, $T - p = T_1 \cup T_2$, where T_1 and T_2 are disjoint nonempty open subsets of T. The continuity of f says that $f^{-1}(T_1)$ and $f^{-1}(T_2)$ are open, and they are certainly disjoint. By easy computation we have

$$f^{-1}(T - p) = f^{-1}(T) - f^{-1}(p) = S - f^{-1}(p) = f^{-1}(T_1) \cup f^{-1}(T_2),$$

which is the desired separation of S. □

THEOREM 2–18. Every nondegenerate T_1 continuum S has at least two non-cut points.

Proof: Let N be the set of all non-cut points of S, and suppose that N consists of at most one point. Let x_0 be a point of $S - N$. Then $S - x_0 = U \cup V$, where U and V are disjoint nonempty open sets with N contained in, say, V. For each point x in U, select a fixed separation $S - x = U_x \cup V_x$, where x_0 lies in V_x. Now $U_x \cup x$ is connected, by Theorem 1–16, since the map $f:S \to U_x \cup x$ defined by $f(y) = y$ for y in $U_x \cup x$ and by $f(y) = x$ for x in V_x is continuous. Since x_0 is in V_x, we must have $U_x \cup x$ contained in U. Order the sets U_x by inclusion and, using the maximal principle, extract a maximal simply-ordered subcollection $\{U_{x_\alpha}\}$ of $\{U_x\}$.

Now $\cap U_{x_\alpha} = \cap[U_{x_\alpha} \cup x_\alpha]$. For if x_β lies in U_{x_α}, then $V_{x_\alpha} \cup x_\alpha$ lies wholly in either U_{x_β} or V_{x_β}. Since x_0 is in $V_{x_\alpha} \cap V_{x_\beta}$, x_α lies in V_{x_β}. Then $U_{x_\beta} \cup x_\beta$, as a connected subset of $S - x_\alpha$, must lie in U_{x_α}.

Each set $U_{x_\alpha} \cup x_\alpha$ is closed and hence compact. Since these sets satisfy the finite intersection hypothesis, there is a point p in the intersection $\cap U_{x_\alpha}$. But then if q is a point of U_p, then p is not in U_q, and also U_q lies in $\cap U_{x_\alpha}$. Hence $\{U_{x_\alpha}\}$ is not maximal, a contradiction that proves that N contains more than one point. □

THEOREM 2–19. A T_1 continuum S is irreducibly connected about the set of all of its non-cut points.

Proof: Let N be the set of all non-cut points of S. Suppose that there is a proper subcontinuum S' containing N. Let x be a point in $S - S'$. Then x is a cut point of S, and $S - x = U \cup V$, where U and V are disjoint nonempty open sets and S' lies in one of these, say U. Then $V \cup x$ is connected and closed and hence is a continuum. Hence by Theorem 2–18, $V \cup x$ has at least two non-cut points, one of which, call it y, is not the point x. Then $(V \cup x) - y$ is connected, $U \cup x$ is connected, and these sets have the point x in common. Thus $S - y = (U \cup x) \cup [(V \cup x) - y]$ is connected, and y is a non-cut point of S that is not in S', a contradiction. □

COROLLARY 2–20. If x is a cut point of a continuum S, and $S - x = U \cup V$, then U and V each contain at least one non-cut point of S.

Let p and q be points of a connected space S. We denote by $E(p, q)$ the subset of S consisting of the points p and q together with all cut points of

S that separate p and q. [There may be no cut points in S separating p and q, in which case $E(p, q) = p \cup q$.] The *separation order* in $E(p, q)$ is defined as follows. Let x and y be two points in $E(p, q)$. Then x precedes y, $x < y$, in $E(p, q)$ if either $x = p$ or if x separates p and y in S.

THEOREM 2–21. The separation order in $E(p, q)$ is a simple order.

Proof: For each point x in $E(p, q)$, $x \neq p$ or q, there is a separation $S - x = A_x \cup B_x$, where p is in A_x and q is in B_x. We use this notation throughout this proof. By virtue of Corollary 1–36, both the sets $A_x \cup x$ and $B_x \cup x$ are connected. We need the following remark.

Remark. Let r and s be two points of $E(p, q) - p - q$. If s is in B_r, then A_s contains $A_r \cup r$, and B_r contains $B_s \cup s$; if s is in A_r, then $A_s \cup s$ is in A_r, and B_s contains $B_r \cup r$. To see this, note that in the first case, the connected set $A_r \cup r$ contains p but not s, and so lies entirely in A_s. The set $(B_s \cup s) \cap (A_r \cup r)$ is then empty, so $B_s \cup s$ must lie in B_r. The second case is similar.

To return to the proof of the theorem, let r and s be two points of $E(p, q) - p - q$. Then either s is in B_r or s is in A_r. If s is in B_r, then $r < s$ in $E(p, q)$. If s is in A_r, then r is in B_s, so $s < r$. Hence any two elements of $E(p, q)$ are ordered.

No element of $E(p, q)$ precedes itself. And if $r < s$ and $s < t$, then by the above remark we know that B_r contains $B_s \cup s$, and $B_s \cup s$ contains B_s, which in turn contains $B_t \cup t$. It follows that $r < t$, so that we have a simple order. The case $E(p, q) = p \cup q$ is trivial. □

Consider now the set A of positive integers ordered by size and the set B of fractions $\frac{1}{2}, \frac{2}{3}, \frac{3}{4}, \ldots, n/(n+1), \ldots$, ordered by size. The transformation $f:A \to B$, defined by $f(n) = n/(n+1)$, is one-to-one and order-preserving; that is, if $a < b$, then $f(a) < f(b)$. Insofar as their orders are concerned, then, there is no way to distinguish between the sets A and B. More generally, two ordered sets A and B are *of the same order type* if there is a one-to-one, order-preserving transformation between them. We might point out that even if A and B are subspaces of E^1 with the natural order, such an *order-isomorphism* between them need not be continuous. For instance, let A be the set of all real numbers x satisfying either $-1 \leqq x < 0$ or $1 \leqq x \leqq 2$, and let B be the set of all numbers x satisfying either $-1 \leqq x \leqq 0$ or $1 < x \leqq 2$. It is easily shown that A and B are order-isomorphic, but there is no continuous order-isomorphism between them.

We next give an example of a metric continuum M, such that one set $E(p, q)$ is of the order type of the set of rational numbers in I^1. We form M by erecting a perpendicular of length 1 at each point of the Cantor set on I^1, and then joining the midpoint of each interval comple-

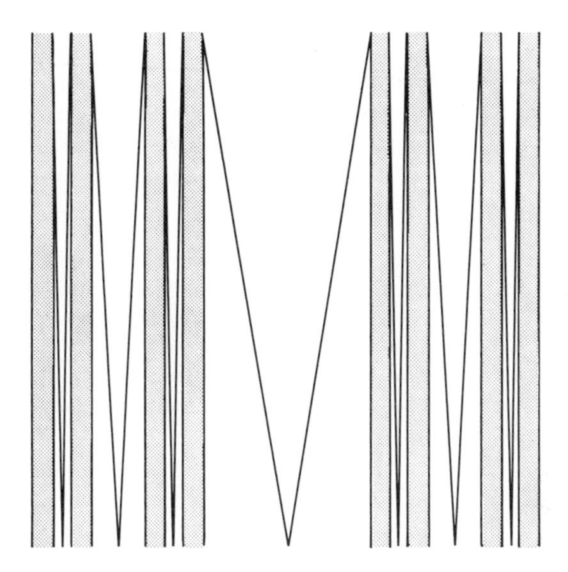

FIGURE 2–3

mentary to the Cantor set to the upper ends of the two nearest perpendiculars, as indicated in Fig. 2–3. The set M consists of I^1 and all the segments just described.

The set of points separating 0 and 1 in the continuum M above is exactly the set of midpoints of the complementary intervals. To prove that this set is order-isomorphic to the rationals, we appeal to the following theorem, which we also use in the next section.

THEOREM 2–22. If A is a countable simply-ordered set such that (1) A has no least element and no greatest element in its order and (2), given two elements a and b of A with $a < b$, there is an element c such that $a < c < b$, then A is of the same order type as the rationals.

Proof: We will actually prove that A is order-isomorphic to the set of proper dyadic fractions, i.e., the set of all numbers $k/2^n$, where k is an integer satisfying $0 < k < 2^n$. Having done so, we will have proof that any two sets that satisfy our hypotheses are order-isomorphic, this being a transitive relation. Since the rationals satisfy the conditions, this will prove the theorem.

Let $A = \{a_1, a_2, a_3, \ldots\}$ be a counting of A, $a_i \neq a_j$ for $i \neq j$. Let $f(a_1) = \frac{1}{2}$. Let n_1 be the first integer such that $a_{n_1} < a_1$ in the order of A, and let n_2 be the first integer such that $a_1 < a_{n_2}$ in A. That these exist follows from condition (1). Let $f(a_{n_1}) = 1/2^2$ and $f(a_{n_2}) = 3/2^2$. Let n_3, n_4, n_5, and n_6 be the first integers such that $a_{n_3} < a_{n_1} < a_{n_4} <$

$a_1 < a_{n_5} < a_{n_2} < a_{n_6}$. The existence of n_3 and n_6 comes from condition (1) again, while the existence of n_4 and n_5 comes from condition (2). Let $f(a_{n_3}) = 1/2^3$, $f(a_{n_5}) = 3/2^3$, $f(a_{n_5}) = 5/2^3$, and $f(a_{n_6}) = 7/2^3$. The remainder of the construction should now be clear. We require each time that the first possible subscript be chosen in order to be certain that we use up all of A in this process. □

EXERCISE 2–11. Construct three continua, each containing two points p and q, such that $E(p, q)$ is order-isomorphic to (a) the Cantor set, (b) the set of numbers $\{0, 1, \frac{1}{2}, \ldots, n/(n+1), \ldots\}$, and (c) the set of numbers $\{0; \ldots, 1/n, \ldots, \frac{1}{4}, \frac{1}{3}, \frac{1}{2}, \frac{2}{3}, \frac{3}{4}, \ldots, n/(n+1), \ldots; 1\}$.

EXERCISE 2–12. Let A be the set obtained from the Cantor set by deleting the points that are right-hand end points of complementary intervals. Prove that A is order-isomorphic to I^1.

EXERCISE 2–13. Let A be the set of all points in the Cantor set that are not end points of complementary intervals. Prove that A is order-isomorphic to the set of irrational numbers.

EXERCISE 2–14. Give constructive proofs for Theorems 2–9 and 2–11 in the case of a compact *metric* space.

2–5 The interval and the circle. In this section, we give topological conditions which when imposed upon a space make it homeomorphic to the interval or to the circle. We begin by defining still another method of introducing a topology into a set.

Let A be a simply-ordered set. The *order topology* in A is the topology given by a basis whose elements are (1) the set A, (2) for each element x in A, the set of all $y < x$, (3) for each x in A, the set of all $y > x$, and (4), for each pair x and y in A with $x < y$, the set of all z satisfying $x < z < y$. (Some of these sets may very well be empty.) We need (1) to take care of the case where A has only one element.

THEOREM 2–23. In its order topology, a simply-ordered set is a Hausdorff space.

Proof: Let x and y be two points of the simply-ordered set A, and suppose that $x < y$. If there is a point z such that $x < z < y$, then the basis elements U consisting of all points $w < z$, and the basis elements V consisting of all points $w > z$, are disjoint and contain x and y respectively. If no such point z exists, then the basis elements U consisting of all points $w < y$, and the basis elements V consisting of all points $w > x$, satisfy the needed conditions. □

Any set $E(p, q)$ in a space S has a simple order, and this order defines an order topology for $E(p, q)$. Is this order topology the same as the subspace topology? We have already seen an example (that following Theorem 2–23) where it is not. Figure 2–4 shows an example in which

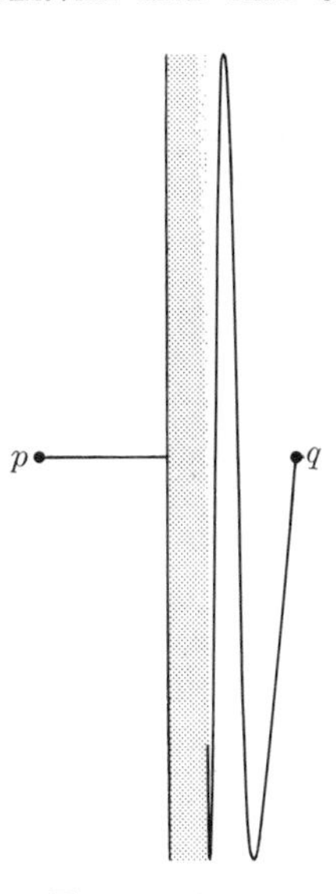

FIGURE 2–4

$E(p, q)$ is connected. Let S be the set of all points (a) on the closure of the graph of $y = \sin \pi/x$, $0 < x \leqq 1$, and (b) on the interval $[-1, 0]$ of the x-axis. For $p = (-1, 0)$ and $q = (1, 0)$, the set $E(p, q)$ is the union of the interval $[-1, 0]$ and the graph of $y = \sin \pi/x$. In the order topology, we have a space homeomorphic to the interval.

THEOREM 2–24. Let S be a connected space, and let p and q be two points of S such that $p \cup q$ is a proper subset of $E(p, q)$. Let $E(p, q)$ have the subspace topology and let E^* denote the set $E(p, q)$ with its order topology. Then the mapping $i: E(p, q) \to E^*$, defined by $i(x) = x$, is continuous.

Proof: It will suffice to prove that every basis element for the order topology in E^* is open in the subspace topology of $E(p, q)$. Reverting to the notation used in the proof of Theorem 2–21, for a point x in $E(p, q) - p - q$ let $S - x = A_x \cup B_x$, where A_x and B_x are disjoint open sets with p in A_x and q in B_x. A basis element for E^* of type (2), determined by the point x, is the intersection of A_x and $E(p, q)$; a basis element of type (3), given by x, is $B_x \cap E(p, q)$; and a basis element of type (4) is of the form $(B_x \cap A_y) \cap E(p, q)$. All these are open in $E(p, q)$. □

THEOREM 2–25. Let S be a compact connected Hausdorff space with just two non-cut points, a and b. Then $S = E(a, b)$ and the order topology defined by the order in $E(a, b)$ is the same as the topology in S.

Proof: Suppose that there is a point x in S that does not separate a from b. Since x is a cut point, $S - x = U \cup V$, where U and V are disjoint nonempty open sets with, say, U containing $a \cup b$. Recall that Corollary 2–20 states that V must contain a non-cut point of S. Thus S must have three non-cut points, a contradiction that proves $S = E(a, b)$.

Since open sets in the order topology were shown to be open in S (Theorem 2–24), we need only show that open sets in S are unions of basis elements of the order topology. If this is not so, there is an open set U in S and a point x in U, such that no order-basis element that contains x lies in U. For verbal simplicity, suppose that x is neither a nor b, so that we need consider only basis elements of type (4). If $y < z$, let (y, z) denote the order-basis element of the type (4) determined by y and z. Using the maximal principle, we obtain a collection of sets (y_α, z_α), which is simply-ordered by inclusion, and which has only x as their intersection. The same is true of the sets $[y_\alpha, z_\alpha] = (y_\alpha, z_\alpha) \cup y_\alpha \cup z_\alpha$, and these are closed sets in S. For each α, $[y_\alpha, z_\alpha] \cap (S - U)$ is nonempty, and these sets are closed in S and simply-ordered by inclusion. Hence there is a point w in $\bigcap[y_\alpha, z_\alpha] \cap (S - U)$. But then w is also in $\bigcap[y_\alpha, z_\alpha]$, which is a contradiction. □

THEOREM 2–26. If S is a connected Hausdorff space which is a set $E(a, b)$, then the Dedekind cut theorem holds in S.

Proof: Let L and R be two nonempty subsets of S such that (1) $S = L \cup R$ and (2), if l is in L and r is in R, then $l < r$ in the cut-point ordering of S. Suppose there is no greatest element in L and no least element in R. Then L is the union of basis elements of type (2), and R is the union of basis elements of type (3) (see the beginning of this section). By Theorem 2–24, L and R are then open in S and give a separation of S. □

The next result is a characterization of the arc.

THEOREM 2–27. If M is a metric continuum with just two non-cut points, then M is homeomorphic to the unit interval I^1.

Proof: We know that M contains a countable dense subset C, and we can assume that C does not contain the non-cut points, a and b, of M. As a subset of $E(a, b)$, C has an order that satisfies the hypotheses of Theorem 2–22, and hence there is an order-isomorphism h of C onto the set R of rationals in I^1. It is easy to see that h is also a homeomorphism of C onto R.

Let x be a point of M other than the two non-cut points. Let C_L be all points of C less than x, and let C_R be all points of C greater than x. The sets $h(C_L)$ and $h(C_R)$ constitute a partition of R that can be extended to a partition of I^1. Such a partition of I^1 determines a unique number y, by the Dedekind cut theorem. We let $h^*(x) = y$. It is easy to show that

$h^*:M \to I^1$ is one-to-one and continuous and hence is a homeomorphism (see Theorem 2–103). □

We recall that any homeomorph of the unit circle S^1 is called a *simple closed curve* (some authors call it a *Jordan curve*). It is clear that the omission of any two distinct points of S^1 separates S^1, and the following theorem proves that this property characterizes a simple closed curve.

THEOREM 2–28. If M is a metric continuum such that for each two points x and y of M, $M - x - y$ is not connected, then M is a simple closed curve.

Proof: (1) No point separates M. For suppose that $M - x = U \cup V$, where U and V are disjoint nonempty open sets. Then $U \cup x$ and $V \cup x$ are both continua, and hence there exist points y in U and z in V such that y does not separate $U \cup x$, and z does not separate $V \cup x$. Then we have $M - y - z = (U \cup x - y) \cup (V \cup x - z)$, which is the union of two connected sets, each containing the point x. Thus $M - y - z$ is connected, contrary to hypothesis.

(2) If $M - a - b = U \cup V$, where U and V are disjoint nonempty open sets, then $U \cup a \cup b$ and $V \cup a \cup b$ are both connected sets. For suppose that $U \cup a \cup b = X \cup Y$, where X and Y are disjoint relatively open nonempty sets. If X contains the point a but not b, then the boundary of X is the point a, so $M - a = (X - a) \cup [Y \cup (V - a)]$ is a separation. This contradicts (1).

(3) Either $U \cup a \cup b$ or $V \cup a \cup b$ is an arc. For if not, then each of these contains a point, say x in $U \cup a \cup b$ and y in $V \cup a \cup b$, distinct from a and b, that is not a cut point of the set. Then we have that $M - x - y = [(U \cup a \cup b) - x] \cup [(V \cup a \cup b) - y]$ is a union of two connected sets having a point in common, which is a contradiction.

(4) Both $U \cup a \cup b$ and $V \cup a \cup b$ are arcs. For by (3) one of them is, say, $V \cup a \cup b$. If $U \cup a \cup b$ is not an arc, then it contains a point $x \neq a, b$ such that x is a non-cut point of $U \cup a \cup b$. Let y be any point of V. Then $V \cup a \cup b - y$ is the union of two connected sets X and Y with, say, a in X and b in Y. Thus $M - x - y = [(U \cup a \cup b) - x] \cup X \cup Y$ is a connected set. This proves that M is the union of two arcs having only their end points in common. □

One of the most instructive examples in topology, the "long line," is of interest here. Consider any uncountable set A and well-order A into a well-ordered sequence $a_1, a_2, \ldots, a_\alpha, \ldots$ Either every element of A is preceded by at most a countable number of elements, or some element has an uncountable number of predecessors. If the second possibility occurs, then the set of all those elements with an uncountable number of predecessors has a first element, say $\bar{a}$, by the well-ordering principle. In this case we let A' denote the set of all predecessors of $\bar{a}$, and in the first case we let $A' = A$. In either case, A' is a well-ordered set with the property that every element has at most countably many predecessors but A' itself is uncountable. (If this seems paradoxical, recall that the positive integers constitute an infinite well-ordered set such that each element has only finitely many predecessors.) Now each element of A' has an immediate successor, but some elements do not have an immediate predecessor. For example, the first element that has infinitely many predecessors has no immediate predecessor.

Now consider a collection $\{I_a\}$ of open intervals indexed by the set A', that is, I_a is an open interval paired with the element a of A'. (How do we know that such a collection exists?) Speaking intuitively, we will insert an open interval between each two elements of A'. More precisely, let $L = A' \cup (\cup_{A'} I_a)$, and order L by the following five conditions. Let x and y be points of L. Then $x < y$ if (1) x and y are in A', and $x < y$ in A', (2) x is in A', and y is in a set I_a, and $x = a$ or $x < a$ in A', (3) x is in a set I_a, and y is in A', and $a < y$ in A', (4) x is in a set I_a, and y is in a set I_b, and $a < b$ in A', or (5) x and y are in the same set I_a, and $x < y$ in I_a. We topologize L by means of the order topology and the resulting space is the *long line*. (It should perhaps be called the *long ray* because it does have a first point.) We leave as an exercise the proof that L satisfies the Dedekind cut theorem. Assuming this, we see that it follows that L is connected in precisely the same way that we proved that the real line is connected.

We note that L is not compact, because the set of all open sets, each of which consists of all predecessors of a point of L, contains no finite subcollection covering L. But, surprisingly, L is countably compact. For suppose that $X = \{b_1, b_2, \ldots\}$ is a countable subset of L consisting entirely of elements of A'. If there is a first element a of A' that follows infinitely many elements of X, then a is a limit point of X. If there is no such element, then given any element a of A', there is an integer n, such that b_n follows a. Then A' is the union of the sets B_n of all elements of A' preceding b_n. But each B_n is countable, so $A' = \cup B_n$ is countable, contrary to hypothesis. Hence X has a limit point. The cases in which X contains infinitely many points not in A' will be left as an exercise. It now follows that the long line is a Hausdorff space that is countably compact but not compact and, as we shall see, is not metric. If we add one more point at the open end of L, we get a compact space with exactly two non-cut points, which is not an arc. This proves the need for metricity in Theorems 2–27 and 2–28.

The reader who is interested in geometry may find it noteworthy that the long line is an example of a non-Archimedean line. That is, each closed interval in L is actually homeomorphic to a straight-line interval, but there is no countable collection of closed intervals that intersect only in end points and fill up L.

We remark for later use that the long line has the fixed point property [134].

2–6 Real functions on a space. Given any space, we can always define real-valued continuous functions over the space, even if only the constant functions. But are there enough such functions to provide useful information about the space? In Example 3 of Section 1-11 we saw that, for any space, the collection of bounded continuous real-valued functions on the space can be made into a metric space whether or not the original space is metric. This might suggest that questions about a given space with some weird topology perhaps can be answered by investigating a rather nice function space. An obvious requirement here would be to have enough continuous real functions to be able to distinguish between the points of the given space. To be precise, if x and y are distinct points of a space S, and if there is a real-valued continuous function $f:S \to E^1$, such that $f(x) \neq f(y)$, then f serves to distinguish between x and y. In fact, the set of points z in S for which $f(z) = \frac{1}{2}[f(x) + f(y)]$ separates x from y in S (see Theorem 2-17), so that *if each two points can be so distinguished by a function, then the space is a Hausdorff space.*

We may also ask for conditions to be placed upon a space S which will allow us to distinguish between closed sets in the same way. *If, for each pair of disjoint closed sets A and B in a space S, there is a real-valued continuous function $F:S \to E^1$, such that $f(A) = a \neq b = f(B)$, then S is normal.* The same argument serves to show this. The converse of this result is also true; that is, the result, together with the following theorem, constitutes a complete characterization of normal spaces.

THEOREM 2-29 (Urysohn's lemma). If S is a normal space and A and B are two disjoint closed subsets of S, then there is a real-valued continuous function $f:S \to I^1$ of S into the unit interval I^1 such that $f(A) = 0$ and $f(B) = 1$.

Proof: Since S is normal, there is a closed set $C(\frac{1}{2})$ separating S into two disjoint open sets $U(\frac{1}{2})$ and $V(\frac{1}{2})$, with A in $U(\frac{1}{2})$ and B in $V(\frac{1}{2})$. We will eventually define f in such a way that $f(x) \leq \frac{1}{2}$ for x in $U(\frac{1}{2})$, $f(x) \geqq \frac{1}{2}$ for x in $V(\frac{1}{2})$, and $f(x) = \frac{1}{2}$ for x in $C(\frac{1}{2})$.

Next we find two closed sets $C(\frac{1}{4})$ and $C(\frac{3}{4})$, on which we will later have $f(x) = \frac{1}{4}$ and $f(x) = \frac{3}{4}$, respectively, as follows. The sets A and $C(\frac{1}{2}) \cup V(\frac{1}{2})$ are disjoint closed sets in S, so normality gives us a closed set $C(\frac{1}{4})$ separating S into disjoint open sets $U(\frac{1}{4})$, containing A, and $V(\frac{1}{4})$, containing $C(\frac{1}{2}) \cup V(\frac{1}{2})$. Also the sets $U(\frac{1}{2}) \cup C(\frac{1}{2})$ and B are disjoint closed sets, so there is a closed set $C(\frac{3}{4})$ separating S into disjoint open sets $U(\frac{3}{4})$ and $V(\frac{3}{4})$, with $U(\frac{1}{2}) \cup C(\frac{1}{2})$ in $U(\frac{3}{4})$ and B in $V(\frac{3}{4})$. Note that the sets $C(\frac{1}{2})$, $C(\frac{1}{4})$, and $C(\frac{3}{4})$ are all disjoint and that $U(\frac{1}{4})$ lies in $U(\frac{1}{2})$, which lies in $U(\frac{3}{4})$, while $V(\frac{3}{4})$ is in $V(\frac{1}{2})$, which lies in $V(\frac{1}{4})$. For uniformity of notation, we set $C(\frac{1}{2}) = C(\frac{2}{4})$, etc.

In general, suppose that we have defined the sets $C(r/2^n)$, $U(r/2^n)$, and $V(r/2^n)$ for a fixed value of n, and for each $r = 1, 2, \ldots, 2^n - 1$, and that these sets have the properties generalized from those above, that is, $U(1/2^n)$ lies in $U(2/2^n)$, etc., and $V((2^n - 1)/2^n)$ lies in $V((2^n - 2)/2^n)$, etc. Now for a fixed $k \leq 2^n - 1$, the sets $C((k-1)/2^n) \cup U((k-1)/2^n)$ and $C(k/2^n) \cup V(k/2^n)$ are disjoint closed sets in S, so there is a closed set $C((2k-1)/2^{n+1})$ separating S into disjoint open sets $U((2k-1)/2^{n+1})$, containing $C((k-1)/2^n) \cup U((k-1)/2^n)$, and $V((2k-1)/2^{n+1})$, containing $C(k/2^n) \cup V(k/2^n)$. Again we agree to take $C(k/2^n) = C(2k/2^{n+1})$. This procedure may be carried on to define the sets $C(k/2^{n+1})$, $U(k/2^{n+1})$ and $V(k/2^{n+1})$, for each $k = 1, 2, \ldots, 2^{n+1} - 1$, with the properties that no two sets $C(k/2^{n+1})$ intersect, and if r_1 and r_2 are two dyadic rationals, $r_1 = k_1/2^{n_1}$ and $r_2 = k_2/2^{n_2}$, with k_1 and k_2 both less than 2^{n+1} and n_1 and n_2 both less than $n + 2$, and if $r_1 < r_2$, then $U(r_1)$ is contained in $U(r_2)$, while $V(r_2)$ is contained in $V(r_1)$.

We will now define the function f as follows. If a point x in S is in a set $U(k_1/2^{n_1})$ but is not in a set $U(k_2/2^{n_2})$, then it must be true that $k_1/2^{n_1} < k_2/2^{n_2}$, and we would want $f(x)$ to lie between these two dyadic rationals. To achieve this condition for all points x at once, we set

$$f(x) = \text{greatest lower bound of } k/2^n \text{ taken over all sets } U(k/2^n) \text{ that contain } x$$

and

$$f(x) = 1, \text{ if } x \text{ lies in no open set } U(k/2^n).$$

Since A is in every set $U(k/2^n)$, it follows that $f(A) = 0$. Similarly, the set B lies in no $U(k/2^n)$, so $f(B) = 1$. All that is left is to show that f is continuous. For the closed unit interval I^1, there is a basis consisting of all intervals of the three forms $[0, k/2^n)$, $(k_1/2^{n_1}, k_2/2^{n_2})$, and $(k/2^n, 1]$. The inverse under f of the first type is an open set $U(k/2^n)$, the inverse of the second type is a set $U(k_2/2^{n_2}) - \overline{U}(k_1/2^{n_1})$, and the inverse of the third type is $S - \overline{U}(k/2^n)$. Each of these is open in S. Thus the inverse of any open set in I^1 is open in S, and f is continuous. □

Remark: The definition of the function f given above does yield the value $f(x) = k/2^n$ for points x in $C(k/2^n)$, but it may happen that f has the value, say, of $\frac{1}{2}$ elsewhere as well as in $C(\frac{1}{2})$. In Fig. 2–5, the sets C are selected so that $C(\frac{1}{2} - 1/2^n)$ approaches the broken segment, which is not in the space. In such a case, $f(x) = \frac{1}{2}$ for all points x between the broken segment and $C(\frac{1}{2})$.

It is easy to give what appears to be greater generality to Theorem 2–29. Suppose that we want a function g mapping S onto the closed interval $[a, b]$, with $f(A) = a$ and $f(B) = b$. The function $h: I^1 \to [a, b]$, defined by $h(x) = a + (b - a)x$, is a homeomorphism. Hence the composite mapping hf, where f is the function given in Theorem 2–29, has the desired property. We state this result explicitly in a theorem equivalent to Theorem 2–29.

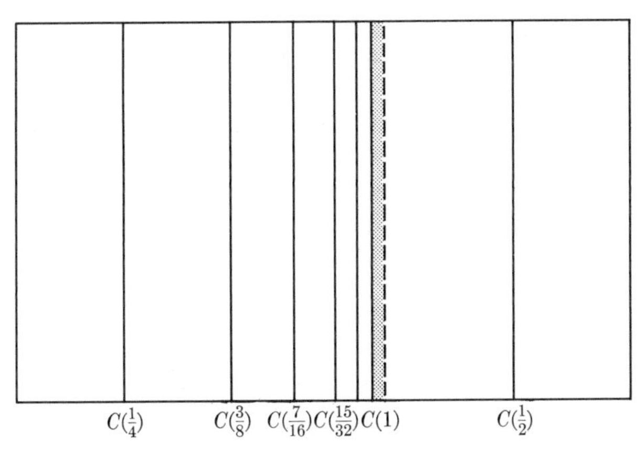

FIGURE 2-5

THEOREM 2-30. If A and B are disjoint closed subsets of a normal space S, then there is a continuous function f of S onto the linear interval $[a, b]$, with $f(A) = a$ and $f(B) = b$.

We may view Theorem 2-29 intuitively as claiming that a normal space S has a topology enough like that of I^1 to permit a very large collection of nonconstant mappings of S onto I^1. On the other hand, the existence of nonconstant mappings of I^1 into S is an entirely different matter! Even if S is quite reasonable, no such mappings need exist. For instance, let S consist of the rational numbers with the subspace topology of E^1. Since I^1 is a continuum, every continuous image of I^1 is a continuum. But the components of S are all single points. Therefore any continuous mapping of I^1 into S must be constant. These considerations will crop up again when we study arcwise connectivity in Section 3-2.

2-7 The Tietze extension theorem. Another way of looking at Theorem 2-29 requires a new definition. Let S and T be spaces, and let S' be a subspace of S. Suppose that $f'{:}S' \to T$ is continuous. Then a continuous mapping $f{:}S \to T$ is an *extension* of f' if $f(x) = f'(x)$ for all points x in S'. Rewording Theorem 2-29, we define the mapping f' of $A \cup B$ into I^1 by setting $f'(x) = 0$ for all points x in A and $f'(y) = 1$ for all points y in B. Clearly f' is continuous in $A \cup B$ if $A \cup B$ is considered as a subspace of S. Theorem 2-29 now asserts that we may extend f' to a mapping f of all of S into I^1.

Such extensions of mappings need not exist even for simple cases. For instance, let $S = E^1$ and $S' = E^1 - 0$. The function $f'{:}S' \to E^1$, defined by $f'(x) = x/|x|$, is continuous on S' but cannot be extended so as to be continuous at $x = 0$ in S. Several more such situations will occur

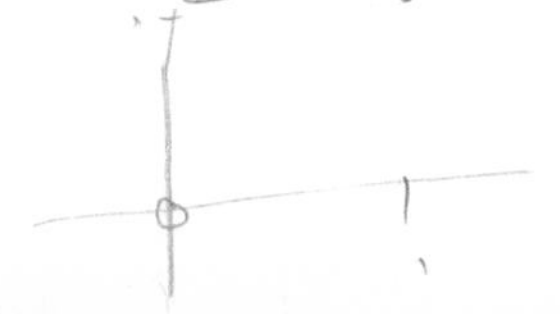

later, and extension of mappings will be discussed at some length. For now, we give a train of results that lead to a metrization theorem.

THEOREM 2–31 (Tietze's extension theorem). Let S be a normal space, and let C be a closed subset of S. Let $f':C \to [a, b]$ be a continuous mapping of C into the linear interval $[a, b]$. Then there exists an extension of f' to a mapping f of S into $[a, b]$.

Before proving this result, we consider infinite series defined on a general space. Suppose S is any topological space and that, for each natural number n, $f_n(x)$ is a real-valued function defined on S. We may form the infinite series $\sum_{n=1}^{\infty} f_n(x)$ just as in the calculus. Convergence of such a series for a fixed point x of S is defined to mean that the partial sums $\sum_{n=1}^{k} f_n(x)$ form a convergent sequence of real numbers, and the value of $\sum_{n=1}^{\infty} f_n(x)$ is taken to be the limit of these partial sums if such exists. The topology of S becomes important here when each $f_n(x)$ is assumed to be continuous and we investigate the continuity of the limit function. Even in this general situation, however, we have access to standard results, such as the following statement of the Weierstrass M-test.

THEOREM 2–32. Let S be a topological space and, for each positive integer n, let $f_n:S \to E^1$ be a real-valued continuous function. Suppose there exists a convergent series of positive numbers, $\sum_{n=1}^{\infty} M_n$, such that for each point x in S and each n, $|f_n(x)| \leqq M_n$. Then for each point x in S, the infinite series $\sum_{n=1}^{\infty} f_n(x)$ converges to a number $f(x)$, and the function f so defined is continuous.

Proof: For any particular point x in S, the series $\sum_{n=1}^{\infty} f_n(x)$ is absolutely convergent by the well-known comparison test from the calculus. Thus for each x, $f(x)$ exists. We remark that (a) if ϵ' is any given positive number, then there is an integer N such that for any point x in S and any integer $k > N$,

$$\left| f(x) - \sum_{n=1}^{k} f_n(x) \right| < \frac{\epsilon'}{3}.$$

For we know that

$$\left| f(x) - \sum_{n=1}^{\infty} f_n(x) \right| = \left| \sum_{n=k+1}^{\infty} f_n(x) \right| \leq \sum_{n=k+1}^{\infty} |f_n(x)| \leq \sum_{n=k+1}^{\infty} M_n,$$

and since $\sum_{n=1}^{\infty} M_n$ converges, we are able to choose N sufficiently large that $\sum_{n=k+1}^{\infty} M_n < \epsilon'/3$ whenever $k > N$. Continuing, we assume that each function $f_n(x)$ is continuous. Thus we may remark further that

(b) for each point x in S, there is an open set U_x in S, such that for any point y in U_x,

$$\left| \sum_{n=1}^{N} f_n(x) - \sum_{n=1}^{N} f_n(y) \right| < \frac{\epsilon'}{3},$$

where N and ϵ' are as above. Now consider

$$\begin{aligned} |f(x) - f(y)| &= \left| \sum_{n=1}^{N} [f_n(x) - f_n(y)] + \sum_{n=N+1}^{\infty} f_n(x) - \sum_{n=N+1}^{\infty} f_n(y) \right| \\ &\leq \left| \sum_{n=1}^{N} [f_n(x) - f_n(y)] \right| + \left| \sum_{n=N+1}^{\infty} f_n(x) \right| + \left| \sum_{n=N+1}^{\infty} f_n(y) \right| \\ &< \frac{\epsilon'}{3} + \frac{\epsilon'}{3} + \frac{\epsilon'}{3} = \epsilon'. \end{aligned}$$

Of course this is a standard proof from analysis (see Kaplan [16], p. 342) and proves that f is continuous. □

Proof of Theorem 2–31 (see Urysohn [127]). We may assume that the interval $[a, b]$ is the interval $[-1, 1]$, without loss of generality. For the mapping $h:[a, b] \to [-1, 1]$, given by $h(x) = (2x - a - b)/(b - a)$, is a homeomorphism and if the mapping hf' of C into $[-1, 1]$ can be extended to a mapping $f:S \to [-1, 1]$, then $h^{-1}f$ is the desired extension of f'.

We prove the theorem by constructing an infinite series that converges to f' on the set C and to some continuous function f on all of S. To start this, let H_1 be the subset of C on which $f'(x) \geqq \frac{1}{3}$, and let K_1 be the subset of C on which $f'(x) \leqq -\frac{1}{3}$. Since H_1 and K_1 are the inverse images of closed intervals under a continuous function f', they are closed in C and hence closed in S. Also H_1 and K_1 are obviously disjoint. By Urysohn's lemma in the form of Theorem 2–30, there is a continuous function $f_1:S \to [-\frac{1}{3}, \frac{1}{3}]$ which has the value $-\frac{1}{3}$ on K_1 and $+\frac{1}{3}$ on H_1. For this function, we have that $|f'(x) - f_1(x)| \leq \frac{2}{3}$ for all points x in C.

Next let H_2 be the subset of C on which $f'(x) - f_1(x) \geqq \frac{2}{9}$, and let K_2 be the subset of C on which $f'(x) - f_1(x) \leqq -\frac{2}{9}$. Again Urysohn's lemma yields a continuous function $f_2:S \to [-\frac{2}{9}, \frac{2}{9}]$, with value $\frac{2}{9}$ on H_2 and $-\frac{2}{9}$ on K_2, and this function satisfies the inequality

$$|f'(x) - f_1(x) - f_2(x)| \leqq \tfrac{4}{9}$$

for each point x in C. Continuing this process with the numbers $\frac{4}{27}, \frac{8}{81}, \ldots, 2^{n-1}/3^n, \ldots$, we obtain a sequence of continuous functions $f_n:S \to [-2^{n-1}/3^n, 2^{n-1}/3^n]$, with the property that

$$|f'(x) - f_1(x) - \cdots - f_n(x)| \leqq (\tfrac{2}{3})^n$$

for each point x in C. Letting $M_n = 2^{n-1}/3^n$ in Theorem 2–32, we see that the functions $f_n(x)$ satisfy the hypothesis of 2–32 and hence have a continuous sum $f:S \to E^1$. Since $\sum_{n=1}^{\infty} 2^{n-1}/3^n = 1$, it follows that $|f(x)| \leq 1$, as required, for all x. Also, $|f'(x) - f(x)| = 0$ for each point x in C, so f is the desired extension of f'. □

A number of important conclusions may be drawn from the Tietze theorem. First, we note that the hypothesis that f' be bounded in this theorem is not necessary. For we know that E^1 is homeomorphic to, say, the *open* interval (0, 1). Thus almost identical arguments suffice to give an extension of any real-valued continuous function on C. Another application of Theorem 2–31 proves that we can also extend some mappings that are not real-valued. We recall that I^n denotes the unit cube in E^n consisting of all points $x = (x_1, \ldots, x_n)$ for which $0 \leqq x_i \leqq 1$.

THEOREM 2–33. Let S be a normal space, and let $f':C \to I^n$ be a continuous mapping defined on the closed subset C of S. Then there is an extension f of f' to all of S.

Proof: For each point x in C, we have $f'(x) = (y_1, \ldots, y_n)$, a point in I^n. Define $f'_i(x) = y_i$. Clearly $f'_i = \pi_i f'$, where π_i is the projection of the product space I^n onto its ith factor. As the composition of two continuous functions, f'_i is certainly continuous. Hence by the Tietze theorem, f'_i can be extended to obtain a continuous mapping $f_i:S \to I^1$. For each point x in S, define $f(x) = (f_1(x), \ldots, f_n(x))$. The mapping f so defined certainly agrees with f' on C, and f is continuous by virtue of Theorem 1–37. Hence f is the desired extension of f'. □

This last result might suggest that for most spaces X, a mapping f' of a closed set C in a normal space S into X can be extended to a mapping $f:S \to X$. *This conjecture is not true!* Such an extension depends as much, or more, upon the topology of X as it does upon the topology of S. In fact, a space X which always permits such extensions is of a highly restricted category known as *absolute retracts.* For instance, the circle does not have this property. We will prove later (Section 6–16) that if D^2 denotes the closed disk in E^2 bounded by the unit circle S^1, then even the identity mapping $i:S^1 \to S^1$, defined by $i(x) = x$, cannot be extended to a continuous mapping $f:D^2 \to S^1$. We may follow this line of thought further by means of several new definitions.

A subset R of a space S is a *retract* of S provided that there is a continuous mapping $r:S \to R$, such that $r(x) = x$ for each point x in R. Such a mapping r is called a *retraction.* Equivalently, R is a retract of S if the identity mapping $i:R \to R$ can be extended to all of S. A space A is an *absolute retract* (often abbreviated AR) provided that if S is any normal space and A' is a closed subset of S that is homeomorphic to A, then A' is a retract of S. (See the definitive paper by Borsuk [67].)

THEOREM 2-34. The unit cube I^n is an absolute retract.

Proof: Let S be any normal space for which there is a homeomorphism h of I^n into S. Since $h(I^n) = A'$ is compact, it is a closed subset of S. By virtue of Theorem 2-33, the mapping $h^{-1}:A' \to I^n$ can be extended to yield a mapping $f:S \to I^n$. Then the composite mapping $hf:S \to A'$ is continuous, and $hf(x) = hh^{-1}(x) = x$ for each point x in A'. □

Although it was said in a different way, we mentioned above that the circle S^1 is not a retract of the disk D^2. However, all the spheres S^n do have an extension property, which is formulated precisely in the following result.

THEOREM 2-35. If C is a closed subset of a normal space S, and $f':C \to S^n$, $n \geqq 0$, is a continuous mapping of C into the n-sphere, then there is an open subset U of S, such that U contains C, and such that there exists an extension $f:U \to S^n$ of f' to all of U.

Proof: The n-sphere S^n may be considered as the boundary of the cube I^{n+1}, so Theorem 2-33 applies to give an extension $g:S \to I^{n+1}$ of f'. Now let p denote the point $(\frac{1}{2}, \frac{1}{2}, \ldots, \frac{1}{2})$ in I^{n+1}. Then there is a retraction r of $I^{n+1} - p$ onto the boundary of I^{n+1}, defined by radial projection from the point p. Now we have that $g^{-1}(I^{n+1} - p) = S - g^{-1}(p)$ is an open set U in S such that U contains C and the composite mapping rg is defined and continuous on U. That $rg(x) = rf'(x) = f'(x)$ for any point x in C is immediate, and hence rg is the desired extension of f'. □

The extension property of S^n expressed in Theorem 2-35 is formulated in retract language as follows. A subset N of a space S is a *neighborhood retract* of S if there is an open set U in S containing N, such that N is a retract of U. Again an equivalent definition is obtained by requiring that the identity mapping $i:N \to N$ be extendable to the open set U. Then a space B is an *absolute neighborhood retract* (often abbreviated ANR) if, for each normal space S and each closed subset B' of S that is homeomorphic to B, B' is a neighborhood retract of S. The reader may follow the proof of Theorem 2-34 to translate Theorem 2-35 into the following form.

THEOREM 2-36. The n-sphere S^n is an absolute neighborhood retract.

Two more theorems on the extension of mappings will be of use later.

THEOREM 2-37. Let A be a closed subset of a normal space S. Let U be an open set in S containing A, and suppose that there is a continuous mapping $f(x, t)$ defined on the subset $(U \times I^1) \cup (S \times 0)$ of $S \times I^1$ and throwing this subset into an arbitrary space Y. Then there exists a mapping $\bar{f}:S \times I^1 \to Y$ that agrees with f on $(A \times I^1) \cup (S \times 0)$.

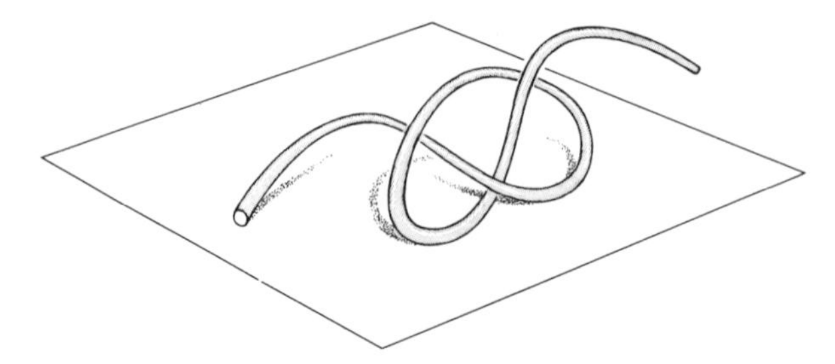

FIGURE 2–6

Proof: The sets A and $S - U$ are disjoint closed subsets of the normal space S. Urysohn's lemma (2–29) provides a mapping $u:S \to I^1$ such that $u(A) = 1$ and $u(S - U) = 0$. Define the mapping

$$\bar{f}(x, t) = f(x, t \cdot u(x)).$$

Clearly $\bar{f}$ is defined if x lies in U, for then the number $t \cdot u(x)$ satisfies the inequality $0 \leq t \cdot u(x) \leq 1$. If x is not in U, then $\bar{f}(x, t) = f(x, t \cdot 0) = f(x, 0)$ and is defined. Thus $\bar{f}$ is defined on all of $S \times I^1$, and $\bar{f}$ is certainly continuous. Moreover, if x lies in A, then $\bar{f}(x, t) = f(x, t \cdot 1) = f(x, t)$ while $\bar{f}(x, 0) = f(x, 0)$ regardless of where x may lie. □

THEOREM 2–38. Let A be a retract of a space X. Then any mapping $f:A \to Y$, where Y is arbitrary, can be extended to all of X.

Proof: By definition, there is a retraction mapping $r:X \to A$, such that $r(x) = x$ for each point x in A. The composite mapping $fr:X \to Y$ is clearly an extension of f. □

EXERCISE 2–15. Let A be an arc in E^3 tied in a simple overhand knot as in Fig. 2–6. Find a retraction of E^3 onto A.

2–8 Completely separable spaces. Many theorems may be extended and improved if we limit consideration to spaces with a countable basis. Such a restriction is in line with our successive specialization of the general topological space, of course, and furthermore it is on the road to a metrization theorem (Section 2–9).

As was remarked in Section 2–4, the existence of a countable basis permits constructive proofs of theorems which require the axiom of choice in more general cases. Before illustrating this, we will need some results about spaces with a countable basis. We recall that a space that contains a countable dense subset is called *separable*. We now introduce the term *completely separable* for a space with a countable basis. (The term *second-countable* is also used for such a space. *First-countable* means that at each point of the space there is a countable basis for the open sets containing that point.) Our first result is merely a restatement of Theorem 1–5.

THEOREM 2–39. Every separable metric space is completely separable.

THEOREM 2–40. Every completely separable space is separable.

Proof: Take one point from each element of a countable basis. □

THEOREM 2–41. If S is a completely separable space, then every subspace of S is completely separable, and hence every subspace is separable. (In other words, the property of being completely separable is *hereditary.*)

Proof: If $\{B_n\}$ is a countable basis for S, and X is any subset of S, then the collection $\{B_n \cap X\}$ is a countable basis for the subspace topology of X. The remainder follows from Theorem 2–40. □

In Theorem 2–41 we have reason for using the term *"completely" separable.* The property of separability alone is not hereditary. At the end of Section 1–4 we gave an example of a separable space having an uncountable subset having no limit point at all. This certainly serves as an example to show that not all subspaces of a separable space are separable. Indeed, the subset given in the example just cited cannot occur in a completely separable space, as the next, slightly startling result shows.

THEOREM 2–42. In a completely separable space, every uncountable subset contains uncountably many limit points of itself.

Proof: If X is an uncountable subset of the completely separable space S, then by Theorem 2–41, X contains a countable dense subset Y of itself. Each point of X is either a point or a limit point of Y. But $X - Y$ is uncountable and, since each point of $X - Y$ is a limit point of Y, the theorem is true. □

Again we point out that the example in Section 1–4 shows that the hypotheses of Theorem 2–42 cannot be weakened to separability (unless something else, like metrizability, is added). We next give an example to prove that the converse of Theorem 2–42 is not true. That is, we construct *a space in which every uncountable subset contains a limit point of itself, but the space is not completely separable.*

Let S be the union of the unit interval I^1 and a point p not on I^1. A basis for a topology in S will consist of all relatively open subsets of I^1 in the usual topology of E^1, together with all sets that are the union of the point p and the complement in I^1 of any finite set. Since in its usual topology I^1 itself has a countable basis, every uncountable subset of S contains a limit point of itself. Suppose that S had a countable basis. Given any countable number of elements $U_1, U_2, \ldots$ of that basis, all containing the point p, the set $I - (I - U_1) - (I - U_2) - \cdots$ is nonempty (in fact, is uncountable) and, for any point y in this set, $S - y$ is an open set containing p but not containing any of the sets $U_1, U_2, \ldots$ This contradicts the assumption that we have a countable basis for S.

The above example is certainly separable. Since it is not completely separable, it cannot be a metric space (Theorem 2–39). In fact, adding the hypothesis that the space be metric allows us to prove a converse of Theorem 2–42.

THEOREM 2–43. If M is a metric space in which every uncountable subset has a limit point, then M is completely separable.

Proof: In view of Theorem 2–39, it suffices to prove that M is separable. The scheme of the proof will be to choose for each integer n a subset X_n in M, with the property that every point in M is at most a distance $1/n$ away from a point of X_n, while no two points of X_n are less than a distance $1/n$ apart. Such a set X_n obviously has no limit points and hence cannot be uncountable by hypothesis. Then the union $\bigcup_{n=1}^{\infty} X_n$ will be dense in M and, as the countable union of countable sets, will be countable.

To prove that such sets X_n exist, let n be a positive integer, and consider all subsets X_α of M having the property that each two points of X_α are not less than $1/n$ apart. Such sets obviously exist, at least for all sufficiently large values of n. Partially-order the collection of all such sets X_α by inclusion. Starting with any fixed X_α, there is a maximal simply-ordered subcollection $\{X_\beta\}$, each X_β containing X_α, by the maximal principle. We let $X_n = \bigcup_\beta X_\beta$. We note that X_n is also a set X_β, for each pair of points x and y of X_n belongs to some X_β and hence x and y are not less than $1/n$ apart. Furthermore, X_n includes each X_β. Now if there were a point p of M at a distance not less than $1/n$ from each point of X_n, then the set $X_n \cup p$ would be an X_β containing X_n, which would contradict the maximality of X_n. □

This is a convenient place to insert two results that will be valuable later.

THEOREM 2–44 (Lindelöf's theorem). Let X be a subset of a completely separable space S, and let $\{U_\alpha\}$ be a collection of open sets covering X. Then some countable subcollection of $\{U_\alpha\}$ also covers X.

Proof: For each point x in X, there is at least one open set $U_\alpha(x)$ containing x. Given a countable basis $\{B_n\}$ for S, the definition of a basis says that there is a basis element B_i containing x and contained in $U_\alpha(x)$. Let $\{B_i\}$ be the subcollection of $\{B_n\}$ consisting of all such sets B_i, each contained in an element of $\{U_\alpha\}$. The collection $\{B_i\}$ covers X. Then for each B_i, let U_i be any one of the sets U_α that contain B_i. It is clear that the countable collection $\{U_i\}$ also covers X. □

The use of open coverings of a space becomes progressively more important as we go deeper into topology. Sections 2–11 and 2–12 and large parts of Chapter 8 will exemplify this.

THEOREM 2–45. A completely separable regular space S is completely normal and hence is normal.

Proof: Let A and B be two separated sets in S. Since $A \cap \overline{B}$ is empty, each point x in A has an open set U_x containing x, such that $\overline{U}_x$ is contained in $S - \overline{B}$ (this follows from the regularity of S). The collection of all such sets U_x covers A and, by Theorem 2–44, there is a countable subcovering $\{U_i\}$ of $\{U_x\}$. In the same way, we find a countable collection $\{V_i\}$ covering B, such that $\overline{V}_i \cap \overline{A}$ is empty for each i. Now let $W_1 = U_1$, $X_1 = V_1 - \overline{U}_1$, and, inductively, $W_{n+1} = U_{n+1} - [\bigcup_{i=1}^{n} \overline{V}_i]$ and $X_n = V_n - [\bigcup_{i=1}^{n} \overline{U}_i]$. The two open sets $U = \bigcup_{n=1}^{\infty} W_n$ and $V = \bigcup_{n=1}^{\infty} X_n$ are then disjoint open sets containing A and B respectively. □

EXERCISE 2–16. Prove that the product of two separable spaces is separable.

EXERCISE 2–17. Prove that the product of two completely separable spaces is completely separable.

EXERCISE 2–18. Give a proof of Theorem 2–43 using well-ordering.

EXERCISE 2–19. Prove that a subset of a completely separable T_2-space is compact if and only if it is countably compact.

(We note that in view of Theorem 2–39, Exercise 2–19 shows that *compact sets and countably compact sets are identical in a separable metric space.* The following exercises pursue this further.)

EXERCISE 2–20. Prove that if the metric space M of Theorem 2–43 is countably compact, then the sets of points X_n are finite.

EXERCISE 2–21. Prove that every countably compact metric space is separable.

EXERCISE 2–22. For subsets of any metric space, show that compact sets and countably compact sets are identical.

2–9 Mappings into Hilbert space. A metrization theorem. In Section 2–7 we were concerned with extensions of mappings of a normal space into E^n. As we pointed out then, the theorems in Section 2–7 assure us of a large collection of nonconstant mappings of a normal space into E^n. But if we ask for more than this, the situation changes quite drastically. A natural desire would be to ask for a homeomorphism of a space S into E^n. For such a homeomorphism to exist, a number of conditions on S are easily seen to be necessary. Since S would be homeomorphic to a subset of E^n, it follows that S must have been metrizable, must have had a countable basis, and so on. However, conditions on S that are both necessary and sufficient for S to be homeomorphic to a subset of E^n are not yet known.

In certain special instances, questions about the existence of homeomorphisms of a particular space into E^n have been answered. For instance, if $m > n$, then there is no homeomorphism of E^m into E^n. This is the Brouwer theorem on the invariance of domain, which will be met again in Section 6–17. Such a result is intuitively obvious but, although E^{827} and E^{819}, say, are defined differently, it is not easy to show that E^{827} cannot be parametrized with 819 coordinates. As another example, we state in Section 3–9 that if a separable metric space has dimension n, in a sense to be defined, then it *can* be mapped homeomorphically into E^{2n+1}.

As a natural generalization of Euclidean space, we define next the *Hilbert coordinate space.* The points of this space are sequences $x = \{x_1, x_2, \ldots\}$ of real numbers satisfying the condition that $\sum_{i=1}^{\infty} x_i^2$ is convergent. This collection is topologized by means of the metric

$$d(x, y) = \left[\sum_{i=1}^{\infty} (x_i - y_i)^2\right]^{1/2}.$$

This is the natural generalization of the Euclidean metric.

Intuitively, Hilbert space is a Euclidean space of infinitely many dimensions and has more room in it than any space E^n. It is true that every separable metric space can be mapped homeomorphically into Hilbert space. In fact, we show next that an apparently larger class of spaces can be so imbedded in Hilbert space, namely, the class of all normal spaces with a countable basis. We say "apparently larger" because, as is obvious, if a space may be imbedded in Hilbert space, then we can utilize the imbedding homeomorphism to apply the Hilbert metric to the original space. In this sense, our next result is a metrization theorem, and proves that every normal space with a countable basis may be assigned a metric equivalent to the original topology. We should point out that the existence of nonseparable metric spaces proves that the conditions used here are sufficient but not necessary for metrizability. Such necessary and sufficient conditions are stated in Section 1–12.

THEOREM 2–46. Every completely separable normal space S can be imbedded in Hilbert coordinate space. (See Urysohn [127].)

Proof: Let $B_1, B_2, \ldots, B_n, \ldots$ be a countable basis for S. In view of Theorem 2–4, there are pairs B_i, B_j, such that $\overline{B}_i$ is contained in B_j; in fact, each point of S lies in infinitely many such pairs, or is itself an open set. However, there are at most a countable number of pairs for each point of S. For each pair B_i, B_j with $\overline{B}_i$ contained in B_j, Urysohn's lemma

(Theorem 2-29) provides a function f_n of S into I^1 with the property that $f_n(\overline{B}_i) = 0$ and $f_n(S - B_j) = 1$. (If the point p forms an open set, then we take $f_n = 0$ for large n.) Letting H denote Hilbert coordinate space, we define a mapping f of S into H by setting

$$f(x) = \left\{f_1(x), \frac{f_2(x)}{2}, \frac{f_3(x)}{3}, \ldots, \frac{f_n(x)}{n}, \ldots\right\}$$

for each point x in S. Since the series $\sum_{n=1}^{\infty} (f_n(x)/n)^2$ is dominated by the convergent series $\sum_{n=1}^{\infty} 1/n^2$, this definition of $f(x)$ does yield a point of H. It remains to prove that the function f so defined is continuous, one-to-one, and interior (see Theorem 1-9).

To establish continuity, let p be any point in $f(S)$, and let $S(p, \epsilon)$ be a spherical neighborhood of p in H. We show that $f^{-1}(S(p, \epsilon))$ is open. First, it is easily seen that there exists an integer N sufficiently large so that if y is any point in S, then we have

$$\sum_{n=N}^{\infty} \frac{[f_n(p) - f_n(y)]^2}{n^2} < \frac{\epsilon}{2}.$$

Since the functions f_n are continuous, there is an open set U_j in S for each integer $j < N$ such that U_j contains p and, if y is any point in U_j,

$$\frac{[f_n(p) - f_j(y)]^2}{j^2} < \frac{\epsilon}{2N}.$$

The intersection $\bigcap_{j=1}^{N} U_j$ is an open set containing p and such that, if y is any point in this intersection, we have

$$\sum_{n=1}^{\infty} \frac{[f_n(p) - f_n(y)]^2}{n^2} = \sum_{n=1}^{N-1} \frac{[f_n(p) - f_n(y)]^2}{n^2} + \sum_{n=N}^{\infty} \frac{[f_n(p) - f_n(y)]^2}{n^2}$$

$$< \frac{(N-1)\epsilon}{2N} + \frac{\epsilon}{2} < \epsilon.$$

This implies that $f(\bigcap_{j=1}^{N} U_j)$ is contained in $S(p, \epsilon)$ so that $\bigcap_{j=1}^{N} U_j$ lies in $f^{-1}(S(p, \epsilon))$.

That f is one-to-one is also easy to prove. For if x and y are distinct points of S, then there is a pair B_i, B_j, with x in B_i, $\overline{B}_i$ contained in B_j, and y in $S - \overline{B}_j$. This comes from the normality of S, of course. Thus there is an integer n such that $f_n(x) = 0$ and $f_n(y) = 1$, so $f(x)$ and $f(y)$ are distinct also.

It remains to show that f is interior. To this end, let x be a point in an open set U in S. There is a pair B_i, B_j such that x lies in B_i, $\overline{B}_i$ is contained in B_j, and B_j is contained in U. Therefore there is an integer n for which $f_n(x) = 0$, while $f_n(S - U) = 1$. It follows that for any point y in $S - U$, we have $d(f(x), f(y)) \geqq 1/n$. For

$$d(f(x), f(y)) \geqq \left(\frac{[f_n(x) - f_n(y)]^2}{n^2}\right)^{1/2} = \frac{1}{n}.$$

Thus the spherical neighborhood $S(f(x), 1/2n)$ lies entirely in $f(U)$, and $f(U)$ is open. □

In view of Theorem 2–45, we may give the following generalization of Theorem 2–46.

THEOREM 2–47. Every completely separable regular space can be imbedded in Hilbert coordinate space.

Since we know that a metric space is normal (Theorem 2–7), we may state the following characterization of metrizable spaces with a countable basis.

THEOREM 2–48. A necessary and sufficient condition for the metrizability of a completely separable space is regularity.

Several comments deserve mention at this point. In the above proof of Theorem 2–46, the space S was actually mapped into the subset I^ω of Hilbert space consisting of all points $x = (x_1, x_2, \ldots, x_n, \ldots)$ satisfying the inequality $0 \leq x_j \leq 1/j$ for each integer j. The subset I^ω is called the *Hilbert cube* or the *Hilbert parallelotope*. Next we note that Theorem 2–46 is a weaker result than we may have been led to expect from analogy to the theorems in Section 2–7. A strictly analogous result would be concerned with the extension of a homeomorphism of a closed subset of S into H. In general, however, no such extension is possible. For example, the set X of all points in H whose first coordinate is 0 is clearly homeomorphic to H by a homeomorphism $h:X \to H$. Considering X as a closed subset of H, there is no extension of the homeomorphism h to a homeomorphism $h^*:H \to H$, since h is already a map onto H.

It is perhaps more important to discuss the motivation behind such imbedding theorems. Theorem 2–46 may be considered as an example of a very general technique (the method of representations) found throughout mathematics. A representation theorem permits the study of certain invariant properties by noninvariant means. For instance, the basic concepts of "coordinates," "slope of a line," and "equation of a curve" in analytic geometry are not invariant under rigid motions. Still we use these tools to prove propositions in Euclidean geometry. Another example is found in the representation of an abstract group as a group of matrices. In a similar way, Theorem 2–46 allows us to prove topological theorems about completely separable normal spaces by utilizing coordinates, straight lines, etc., in Hilbert coordinate space, although all these concepts are of a nontopological nature.

EXERCISE 2–23. Prove that Hilbert coordinate space is separable.

EXERCISE 2–24. Prove by some direct method that I^ω is compact.

EXERCISE 2–25. Prove that I^ω is homeomorphic to $\mathsf{P}_{n=1}^{\infty} I_n$, where each $I_n = I^1$. (Note that this affords an indirect solution to Exercise 2–24.)

EXERCISE 2–26. The unit sphere in H is the set of all points x for which $\sum x_i^2 = 1$. Is this subset compact? If so, prove it, and if not, show why it is not.

2–10 Locally compact spaces. We exhibit here one form of a general concept called *localization*. Speaking loosely, localization of a topological property, such as compactness or connectedness, is the requirement that "small" open sets have the desired property even though the space as a whole may not.

A space is said to be *locally compact at a point* p if there is some open set U containing p whose closure $\overline{U}$ is compact. The space is locally compact if it is *locally compact* at each of its points. Notice that any other open set in U will also have compact closure, and hence no space can be locally compact at just one point unless that point is an isolated point. (This is in contradistinction to the local connectivity property in Section 3–1.)

If the reader returns to the examples in Section 2–4, he will find that although the spaces given there (particularly in Fig. 2–3) are not compact, they are locally compact. Hence, if we restrict our attention to locally compact spaces, we are certain to lose some of the power of compactness. But much still remains, and we establish some important results here. The first theorem simply shows that we are dealing with a valid generalization of compactness.

THEOREM 2–49. Every compact space is locally compact.

Proof: Every point is contained in an open set, namely the entire space, with compact closure. □

THEOREM 2–50. Each closed subspace of a locally compact space is locally compact.

Proof: Let S be a locally compact space and C be a closed subset of S. Every open set in S which intersects C yields an open set in the subspace topology of C. Thus if x is a point of C, and U is an open set in S with compact closure containing x, then $U \cap C$ is open in C, and $\overline{U \cap C}$ is a closed subset of $\overline{U}$ and is therefore compact. □

We know that Euclidean n-space E^n is not compact; however, E^n is *locally compact,* because every spherical neighborhood $S(x, r)$ in E^n has a closure homeomorphic to the compact cube I^n. Indeed, any *open* subset of E^n is locally compact, for each point of such an open set lies in a spherical neighborhood whose closure is contained in the open set. From this, one sees that the locally compact spaces include some special cases of importance in analysis.

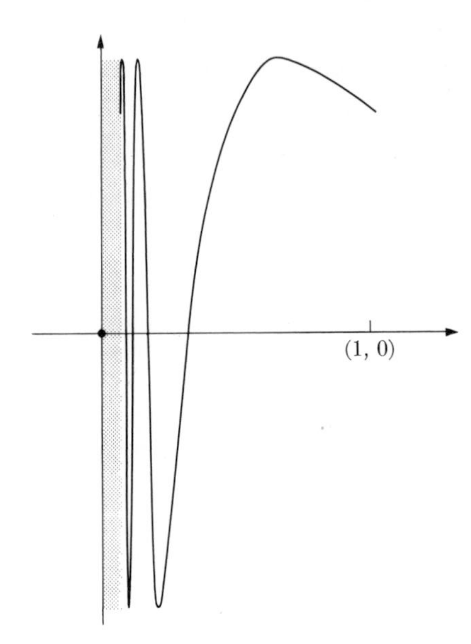

FIGURE 2–7

Compactness is a continuous invariant, but this is not true of local compactness. For consider the space S consisting of the point -1 and the open interval $0 < x < 1$ in E^1 with the subspace topology of E^1. Let T be the subspace of E^2 consisting of the point $(0, 0)$ and the graph of the function $y = \sin(1/x)$, $0 < x < 1$. Define the mapping $f:S \to T$ that carries the point -1 onto $(0, 0)$ and that carries a point x, $0 < x < 1$, onto the point $(x, \sin(1/x))$. Figure 2–7 illustrates the space T. It is easy to see that no open set containing $(0, 0)$ has compact closure in T and hence that T is not locally compact. Certainly S *is* locally compact, so this is an example of a continuous function destroying the local compactness property.

The mapping f in the above example has the property that the open set in S consisting of the single point -1 is mapped onto the nonopen set $(0, 0)$ in T. In brief, the mapping f is not *open*. For open mappings we have the following result which, incidentally, proves that local compactness is a topological invariant.

THEOREM 2–51. Local compactness is invariant under open mappings.

Proof: Let S be a locally compact space, and let f be an open mapping of S onto a space T. For any point p in S, there is an open set U containing p and having compact closure. Then the image $f(U)$ is an open set in T containing $f(p)$. Now $f(\overline{U})$ is compact since f is continuous. Hence the interior of $f(\overline{U})$ is an open set containing $f(p)$ and having compact closure.

Since every point of T is the image of some point of S, this proves that T is locally compact. □

The next pair of theorems are in analogy to the Tychonoff theorem (1–28).

THEOREM 2–52. The product of finitely many locally compact spaces is a locally compact space.

Proof: Let S_1, $S_2, \ldots,$ and S_n each be locally compact, and let $p = (p_1, \ldots, p_n)$ be any point in $S_1 \times S_2 \times \cdots \times S_n$. In each space S_i there is an open set U_i containing p_i and having compact closure. Then the basis element $U_1 \times U_2 \times \cdots \times U_n$ contains p and has closure $\overline{U}_1 \times \overline{U}_2 \times \cdots \times \overline{U}_n$, which is compact by the Tychonoff theorem. Hence $S_1 \times S_2 \times \cdots \times S_n$ is locally compact. □ (Why does this proof fail for an infinite product?)

THEOREM 2–53. If a product space $\mathbb{P}_A S_x$ is locally compact, then each factor space S_α is locally compact, and all but a finite number of factors are compact.

Proof: We know that the projection mappings $\pi_\alpha\colon \mathbb{P}_A S_\alpha \to S_\alpha$ are open. Hence Theorem 2–49 applies to prove that each factor S_α is locally compact. But furthermore, if p is any point of $\mathbb{P}_A S_\alpha$, and U is an open set containing p and having compact closure, then there is an element $\mathbb{P}_A U_\alpha$ of the basis which contains p and lies in U. Hence each U_α has compact closure. By definition, all but a finite number of sets U_α are the entire spaces S_α, so $\overline{U}_\alpha = S_\alpha$ for all but a finite number of factors S_α. □

It is usually easier to deal with a compact space than with a noncompact space. For this reason the topologist often imbeds a noncompact space in a compact space as an aid to proving theorems. (A homeomorphism of a space S *into* a space T is called an *imbedding* of S in T, and S is said to be *imbedded* in T.) For locally compact Hausdorff spaces, this can be done in a very simple manner. The process is a generalization of the familiar process in which we add a "point at infinity" to the plane of complex numbers to obtain the complex sphere.

Let S be a T_1-space that is not compact, and let ω be any abstract element not in S. The *one-point compactification* $\widetilde{S}$ of the space S consists of the points of $S \cup \{\omega\}$ with a basis for a topology of $\widetilde{S}$ consisting of (a) all open sets of S and (b) all subsets U of $\widetilde{S}$ such that $\widetilde{S} - U$ is a closed compact subset of S. Of course, a single point of any space whatsoever is a compact set, but a single point need not be closed. Hence we make this definition only for T_1-spaces, so that (b) is not satisfied vacuously.

THEOREM 2–54. The one-point compactification $\widetilde{S}$ of a T_1-space S is a compact T_1-space. And $\widetilde{S}$ is Hausdorff if and only if S is Hausdorff and locally compact.

Proof: We first show that $\tilde{S}$ is a T_1-space by proving that points of $\tilde{S}$ are closed. That the point ω is closed follows immediately because $\tilde{S} - \omega = S$ is open in $\tilde{S}$ by definition. If x is any point of S, then x has no limit points in S, for the topology in S has not been changed. Since x is a closed compact set in S, $\tilde{S} - x$ is an open set in $\tilde{S}$ that contains ω but not x. Therefore ω is not a limit point of x, either. Each point of $\tilde{S}$ is closed, and $\tilde{S}$ is a T_1-space.

To prove that $\tilde{S}$ is compact, let $\mathfrak{U}$ be any covering of $\tilde{S}$ by open sets. Then there is at least one element U_ω of $\mathfrak{U}$ containing ω. By definition, there is an element of the basis, say V, that contains ω and lies in U_ω. Then $\tilde{S} - V$ is a compact set in S that is covered by $\mathfrak{U}$. Hence there is a finite subcovering $\{U_1, U_2, \ldots, U_n\}$ in $\mathfrak{U}$ such that $\cup_{i=1}^n U_i$ contains $\tilde{S} - V$. It follows that $\{U_\omega, U_1, \ldots, U_n\}$ is a finite covering of $\tilde{S}$. This proves the first part of the theorem.

Next suppose that $\tilde{S}$ is a Hausdorff space. We show that S is locally compact. Given any point x in S, there are disjoint open sets U and V in $\tilde{S}$, with x in U and ω in V. There is no loss of generality in assuming V to be an element of the basis so that $\tilde{S} - V$ is compact. Thus U is an open set in S contained in the compact set $\tilde{S} - V$, and hence $\overline{U}$ is compact. Therefore S is locally compact. To see that S is Hausdorff, we need only note that S is a subspace of a Hausdorff space.

Finally, suppose that S is a locally compact Hausdorff space. Let x and y be two points of $\tilde{S}$. If x and y both lie in S, then there are disjoint open sets U and V containing x and y respectively, and U and V are also open in $\tilde{S}$. If y, say, is the point ω, then we know that there is an open set U in S such that U contains x and $\overline{U}$ is compact. The set $\tilde{S} - \overline{U}$ is an open set V in S that contains ω, and U and V are disjoint open sets containing x and ω respectively. □

We give some important properties of locally compact Hausdorff spaces as examples of the usefulness of the one-point compactification. The first of these is a direct generalization of Theorem 2–1.

THEOREM 2–55. Every locally compact Hausdorff space is regular.

Proof: Let S be a locally compact Hausdorff space, and let $\tilde{S}$ be its one-point compactification. Then $\tilde{S}$ is regular by Theorem 2–1 and, as a subspace of a regular space, S is also regular. □

A space S is said to be *completely regular* (also called a *Tychonoff space*) if for every point p of S and for any open set U containing p, there is a continuous function of S into I^1 such that $f(p) = 0$ and $f(x) = 1$ for all points x in $S - U$.

THEOREM 2–56. Every locally compact Hausdorff space is completely regular.

Proof: Again let $\tilde{S}$ be the one-point compactification of a locally compact Hausdorff space S. Then $\tilde{S}$ is normal by Theorem 2–3. Let p be any point of S, and let U be an open set in S containing p. Since S is locally compact, there is an open set W in S containing p, such that $\overline{W}$ is compact. Then $\tilde{S} - \overline{W}$ is an open set V in $\tilde{S}$. Now $\overline{V}$ and p are disjoint closed sets in the normal space $\tilde{S}$, and Urysohn's lemma applies to give a real-valued function $f:\tilde{S} \to I^1$, such that $f(p) = 0$ and $f(\overline{V}) = 1$. Since $\overline{V}$ contains $S - U$, this completes the proof. □

Then, in analogy to Theorem 2–14, the reader may apply the same sort of argument to prove the next result.

THEOREM 2–57. In a locally compact Hausdorff space, every compact quasicomponent is a component, and every compact component is a quasicomponent.

THEOREM 2–58. If the space S is a locally compact, completely separable Hausdorff space, then the one-point compactification $\tilde{S}$ of S is metrizable.

Proof: Let $\{U_n\}$ be a countable basis for S. Clearly there is no loss of generality in assuming that each set $\overline{U}_n$ is compact. Let $V_n = \tilde{S} - \bigcup_{j=1}^{n} \overline{U}_j$. If V is an open set in $\tilde{S}$ that contains ω, then $S - V$ is compact and so is covered by a finite number of basis elements $U_{n_1}, \ldots, U_{n_k}$; let $n = \max(n_1, \ldots, n_k)$. Then $S - V$ is contained in $\bigcup_{j=1}^{n} U_j$, so V_n lies in V. This (and the fact that $\{U_n\}$ is a basis for S) establishes that the collection of all sets V_n and of all sets U_n is a countable basis for the normal space $\tilde{S}$. □

COROLLARY 2–59. Every locally compact, completely separable Hausdorff space is metrizable.

Proof: A subspace of a metrizable space is metrizable. □

Euclidean n-space E^n is the union of all spherical neighborhoods $S(O, n)$ of the origin with positive integral radii. Clearly, the closure of each $S(O, n)$ is compact and is contained in $S(O, n + 1)$. This situation is not peculiar to E^n; it actually characterizes the locally compact separable metric spaces.

THEOREM 2–60. A separable metric space M is locally compact if and only if M is the union of a countable number of open sets $U_1, U_2, \ldots$ such that for each n, $\overline{U}_n$ is a compact subset of U_{n+1}.

Proof: Clearly, if M has such a sequence of open sets, then every point of M lies in some set U_n with compact closure, and hence M is locally compact. Conversely, if M is a locally compact separable metric space, then $\tilde{M}$ is a compact metric space by Theorem 2–58. Define $U_n = \tilde{M} - \overline{S}(\omega, 1/n)$, $n = 1, 2, 3, \ldots$ Then each U_n is open in M, and $\overline{U}_n$ is compact both in M and in $\widetilde{M}$. The set $\overline{U}_n - U_n$ is contained in $\overline{S}(\omega, 1/n)$, so U_{n+1} contains not only U_n but also $\overline{U}_n$. It is possible that if $\widetilde{M}$ happens to have a small diameter, the sets U_n may be empty for the first few values of n, but this does not change the statements. □

Euclidean spaces have the property that every bounded set has compact closure. This is not true of every locally compact separable metric space. That is, it may not be true of such a space in its given metric. For instance, if we use the subspace metric, the open interval is a locally compact separable metric space that is itself bounded but not compact. However, every locally compact separable metric space has *a* metric such that each set bounded in this new metric has a compact closure. This is the essence of the next result.

THEOREM 2–61. Let M be a locally compact separable metric space. Then M may be imbedded in Hilbert coordinate space H in such a way that every subset of M that is bounded in the Hilbert space metric has a compact closure.

Proof: From Theorem 2–58, M is contained in a compact metric space $\widetilde{M}$, such that $\widetilde{M} - M$ is a single point ω. From Theorem 2–46, we know there is a homeomorphism h of $\widetilde{M}$ into H. Now H is a vector space over the real numbers (see Chapter 5), and the translation φ of H onto itself, defined by $\varphi(p) = p - h(\omega)$, is a homeomorphism. Clearly φ carries $h(\omega)$ onto the origin in H, so the composite mapping φh is an imbedding of $\widetilde{M}$ into H such that $\varphi h(\omega) = O$, the origin in H. We next define an "inversion" in H. Given any point $p = (p_1, p_2, \ldots)$ in $H - O$, there is a real number t_p such that $t_p^2 \cdot \sum_{n=1}^{\infty} p_n^2 = 1$. Let j be the mapping of $H - O$ onto itself, defined by

$$j(p) = \frac{1}{t_p} \cdot p = \left(\frac{p_1}{t_p}, \frac{p_2}{t_p}, \ldots\right).$$

Then j is a homeomorphism of $H - O$ onto itself that cannot be extended to be continuous at the origin. The composite mapping $j\varphi h|M$ ($j\varphi h$ restricted to M) of M into $H - O$ is the desired imbedding of M in H. The reader may prove this last statement as an exercise. □

One might suppose that if M is any locally compact metric space, then the one-point compactification $\widetilde{M}$ could also be taken to be metric. *This is not true for nonseparable metric spaces* (which is why we considered only the separable case in Theorem 2–60 and 2–61). For if $\widetilde{M}$ were a compact metric space, then it would be completely separable, as is easily seen. Since complete separability is hereditary, this would imply that the subspace M of $\widetilde{M}$ would be separable, counter to hypothesis. The space $\widetilde{M}$ in this case would be compact and Hausdorff but not metric.

The one-point compactification is only one of many ways of compactifying a space. In general, a compactification of a space S may be defined to be a pair $(\widetilde{S}, f)$ where $\widetilde{S}$ is a compact space and f is a homeomorphism of S into $\widetilde{S}$. We mention briefly the *Stone-Čech compactification*. This is currently finding application in functional analysis, rings of continuous functions, and similar studies. It is defined as follows. For a given topological space S, let C denote the collection of all continuous functions of S into the closed unit interval I^1. Consider the product space $\mathbb{P}_C I^1_f$, the indexing set being the collection C; that is, we have one factor I^1 for each mapping f in C. This is a compact space by the Tychonoff theorem. Next, the *evaluation mapping* e of S into $\mathbb{P}_C I^1_f$ is the mapping that carries each point x of S onto the point $e(x)$ in $\mathbb{P}_C I^1_f$, where the coordinate in I^1_f of $e(x)$ is the functional value $f(x)$. It can be shown that e is continuous and that, if S is a completely regular T_1-space, then e is a homeomorphism of S into $\mathbb{P}_C I^1_f$. If we take $\widetilde{S}$ to be the closure of $e(S)$ in $\mathbb{P}_C I^1_f$, then the pair $(\widetilde{S}, e)$ is the Stone-Čech compactification of S. We leave the topic with this definition and a reference to the recent book by J. L. Kelley [17].

2–11 Paracompact spaces. The concept of a paracompact space was introduced in 1944 by Dieudonne [73] as a generalization of certain compact spaces. We insert mention of this topic here for two reasons. First, it affords another example of the widespread use of open coverings of a space. Second, a most definitive work on the metrization problem (see Section 2–12) is couched in the language of paracompactness.

A covering $\{V_\beta\}$ of a space S is said to be a *refinement of a covering* $\{U_\alpha\}$ if for each element V_β of $\{V_\beta\}$ there is an element U_α of $\{U_\alpha\}$ such that U_α contains V_β. We write this as $\{U_\alpha\} < \{V_\beta\}$. It is readily proved that the collection of all coverings of S is a partially-ordered system under this relation.

A covering $\{U_\alpha\}$ of a space S is a *locally finite covering* if for each point x in S there is an open set in S containing x and intersecting only a finite number of elements of $\{U_\alpha\}$.

A space S is *paracompact* if S is a Hausdorff space and if every open covering of S has an open, locally finite refinement.

First, notice that *paracompactness is a topological invariant.* As usual, this means that any homeomorphism on a paracompact space yields a paracompact image space. Second, since every open covering of a compact space contains a finite open covering (which is a refinement by definition) and since a finite covering is *a fortiori* locally finite, we see that *every compact Hausdorff space is paracompact.* The converse is not true, but paracompactness has some of the force of compactness, as the following two results of Dieudonne will indicate. The first of these is an analogue of Theorem 2–1 and the second of Theorem 2–3.

THEOREM 2–62. Every paracompact space S is regular.

Proof: Let p be any point of S, and let C be a closed subset of S not containing the point p. Because S is Hausdorff, for every point x in C there are disjoint open sets U_x and V_x, with p in U_x and x in V_x. Consider the covering of S consisting of $S - C$ and all the sets V_x, x in C. By paracompactness, there is an open, locally finite refinement $\{V_\alpha\}$ of this covering. Let V be the union of all those elements of $\{V_\alpha\}$ that intersect C so that V is an open set containing C. By hypothesis, there is an open set W in S such that W contains p and meets only a finite number $V_1, \ldots, V_n$ of elements of $\{V_\alpha\}$. Each such V_i that meets C must lie in some V_{x_i}, x_i in C. If we now take the intersection $W \cap (\bigcap_{i=1}^n U_{x_i})$ of W and the sets U_{x_i} corresponding to the points x_i, we obtain an open set U containing p that does not meet the sets V_{x_i}, and hence does not meet the set V. □

THEOREM 2–63. Every paracompact space S is normal.

Proof: Let A and B be disjoint closed subsets of S. For every point x in A, the regularity of S established in Theorem 2–62 provides disjoint open sets U_x and V_x, with x in U_x, and B contained in V_x. Consider the open covering of S consisting of $S - A$ and of all sets U_x, x in A. Paracompactness yields an open, locally finite refinement $\{U_\alpha\}$ of this covering. Let U be the union of all members of $\{U_\alpha\}$ that intersect A; certainly U is an open set containing A. Then for each point y in B, there is an open set $W(y)$ that meets only a finite number $U_1(y), \ldots, U_{n(y)}(y)$ of elements of $\{U_\alpha\}$. Each of these sets $U_i(y)$ that meets A is, by definition, contained in some set U_{x_i} for a point x_i in A. Let $X_y = W(y) \cap V_{x_i}$ be the intersection of $W(y)$ and the finitely many V_{x_i} corresponding to the points x_i. Then X_y is an open set containing the point y and not meeting U. Letting V be the union of all such sets X_y, y in B, we have an open set containing B and not meeting U. □

We know that compact spaces have noncompact subspaces, and a similar remark is true of paracompactness, i.e., neither compactness nor paracompactness is hereditary. However, we do have the following analogue of Theorem 1–25.

THEOREM 2–64. Every closed subspace of a paracompact space is paracompact.

Proof: Let A be any closed subset of a paracompact space S, and apply the subspace topology to A. By definition, an open set in the subspace A is the intersection of A with some open set of S. Thus if $\{V_\alpha\}$ is an open covering of A (by subsets of A that are open in the subspace topology), then each $V_\alpha = A \cap U_\alpha$, where U_α is open in S. The open covering of S consisting of $S - A$ and the sets U_α has a locally finite refinement $\{X_\beta\}$.

Then the collection $\{A \cap X_\beta\}$ is a locally finite refinement of $\{V_\alpha\}$ by sets that are open in A. $\square$

Local compactness relates to paracompactness with the following two results.

THEOREM 2-65. Any locally compact Hausdorff space that is the union of a countable number of compact sets is paracompact.

Proof: Let the locally compact Hausdorff space S be the union of countably many compact sets C_n. We may assume that each C_n lies in C_{n+1} (for if not, we can set $C_n' = \bigcup_{i=1}^n C_n$). We first show that S is a union of open sets W_n such that each $\overline{W}_n$ is compact and lies in W_{n+1}. For each point x of C_1, let U_x be an open set containing x and such that $\overline{U}_x$ is compact. The compact set C_1 is covered by this collection, and we have a finite number of sets $U_{x_1}, \ldots, U_{x_n}$ that cover C_1. Let $W_1 = \bigcup_{i=1}^n U_{x_i}$. Then suppose W_m has been defined for each $m < n$ such that C_m lies in W_m, and such that $\overline{W}_m$ is compact and lies in W_{m+1}. Cover the compact set $\overline{W}_{n-1} \cup C_n$ as we just did for C_1 and so obtain an open set W_n containing $W_{n-1} \cup C_n$ and having compact closure.

Let $\mathfrak{U} = \{U_\alpha\}$ be any open covering of S, and define the compact sets $K_n = \overline{W}_n - W_{n-1}$. For each point x in K_x, there is an open set V_x containing x and lying in one of the open sets U_α containing x. Also V_x can be chosen to lie in W_{n+1} (since $\overline{W}_n$ does) and can be chosen so as to be disjoint from $\overline{W}_{n-2}$ (since $\overline{W}_{n-2}$ lies in W_{n-1}). Then K_n may be covered by a finite number of these sets V_x. Doing the same for each integer n, we let $\mathfrak{V}$ be the covering of S so obtained. Now $\mathfrak{V}$ refines $\mathfrak{U}$ by construction. If y is any point of S, then there is a smallest integer n such that y lies in $\overline{W}_n$. Since y is not in W_{n-1}, there is an open set V in $\mathfrak{V}$ containing y, and V can only meet the finite number of elements of $\mathfrak{V}$ that cover K_{n-2}, K_{n-1}, K_n, and K_{n+1}. Hence $\mathfrak{V}$ is locally finite. $\square$

THEOREM 2-66. Any locally compact completely separable Hausdorff space is paracompact.

Proof: Clearly, the countable basis of such a space may be taken to be composed of open sets with compact closures. Then the space is a countable union of compact sets and Theorem 2-65 applies. $\square$

One might conjecture that every locally compact space is the union of a countable number of compact sets, but this is false. Every discrete space is locally compact and metric, but only the countable discrete spaces are unions of a countable number of compact sets. More complicated examples can be given, for example by taking the product of a discrete space and any locally compact space. The "long line" of Section 2-5 is an example of a connected nonmetric locally compact space without this property. It is a theorem of Alexandroff [47] that a locally separable connected metric space has a countable basis and that local compactness implies local separability. (Also see Jones [87] and Treybig [125(a)].)

Unlike compact spaces, the product of two paracompact spaces need not be paracompact. A space may be normal and first-countable (see Section 2–6), and even more, and still fail to be paracompact. A surprising result of A. H. Stone [124] is that *every metric space is paracompact.* This last holds whether or not the metric space is separable. We finish this section by remarking that paracompactness is still under intensive investigation, as one may discover by consulting the current literature.

EXERCISE 2–27. Prove that the product of a paracompact space and a compact Hausdorff space is paracompact.

EXERCISE 2–28. Construct an example to prove that the product of two paracompact spaces need not be paracompact.

EXERCISE 2–29. Find a normal first-countable space that is not paracompact.

2–12 A general metrization theorem. In Theorem 2–48 we saw the classic metrization theorem of Urysohn, which characterizes those completely separable spaces that are metrizable as being the regular spaces. In 1951, Smirnov [123] gave a complete characterization of metric spaces, separable or not. We review his results very briefly in this section.

Smirnov begins by defining any system γ of sets in a space S to be a *locally finite system* if every point of S lies in an open set that meets at most a finite number of sets in γ. (This is only a slight generalization of a locally finite covering.) His principal result is stated next.

THEOREM 2–67. A space S is metrizable if and only if it is regular and has a basis that is the union of at most countably many locally finite systems of open sets.

As was remarked in Section 2–11, Stone had already proved the necessity of the conditions given in Theorem 2–67 by showing that every metric space is paracompact (a simple argument is needed here). The sufficiency of the conditions depends upon proving that the space S is normal and then imbedding S in a generalized Hilbert space H^τ, where τ is an infinite cardinal number. The method is analogous to the proof of Theorem 2–46, and the details are available in the translation of the Smirnov paper cited above. This paper also gives a succinct historical review of the metrization problem. (Also see Bing [57].) We close this section with a statement of another result from the same paper.

A space is said to be *locally metrizable* if every point of the space lies in an open set that, as a subspace, is metrizable. Combining the result of Dieudonne, which we stated as Theorem 2–63, with his own work, Smirnov gives the most natural metrization theorem we have seen. (See also Stone [125].)

THEOREM 2–68. A locally metrizable Hausdorff space is metrizable if and only if it is paracompact.

2–13 Complete metric spaces. The Baire-Moore theorem. We conclude this chapter with several special topics, of which this section is the first. Our considerations here are limited to metric spaces. The results find frequent application in analysis.

Let M be a metric space with metric d. Precisely as is done in the theory of real numbers, a sequence $\{x_n\}$ of points in M is called a *Cauchy sequence* provided that for any positive number ϵ, there is an integer N_ϵ sufficiently large that $d(x_m, x_n) < \epsilon$ whenever m and n exceed N_ϵ. In the real numbers, this *Cauchy condition* is necessary and sufficient for the convergence of the sequence $\{x_n\}$.

A metric space M is *complete* if every Cauchy sequence of points in M has a limit point in M. Thus the real numbers are complete (in the usual metric), but the rational numbers are not. (Indeed, the reals are often defined as a completion of the rationals in the sense of Theorem 2–72 below.) It should be noted immediately that *completeness is not a topological invariant;* it depends upon the chosen metric in the space M. For instance, let $|x - y|$ be the usual metric for the reals E^1, and define the new (but equivalent) metric

$$\rho(x, y) = \frac{|x - y|}{1 + |x - y|^2}.$$

Each sequence $\{x_n\}$ that satisfied the Cauchy condition in terms of the old metric still does, but the sequence of numbers $\sum_{n=1}^{n} 1/k$ forms a Cauchy sequence in terms of the new metric and, of course, does not converge. A space that is homeomorphic to a complete metric space is called *topologically complete* by some authors.

Our first few theorems relate the property of completeness to matters already familiar.

THEOREM 2–69. Every compact metric space is complete.

Proof: By Theorem 1–23, every infinite subset of a compact space has a limit point. □

THEOREM 2–70. Every closed subspace of a complete metric space is complete.

Proof: Let X be a closed subset of a complete metric space M. Then every Cauchy sequence of points in X has a limit point in M but, since X is closed, the limit point must be in X. □

THEOREM 2–71. If M and N are complete metric spaces, then the product $M \times N$ is complete in the product metric.

Proof: Let d_1 and d_2 be the metrics in M and N, respectively. Then if (x_1, y_1) and (x_2, y_2), x_i in M, y_i in N, are two points in $M \times N$, the *product metric* is given by

$$d[(x_1, y_1), (x_2, y_2)] = [d_1^2(x_1, x_2) + d_2^2(y_1, y_2)]^{1/2}.$$

Now let $\{(x_n, y_n)\}$ be a Cauchy sequence in $M \times N$ (in terms of the product metric). It is easily seen that this implies that $\{x_n\}$ and $\{y_n\}$ are Cauchy sequences in M and N respectively and hence converge to points x and y. The point (x, y) in $M \times N$ is then the limit point of the sequence $\{(x_n, y_n)\}$. The details are left as an exercise. □

A metric space M is said to be *isometrically imbedded* in a metric space N if there is a distance-preserving homeomorphism of M into N. In this language, we can state a generalization of the process of completing the rationals by means of Cauchy sequences.

THEOREM 2–72. Any metric space M can be isometrically imbedded in a complete metric space N in such a way that M is dense in N.

Proof: Consider the collection of all Cauchy sequences $\{x_n\}$ in M. Two such sequences $\{x_n\}$ and $\{y_n\}$ will be said to be equivalent if $\lim_{n\to\infty} d(x_n, y_n) = 0$. (It is easy to see that this is a true equivalence relation.) The equivalence classes of Cauchy sequences in M so obtained form the points of the space N, and we denote such a class by $[\{x_n\}]$. A metric for N may be defined as

$$\rho([\{x_n\}], [\{y_n\}]) = \lim_{n\to\infty} d(x_n, y_n),$$

where $\{x_n\}$ and $\{y_n\}$ are any representatives of $[\{x_n\}]$ and $[\{y_n\}]$ respectively. To prove that this definition of ρ is independent of the choice of these representations, let $\{x_n\}$ and $\{x'_n\}$ represent $[\{x_n\}]$, and let $\{y_n\}$ and $\{y'_n\}$ represent $[\{y_n\}]$. Then

$$\lim_{n\to\infty} d(x_n, y_n) \leqq \lim_{n\to\infty} [d(x_n, x'_n) + d(x'_n, y'_n) + d(y'_n, y_n)] = \lim_{n\to\infty} d(x'_n, y'_n)$$

and

$$\lim_{n\to\infty} d(x'_n, y'_n) \leqq \lim_{n\to\infty} [d(x'_n, x_n) + d(x_n, y_n) + d(y_n, y'_n)] = \lim_{n\to\infty} d(x_n, y_n).$$

A verification that ρ is indeed a metric is left as an exercise.

Next, define the mapping h that carries a point x in M onto the equivalence class of all Cauchy sequences in M that converge to x. This class is not empty, for if we set $x_n = x$ for all n, then $\{x\}$ is such a sequence. It is easily seen that h is an isometry of M into N, as required. That $h(M)$ is dense in N will follow from the arguments below.

We show that N is complete. To do so, let $\{[\{x_{m,n}\}]_m\}$ be a Cauchy sequence in N, and choose a representative $\{x_{m,n}\}$ for each "point" of the sequence. We obtain an array of sequences

$$\begin{aligned}&x_{1,1}, x_{1,2}, x_{1,3}, \ldots\\&x_{2,1}, x_{2,2}, x_{2,3}, \ldots\\&x_{3,1}, x_{3,2}, x_{3,3}, \ldots\\&\vdots\end{aligned}$$

For the kth sequence there is, by definition, an integer n_k such that $d(x_{k,n_k}, x_{k,i}) < 1/k$ whenever $i > n_k$. We can then define a Cauchy sequence $\{y_n\}_k$ in M where each $y_n = x_{k,n_k}$. From the definition of ρ we see that

$$\rho([\{x_{k,n}\}], [\{y_n\}_k]) < \frac{1}{k}.$$

Therefore

$$\lim_{k\to\infty} \rho([\{x_{k,n}\}], [\{y_n\}_k]) = 0,$$

and the two sequences are equivalent in N.

But since $\{[\{y_n\}_k]\}$ is a Cauchy sequence in N, it follows that, given ϵ, there is an integer K such that $\rho([\{y_n\}_k], [\{y_n\}_l]) < \epsilon$ whenever k and l exceed K. But this implies that $d(x_{k,n_k}, x_{l,n_l}) < \epsilon$ whenever $k, l > K$. Thus the sequence $\{x_{k,n_r}\}$ is a Cauchy sequence in M. That the sequences $\{x_{k,n_k}\}_k$ of constants converge to this diagonal sequence is immediate. Therefore the sequence $[\{x_{k,n_k}\}]_k$ in N has a limit point in N, and so does the equivalent sequence $\{[\{x_{m,n}\}]_m\}$. This proves that N is complete and moreover that every point of N is the limit of a Cauchy sequence (in N) of constant sequences (in M). It follows that $h(M)$ is dense in N. □

Most of the results of this section find their primary use in analysis. However, the next result, together with Theorem 2–79, provides the basis for an important imbedding property in topology (see Theorem 3–62).

THEOREM 2–73. If M and N are metric spaces, and if N is bounded and complete, then the function space N^M of all continuous mappings of M into N is complete in the metric $\rho(f, g) = \sup_x d(f(x), g(x))$, where d is the metric in N.

Proof: Let $\{f_n\}$ be a sequence of continuous mappings of M into N that have the property that, given $\epsilon > 0$, there is an integer K such that $\rho(f_n, f_m) < \epsilon$ whenever $m, n > K$. For a fixed point x in M, the sequence of points $\{f_n(x)\}$ then forms a Cauchy sequence in N since $d(f_n(x), f_m(x)) \leqq \rho(f_n, f_m)$. Because N is assumed to be complete, there is a point $f(x)$ in N such that $\lim_{n\to\infty} f_n(x) = f(x)$. Therefore we have a function f of M into

N defined by

$$f(x) = \lim_{n\to\infty} f_n(x)$$

for all x in M.

To complete the proof, we must show that f is continuous. To do so, we use the customary $(\epsilon - K)$-argument. That is, given $\epsilon > 0$, there is an integer K such that $\rho(f_n, f) < \epsilon/3$ whenever $n > K$. For such a value of n, there is a positive number such that $d(f_n(x), f_n(y)) < \epsilon/3$ whenever $d_1(x, y) < \delta$ (d_1 is the metric in M). Hence we have

$$\begin{aligned} d(f(x), f(y)) &\leqq d(f(x), f_n(x)) + d(f_n(x), f_n(y)) + d(f_n(y), f(y)) \\ &\leqq \rho(f, f_n) + d(f_n(x), f_n(y)) + \rho(f_n, f) < \epsilon \end{aligned}$$

whenever $d_1(x, y) < \delta$. This proves that f is continuous. □

The requirement that N be bounded in Theorem 2–73 is needed only to show that $\rho(f, g)$ exists. Our argument above actually proves the following.

Corollary 2–74. If N in Theorem 2–73 is complete (but not necessarily bounded), then the space of bounded continuous mappings of M into N is complete.

We remark that *every metric space with metric d has a metric d' that is bounded and that does not alter Cauchy sequences.* One such metric may be obtained by replacing the original values $d(x, y)$ by values $d'(x, y)$, defined by $d'(x, y) = d(x, y)$ if $d(x, y) < 1$ and by $d'(x, y) = 1$ if $d(x, y) \geqq 1$. We leave it to the reader to verify that d' is a metric.

A metric space M with metric d is said to be *totally bounded* if, given any positive number r, M is the union of finitely many sets of d-diameter less than r.

Theorem 2–75. A metric space is compact if and only if it is complete and totally bounded.

Proof: From Theorem 2–69 we know that a compact metric space is complete. And such a space must also be totally bounded or else the covering by open spherical neighborhoods of some radius r would not have a finite subcovering. Hence the condition is necessary.

To prove sufficiency, we take advantage of Exercise 2–21 and prove that a complete and totally bounded metric space M is countably compact. To do so, let $\{x_n\}$ be any sequence of points of M. Now M is a union of a finite number of sets $X_{1,1}, \ldots, X_{1,n_1}$ of diameter < 1. At least one of these sets, say $X_{1,1}$, contains an infinite number of points x_n. Let x_{k_1} be the first point of $\{x_n\}$ in $X_{1,1}$. Again, M is a union of a finite number of sets $X_{2,1}, \ldots, X_{2,n_2}$ of diameter $< \frac{1}{2}$, and one of these, say $X_{2,1}$, has the property that $X_{1,1} \cap X_{2,1}$ contains infinitely many points of the sequence $\{x_n\}$. Choose x_{k_2} as the first point of $\{x_n\}$, with $k_2 > k$, and lying in $X_{1,1} \cap X_{2,1}$. In general, we consider M as a finite union of sets of diameter $< 1/i$ and choose a new point x_{k_i}, $k_i > k_{i-1} > \cdots > k_2 > k_1$, of the sequence $\{x_n\}$ lying in $\bigcap_{j=1}^{i} X_{i,1}$. Since for any $k_j > k_i$, the points x_{k_i} and x_{k_j} lie together in a set of diameter $< 1/i$, the subsequence $\{x_{k_i}\}$ which we have extracted is a Cauchy sequence. Since M is assumed to be complete, this subsequence converges to a point of M and hence the sequence $\{x_n\}$ has a limit point. □

Some new (to us) terminology is often seen in analysis. A subset of a space S is called a *G_δ-set* if it is the countable intersection of open sets, and is called an *F_σ-set* if it is the countable union of closed sets. It is obvious that a subset is a G_δ-set if and only if its complement is an F_σ-set. As a point of interest, the genesis of these terms is as follows. The G in G_δ stands for the German word *Gebiet* (open set), and the δ means *Durchschnitt* (intersection). The F in F_σ comes from the French word *fermé* (closed), and the σ stands for *sum*, which many authors use in place of *union*.

THEOREM 2–76 (Alexandroff). Every G_δ-set in a complete metric space is homeomorphic to a complete space (or is topologically complete).

Proof: Let Q be a G_δ-set in the complete metric space M. We show that a new (but equivalent) metric can be placed upon Q so that Q is complete in terms of the new metric. By definition, $Q = \bigcap_{i=1}^{\infty} U_i$, where each U_i is open in M. As in Section 2–3, we consider the distance $d(x, M - U_i)$ for each point x in U_i and define a function $f_i\colon U_i \to E'$ by

$$f_i(x) = \frac{1}{d(x, M - U_i)}.$$

Now let $\varphi_i(x, y)$ be the real function defined on $U_i \times U_i$ by

$$\varphi_i(x, y) = \frac{|f_i(x) - f_i(y)|}{1 + |f_i(x) - f_i(y)|}.$$

The function φ_i will in general not be a metric for U_i, because it is possible to have $\varphi_i(x, y) = 0$ without having $x = y$. However, we do have

$$\varphi_i(x, y) + \varphi_i(y, z) \geqq \varphi_i(x, z),$$

for all x, y, z in U_i. Since

$$|f_i(x) - f_i(y)| + |f_i(y) - f_i(z)| \geqq |f_i(x) - f_i(z)|,$$

the inequality will follow if we can show that $a + b \geqq c > 0$ implies

$$\frac{a}{1+a} + \frac{b}{1+b} \geqq \frac{c}{1+c}.$$

But this is equivalent to

$$1 - \frac{1}{1+a} + 1 - \frac{1}{1+b} \geqq 1 - \frac{1}{1+c}$$

or to

$$1 + \frac{1}{1+c} \geqq \frac{1}{1+a} + \frac{1}{1+b}.$$

From $a + b \geqq c$, we have

$$\begin{aligned} 1 + \frac{1}{1+c} &\geqq 1 + \frac{1}{1+a+b} \\ &= \frac{2+a+b}{1+a+b} \\ &\geqq \frac{2+a+b}{1+a+b+ab} \\ &= \frac{1}{1+a} + \frac{1}{1+b}, \end{aligned}$$

which proves the inequality.

The desired metric for Q is defined by

$$\rho(x, y) = d(x, y) + \sum_{n=1}^{\infty} 2^{-n} \varphi_n(x, y),$$

x, y in Q. The series for ρ converges uniformly in Q, and it is easy to verify that ρ is indeed a metric for Q. To see that Q is complete in this metric, note first that if $\{x_n\}$ is a Cauchy sequence in the metric ρ, then it is a Cauchy sequence in the metric d. Hence it has a limit, x, in the metric d for M. If x belongs to Q it is easy enough to verify that $x_n \to x$ in the metric ρ. It follows that every ρ-Cauchy sequence converges, and also that the topologies for Q given by the metrics ρ and d are the same.

If x does not belong to Q, there is an integer N such that for all $n > N$, x is in $M - U_n$. Select a term x_k of the sequence $\{x_n\}$, and consider $\varphi_i(x_k, x_{k+j})$, $i > N$. As j increases $\varphi_i(x_k, x_{k+j}) \to 1$, since $x_{k+j} \to x$, so that $d(x_{k+j}, M - U_i) \to 0$. But then $\rho(x_k, x_{k+j})$ has a limit not less than $\sum_{n=N}^{\infty} 2^{-n}$, and $\{x_n\}$ cannot be a Cauchy sequence. □

It will follow from Theorem 2–79 below that the rationals cannot be assigned an equivalent metric in which they form a complete space. Intuitively, this seems quite plausible, since there are so many "holes," dense in the rationals. But the irrationals also have many "holes," dense in the irrationals. However, the irrationals form a G_δ-set (the rationals are an F_σ-set), and hence, by Theorem 2–76, there is a metric in which the irrationals form a complete space.

We continue this section with a theorem due to Baire [51], which we will give in several forms. The first form is due to R. L. Moore [105].

THEOREM 2–77. Let S be a compact Hausdorff space. Then S is not the union of a countable number of closed subsets, no one of which contains an open subset of S.

Proof: If S is a union $\bigcup_{i=1}^{\infty} C_i$, where each C_i is closed and contains no open set, then each C_i lies in $\overline{S - C_i}$. For if C_i contains no open sets, each point of C_i is a limit point of its complement. Let p_1 be a point of $S - C_1$, and let U_1 be an open set containing p_1 whose closure does not meet C_1 (Theorem 2–1). Let n_1 be the first integer such that $C_{n_1} \cap U_1$ is not empty.

There is a point p_2 in $U_1 - C_{n_1}$, for otherwise C_{n_1} would contain the open set U_1. Let U_2 be an open set containing p_2 whose closure does not meet the closed set $(S - U_1) \cup C_{n_1}$. Let n_2 be the first integer such that C_{n_2} meets U_2. In general, if C_{n_j}, U_j, and p_j have been defined, let p_{j+1} be a point of $U_j - C_{n_j}$. Let U_{j+1} be an open set containing p_{j+1} whose closure does not meet $(S - U_j) \cup C_{n_j}$, and let n_{j+1} be the first integer such that $C_{n_{j+1}}$ meets U_{j+1}.

Now consider the sets $\overline{U}_1, \overline{U}_2, \overline{U}_3, \ldots$ We have that for each n, $\overline{U}_n$ contains U_n, and U_n contains $\overline{U}_{n+1}$, so these sets have the finite intersection property and the set $\bigcap_{j=1}^{\infty} \overline{U}_j$ is not empty. But if p is a point in this intersection, then p must lie in some set C_k. However U_{nk+1} cannot meet C_k, which contradiction proves the theorem. □

THEOREM 2–78. Let S be a compact Hausdorff space, and let $\{U_n\}$ be a countable collection of open sets, each U_n being dense in S. Then the intersection $\bigcap_{n=1}^{\infty} U_n$ is not empty. Indeed, $\bigcap_{n=1}^{\infty} U_n$ is dense in S.

Except for the last sentence, this result is a dual to Theorem 2–77, as is easily seen. The details of the duality, and hence a proof, are left as an exercise. The last sentence follows from applying the rest of the theorem to the closures of open sets.

Our next result relates the Baire-Moore theorems above to the property of completeness.

THEOREM 2–79. Let M be a complete metric space. Then M is not the union of a countable number of closed subsets, no one of which contains an open subset of M.

Proof: In the proof of Theorem 2–77, we put no condition of size upon the sets U_n. However, if M is metric and we apply the construction of Theorem 2–77, we could obviously require that each U_n satisfy the additional condition that its diameter be less than $1/n$. If for each n we select a point q_n in U_n, then $\{q_n\}$ is a Cauchy sequence, since for any two integers m and n, with $m < n$, q_n and q_m are both in U_m, so $d(q_n, q_m) < 1/m$. It follows that $\{q_n\}$ converges to a point q. For each n, all but a finite number of points of this sequence belong to U_n, so q is in $\overline{U}_n$. This shows that $\bigcap_{n=1}^{\infty} \overline{U}_n$ is not empty, which is all we need to reach the contradiction of Theorem 2–77. □

This last result, combined with Theorem 2–73, provides some interesting existence proofs. For instance, it is possible to show [52] that the collection of all real-valued continuous functions on the unit interval I^1 that have a derivative at at least one point of I^1 consists of a countable number of closed sets, each containing no open set in the complete metric space C of all real-valued continuous functions on I^1. Therefore it follows from Theorem 2–79 that there are real-valued continuous functions on I^1 having no derivative anywhere.

A subset X of a space S is said to be *perfect* if X is closed and if every point of X is a limit point of X. This latter property is sometimes called (misleadingly) *dense in itself*. Restating Theorems 2–77 and 2–79 in these terms, we have the next result.

THEOREM 2–80. No compact Hausdorff space and no complete metric space is both countable and perfect.

COROLLARY 2–81. No compact Hausdorff space is countable and connected.

Note. There exist countable connected Hausdorff spaces. One example has been given by Bing [61].

EXAMPLE. We construct a closed, totally disconnected set K in I^1 such that K has Lebesgue measure $\frac{9}{10}$ (see below). To begin with, let $U_{1,1}$ be the open interval centered at the point $\frac{1}{2}$ and having length $(\frac{1}{2})(\frac{1}{10})$. About the midpoint of each closed interval in $I^1 - U_{1,1}$, take open intervals $U_{2,1}$ and $U_{2,2}$, each of length $(\frac{1}{8})(\frac{1}{10})$. Then $I^1 - U_{1,1} - U_{2,1} - U_{2,2}$ consists of four closed intervals, and about the midpoint of each we take an open interval of length $(\frac{1}{32})(\frac{1}{10})$. At the $(n+1)$th step, we will have 2^n closed intervals about whose midpoints we take open intervals of length $(1/2^{2n+1})(\frac{1}{10})$. We then define

$$K = I^1 - U_{1,1} - U_{2,1} - U_{2,2} - U_{4,1} - \cdots - U_{4,4} - \cdots$$

Now the *Lebesgue measure* $\mu(U)$ of an open set U in E^1 is defined to be the sum of the lengths of the disjoint open intervals composing U, and the measure

$\mu(X)$ of a closed set in any interval $[a, b]$ is defined by

$$\mu(X) = b - a - \mu([a, b] - X).$$

It is readily seen that

$$\mu\left(\bigcup_{i,j} U_{k,j}\right) = \frac{1}{2}\left(\frac{1}{10}\right) + 2\left(\frac{1}{8}\right)\left(\frac{1}{10}\right) + 4\left(\frac{1}{32}\right)\left(\frac{1}{10}\right) + \cdots$$

$$+ 2^n\left(\frac{1}{2^{2n+1}}\right)\left(\frac{1}{10}\right) + \cdots$$

$$= \frac{1}{10}\left(\frac{1}{2} + \frac{1}{4} + \frac{1}{8} + \cdots + \frac{1}{2^{n+1}} + \cdots\right) = \frac{1}{10},$$

and hence $\mu(K) = \frac{9}{10}$ as claimed. The set K is totally disconnected, for it can contain no interval whatsoever. A startling remark is that I^1 is not a countable union of such closed sets as K, for no such set contains an open interval, and Theorem 2–77 applies!

The results above are often expressed in other terms, with which the reader should be acquainted. A set X in a space S is said to be of *first category* in S if X is the union of a countable number of sets, no one of which is dense in any open subset of S. (A set that is dense in no open set is said to be *nowhere dense.*) A set is of *second category* if it is not of first category. We may rephrase Theorem 2–79 in these terms.

THEOREM 2–82. No compact Hausdorff space and no complete metric space is of first category.

In a sense, the notion of category is not topological. That is, we can have a set X of first category in E^1, say, and a set Y in E^1 of second category, and have a homeomorphism of X onto Y. On the other hand, there is no homeomorphism of E^1 onto itself carrying a set of first category onto one of second category. To prove the first statement, let the points of I^1 be given in ternary notation (to the base 3). Let X be all points of I^1 that have a ternary expansion involving only the digits 0 and 2, but that do not end in all 0 or all 2. The set X is the subset of the Cantor set consisting of all points that are not end points of deleted intervals. In the expansion of an element x of X, replace each digit 2 by a digit 1, and consider the resulting number in the binary scale. The mapping thus defined maps X homeomorphically onto the set Y of all numbers on I^1 that are not of the form $k/2^m$, k and m positive integers. The complement of Y is of first category, and clearly the union of two sets of first category is of first category. Thus if Y were of first category, so would I^1 be. The reason for the existence of such examples is that the sets in the definition of first category are not required to be closed. If they were, and S were compact, then the property would be topological in every sense.

If we consider a set to be "important" only if it is dense in some open set, then a set of first category is a countable union of "unimportant" sets. In some sense, a set of first category is analogous to a set of measure zero in measure theory. Indeed, Oxtoby and Ulam [112] have proved, assuming the continuum hypothesis, that for each set of first category in I^n there is a homeomorphism of I^n onto itself carrying that set onto a set of measure zero, and that each set of measure zero may be so obtained. One also finds such statements as, "Almost all continuous functions are not polynomials," meaning that, although the polynomials are dense in the function space $(E^1)^{I^1}$ by the Weierstrass approximation theorem, the set of polynomials is of first category.

Our final result in this section concerns the extension of mappings and has been a motivating result in the study of *uniform spaces* (see Section 1–12). We will not follow in this direction, however.

THEOREM 2–83. If f is a uniformly continuous function on a subset A of a metric space M into a complete space N, then f has a unique uniformly continuous extension to the closure $\overline{A}$ of A.

Proof: For each point x in $\overline{A}$, select a sequence $\{x_n\}$ of points in A such that $d(x_n, x)$ approaches zero as n increases indefinitely. If x is in A, we agree to let each $x_n = x$ so that we have a constant sequence in this case. Now by the uniform continuity of f, given $\epsilon > 0$, there is an integer K such that $\rho[f(x_m), f(x_n)] < \epsilon$ whenever $m, n > K$ (ρ is the metric in N). It follows that $\{f_n(x)\}$ is a Cauchy sequence in N and hence converges to a point in N. We define

$$\bar{f}(x) = \lim_{n\to\infty} f(x_n).$$

To show that $\bar{f}(x)$ is actually independent of the sequence used in its definition, let $\{y_n\}$ be another sequence in A converging to the point x.

Then clearly the sequence $x_1, y_1, x_2, y_2, \ldots, x_n, y_n, \ldots$ also converges to x. It follows that $f(x_1), f(y_1), f(x_2), f(y_2), \ldots$ converges to $\bar{f}(x)$. Therefore $\bar{f}(x)$ does not depend upon the sequence.

To prove that $\bar{f}$ is uniformly continuous on $\overline{A}$, let $\epsilon > 0$ be given, and choose $\delta > 0$ by the uniform continuity of f on A in such a way that $\rho[f(x), f(y)] < \epsilon/3$ whenever $d(x, y) < \delta$ (d is the metric in M). Let x and y be points of $\overline{A}$ such that $d(x, y) < \delta/2$. Choose an integer K sufficiently large that $d(x_n, x) < \delta/4$ and $d(y_n, y) < \delta/4$ whenever $n > K$, where $\{x_n\}$ and $\{y_n\}$ are sequences in A converging to x and y, respectively. Then $d(x_n, y_n) < \delta$, and $\rho[f(x_n), f(y_n)] < \epsilon/3$. Hence we have

$$\rho[\bar{f}(x), \bar{f}(y)] \leqq \rho[\bar{f}(x), f(x_n)] + \rho[f(x_n), f(y_n)] + \rho[f(y_n), \bar{f}(y)] < \epsilon.$$

Thus $\bar{f}$ is uniformly continuous on $\overline{A}$, and uniqueness of $\bar{f}$ follows from its continuity since A is dense in $\overline{A}$. □

EXERCISE 2–30. Use the Baire-Moore theorem to show that the real numbers are uncountable.

EXERCISE 2–31. Prove that the set of irrational numbers is not the union of a countable number of closed sets.

EXERCISE 2–32. Prove that if X is a set of first category in a compact Hausdorff space S and if X is a countable union of closed sets, then no one of these closed sets contains an open set.

EXERCISE 2–33. Prove Theorem 2–76 for the case of a countably compact Hausdorff space.

EXERCISE 2–34. Construct a function that is continuous at each irrational point of E^1 and is discontinuous at each rational point. Prove that there is no function that is continuous at each rational and discontinuous at each irrational. More generally, prove that the set of points of discontinuity of a real-valued function on a space is an F_σ-set.

2–14 Inverse limit systems. The concepts discussed in this section are of recent importance in topology. Applications of these ideas are found in the next section and in Chapter 8. We begin with a special case of an inverse limit system, study that special case, and then indicate how the general concept is defined.

Let $X_0, X_1, X_2, \ldots$ be a countable collection of spaces, and suppose that for each $n > 0$, there is a continuous mapping $f_n: X_n \to X_{n-1}$. The sequence of spaces and mappings $\{X_n, f_n\}$ is called an *inverse limit sequence* and may be represented by means of the diagram

$$\cdots \xrightarrow{f_{n+1}} X_n \xrightarrow{f_n} X_{n-1} \xrightarrow{f_{n-1}} \cdots \xrightarrow{f_3} X_2 \xrightarrow{f_2} X_1 \xrightarrow{f_1} X_0.$$

Clearly, if $n > m$, there is a continuous mapping $f_{n,m}: X_n \to X_m$ given by the composition $f_{n,m} = f_{m+1} \cdot f_{m+2} \cdots f_{n-1} \cdot f_n$.

Consider a sequence $(x_0, x_1, \ldots, x_n, \ldots)$ such that each x_n is a point of the space X_n and such that $x_n = f_{n+1}(x_{n+1})$ for all $n \geqq 0$. Such a sequence can be identified with a point in the product space $\mathsf{P}_{n=0}^{\infty} X_n$ by considering the function φ from the nonnegative integers into $\bigcup_{n=0}^{\infty} X_n$, given by $\varphi(n) = x_n$. The set of all such sequences is in this way a subset of $\mathsf{P}_{n=0}^{\infty} X_n$ and has a topology as a subspace. This topological space is *the inverse limit space of the sequence* $\{X_n, f_n\}$. We will denote it by X_∞.

Our first result concerns a condition that permits us to specify some of the coordinates of a point in X_∞ and to "fill in" the rest.

LEMMA 2–84. If $\{X_n, f_n\}$ is an inverse limit sequence, if each f_n is a mapping onto, and if $x_{n_1}, x_{n_2}, \ldots, x_{n_k}, \ldots$ is a set of points with x_{n_i} in X_{n_i} for $i = 1, 2, 3, \ldots$ and such that if $i < j$, then $f_{n_j\, n_i}(x_{n_j}) = x_{n_i}$, then there is a point in X_∞ whose coordinate in X_{n_i} is x_{n_i}, $i = 1, 2, 3, \ldots$

Proof: There are two cases, the first in which the set $\{x_{n_i}\}$ is infinite, and the second in which it is finite. The infinite case is easy. For arbitrary n, there is a least integer n_j with $n_j \geqq n$. If $n_j = n$, set $x_n = x_{n_j}$; if $n_j > n$, define $x_n = f_{n_j,n}(x_{n_j})$. Clearly the sequence $(x_0, x_1, \ldots, x_n, \ldots)$ so defined is a point of X_∞.

In the finite case, there is a greatest n_i, say n_k. For $n < n_k$, we may define $x_n = f_{n_k,n}(x_{n_k})$. Now suppose that for some $m \geqq n_k$, we have already defined x_m. Since f_{m+1} is onto, there is at least one point x in X_{m+1} such that $f_{m+1}(x) = x_m$. Choose any such point as x_{m+1}. The existence of our desired sequence now follows by induction. □

Lemma 2–84 is an existence theorem. It illustrates a characteristic feature of an inverse limit space, namely, from any coordinate x_n toward the "front" of the sequence, the coordinates of a point are controlled absolutely by x_n, but there is room for some choice from x_{n+1} on in the sequence. In fact, if the mappings f_n are not onto, the space X_∞ may be empty. This is true even if each f_n is a homeomorphism into. For instance, consider a sequence of countable discrete spaces $X_n = \bigcup_{m=1}^{\infty} x_{n,m}$, $n = 0, 1, 2, \ldots$. Define the mappings $f_n : X_n \to X_{n-1}$ by $f_n(x_{n,m}) = x_{n-1,m+1}$. Then we have an inverse limit sequence. But if we begin with a point $x_{0,j}$ of X_0 and attempt to form a point of X_∞, we can only construct the first j coordinates and then are forced to stop. Therefore X_∞ has no points in it at all.

We can prove another existence theorem also, in case all the spaces X_n are compact.

THEOREM 2–85. Suppose that each space X_n in the inverse limit sequence $\{X_n, f_n\}$ is a compact Hausdorff space. Then X_∞ is not empty.

Proof: For each integer $n \geqq 1$, let Y_n be the set of all sequences $p = (p_0, p_1, \ldots)$ such that for $1 < j < n$, $p_{j-1} = f_j(p_j)$. Each Y_n is a subset of $\mathsf{P}_{n=0}^{\infty} X_n$, and we will show that Y_n is closed in $\mathsf{P}_{n=0}^{\infty} X_n$. Suppose that for a given n, q is not a point of Y_n. If $q = (q_0, \ldots, q_n, \ldots)$, then for some $j < n$, we have $q_j \neq f_{j+1}(q_{j+1})$. Now X_j is a Hausdorff space, so there exist disjoint open sets U_j and V_j in X_j, with q_j in U_j and $f_{j+1}(q_{j+1})$ in V_j. Define $V_{j+1} = f_{j+1}^{-1}(V_j)$. Let U_q denote any basis element in $\mathsf{P}_{n=0}^{\infty} X_n$ containing q and having U_j and V_{j+1} as factors. Then no point of Y_n lies in U_q. For if $p = (p_0, p_1, \ldots, p_n, \ldots)$ were in Y_n and in U_q, then p_{j+1} lies in V_{j+1} and p_j in V_j, and not in U_j as required. Thus the complement of Y_n is a union of such open sets U_q, and hence Y_n is closed. Since the collection $\{Y_n\}$ obviously satisfies the finite intersection hypothesis and $\mathsf{P}_{n=0}^{\infty} X_n$ is compact, the intersection $\bigcap_{n=1}^{\infty} Y_n$ is not empty. But also each point in $\bigcap_{n=1}^{\infty} Y_n$ satisfies the condition for being a point in X_∞. Hence X_∞ is not empty. □

Closely related to the concept of an inverse limit sequence of spaces is an *inverse limit sequence of algebraic groups*. This is a sequence of groups $G_0, G_1, G_2, \ldots$ and homomorphisms $\varphi_n : G_n \to G_{n-1}$, $n \geqq 1$. The *inverse limit group* G_∞ of such a sequence is the collection of all sequences $(g_0, g_1, g_2, \ldots)$ with g_i in G_i and such that $g_i = \varphi_{i+1}(g_{i+1})$ for all i. The product of two such elements, say $(f_0, f_1, f_2, \ldots)$ and $(g_0, g_1, g_2, \ldots)$, is given by the formula

$$\{f_i\} \cdot \{g_i\} = \{f_i \cdot g_i\},$$

where the dot on the right indicates the group operation in G_i. The reader may prove that G_∞ is indeed a group. Note that G_∞ always contains at least one element, namely $(e_0, e_1, e_2, \ldots)$, where e_i denotes the identity element of G_i.

There is a natural way to map one inverse limit sequence into another. Let $\{A_n, f_n\}$ and $\{B_n, g_n\}$ be two inverse limit sequences of spaces. A mapping $\Phi:\{A_n, f_n\} \to \{B_n, g_n\}$ is a collection $\{\varphi_n\}$ of continuous mappings $\varphi_n:A_n \to B_n$ satisfying the condition $g_n\varphi_n = \varphi_{n-1}f_n$, $n \geqq 1$. This condition may be given by saying that we have *commutativity in the diagram* below.

$$\begin{array}{ccccccccccc}
\cdots \longrightarrow & A_n & \xrightarrow{f_n} & A_{n-1} & \xrightarrow{f_{n-1}} & A_{n-2} & \longrightarrow \cdots \longrightarrow & A_1 & \xrightarrow{f_1} & A_0 \\
& \downarrow{\scriptstyle \varphi_n} & & \downarrow{\scriptstyle \varphi_{n-1}} & & \downarrow{\scriptstyle \varphi_{n-2}} & & \downarrow{\scriptstyle \varphi_1} & & \downarrow{\scriptstyle \varphi_0} \\
\cdots \longrightarrow & B_n & \xrightarrow[g_n]{} & B_{n-1} & \xrightarrow[g_{n-1}]{} & B_{n-2} & \longrightarrow \cdots \longrightarrow & B_1 & \xrightarrow[g_1]{} & B_0
\end{array}$$

This means that we may pass from A_n to B_{n-1} in two ways but the result is the same. Such a mapping Φ *induces* a mapping $\varphi:A_\infty \to B_\infty$ of the inverse limit spaces as follows. For each point $a = (a_0, a_1, \ldots)$ in A_∞, let $\varphi(a) = (\varphi_0(a_0), \varphi_1(a_1), \ldots)$. That $\varphi(a)$ is indeed a point of B_∞ follows immediately from the equations

$$g_n[\varphi_n(a_n)] = \varphi_{n-1}[f_n(a_n)] = \varphi_{n-1}(a_{n-1}).$$

Theorem 2–86. The mapping $\varphi:A_\infty \to B_\infty$ induced by the mapping $\Phi:\{A_n, f_n\} \to \{B_n, g_n\}$ is continuous.

Proof: The mapping φ may be regarded as a mapping of A_∞ into $\mathbb{P}_{n=0}^{\infty} B_n$ since B_∞ is contained in $\mathbb{P}_{n=0}^{\infty} B_n$. Now if a is any point of A_∞, each coordinate of $\varphi(a) = (\varphi_0(a_0), \varphi_1(a_1), \ldots)$ is defined by a continuous mapping of A_n into B_n. By Theorem 1–37, this implies the continuity of φ. □

Rather than give an application of this result here, we merely refer to Section 1–15 and go on to a brief discussion of a generalization of the inverse limit sequence. Consider a set Γ partially-ordered by a relation $<$. If for any pair of elements α, β in Γ there exists an element γ in Γ such that both $\alpha < \gamma$ and $\beta < \gamma$, then Γ is called a *directed set*.

Suppose that for each element α of Γ there is a unique set A_α in a collection $\mathcal{A}$ of sets (we say that $\mathcal{A}$ *is indexed by* Γ), and suppose that whenever $\alpha < \beta$ in Γ, there is a transformation $f_{\alpha\beta}:A_\beta \to A_\alpha$ of A_β into A_α. Note that $f_{\alpha\beta}$ acts *against* the order relation. Assume further that these transformations satisfy

(i) $f_{\alpha\alpha}$ is the identity transformation for each α in Γ,

and

(ii) $f_{\beta\gamma}\, f_{\alpha\beta} = f_{\alpha\gamma}$ whenever $\alpha < \beta < \gamma$.

If F denotes the collection $\{f_{\alpha\beta}\}$ of all such transformations, the pair $\{\mathcal{A}, F\}$ is called an *inverse limit system over the directed set* Γ. It is clear that an inverse limit sequence is merely an inverse limit system over the directed set of all nonnegative integers.

We are interested in two particular instances of such systems. The first of these is the case where each A_α in $\mathcal{A}$ is a topological space and each $f_{\alpha\beta}$ in F is a continuous mapping. Just as for an inverse sequence, we define the *inverse limit space* A_∞ *of the system* $\{\mathcal{A}, F\}$ as follows. Let $\{x_\alpha\}$ be a set consisting of one point x_α from each space A_α in $\mathcal{A}$ and satisfying the condition that if $\alpha < \beta$ in Γ, then $f_{\alpha\beta}(x_\beta) = x_\alpha$. Such a set $\{x_\alpha\}$ may be identified with the point ψ in the product space $\mathbb{P}_\Gamma A_\alpha$ having coordinates $\psi(\alpha) = x_\alpha$. Hence the collection of all such sets $\{x_\alpha\}$ constitutes a subspace of $\mathbb{P}_\Gamma A_\alpha$, and this subspace is the inverse limit space A_∞ of $\{\mathcal{A}, F\}$.

We use only an inverse limit *sequence* of spaces (in Section 2–15). Therefore we merely quote a few results. For a comprehensive treatment, the reader is referred to Chapter VIII of Eilenberg and Steenrod [7]. Before quoting results, note that if A_∞ is the limit space of an inverse system $\{\mathcal{A}, F\}$, then for each β in Γ there is a natural *projection* $\pi_\beta:A_\infty \to A_\beta$, defined by $\pi_\beta(\{x_\alpha\}) = x_\beta$.

LEMMA 2–87. If A_∞ is the inverse limit space of an inverse limit system $\{\mathcal{A}, F\}$, then each projection π_α of A_∞ into A_α is continuous.

LEMMA 2–88. The inverse limit space A_∞ of the system $\{\mathcal{A}, F\}$ is a closed subspace of the product space $\mathbb{P}_\Gamma A_\alpha$.

In analogy to Theorem 2–85 we have the existence theorem.

THEOREM 2–89. The inverse limit space of an inverse limit system of compact Hausdorff spaces is a compact Hausdorff space, and if each space of the system is nonempty, then the limit space is also nonempty.

The second instance of an inverse limit system we will consider requires some algebraic preparation. Let each set A_α in the collection $\mathcal{A}$ be either a module over a ring with unit or a topological group, and let each $f_{\alpha\beta}$ in F be a homomorphism or a continuous homomorphism. The *inverse limit group* A_∞ *of the system* $\{\mathcal{A}, F\}$ is that subgroup of the direct sum

$\sum_\Gamma A_\alpha$ consisting of all sets $\{x_\alpha\}$, one element from each group A_α, for which $f_{\alpha\beta}(x_\beta) = x_\alpha$ whenever $\alpha < \beta$ in Γ. The group operation in A_∞ is defined naturally by the formula

$$\{x_\alpha\} + \{y_\alpha\} = \{x_\alpha + y_\alpha\},$$

where the sum on the right indicates the group operation in each A_α. Again we have the projections $\pi_\beta : A_\infty \to A_\beta$ given by $\pi_\beta(\{x_\alpha\}) = x_\beta$.

Note that the direct sum $\sum_\Gamma A_\alpha$ is used above. We use the *weak direct sum* shortly.

LEMMA 2–90. If A_∞ is the inverse limit group of a system of topological groups or modules over a ring (or of topological spaces), each projection π_α is a homomorphism (or a continuous homomorphism).

Here there is no question about the existence of A_∞ because the element $\{e_\alpha\}$ consisting of the set of all identities e_α in A_α is obviously an element of A_∞. We will see this concept in use in Section 8–3.

Let $\{\mathfrak{A}, F\}$ and $\{\mathfrak{A}', F'\}$ be inverse limit systems over directed sets Γ and Γ', respectively. We define a transformation Φ of $\{\mathfrak{A}, F\}$ into $\{\mathfrak{A}', F'\}$ to consist of an order-preserving transformation ϕ of Γ' into Γ (note the direction) and for each α' in Γ', a transformation $\varphi_{\alpha'}$ of $A_{\phi(\alpha')}$ into $A'_{\alpha'}$. Furthermore, we require that whenever $\alpha' < \beta'$ in Γ', we have the commutative relation

$$\varphi_{\alpha'} f_{\phi(\alpha')\phi(\beta')} = f'_{\alpha'\beta'}\varphi_{\beta'}.$$

Again, this is more easily envisioned by requiring "commutativity in the diagram"

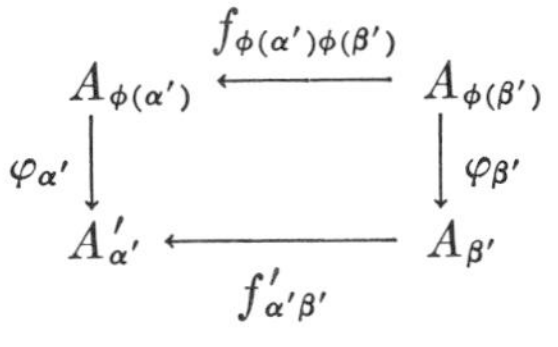

$$\begin{array}{ccc} A_{\phi(\alpha')} & \xleftarrow{f_{\phi(\alpha')\phi(\beta')}} & A_{\phi(\beta')} \\ \varphi_{\alpha'}\Big\downarrow & & \Big\downarrow\varphi_{\beta'} \\ A'_{\alpha'} & \xleftarrow[f'_{\alpha'\beta'}]{} & A_{\beta'} \end{array}$$

Such a transformation Φ of $\{\mathfrak{A}, F\}$ into $\{\mathfrak{A}', F'\}$ induces a transformation φ_∞ of the inverse limits A_∞ into A'_∞ as follows. If $\{x_\alpha\}$ is an element of A_∞ and α' in Γ' is given, set $x'_{\alpha'} = \varphi_{\alpha'}(x_{\phi(\alpha')})$. Note that if $\alpha' < \beta'$, the above commutative relation tells us that

$$f'_{\alpha'\beta'}(x'_{\beta'}) = x'_{\alpha'}.$$

Thus $\{x'_{\alpha'}\}$ is an element of A'_∞. We define $\varphi_\infty(\{x_\alpha\}) = \{x'_{\alpha'}\}$. It is not difficult to show that *the induced transformation φ_∞ also commutes with the projections.* Of course, in the case where we have two inverse limit systems of groups, we would require the $\varphi_{\alpha'}$'s to be homomorphisms and show that the induced φ_∞ is a homomorphism too. This idea will also be found in Section 8–4.

While we are about it, we will also mention another concept, which will be of use in Chapter 8. Let $\mathcal{A}$ be a collection of sets A^α indexed by a directed set Γ, and for each $\alpha < \beta$ in Γ let $g^{\alpha\beta}$ be a transformation of A^α into A^β. Note that now $g^{\alpha\beta}$ acts *with* the order relation. Again we will assume that

$$\text{(i)} \qquad g^{\alpha\alpha} \quad \text{is the identity for each } \alpha$$

and that

$$\text{(ii)} \qquad g^{\beta\gamma}g^{\alpha\beta} = g^{\alpha\gamma} \quad \text{whenever} \quad \alpha < \beta < \gamma.$$

If we let the collection of all such $g^{\alpha\beta}$ be denoted by G, the pair $\{\mathcal{A}, G\}$ is a *direct limit system over the directed set* Γ. Our interest here will be confined to the case in which each A^α in $\mathcal{A}$ is a module, all over the same ring with unit, and where each $g^{\alpha\beta}$ is a homomorphism.

Let $\sum_\Gamma^f A^\alpha$ denote the weak direct sum of the modules in $\mathcal{A}$, that is, all sets $\{x^\alpha\}$, one element x^α from each A^α, where only a finite number of x^α are different from the identity element e^α. Note that $\sum_\Gamma^f A^\alpha$ is a subgroup of the direct sum $\sum_\Gamma A^\alpha$. Now if x^β is an element of A^β, there is an element $\{x^\alpha\}$ in $\sum_\Gamma^f A^\alpha$ with coordinates

$$x^\alpha = x^\beta \qquad \text{if} \qquad \alpha = \beta,$$

$$x^\alpha = e^\alpha \qquad \text{if} \qquad \alpha \neq \beta.$$

This obviously defines an isomorphism i_β, called an *injection* of A^β onto a subgroup of $\sum_\Gamma^f A^\alpha$ and allows us to identify x^β with an element of the weak direct sum. Hence we can and do retain the same notation for x^b as an element of A^β and as an element of $\sum_\Gamma^f A^\alpha$.

Whenever $\alpha < \beta$ in Γ, there is an element $g^{\alpha\beta}(x^\alpha) - x^\alpha$ in $\sum_\Gamma^f A^\alpha$. [Actually this is $i_\beta[g^{\alpha\beta}(x^\alpha)] - i_\alpha(x^\alpha)$, but we are using the identification just mentioned.] Such an element $g^{\alpha\beta}(x^\alpha) - x^\alpha$ is called a *relation*. Now the collection of all relations generates a subgroup B of $\sum_\Gamma^f A^\alpha$, as is easily verified. Then the (additively written) factor group $(\sum_\Gamma^f A^\alpha) - B$ is the *direct limit group* A^∞ *of the system* $\{\mathcal{A}, G\}$. The canonical homomorphism of $\sum_\Gamma^f A^\alpha$ onto the factor group A^∞ clearly defines *projection homomorphisms* $\pi^\alpha : A^\alpha \to A^\infty$ for each α in Γ.

LEMMA 2–91. If $\alpha < \beta$, then $\pi^\beta g^{\alpha\beta} = \pi^\alpha$.

The next lemma is important in that it claims that elements of a direct limit group are easily constructed, which is not necessarily true of an inverse limit group.

LEMMA 2–92. If x is any element of A^∞, then there is an α in Γ and an x^α in A^α such that $\pi^\alpha(x^\alpha) = x$.

Again we point out that this material is a bare introduction to topics of importance in our Chapter 8 and is to be found in detail in Chapter VIII of Eilenberg and Steenrod.

EXERCISE 2–35. Modify the example following Lemma 2–84 to show that all the spaces X_n can be taken to be connected and still have X_∞ be empty.

EXERCISE 2–36. Modify the same example again to show that each X_n can be infinite and compact and have X_∞ consist of a single point.

2–15 A characterization of the Cantor set. We recall that the *Cantor set* (or the *Cantor ternary set;* or the *Cantor middle-third set;* or the *Cantor discontinuum*) consists of all points in the closed unit interval I^1 that, when expressed to the base 3, have no units in their ternary expansion. We know that the Cantor set is totally disconnected, compact, perfect, and metric. We will make use of the results of Section 2–14 to prove that every two such totally disconnected compact perfect metric spaces are homeomorphic and hence will have a topological characterization of the Cantor set.

LEMMA 2–93. *If $\mathfrak{U}$ is any covering by open sets of a metric space M, and if n is any integer, then there is a refinement $\mathfrak{V}$ of $\mathfrak{U}$ composed of open sets of diameter $< 1/n$. If M is compact, then $\mathfrak{V}$ can be taken to be finite.*

The proof is left as an exercise.

THEOREM 2–94. *Let M be a compact totally disconnected metric space. Then M has a sequence $\mathfrak{U}_1, \mathfrak{U}_2, \ldots$ of finite coverings, each $\mathfrak{U}_n$ being a collection of disjoint sets of diameter $< 1/n$ that are both open and closed and $\mathfrak{U}_{n+1}$ being a refinement of $\mathfrak{U}_n$ for each n.*

Proof: From Theorem 2–15 we know that if C is a component (a single point in this case) of M, and if U is any open set containing C, then there is an open and closed set V lying in U and containing C. Begin with a covering $\mathfrak{U}_0$ of M. Each point x of M lies in an open set U_x of $\mathfrak{U}_0$; there is an open and closed set V_x of diameter < 1 containing x and lying in U_x. By compactness, a finite number $V_1, \ldots, V_n$ of these sets covers M. However, the set V_i need not be disjoint. Consider the sets $U_1 = V_1$, $U_2 = V_2 - V_1, \ldots, U_j = V_j - (\cup_{i=1}^{j-1} V_i)$. Each of these is an open set minus a closed set and is open, but also each is a closed set minus an open set and is closed. No two intersect, for given U_k and U_j, $i < j$, U_i is a subset of V_i, and U_j is a subset of $M - V_i$. We have diameter $U_i \leqq$ diameter $V_i < 1$. We let $\mathfrak{U}_i = \{U_i\}$. The general inductive step should now be obvious. □

Next let us take a sequence $\mathfrak{U}_1, \mathfrak{U}_2, \ldots$ of coverings of the space M as described in Theorem 2–94 and construct an inverse limit sequence. We take each covering $\mathfrak{U}_n$ to be a space, with the "points" being the open sets in $\mathfrak{U}_n$ and using the discrete topology. The exact process is described in the next proof. These *spaces* $\mathfrak{U}_n$ are examples of a more general concept, the *nerve of a covering*, to be found in Section 5–7 and again in Chapter 8.

THEOREM 2–95. Let M be a compact, totally disconnected metric space. Then M is homeomorphic to the inverse limit space of an inverse limit sequence of finite, discrete spaces.

Proof: Let $\mathfrak{U}_1, \mathfrak{U}_2, \ldots$ be a sequence of coverings of M as given in Theorem 2–94. For each n, let $\mathfrak{U}_n^*$ denote the space whose points are the open sets of $\mathfrak{U}_n$ and which has the discrete topology. We will use the same notation for an element of $\mathfrak{U}_n$ and the corresponding point of $\mathfrak{U}_n^*$. A continuous mapping $f_n:\mathfrak{U}_n^* \to \mathfrak{U}_{n-1}^*$, $n > 1$, may be defined as follows. If $U_{n,i}$ is an element of $\mathfrak{U}_n$, then there is a unique element $U_{n-1,j}$ of $\mathfrak{U}_{n-1}$ containing $U_{n,i}$ because the elements of $\mathfrak{U}_{n-1}$ are disjoint. We set $f_n(U_{n,i}) = U_{n-1,j}$, thinking now of these sets as points in $\mathfrak{U}_n^*$ and $\mathfrak{U}_{n-1}^*$. The mappings f_n are continuous in a trivial manner since each $\mathfrak{U}_n^*$ is discrete. With these definitions, it is obvious that $\{\mathfrak{U}_n^*, f_n\}$ is an inverse limit sequence of compact spaces. Hence by Theorem 2–85, the inverse limit space U_∞ is nonempty.

We next define a mapping $h:U_\infty \to M$. If $p = (U_{1,n_1}, U_{2,n_2}, \ldots)$ is a point of U_∞, then the *sets* $U_{1,n_1}, U_{2,n_2}, \ldots$ in M form a sequence of closed sets, each containing the succeeding one. Thus the compactness of M assures us that the intersection $\bigcap_{j=1}^{\infty} U_{j,n_j}$ is not empty. Since diameter $U_{j,n_j} < 1/j$, there can be at most one point q of M in this intersection, and we let $h(p) = q$. Our proof will show that h is a homeomorphism onto.

First, *h is one-to-one*, for if p is a point of U_∞, then $h(p)$ is in each of the point sets in M that are coordinates of p. Hence if two points p and p' of U_∞ differ in the nth coordinate, then $h(p) \neq h(p')$ because the elements of $\mathfrak{U}_n$ are disjoint. Second, *h is onto*, for each point q of M lies in the intersection of such a sequence of sets. Third, *h is continuous*. To see this, note first that the collection of all sets $U_{j,i}$ is a basis for the topology of M. Thus if we prove that for each $U_{j,i}$ in $\mathfrak{U}_j$, $h^{-1}(U_{j,i})$ is open in U_∞, we are finished. But $h^{-1}(U_{j,i})$ consists of all points of U_∞ having $U_{j,i}$ for their jth coordinate, and the *point* $U_{j,i}$ of $\mathfrak{U}_j^*$ is open in $\mathfrak{U}_j^*$. Hence $h^{-1}(U_{j,i})$ is open in U_∞.

From Theorem 2–85 we know that U_∞ is a compact Hausdorff space, and hence from Exercise 2–43 we know that h is a homeomorphism. □

Concerning Theorem 2–95, we should point out that this is a case in which the statement of the theorem is inadequate. A complete statement of the theorem would include most of the proof in that it would not only state that M is homeomorphic to an inverse limit of discrete spaces but would also tell just *which* discrete spaces and how the homeomorphism is defined. In applications, it is these facts, not merely the existence, that are used. The reader will find that this is a common occurrence in mathematical writing.

We are now ready to prove the characterization of the Cantor set. We have seen how to obtain compact, totally disconnected metric spaces as the inverse limits of nerves of coverings. Our next task is to show that if two such spaces are perfect, then the two inverse limit sequences can be chosen so that their inverse limit spaces are homeomorphic.

Theorem 2–96. If U is an open set in a totally disconnected perfect topological space, and n is an integer, then U is a union of n disjoint nonempty open sets.

Proof: For $n = 1$, U itself satisfies the condition. Suppose that for $n = k$ we have $U = U_1 \cup \cdots \cup U_k$, where the U_i are open, disjoint, and nonempty. The set U_k is not connected because the space is totally disconnected, and a single point is not open. Thus $U_k = U_{k,1} \cup U_{k,2}$, where $U_{k,1}$ and $U_{k,2}$ are disjoint. Each of these sets is open in U_k and hence in the space. Then $U_1, \ldots, U_{k-1}, U_{k,1}, U_{k,2}$ is a desired decomposition of U for $n = k + 1$. □

Theorem 2–97. Any two totally disconnected, perfect, compact metric spaces are homeomorphic.

Proof: Let S and T be two such spaces, and let $\mathfrak{U}_1, \mathfrak{U}_2, \ldots$ and $\mathfrak{V}_1, \mathfrak{V}_2, \ldots$ be sequences of open coverings of S and T, respectively, where $\mathfrak{U}_k = \{U_{k,1}, \ldots, U_{k,n_k}\}$ and $\mathfrak{V}_k = \{V_{k,1}, \ldots, V_{k,m_k}\}$, as produced in the proof of Theorem 2–95. If $\mathfrak{U}_1$ and $\mathfrak{V}_1$ have the same number of elements, we set $\mathfrak{U}'_1 = \mathfrak{U}_1$ and $\mathfrak{V}'_1 = \mathfrak{V}_1$. If $n_1 > m_1$, then by Theorem 2–96, $V_{1,1}$ is the union of $n_1 - m_1 + 1$ disjoint open (and closed) sets. Take $\mathfrak{U}'_1 = \mathfrak{U}_1$, and let $\mathfrak{V}'_1$ consist of $V_{1,2}, \ldots, V_{1,m_1}$ together with the sets into which $V_{1,1}$ has been decomposed. If $m_1 > n_n$, then the roles of $\mathfrak{U}_1$ and $\mathfrak{V}_1$ are interchanged.

Now suppose that $\mathfrak{U}'_j$ and $\mathfrak{V}'_j$ have been defined so as to have the same number of elements. Since the elements of $\mathfrak{U}'_j = \{U'_{j,1}, \ldots, U'_{j,n_j}\}$ are disjoint closed sets, there is an integer $m > j$ such that no set of diameter $< 1/m$ intersects any two $U'_{j,i}$, and there is a similar integer m' for $\mathfrak{V}'_j$. Let m denote the larger of these two integers. Then $\mathfrak{U}_m$ refines $\mathfrak{U}'_j$, and $\mathfrak{V}_m$ refines $\mathfrak{V}'_j$. Consider the elements of $\mathfrak{U}_m$ in $U'_{j,i}$ and the elements of $\mathfrak{V}_m$ in $\mathfrak{V}'_{j,i}$ for each i. If there are the same number of these elements for a given i, we leave them unaltered. If, for example, there are more elements of $\mathfrak{U}_m$ in $U'_{j,i}$ than elements of $\mathfrak{V}_m$ in $V'_{j,i}$, then we use Theorem 2–96 again to decompose one of the elements of $\mathfrak{V}_m$. Carrying out this process for each $i \leq n_j$ yields coverings $\mathfrak{U}'_{j+1}$ and $\mathfrak{V}'_{j+1}$, which refine $\mathfrak{U}'_j$ and $\mathfrak{V}'_j$, respectively, and which have the property that for each i, $U'_{j,i}$ and $V'_{j,i}$

contain the same number of elements of $\mathfrak{U}'_{j+1}$ and $\mathfrak{V}'_{j+1}$, respectively. The inductive definition of sequences $\mathfrak{U}'_1, \mathfrak{U}'_2, \ldots$ and $\mathfrak{V}'_1, \mathfrak{V}'_2, \ldots$ is complete, and we let $\mathfrak{U}^*_1, \mathfrak{U}^*_2, \ldots$ and $\mathfrak{V}^*_1, \mathfrak{V}^*_2, \ldots$ be the associated sequences of discrete spaces as defined in the proof of 2–95. We define a mapping $\Phi:\{\mathfrak{U}^*_n\} \to \{\mathfrak{V}^*_n\}$ by induction also. For $n = 1$, let $\varphi_1:\mathfrak{U}^*_1 \to \mathfrak{V}^*_1$ be any arbitrary one-to-one correspondence. Supposing that φ_{n-1} has been defined, let $\varphi_n:\mathfrak{U}^*_n \to \mathfrak{V}^*_n$ be given by assigning to each $U^*_{n,j}$ in $\mathfrak{U}^*_n$ an element of $\mathfrak{V}^*_n$ in $\varphi_{n-1}(f^{-1}(U'_{n,j}))$, where f is the projection of $\mathfrak{U}'_n$ into $\mathfrak{U}'_{n-1}$. This assignment is made in such a way that φ_n is a one-to-one correspondence also. It is now easy to verify that $\Phi = \{\varphi_n\}$ is a mapping of the inverse limit sequences $\{\mathfrak{U}^*_n, f_n\}$ and $\{\mathfrak{V}^*_n, f_n\}$ and that each φ_n is a homeomorphism onto. Then the inverse limit space U_∞ is mapped onto the inverse limit space V_∞ by the induced mapping φ. It is easily seen that φ is a homeomorphism of U_∞ onto V_∞ and since, by Theorem 2–95, U_∞ and V_∞ are homeomorphic to S and T, respectively, it follows that S and T are homeomorphic. □

COROLLARY 2–98. Any compact totally disconnected perfect metric space is homeomorphic to the Cantor set.

EXERCISE 2–37. Show that the homeomorphism h of Theorem 2–97 can be required to have the property that if $x_1, \ldots, x_n$ and $y_1, \ldots, y_n$ are points of S and T, respectively, and if we define a one-to-one correspondence $f(x_i) = y_i$, then the homeomorphism h is an extension of f. This points up the fact that the Cantor set is *homogeneous*. A space S is called *homogeneous* provided that if we are given any two points a and b in S, then there is a homeomorphism h of S onto itself such that $h(a) = b$. In other words, the apparent distinction between those points of the Cantor set that are end points of deleted intervals and those that are not is not an intrinsic topological property but is merely an accidental result of the particular imbedding of the Cantor set in the real line.

A further corollary can also be proved.

COROLLARY 2–99. Any compact totally disconnected metric space is homeomorphic to a subset of the Cantor set.

Proof: Let C be the Cantor set, and let M be a compact totally disconnected metric space. The space $M \times C$ is compact, totally disconnected, perfect, and metric, so there is a homeomorphism h of $M \times C$ onto C. Let i be the homeomorphism of M into $M \times C$, defined by $i(x) = (x, 0)$. Then $hi:M \to C$ is the desired homeomorphism. □

2–16 Limits inferior and superior. Suppose that $\{X_n\}$ is a sequence of subsets of a space S. The set of all points x in S such that every open set containing x intersects all but a finite number of the sets X_n is called the *limit inferior* of the sequence $\{X_n\}$ and is abbreviated "lim inf X_n"; the

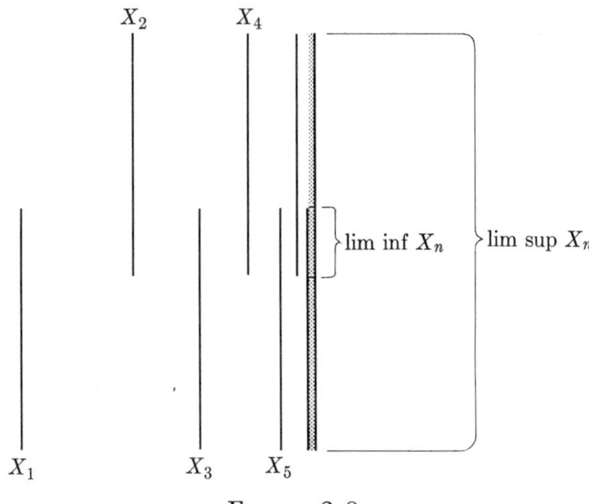

FIGURE 2–8

set of all points y in S such that every open set containing y intersects infinitely many sets X_n is called the *limit superior* of $\{X_n\}$ and is abbreviated "lim sup X_n." If these two sets coincide (so that lim inf $X_n = L =$ lim sup X_n), we say that $\{X_n\}$ is a *convergent sequence* of sets and that L is the *limit* of $\{X_n\}$, which is abbreviated "$L = \lim X_n$."

It is obvious from the definition that lim inf X_n is contained in lim sup X_n. But the two need not coincide. The sequence of sets in E^2 shown in Fig. 2–8 is an example in which lim inf $X_n \neq$ lim sup X_n.

EXERCISE 2–38. Give an example in which lim sup $X_n = \phi$.

EXERCISE 2–39. Give an example in which lim inf $X_n = \phi \neq$ lim sup X_n.

EXERCISE 2–40. In Fig. 2–8, every subsequence of $\{X_n\}$ contains a convergent subsequence. Give an example with lim inf $X_n \neq \phi$ which lacks this property.

LEMMA 2–100. *If $\{X_n\}$ is a sequence of sets in a space S, then lim inf $X_n =$ lim inf $\overline{X}_n$ and lim sup $X_n =$ lim sup $\overline{X}_n$. Furthermore, both lim inf X_n and lim sup X_n are closed, and so is the set $\bigcup_{n=1}^{\infty} \overline{X}_n \cup$ lim sup X_n.*

The proof is left as an exercise.

There is a high probability that the example that the reader produced for Exercise 2–39 has the property that each X_n is connected but that lim sup X_n is not. (If this is not the case, try again!) We next give a theorem that will control this situation.

THEOREM 2–101. *If $\{X_n\}$ is a sequence of connected sets in a compact Hausdorff space S, and if lim inf X_n is not empty, then lim sup X_n is connected.*

Proof: Suppose to the contrary that lim sup X_n is the union of two separated sets M and N; each is closed by 2–100. Since S is normal (Theorem 2–3), there exist disjoint open sets U and V containing M and N, respectively. Then there is an integer j such that for $n > j$, X_n lies in $U \cup V$. For if not, then there would be an infinite sequence $x_{n_1}, x_{n_2}, \ldots$ of points such that x_{n_i} lies in $X_{n_i} - (U \cup V)$. The set $\{\cup x_{n_i}\}$ is either finite or has a limit point, so there is a point x such that every open set containing x contains infinitely many of the points x_{n_i}. It follows that x is in lim sup X_n, and this contradicts the fact that x is in $S - (U \cup V)$.

The set lim inf X_n intersects one of the sets U and V; suppose that it intersects U. Then all but a finite number of the sets X_n intersect U. But if $X_n \cap U$ is not empty, then $X_n \cap V$ is empty; otherwise X_n is not connected. Therefore only a finite number of the sets X_n meet V, so $V \cap$ lim sup X_n is empty, a contradiction. □

Next we are interested in giving an analogue to the theorem that compactness implies countable compactness. The analogy is not quite perfect, for in a compact Hausdorff space, a point may be a limit of a set X but not be the limit of any sequence of points in X. An example of such a space is the *long line*, at the end of Section 2–5.

THEOREM 2–102. If M is a compact metric space, then every sequence of subsets of M contains a convergent subsequence.

Proof: By Exercise 2–21, there is a countable basis $\{B_n\}$ for M. Let $\{X_n\}$ be a sequence of subsets of M. We will define a collection of subsequences $\{X_n^k\}$ of $\{X_n\}$, one for each integer k, such that $\{X_n^{k+1}\}$ is a subsequence of $\{X_n^k\}$. Let $\{X_n^1\} = \{X_n\}$. Now if $\{X_n^1\}$ contains a subsequence whose limit superior contains no point of B_1, we take $\{X_n^2\}$ to be such a subsequence; if there is no such subsequence, we set $\{X_n^2\} = \{X_n^1\}$. If $\{X_n^2\}$ contains a subsequence whose limit superior contains no point of B_2, let $\{X_n^3\}$ be such a subsequence; if no such exists, let $\{X_n^3\} = \{X_n^2\}$. The general inductive step in the definition is now easy to formulate. We have the array

$$\begin{array}{llll} X_1^1, & X_2^1, & X_3^1, & X_4^1, \ldots \\ X_1^2, & X_2^2, & X_3^2, & X_4^2, \ldots \\ X_1^3, & X_2^3, & X_3^3, & X_4^3, \ldots \\ \vdots & & & \end{array}$$

where each row is a subsequence of the row above. Now consider the "diagonal sequence" $X_1^1, X_2^2, X_3^3, \ldots$ For each m, the terms from X_m^m on constitute a subsequence of $\{X_n^m\}$.

We assert that the sequence $\{X_m^m\}$ converges. Suppose to the contrary that there is a point p in lim sup X_m^m $-$ lim inf X_m^m. Then there is an open set U containing p, and a subsequence $\{Y_n\}$ of $\{X_m^m\}$ such that no Y_n intersects U. In U there is a basis element B_j containing p. Then, the first few terms excepted, $\{Y_n\}$ is a subsequence of $\{X_n^j\}$ such that $(\limsup Y_n) \cap B_j$ is empty. Since such a subsequence exists, $\{X_m^{j+1}\}$ is such a subsequence. But since $\{X_m^m\}$, $m \geqq j + 1$, is a subsequence of $\{X_m^{j+1}\}$, we have that $\limsup X_m^{j+1}$ contains $\limsup X_m^m$, which therefore cannot contain the point p, a contradiction. □

The reader will notice the resemblance between Theorem 2–102 and the familiar theorem in analysis that states that every bounded sequence of real (or complex) numbers contains a convergent subsequence.

Exercise 2–41. Prove that the intersection of all open sets containing a subset X of a T_1-space is X itself.

Exercise 2–42. Prove that the Kuratowski closure operation (Section 1–13) yields a T_1-space and conversely.

Exercise 2–43. Show that a subset A is nowhere dense in a space S if and only if $S - \overline{A}$ is dense in S.

Exercise 2–44. Prove the following theorem.

Theorem 2–103. Let X and Y be compact Hausdorff spaces and $f:X \to Y$ be a one-to-one continuous mapping of X onto Y. Then f is a homeomorphism.

Exercise 2–45. Prove the following theorem.

Theorem 2–104. Let Y be a Hausdorff space and Y^X be assigned the compact-open topology. Then Y^X is also a Hausdorff space.

Exercise 2–46. Prove that every perfect set in a complete metric space contains a compact perfect set.

Exercise 2–47. If S is a normal space that is separable, then show that every subset of cardinality c (the power of the continuum) has a limit point. (See Jones [88].)

Exercise 2–48. A transformation $f:X \to Y$ is said to be *arc-preserving* if the image of every arc in X is either an arc or a point. Show that if X is a space such that every infinite subset of X intersects some arc in X in an infinite set and if $f:X \to X$ is an arc-preserving transformation of X into itself, then f is continuous. (See Hall and Puckett [81].)

Exercise 2–49. Prove the following theorems concerning topological groups. (See Section 1–14.)

Theorem 2–105. If a topological group G satisfies Axiom T_0, then it also satisfies Axiom T_2.

Theorem 2–106. A T_0 topological group is completely regular (is a *Tychonoff space*).

Theorem 2–107. A topological group G is locally compact if and only if there is an open set U containing the identity e such that $\overline{U}$ is compact.

EXERCISE 2–50. A space is *rim-compact*, or *locally peripherally compact* if each point has "arbitrarily small" open sets containing it with compact boundaries. Show that if S is a rim-compact Hausdorff space, Theorem 2–1 holds without the requirement that C be compact. Are there other theorems in this chapter which can be altered in a similar way?

EXERCISE 2–51. A space S is *pseudo-compact* if every continuous real-valued function defined on it is bounded. Show that the "long line" is pseudo-compact. Show that a metric space is pseudo-compact if and only if it is bounded.

EXERCISE 2–52. Find a connected metric space that contains an open set U such that no component of U has a point of $\overline{U} - U$ as limit point.

CHAPTER 3

FURTHER TOPICS IN POINT-SET TOPOLOGY

Many interesting and useful topics were necessarily omitted from the first two chapters. Now that a background has been given, further developments may be made. Some of these may be considered as "classic," and others touch upon the frontier of our current knowledge.

3–1 Locally connected spaces. In Section 2–10 we exhibited one technique for localizing a topological property. This may be stated in general terms as follows. Let P denote a topological property. Then a space has property P at a point x if there is an open set containing x and having property P (or whose closure has property P). This gives a useful meaning to local compactness, for instance, but is not satisfactory for some other properties. To truly "localize" a given property, we should ask for "arbitrarily small" open sets with this property. That is, we want an "epsilon-delta" definition without restricting ourselves to metric spaces. This is precisely the content of the following formulation of local connectivity.

A space S is said to be *locally connected at a point* x if for every open set U containing x there is a connected open set V containing x and contained in U. The space S is *locally connected* if it is locally connected at each point. A few words about this property are in order before the precise study begins. First, a space may be locally connected at all but one point. (This is in contradistinction to local compactness.) For, consider the graph of the function $y = \sin(1/x)$, $0 < x \leqq 1$, together with the origin, in E^2 (see Fig. 3–1).

Any small circle, such as C in the figure, defines an open set containing the origin. But the only connected set containing the origin and lying within C is the origin itself, and this one-point set is not open. (Why?) Any other point in this space lies in arbitrarily small open arcs, however, so the space fails to be locally connected at just the one point. Where is the space locally compact?

Every compact space is locally compact, but not every connected space is locally connected. Indeed, the example just given above is connected. A more widely used example is the *compact*, connected, but not locally connected, set often called "the topologist's sine curve." It is the graph of the function $y = \sin(1/x)$, $0 < x \leq 1$, together with the interval $-1 \leqq y \leqq 1$ on the y-axis in E^2 (Fig. 3–2). Again, a small circle C about a point p on the segment $-1 \leq y \leq 1$ defines an open set containing p.

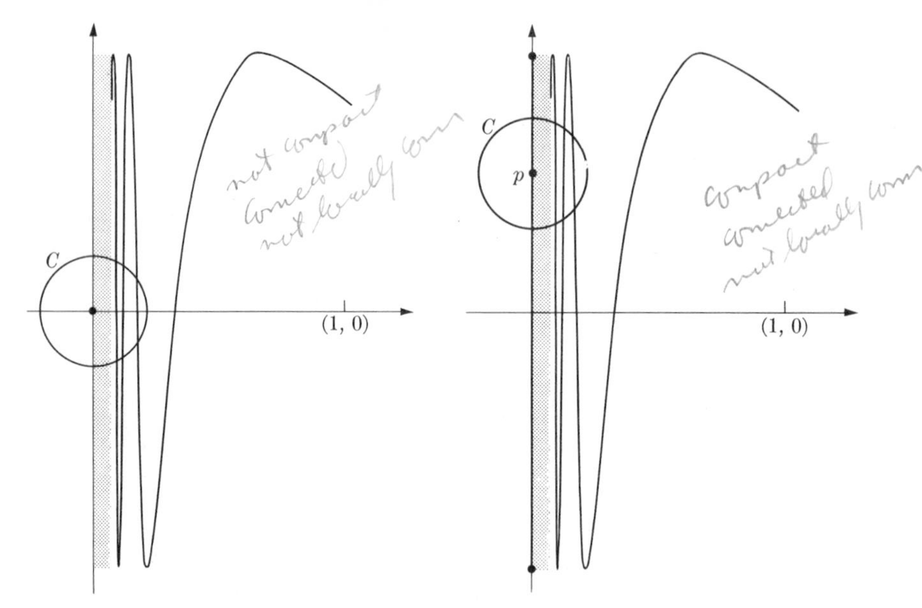

FIGURE 3–1

FIG. 3–2. The topologist's sine curve.

The only connected set lying within C and containing p is the segment on the interval $-1 \leq y \leq 1$ which lies within C. But this open segment is *not* open in the space. (Why?) It follows that the topologist's sine curve fails to be locally connected at each point of the interval $-1 \leqq y \leqq 1$. The reader may supply the arguments to show that this set is compact and connected.

We may now re-examine the "one-neighborhood" definition of local compactness for comparison purposes. The use of just one set is possible, for if U is an open set containing a point p and having a compact closure, then any open set contained in U also has a compact closure. Thus the "two-neighborhood" type of definition would follow as an easy theorem. Other local properties that may be given in a one-neighborhood definition are "locally separable," "locally countable," and "locally perfect."

The proof of the first result is left as an easy exercise.

LEMMA 3–1. An open subset of a locally connected space is locally connected.

THEOREM 3–2. For a space to be locally connected, it is necessary and sufficient that each component of an open set be open.

Proof: Suppose that S is a locally connected space, U is an open set in S, and C is a component of U. For each point x in C, there is an open connected set V_x containing x and lying in U. Then $C \cup V_x$ is connected and lies in U, and V_x lies in C, by the maximality of C. It follows that $C = \bigcup_C V_x$, so C is a union of open sets.

Conversely, if U is an open set containing a point x and C is the component of U that contains x, and if every component of an open set is open, then C itself is the set V of the definition. □

We note that *a space is locally connected if it has a basis of connected open sets.* This is in contradistinction to the higher-dimensional local connectivity properties to be seen in Sections 4–9 and 8–7.

A frequently used abbreviation for the phrase *locally connected* are the letters "lc." We shall often use it, too.

THEOREM 3–3. If S is a connected lc space, and C is a component of an open set in S such that $S - \overline{C}$ is not empty, then $\overline{C} - C$ is not empty and separates C and $S - \overline{C}$ in S.

Proof: If $\overline{C} - C$ is empty, then C is closed. We know from Theorem 3–2 that C is open, so $S - \overline{C} = S - C$ is also both open and closed. It follows that S is not connected, counter to hypothesis. Therefore $\overline{C} - C$ cannot be empty. Since $S - (\overline{C} - C) = C \cup (S - \overline{C})$, it also follows that $\overline{C} - C$ separates S as asserted. □

This result is not true without the assumption of local connectedness, but it should be compared with Theorem 2–16, where under the alternative hypothesis that S be compact, it is shown that each component of an open set has limit points in its boundary.

Given two points a and b of a space S, a collection $A_1, \ldots, A_n$ of sets is a *simple chain from a to b* provided that A_1 (and only A_1) contains a, A_n (and only A_n) contains b, and $A_i \cap A_j$ is nonempty if and only if $|i - j| \leqq 1$; that is, each *link* intersects just the one before it and the one after it (and itself). Figure 3–3(a) illustrates a simple chain of regions from a to b.

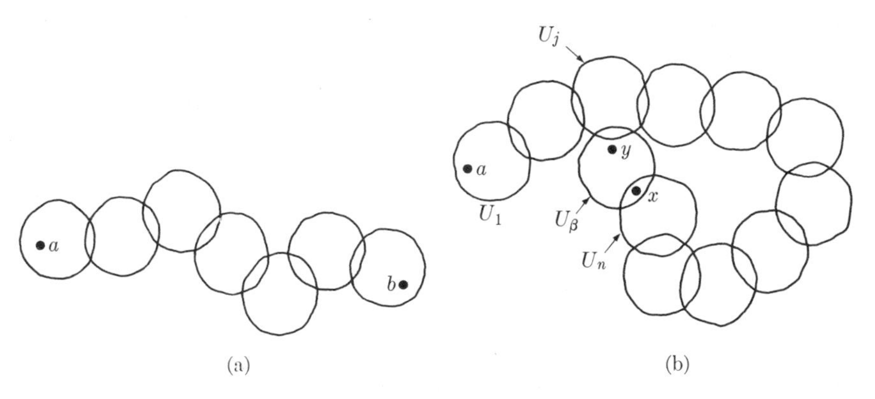

FIGURE 3–3

THEOREM 3–4. If a and b are two points of a connected space S, and $\{U_\alpha\}$ is a collection of open sets covering S, then there is a simple chain of elements of $\{U_\alpha\}$ from a to b.

Proof: Let X denote the set of all points x in S such that there is a simple chain of elements of $\{U_\alpha\}$ from a to x. Then X is open. For if x is a point in X, and $U_1, \ldots, U_n$ is a simple chain of elements of $\{U_\alpha\}$ from a to x, then for each point y in U_n, either $U_1, \ldots, U_n$ or $U_1, \ldots, U_{n-1}$ is a simple chain from a to y (it may happen that y is in $U_{n-1} \cap U_n$). It follows that y is in X, and hence all of U_n is in X. Thus X is a union of open sets.

Now X is also closed. For suppose that y is a point of $\overline{X} - X$. Then there is an element U_β of $\{U_\alpha\}$ that contains y. Since y is a limit point of X, U_β also contains a point x of X. There is a simple chain $U_1, \ldots, U_n$ from a to x, and the collection $U_1, U_2, \ldots, U_n, U_\beta$ contains a simple chain from a to y. (Figure 3–3b indicates that this simple chain may have less than $n + 1$ links.) Since S is connected and X is both open and closed in S, we have $X = S$. □

The above result is based upon Cantor's first definition of connectedness for metric spaces. A metric space M is connected in the sense of Cantor, provided that given two points a and b and any positive number ϵ, there is a sequence $a = x_1, x_2, \ldots, x_n = b$ of points of M such that $d(x_i, x_{i+1}) < \epsilon, i = 1, \ldots, n - 1$. This definition agrees with the more general definition that we have adopted in *compact* metric spaces. But the rationals in E^1 are connected in the sense of Cantor, and the reader will easily see that Theorem 3–4 does not hold for the rationals.

A simple consequence of Theorem 3–4 is a theorem that may be familiar to the reader from his studies in analysis.

THEOREM 3–5. Each two points of a connected open set U in E^n can be joined by a polygonal arc in U.

Proof: Let $\{S(x_\alpha, r_\alpha)\}$ be a collection of spherical neighborhoods covering U and such that each $S(x_\alpha, r_\alpha)$ lies in U. (Such a covering of U exists because the collection of all spherical neighborhoods is a basis for E^n.) If a and b are two points of U, then by Theorem 3–4 there is a simple chain $S(x_1, r_1), \ldots, S(x_n, r_n)$ from a to b. Then the union of the straight-line segments $\overline{ax_1}, \overline{x_1x_2}, \ldots, \overline{x_{n-1}x_n}, \overline{x_nb}$ certainly contains a polygonal arc from a to b. □

It is interesting to note that if U is the open set of Theorem 3–5, it is *not* true that each two points of the closure $\overline{U}$ can be joined by a polygonal arc. Even if U is a plane region bounded by a simple closed curve so that U is homeomorphic to the unit disc under a conformal mapping (the

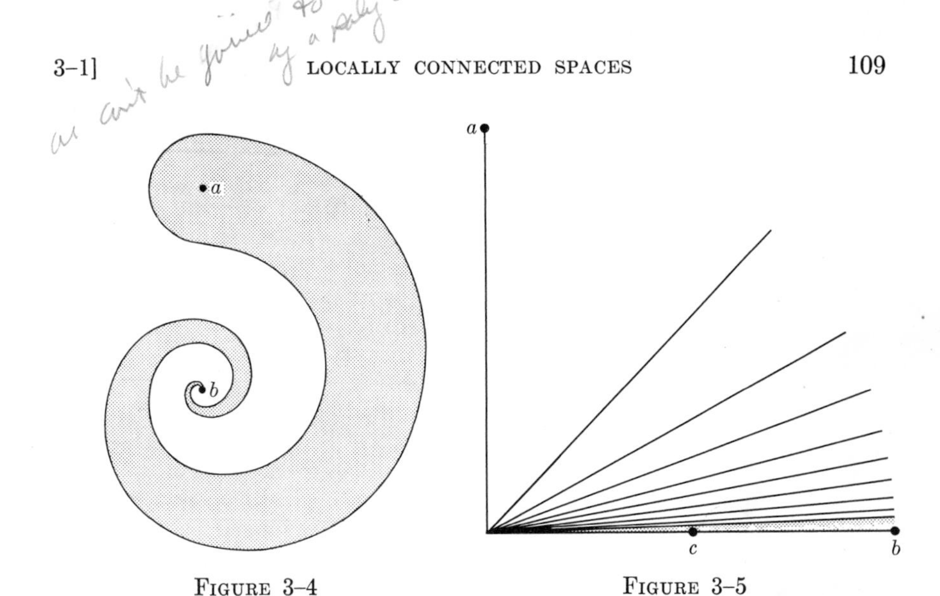

FIGURE 3–4 FIGURE 3–5

Riemann mapping theorem), the conclusion of Theorem 3–5 may be false. Figure 3–4 shows such a region. Indeed, the spiral can be made "infinitely long," so that no rectifiable arc in $\overline{U}$ joins a to b.

There is a useful property called *cutting*, which is a weak form of separation. A set X in a connected space S cuts S between two points a and b of $S - X$ if X intersects every closed connected subset that contains both a and b. Clearly, if X separates a and b, then X cuts S between a and b, but the converse may fail to hold. In Fig. 3–5, the point c cuts the set between a and b but does not separate these points. For locally connected spaces, however, the two concepts, separation and cutting, agree.

THEOREM 3–6. A closed set X cuts a connected lc regular space S between points a and b if and only if X separates a from b.

Proof: Suppose X cuts S between a and b. The components of $S - X$ are all open in view of Theorem 3–2, so that if a lies in a component U, and if b does not lie in U, then $S - X = U \cup (S - X - U)$ is a separation of a and b. Suppose, then, that a and b both lie in the same component U of $S - X$. Each point x of U lies in an open connected set U_x whose closure lies in U (Theorem 2–4). If we now apply Theorem 3–4 to U as a subspace of S and take the covering $\{U_x\}$, we find a simple chain $U_1, \ldots, U_n$ of sets in $\{U_x\}$ from a to b. The set $\bigcup_{i=1}^{n} \overline{U}_i$ is then a closed connected subset of $S - X$ that contains a and b. Hence X could not cut S between a and b. □

If the set X in Theorem 3–6 is not required to be closed, the conclusion may be quite false. Indeed we have the following modification of a theorem due to E. Bernstein [54].

THEOREM. The plane E^2 is the union of two disjoint sets each of which cuts E^2 between each two points of the other.

We will only indicate the proof. First we remark that there are only c open sets in E^2 (c is the cardinal number of the real numbers). For if $\{V_1, V_2, \ldots\}$ is a countable basis for E^2, then each open set U is the union of all basis elements that are contained in U. Different open sets are composed of different collections of basis elements, so the number of open sets is the same as the number of subcollections of $\{V_1, V_2, \ldots\}$. There are only c such subcollections. By complementation there are c closed subsets in E^2 and hence only c closed connected subsets. We well-order the collection of all nondegenerate closed connected sets into a sequence $C_1, C_2, \ldots, C_\alpha, \ldots$ such that each element has less than c predecessors. Let A_1 be a set consisting of one point of C_1, and let B_1 be a set consisting of some other point of C_1. Suppose that A_β and B_β have been defined for all $\beta < \alpha$. Then $\bigcup_{\beta<\alpha}(A_\beta \cup B_\beta)$ has less than c points, whereas C_α has c points. Hence there are many points in $C_\alpha - \bigcup(A_\beta \cup B_\beta)$. Let a_α and b_α be two of these points, and define $A_\alpha = a_\alpha \cup \bigcup_{\beta<\alpha} A_\beta$ and $B_\alpha = b_\alpha \cup \bigcup_{\beta<\alpha} B_\beta$. This defines A_α and B_α for all α. Now let $A = \bigcup_\alpha A_\alpha$ and $B = E^2 - A$. Then B contains $\bigcup_\alpha B_\alpha$. By construction, any closed connected set C meets both A and B. It follows that any closed connected set containing two points of A (or B) meets B (or A).

Bernstein's statement was that the plane is a union of two disjoint *connected* sets, etc. Our sets have this property. Suppose that A, for instance, were not connected, that is, $A = A' \cup A''$, where A' and A'' form a separation. By the complete normality of E^2, there would be a closed set X separating A' from A''. We will see later that X must contain a nondegenerate closed connected set C separating some point in A' from some point in A''. But C meets A and cannot lie in $E^2 - A$. Thus A is connected. □

As a corollary to the proof of Theorem 3–6, we have the following result.

THEOREM 3–7. If S is a locally connected regular space, and if U is a connected open set in S, then each two points of U lie in a closed connected subset C of S such that C is contained in U.

EXERCISE 3–1. Show that if S is lc and Hausdorff, every quasicomponent is a component.

EXERCISE 3–2. Show that if a and b are two points in an lc Hausdorff space S, then a necessary and sufficient condition that a point p separate a from b is that every simple chain of open sets from a to b have a link containing p.

THEOREM 3–8. If a and b are two points in a connected lc Hausdorff space S, then the set $E(a, b)$ of cut points separating a and b is closed (see Theorem 2–21).

Proof: Suppose that there is a limit point p of $E(a, b)$ that is not in $E(a, b)$. Since p does not separate a from b, the points a and b lie in the

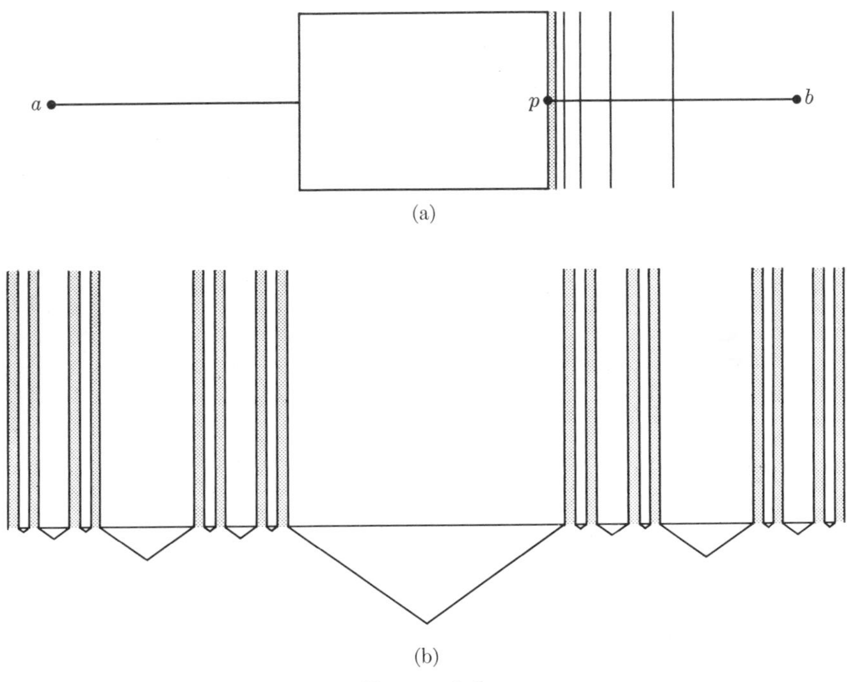

(a)

(b)

FIGURE 3–6

same component U of $S - p$. Applying Theorem 3–7, we find a closed connected subset C of U that contains a and b. But C must contain each point of $E(a, b)$ by Theorem 3–6, and p is not a limit point of C. This is a contradiction. □

Figure 3–6 gives two examples of non-locally connected plane continua for which the conclusion of Theorem 3–8 is false. In Fig. 3–6(a) the point p is a limit point of $E(a, b)$ but is not in $E(a, b)$. Figure 3–6(b) pictures an even more drastic case, for there we have only a countable number of points in $E(a, b)$, but every point on the Cantor set *cuts* the continuum between a and b.

A useful criterion for local connectedness is stated next.

THEOREM 3–9. A necessary and sufficient condition that a locally compact connected Hausdorff space S be locally connected is that if C is a compact subset of S and U is an open set containing C, then all but a finite number of components of $S - C$ lie in U.

Proof: We prove sufficiency first. Suppose that S is not locally connected at a point p. Then there is an open set V containing p, such that the component K of V that contains p is not open. There is no loss of generality in supposing that $\overline{V}$ is compact. Since K is not open, it contains a limit point q of the union of the remaining components of V, but of no finite number of these. By Theorem 2–56, there is an open set U containing $\overline{V} - V$, such that $\overline{U}$ does not contain q. Every component of V is a component of $S - (\overline{V} - V)$, so all but a finite number of these lie in U by the condition we are assuming. This is a contradiction, since infinitely many of these components intersect every open set containing q and, in particular, the open set $S - \overline{U}$.

The necessity part is somewhat easier. Suppose that C is a compact subset of S, and that V is an open set containing C. Let $\{O_\alpha\}$ be the collection of all components of $S - C$ that intersect $S - V$. Since S is lc, each such component is open, and they are clearly disjoint. By Theorem 3–3, each $\overline{O}_\alpha$ meets C. Now in V, there is an open set U with compact closure, and each O_α meets $\overline{U} - U$. Also each point of $\overline{U} - U$ is in some O_α. By compactness of $\overline{U} - U$, a finite number of sets O_α covers $\overline{U} - U$. But this means that there are only a finite number of the O_α altogether. □

For a metric space, the above theorem may of course be formulated in terms of distance.

THEOREM 3–10. A necessary and sufficient condition that a locally compact connected metric space M be locally connected is that if C is a compact subset of M and $\{x_n\}$ is a sequence of points from different components of $M - C$, then $\lim_{n\to\infty} d(x_n, C) = 0$.

The modifications required to prove Theorem 3–10 are left as an exercise.

One might suppose at first that Theorem 3–10 is equivalent to saying that the diameters of the components of $M - C$ approach zero. Figure 3–7 shows that this is not true, even in the plane.

If C were locally connected as well as compact, the conjecture that the components of $E^2 - C$ have diameters approaching zero would be true. But even adding local connectedness to C does not establish the

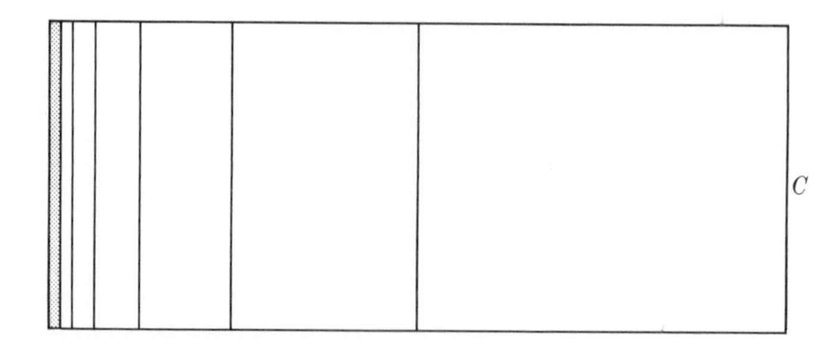

FIGURE 3–7

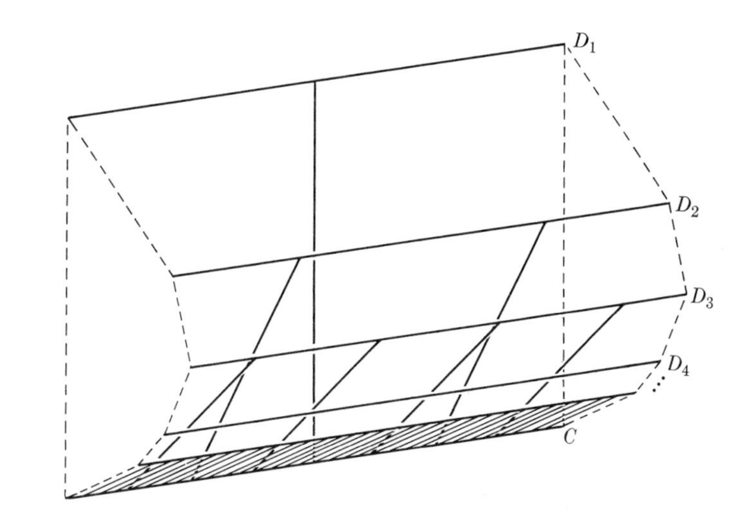

FIGURE 3–8

conjecture in general. Figure 3–8 is a locally connected continuum P in E^3 that is separated by an arc C. Each set D_n has the same diameter. If one imagines that each D_n is a hollow pipe tapering to a point as it approaches C, then removing this modified continuum P' from E^3 gives an example in E^3, the diameters of the components of $E^3 - P'$ not approaching zero.

Closely related to local connectedness is the concept of connectedness *im kleinen*. A space S is *connected im kleinen at a point* x provided that for each open set U containing x, there is an open set V containing x and lying in U, such that if y is any point in V, then there is a connected subset of U containing $x \cup y$. It is obvious that local connectedness at the point x implies connectedness *im kleinen*, but a space may be connected *im kleinen* at a point x and yet not be locally connected at x. For example, consider the point x in the set illustrated in Fig. 3–9.

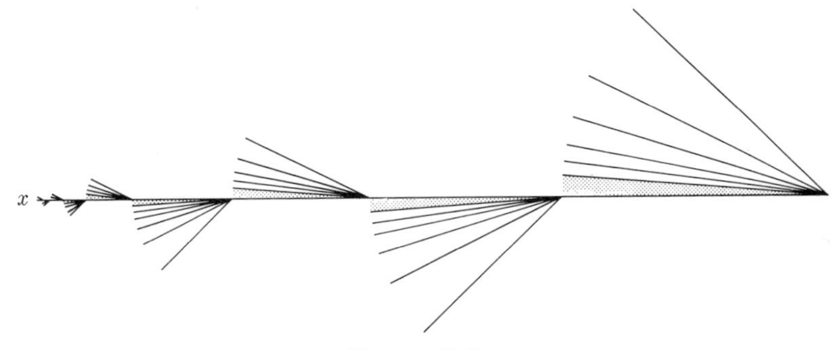

FIGURE 3–9

Despite this example, however, we do have the following result.

THEOREM 3–11. If a space S is connected *im kleinen* at each point, then it is locally connected.

Proof: Let U be an open subset of S, and let C be a component of U. Let x be a point of C. Then there is an open set V_x containing x and lying in U, such that each point y of V_x is in a connected set C_{xy} lying in U. Then C_{xy} is a subset of C, so V_x lies in C. Thus $C = \cup_x V_x$ is open, and Theorem 3–2 applies. □

The next result is allied to Theorem 3–9 and is a useful criterion for the *failure* of the local connectivity property.

THEOREM 3–12. If a locally compact connected metric space M is not connected *im kleinen* at a point p, then there exist an open set U containing p, a continuum K that contains p, lies in $\overline{U}$, and meets $\overline{U} - U$, and a sequence of distinct components $\{C_n\}$ of U such that $K = \lim C_n$.

Proof: Suppose that V is an open set containing p, such that $\overline{V}$ is compact and such that there is no open set U containing p and lying in V, with the property that every point in U can be joined to p by a connected subset of V. In particular, for any positive n, we can choose an open set U_n with diameter $< 1/n$ and a point x_n in U_n so as to obtain a sequence $x_1, x_2, \ldots$ of points of V converging to p, none of which lies in the same component of V as does p. Let K_n be the component of V containing x_n. By Theorem 2–16, the closure of K_n meets $\overline{V} - V$. It should be evident that no component of V can contain more than a finite number of points x_n, so we may assume that the K_n are all distinct. Now by Theorem 2–102, some subsequence of $\{\overline{K}_n\}$ converges to a continuum K. Let $C_1, C_2, \ldots$ be that subsequence. Then K contains p, and the conclusion is satisfied. □

The chief reason for our introducing this other form of local connectedness is that it lends itself most readily to being altered into a uniform local connectedness analogous to uniform continuity. We will do this only for metric spaces and hence will first rephrase the definition of *connected im kleinen* as follows. A metric space is connected *im kleinen* at a point x provided that, given $\epsilon > 0$, there is a number $\delta = \delta(x, \epsilon) > 0$ such that if $d(x, y) < \delta$, then $x \cup y$ lies in a connected set of diameter $< \epsilon$. (It follows that $\delta \leqq \epsilon$.) We now say that a metric space is *uniformly connected im kleinen* or *uniformly locally connected* provided that, given $\epsilon > 0$, there is a number $\delta = \delta(\epsilon)$, independent of position, such that any two points x and y, with $d(x, y) < \delta$, lie in a connected set of diameter $< \epsilon$.

In analogy to Theorem 1–31, we have the final result of this section.

THEOREM 3–13. If a compact metric space M is locally connected, then it is uniformly locally connected.

Proof: Given $\epsilon > 0$, each point x of M lies in a connected open set V_x of diameter $< \epsilon$, since M is locally connected. A finite number of these sets V_x, say $V_1, \ldots, V_n$, covers M. Let δ be the Lebesgue number of this covering (see Theorem 1–32). Then if $d(x, y) < \delta$, x and y lie in some V_j. This V_j is the desired connected set. □

EXERCISE 3–3. Prove that the two definitions of connected *im kleinen* agree on metric spaces.

EXERCISE 3–4. Define uniform local connectedness for topological spaces, and use your definition to prove the analogue of Theorem 3–13.

EXERCISE 3–5. Show that a uniformly locally connected metric space is locally connected.

EXERCISE 3–6. A metric space has *property S* if, for every $\epsilon > 0$, it is the union of a finite number of connected sets, each of diameter $< \epsilon$. Prove that a space having property S is connected *im kleinen* at each of its points and hence is locally connected.

EXERCISE 3–7. Show that if a metric space has property S, each of its points lies in arbitrarily small open sets having property S.

EXERCISE 3–8. Show that property S is not equivalent to uniform local connectedness.

EXERCISE 3–9. Prove that a compact locally connected metric space has property S.

3–2 Arcs, arcwise connectivity, and accessibility. In this section, we give some further characterization of the unit interval and show that locally connected spaces with a compactness or a completeness condition have the added property that each two points can be joined by an arc in the space.

A locally connected and connected space has the property that each two points can be joined by a simple chain of connected sets (Theorem 3–4). Such a simple chain may be regarded as a sort of approximation to an arc. By joining two points with finer and finer simple chains, we should come closer and closer to an arc. There are three reasons why such a construction may fail. If the simple chains are not related in some way, their "limit" may be almost any kind of a continuum. To avoid this, we can require that the links of each successive chain be contained in the links of its predecessor. Even with this precaution, however, we will see in Section 3–8 that the intersection of all the chains need not be an arc. This kind of behavior will have to be ruled out. And finally, even after doing this, it may happen that the intersection of the simple chains may lack some of the points necessary to form an arc. To illustrate this last point, let S denote the set of all points in E^2 except those on the x-axis having rational coordinates. The sequence of simple chains indicated by Fig. 3–10 will have an intersection in S that is not an arc because of the omission of the rational points.

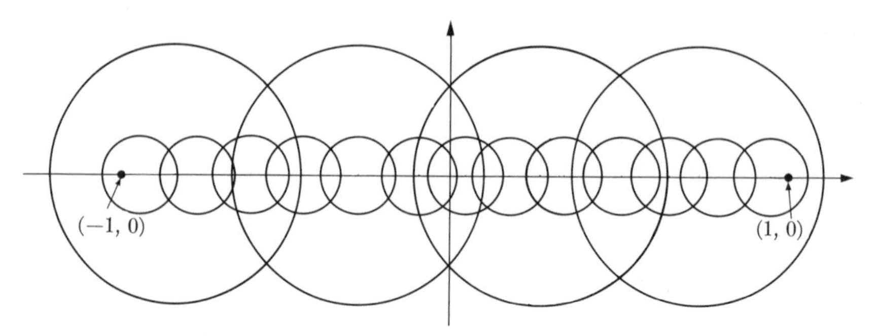

FIGURE 3–10

Let $C_1 = \{U_{11}, \ldots, U_{1n_1}\}$ and $C_2 = \{U_{21}, \ldots, U_{2n_2}\}$ be simple chains from a point a to a point b in a space S. The chain C_2 will be said to go straight through C_1 provided that (a) every set U_{2i} is contained in some set U_{1j} and (b) if U_{2i} and U_{2k}, $i < k$, both lie in a set U_{1r}, then for every integer j, $i < j < k$, U_{2j} also lies in U_{1r}. The finer chain in Fig. 3–10 goes straight through the other, for example. To see a counterexample, look at Fig. 3–21.

THEOREM 3–14. Suppose that S is a locally connected and connected Hausdorff space and that C is a simple chain of connected open sets $U_1, \ldots, U_n$ from a point a to a point b. Suppose that $\mathcal{V}$ is a collection of open sets such that each link U_i is a union of elements of $\mathcal{V}$. Then there is a simple chain of elements of $\mathcal{V}$ from a to b that goes straight through C.

Proof: Let $x_0 = a$, $x_n = b$, and for $i = 1, 2, \ldots, n - 1$, take x_i to be a point of $U_i \cap U_{i+1}$. Each U_i is connected and is a union of elements of $\mathcal{V}$, so by Theorem 3–4 there is a simple chain C_i of elements of $\mathcal{V}$ from x_{i-1} to x_i, all links of C_i lying in U_i. The collection of all links of all simple chains C_i is a chain from a to b, but it need not be a simple chain; for instance, the situation pictured in Fig. 3–11 might arise. However, this collection of all links contains a simple chain going straight through C. In C_1 there is a first link (U_i in Fig. 3–11) that intersects a link of C_2, and there is a last link of C_2 (V_j in Fig. 3–11) that meets U_i. We omit the links of C_1 following U_i and those of C_2 preceding V_j. Repeating this process for each i, we easily obtain the desired simple chain. □

THEOREM 3–15. Each two points of a compact, connected, and locally connected metric space S can be joined by an arc in S.

Proof: Let a and b be two points of S. There is a simple chain of connected open sets, $C_1 = \{U_{11}, \ldots, U_{1n_1}\}$, joining a to b, each U_{1i} having diameter < 1. About each point of each U_{1i} there is a connected open set

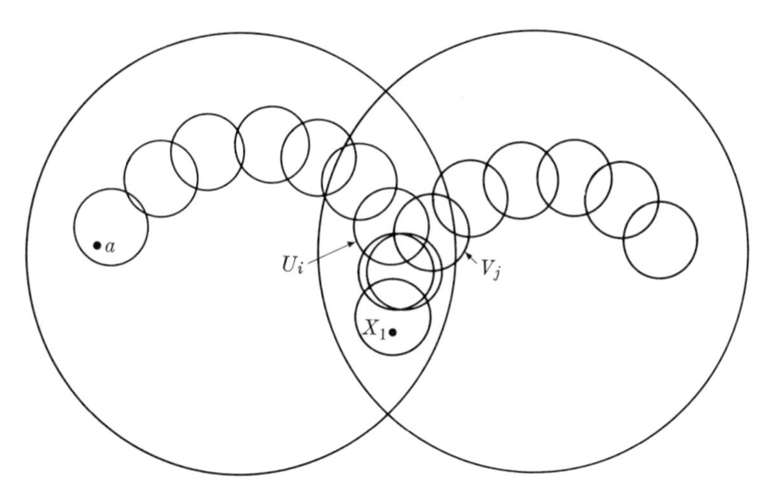

FIGURE 3–11

of diameter $< \frac{1}{2}$ with its closure lying in U_{1i}; if the point in question lies in two sets U_{1i} and U_{1i+1}, the open set of diameter $< \frac{1}{2}$ can be chosen so that its closure is in $U_{1i} \cap U_{1i+1}$. By Theorem 3–14, there is a simple chain $C_2 = \{U_{2i}, \ldots, U_{2n_2}\}$ of these open sets of diameter $< \frac{1}{2}$ joining a to b and going straight through C_1. Similarly, we construct a simple chain $C_3 = \{U_{31}, \ldots, U_{3n_3}\}$ from a to b such that each U_{3j} has diameter $< \frac{1}{3}$ and has closure lying in a link of C_2 and such that C_3 goes straight through C_2. It is now evident how we will construct $C_4, C_5, \ldots$. Let $K_j = \overline{U}_{j1} \cup \cdots \cup \overline{U}_{jn_j}$. For each j, K_j is a continuum containing $a \cup b$ and also containing K_{j+1}. By Theorem 2–8, $K = \bigcap_{j=1}^{\infty} K_j$ is also a continuum containing $a \cup b$. Note that each point of K is also in $\bigcap_{j=1}^{\infty} (U_{j1} \cup \cdots \cup U_{jn_j})$.

Now let x be a point of $K - a - b$. For each integer j, let P_j be the union of all links U_{ji} in C_j that precede the one or two links in C_j that contain x, and let F_j be the union of all links U_{jk} in C_j that follow the one or two links containing x. Let $P = \bigcup_{j=1}^{\infty} P_j \cap K$, and let $F = \bigcup_{j=1}^{\infty} F_j \cap K$. Then P and F are disjoint relatively open nonempty subsets of K, and each point of $K - x$ lies in one or the other. Therefore x is a cut point of K, and K has only two non-cut points, a and b. In view of Theorem 2–27, K is then an arc from a to b. □

Two definitions will shorten our statements as well as introduce two commonly used terms. A compact, connected, and locally connected metric space is called a *Peano space* or a *Peano continuum*. A space S is *arcwise connected* if each two points of S are the end points of an arc in S. In these terms, Theorem 3–15 may be stated as follows. *Every Peano space is arcwise connected.*

We will profit by a few comments upon the hypotheses of Theorem 3–15. First, it is evident that if any one of the sets K_j above is compact, then all its successors will also be compact, and the proof will go through without alteration. This means that the requirement of compactness for S may be replaced by local compactness. In particular, an open subset of a compact T_2 space is locally compact, so we may state the following generalization of Theorem 3–15.

THEOREM 3–16. A connected open subset of a Peano space is arcwise connected.

A second comment on Theorem 3–15 is this: the primary use of compactness in the proof of Theorem 3–15 is to establish that $\cap K_j$ is a continuum. This can also be established by requiring only that the connected and locally connected space S be complete and metric. The construction is exactly the same in this case; the only change occurs in the proof that K is an arc. We prove first that K is countably compact. Suppose that X is an infinite subset of K. Some set U_{1i_1} then contains an infinite subset X_1 of X such that diameter $X_1 < 1$. That part of K lying in $\overline{U}_{1i_1}$ also lies in only a finite number of sets $\overline{U}_{2j}$, so some set U_{2j_2} contains an infinite subset X_2 of X_1 with diameter $X_2 < \frac{1}{2}$. Similarly, we obtain $X_3, X_3, \ldots,$ with X_{j+1} contained in X_j and diameter $X_j < 1/j$. It follows from the completeness of S that $\cap \overline{X}_j$ is not empty and hence that X has a limit point. To see that K is connected, we proceed as follows. Suppose that $K = A \cup B$, where A and B are disjoint, closed, and nonempty subsets of K. Then the distance $d(A, B)$ is a positive number ϵ. Now there is a subchain C_1' of C_1 connecting A and B; there is a subchain C_2' of C_2 connecting A and B, each link of C_2' lying in a link of C_1'; there is a subchain C_3' of C_3 connecting A and B, each link of C_3' lying in a link of C_2'; etc. The argument used to prove that K is compact also shows that if $\overline{U}_{1i_1}$ contains $\overline{U}_{2i_2}$ contains $\overline{U}_{3i_3}$, etc., then $\cap U_{ji_j}$ is not empty, and hence each $\overline{U}_{ji_j}$ meets K. But if n is so large that $1/n < \epsilon/2$, there is a link of C_n' whose closure does not meet A or B. This link then fails to meet K, and this contradiction proves that K is connected. Then the same argument as before applies to show that K is an arc. We have proved the following result.

THEOREM 3–17. A connected, locally connected complete metric space is arcwise connected.

The hypothesis of completeness in Theorem 3–17 is vital. R. L. Moore [107] has given an example of a locally connected subset S of the plane such that each two points of S lie in a continuum in S but S contains no arcs. On the other hand, Knaster and Kuratowski [92] have given an example of a connected and locally connected subset of the plane that

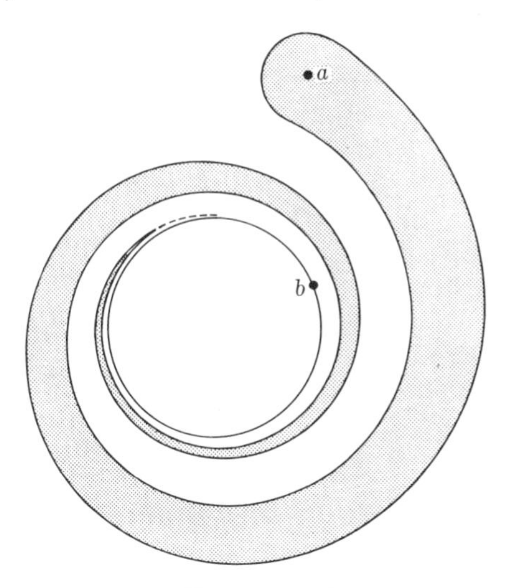

FIGURE 3-12

contains no continuum whatsoever. The example following Theorem 3-6 is actually another such subset, although we do not prove this.

EXERCISE 3-10. Prove that in a locally compact, but not compact, connected and locally connected metric space, each point is the end point of a closed set that is homeomorphic to a closed half-line (ray). [*Hint:* There is a very short proof.]

In Theorem 3-16, we have seen that connected open sets in a Peano space are arcwise connected. Now consider such an open set U. Is the closure $\overline{U}$ necessarily arcwise connected? The answer to this question must be negative. For example, the spiral region depicted in Fig. 3-12 does not have in its closure an arc joining the indicated points a and b.

It is even less reasonable to expect that each point on the boundary of a connected open set U of a Peano space is *arcwise accessible* from U; that is, it is not always true that given a point x on $\overline{U} - U$ there is an arc lying in $U \cup x$ and having x as an end point. Figure 3-13 shows an open set U in the plane such that $\overline{U}$ is arcwise connected, but the point a is not arcwise accessible from U.

In analogy to the definition of local connectedness, one says that a space S is *locally arcwise connected* if S has a basis of arcwise-connected open sets. Adding this property to a space permits us to give the following result concerning accessibility of boundary points.

THEOREM 3-18. In a locally connected and locally arcwise-connected space S, the set of all points on the boundary of an open set U that are accessible from U is dense in the boundary of U.

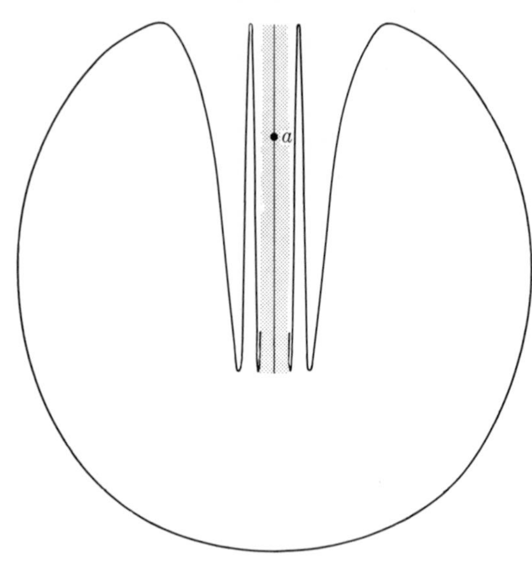

FIGURE 3–13

Proof: Let x be a point on the boundary of U, and let W be an open set containing x. In W there is an open set V that contains x and that is arcwise connected. Let y be a point of $V \cap U$. There is an arc A from y to x in V. Let z be the first point of the closed set $A \cap (\overline{U} - U)$ in the natural ordering of A from y to x. Then the segment $[y, z]$ of A lies in $U \cup z$, so z is accessible from U. This shows that x is a limit point of accessible points. □

Suppose that U is a bounded open set in E^2, or more generally, in E^n. Then the family of all lines parallel to a given line has uncountably many lines meeting U. For each such line l, each component of $l \cap U$ is an interval, both end points of which are accessible from U. Therefore, the boundary of U contains uncountably many points that are accessible from U and by disjoint arcs. This is not true in general. Figure 3–14 depicts a Peano space and a connected open set U with uncountable boundary, each point of which is accessible from U. But any set of disjoint arcs in $\overline{U}$ is countable. Note, incidentally, that $\overline{U}$ is locally connected but that the boundary of U is not.

The last example has uncountably many accessible boundary points. The next example, Fig. 3–15, is a connected open set in a Peano space with an uncountable boundary, and in fact, both $\overline{U}$ and $\overline{U} - U$ are Peano continua. However, only a countable number of points of the boundary of U are accessible from U. Note that this is the same example as pictured in Fig. 3–8.

EXERCISE 3–11. Construct an example of a connected open set whose closure is a Peano continuum, but whose boundary, while connected, is not locally connected.

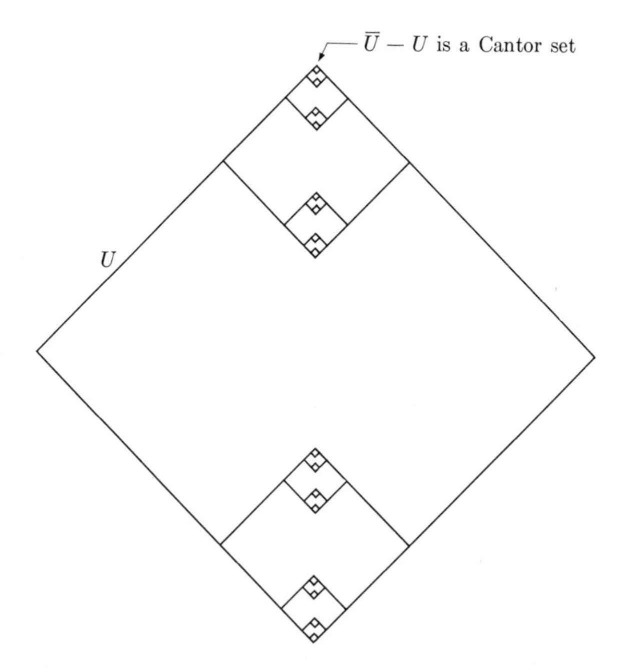

FIGURE 3–14

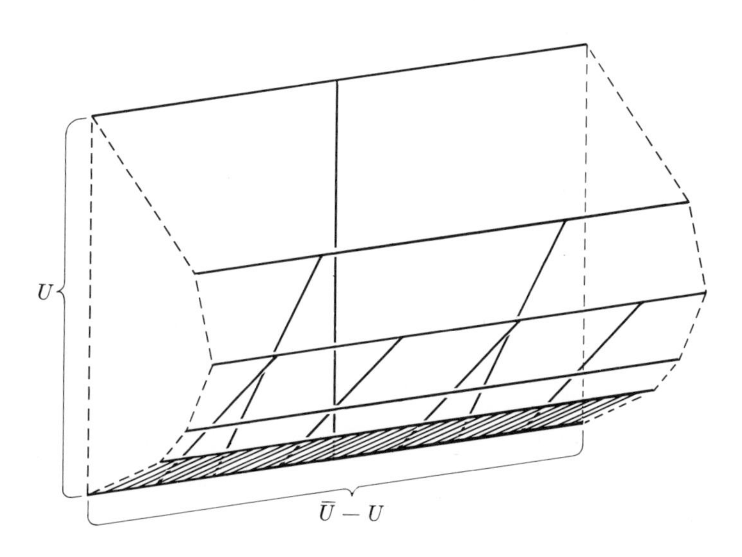

FIGURE 3–15

Of interest to the analyst is such a property as accessibility by rectifiable arcs. It may easily happen that a point is accessible from an open set but not accessible by an arc of finite length. A spiral region similar to Fig. 3–4 can be constructed by imagining the polar coordinate curve $r = e^{-\theta}$ expanded into a long tapering region. It was remarked earlier that there will be no rectifiable arc from an interior point to the center of the spiral.

EXERCISE 3–12. Show that an arc with the same end points as a straight-line interval can intersect that interval in a set of points of positive Lebesgue measure but containing no interval.

EXERCISE 3–13. Is the boundary of a connected open subset of a locally connected space necessarily connected? locally connected? compact? locally compact?

3–3 Mappings of the interval. As was pointed out in Section 2–6, there are always nontrivial mappings of a normal space into the unit interval I^1. We now want to look in the other direction and investigate the question of what kind of a space is a continuous image of the unit interval. Since I^1 is both connected and compact, it is obvious that any such continuous image will be a continuum (Theorems 1–16 and 1–24). But there are continua so "pathological" (we shall see one in Section 3–8) that the only mappings of I^1 into these continua are the (trivial) constant mappings. The missing property in such continua is local connectedness. We will show in this section that, in the class of Hausdorff spaces, every continuous image of the unit interval I^1 is a locally connected metric continuum. Then in Section 3–5, we will establish the converse, namely, every locally connected metric continuum is a continuous image of I^1. This characterizes the Peano spaces. Incidentally, instead of *Peano space* many authors use the term *continuous curve*, which is a more logical name but is less often used.

The Peano spaces have an interesting history. During the last century, when mathematicians were first formulating concepts with a careful regard for rigor, the notion of a "curve" caused considerable difficulty. A curve in E^2 was taken to be the graph of a pair of parametric functions, $x = f(t)$ and $y = g(t)$, with, say, $0 \leqq t \leqq 1$. The question arises as to what conditions should be placed upon the functions f and g. To require differentiability would be too much; it would bar such configurations as a polygon, for instance. Jordan proposed that only continuity be required of the functions f and g. This definition seemed acceptable until Peano found a pair of continuous functions f and g whose graph is 2-dimensional, filling up the square and its interior. This example, surprising and almost paradoxical at the time, is commemorated in the term *Peano space*. It is of interest to examine Peano's example.

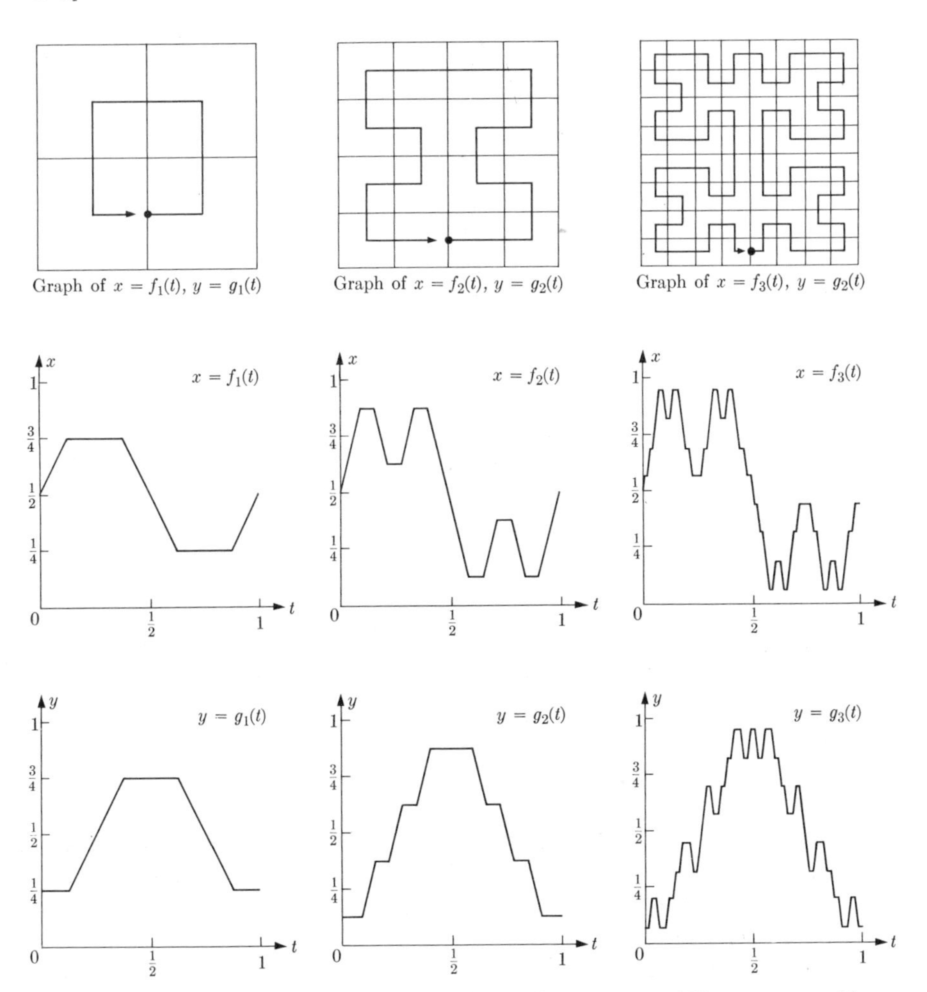

FIG. 3-16. Three stages in constructing a Peano space-filling curve, with graphs of the equations that generate each stage.

In Fig. 3-16 we show three stages in the construction of a Peano "space-filling curve," together with graphs of the parametric equations $x = f_n(t)$, $y = g_n(t)$, which generate each stage. From the form of these functions, it is not difficult to see that they have continuous limit functions $f(t) = \lim_{n\to\infty} f_n(t)$ and $g(t) = \lim_{n\to\infty} g_n(t)$ and that the graph of the pair $x = f(t)$, $y = g(t)$ does indeed fill the unit square. This construction is due to Hilbert.

The modern theory of curves has absorbed this phenomenon and carried on. For a comprehensive treatment of the subject, see Rado's *Length and Area* [28].

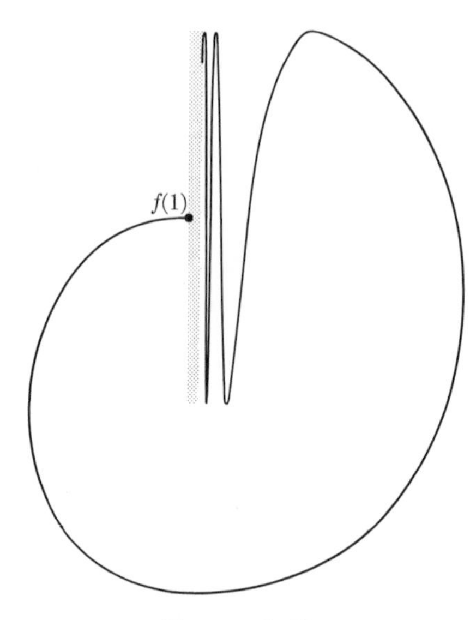

FIGURE 3–17

We point out that local connectedness is obviously a topological invariant. On the other hand, the local connectedness property is not invariant under all continuous mappings; even those that are one-to-one may fail to preserve this property. For instance, consider a mapping f of the half-open interval $0 < t \leqq 1$ onto the curve shown in Fig. 3–17. The image fails to be locally connected at $p = f(1)$ although the half-open interval is locally connected. This example shows that we can expect very little from the general continuous function on lc spaces. However, there is a type of mapping, more general than homeomorphisms, which preserves the lc property.

A mapping is said to be *closed* if it carries closed sets onto closed sets. This is in analogy to the open or interior mappings (see Section 1–5). One might ask, if closed sets are carried onto closed sets, why are not open sets carried onto open sets? That is, why is not a closed mapping also open and *vice versa*? An answer to this question is furnished by the following example.

Map the line E^1 onto the circle S^1 by sending each point x onto the point $(\cos \pi x, \sin \pi x)$. Geometrically, we are simply wrapping the line around the circle infinitely many times. Clearly, this mapping is open. The set of points $\{2n + 1/n\}$ in E^1 is closed, for it has no limit point. But the image set $\{(\cos \pi(2n + 1/n), \sin \pi(2n + 1/n)\}$ has $(1, 0)$ as a limit point, so is not closed. And for a converse situation, consider the mapping of the closed unit interval $-2 \leqq t \leqq 2$ onto itself, given by $f(t) = \frac{1}{18}(t^5 + 7t^3 - 26t)$, whose graph is shown in Fig. 3–18. This

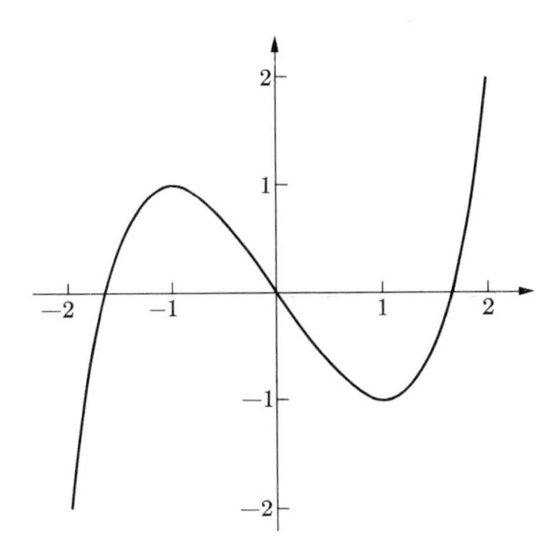

FIGURE 3–18

mapping is closed, as we shall see shortly, but it is *not* open. The (relatively) open interval $0 < t \leq 2$ maps onto the closed interval $-1 \leqq t \leqq 2$.

The following sequence of results yields the theorem that we mentioned was our goal in this section.

LEMMA 3–19. Any mapping of a compact space into a Hausdorff space is closed.

Proof: A closed subset of a compact space is compact. Therefore its continuous image is compact. Being in a Hausdorff space, the continuous image is closed (Corollary 2–2). □

LEMMA 3–20. If $f:S \to T$ is continuous, and if C is a component of T, then $f^{-1}(C)$ is a union of components of S.

Proof: Let B be a component of S. Then $f(B)$ is connected, and if $f(B)$ meets C, it must lie entirely in C. □

LEMMA 3–21. If S is a locally connected space, and if f is a closed mapping of S, then the image $f(S)$ is locally connected.

Proof: Let $f(S) = T$, and suppose that C is a component of an open set U of T. Since f is continuous, $f^{-1}(U)$ is open and, by Lemma 3–20, $f^{-1}(C)$ is a union of components of $f^{-1}(U)$. The components of $f^{-1}(U)$ are open by Theorem 3–2, so $f^{-1}(C)$ is open. Since $ff^{-1}(C) = C$, and since f is closed, it follows that $ff^{-1}(T - C)$ is closed, and hence that C is open. Then, by Theorem 3–2, T is locally connected. □

Combining Lemmas 3–19 and 3–21, we immediately state the following result.

THEOREM 3–22. The continuous image in a Hausdorff space of a compact locally connected space is again a compact locally connected space.

The next theorem could have been proved in Section 2–9, but we did not need it until now.

THEOREM 3–23. The continuous image of a compact metric space in a Hausdorff space is a compact metrizable space.

Proof: Let $f:S \to X$ be a mapping of a compact metric space S into a Hausdorff space X, and let $f(S) = T$. Since f is continuous, T is compact and, as a compact Hausdorff space, T is normal (Theorem 2–3). We need only show that T is completely separable for Theorem 2–46 to apply to give metrizability. To this end, let $\{U_i\}$ be a countable basis for S. This exists by virtue of Theorem 1–5. Each set $S - U_i$ is closed, so $f(S - U_i)$ is closed by Lemma 3–19. We show that $\{T - f(S - U_i)\}$ forms a countable basis for T.

Given any point x in T and an open set V containing x, then $f^{-1}(x)$ is a closed set contained in the open set $f^{-1}(V)$. Since $f^{-1}(x)$ is also compact, it is covered by a finite number of sets U_i whose union we may call U. Thus $f^{-1}(x)$ lies in U, and U lies in $f^{-1}(V)$. Taking complements, we have that $S - f^{-1}(x)$ contains $S - U$, which contains $S - f^{-1}(V)$. Since for any set X in T, we have $S - f^{-1}(X) = f^{-1}(T - X)$, it follows that $f^{-1}(T - x)$ contains $S - U$, which contains $f^{-1}(T - V)$. Application of f tells us either that $T - x$ contains $f(S - U)$, which contains $T - V$, or that V contains $T - f(S - U)$. □

To obtain the final result, we may combine Theorems 3–22 and 3–23.

THEOREM 3–24. The result of mapping the closed unit interval I^1 into a Hausdorff space is a compact connected, locally connected metric space.

In Theorem 3–24 we could of course replace I^1 by any locally connected metric continuum. Our reason for stating Theorem 3–24 in this way lies in the remarkable fact that its converse is also a true theorem, as we will prove in Section 3–5.

3–4 Mappings of the Cantor set. To prove the main theorem of this section, we need two lemmas. The first of these is an easily established result on product spaces, whose proof is left as an exercise.

LEMMA 3–25. If S is a space, and if for each α in an index set A, $S_\alpha = S$, then the diagonal in $\mathbb{P}_A S_\alpha$ consisting of all constant functions $\psi:A \to \cup S_\alpha$ is homeomorphic to S.

LEMMA 3–26. Let $S_0, S_1, S_2, \ldots$ each be the same space S, and let $f_n:S_n \to S_{n-1}$, $n > 0$, be the identity mapping. Then the inverse limit space S_∞ of the sequence $\{S_n, f_n\}$ is homeomorphic to S.

Proof: The points of S_∞ are sequences $(x, x, x, \ldots)$, x in S, and there is an obvious one-to-one transformation $h:S_\infty \to S$, defined by $h(x, x, x, \ldots) = x$. The only question that remains is whether the topologies of S and S_∞ are equivalent. But S_∞ is the diagonal in $\mathbb{P}S_n$, so Lemma 3–25 applies. □

THEOREM 3–27. Let $\{A_n, f_n\}$ and $\{B_n, g_n\}$ be two inverse limit sequences of compact T_2 spaces, and let $\Phi = \{\varphi_n\}$ be a mapping of $\{A_n, f_n\}$ into $\{B_n, g_n\}$ such that each $\varphi_n:A_n \to B_n$ is onto. Then the induced mapping $\varphi:A_\infty \to B_\infty$ is also onto.

Proof: Let $(b_0, b_1, \ldots)$ be a point in B_∞. For each n, let A'_n be the set $\varphi_n^{-1}(b_n)$. The subsets A'_n exist since φ_n is onto. Define the mappings $f'_n = f_n|A'_n$ (f_n restricted to A'_n). Then $\{A'_n, f'_n\}$ is an inverse limit sequence of compact spaces, for if a_n is in A'_n, then $f'_n(a_n)$ is in $\varphi_{n-1}^{-1}\, g_n\varphi_n(a_n) = \varphi_{n-1}^{-1}\, g_n(b_n) = \varphi_{n-1}^{-1}\,(b_{n-1}) = A'_{n-1}$. Hence A'_∞ exists (Lemma 2–85) and any point in A'_∞ is mapped by φ onto $(b_0, b_1, \ldots)$. □

EXERCISE 3–14. If the spaces A_n in Theorem 3–27 are not required to be compact, show that the theorem may be false.

The chief result of this section is somewhat startling.

THEOREM 3–28. If S is any compact metric space, there is a continuous mapping of the Cantor set onto S.

Proof: There is a sequence $\mathcal{U}_1, \mathcal{U}_2, \ldots$ of coverings of S, each $\mathcal{U}_n$ being a finite covering of S by closures of open sets of diameter $< [1/(n+1)]$, and each $\mathcal{U}_n$ being a refinement of its predecessor. Let $\mathcal{U}_n = \{U_{n,1}, \ldots, U_{n,j_n}\}$. Our method is this: we form spaces V_n, closely related to the coverings $\mathcal{U}_n$, forming an inverse limit system of compact spaces, the inverse limit space V_∞ being totally disconnected. Also we will have a mapping of the system $\{V_n\}$ onto the system $\{S_n, i_n\}$, where $S_n = S$ and each i_n is the identity. The preceding results can then be applied.

Consider $\mathcal{U}_1$. We will form disjoint compact sets $V_{1,1}, \ldots, V_{1,j_1}$, each $V_{1,i}$ being homeomorphic to the corresponding $U_{1,i}$. As a useful device, consider for each $U_{1,i}$ all pairs (u, i), where u is a point of $U_{1,i}$. Let $V_{1,i}$ be the collection of all such pairs. Then no two $V_{1,i}$'s intersect. We topologize $V_1 = \bigcup_1^{j_1} V_{1,i}$ by requiring that the natural mapping $h_{1,i}:V_{1,i} \to U_{1,i}$, defined by $h_{1,i}(u, i) = u$, be a homeomorphism and by requiring that each $V_{1,i}$ be open in V_1. Let the mapping $\varphi_1:V_1 \to S_1$ be defined by $\varphi_1|V_{1,i} = h_{1,i}$.

Now go on to $\mathcal{U}_2$. Each element $U_{2,i}$ of $\mathcal{U}_2$ lies in at least one element $U_{1,j}$ of $\mathcal{U}_1$. For each such $U_{1,j}$, let $V_{2,i,j}$ be the collection of triples (u, i, j), where u is a point in $U_{2,i}$. We let $V_2 = \bigcup_{i,j} V_{2,i,j}$, again topologizing the set by requiring that each $V_{2,i,j}$ be open in V_2 and that each natural mapping $h_{2,i,j}:V_{2,i,j} \to U_{2,i}$, defined by $h_{2,i,j}(u, i, j) = u$, be a homeomorphism. We define a mapping $f_2:V_2 \to V_1$ by setting $f_2(u, i, j) = (u, j)$. Letting $\varphi_2:V_2 \to S_2 = S$ be defined by $\varphi_2|V_{2,i,j} = h_{2,i,j}$, and letting $g_2:S_2 \to S_1$ be the identity mapping, we see that $\varphi_1 f_2 = g_2\varphi_2$, and we have the necessary commutativity in the first rectangle.

Consider $\mathfrak{U}_3$. Each element $U_{3,i}$ of $\mathfrak{U}_3$ lies in at least one element $U_{2,j}$ of $\mathfrak{U}_2$, which is in turn contained in an element $U_{1,k}$ of $\mathfrak{U}_1$. For each such choice of j and k, let $V_{3,i,j,k}$ be the set of all quadruples (u, i, j, k), where u is in $U_{3,i}$, and let $V_3 = \bigcup_{i,j,k} V_{3,i,j,k}$. Topologize V_3 by means of the natural homeomorphisms $h_{3,i,j,k}:V_{3,i,j,k} \to U_{3,i}$ as before. Define $f_3:V_3 \to V_2$ by $f_3(u, i, j, k) = (u, j, k)$, and define $\varphi_3:V_3 \to S_3 = S$ by $\varphi_3|V_{3,i,j,k} = h_{3,i,j,k}$. Letting $g_3:S_3 \to S_2$ be the identity, we have the desired commutativity in the second rectangle. Although it is complicated notationally, the general inductive step should now be clear.

The inverse limit sequence $\{V_n, f_n\}$ has an inverse limit space V_∞. Since the mapping $\Phi:\{V_n, f_n\} \to \{S_n, g_n\}$ defined by $\Phi = \{\varphi_n\}$ is onto, Theorem 3–27 tells us that the induced continuous mapping $\varphi:V_\infty \to S_\infty$ is also onto. Then by Lemma 3–26, there is a homeomorphism h of S_∞ onto S. If we knew that V_∞ were totally disconnected *and perfect*, then by Theorem 2–100 there would be a homeomorphism h' of the Cantor set onto V_∞, and $h\varphi h'$ would be the desired mapping. It is not difficult to show that although V_∞ is totally disconnected, it need not be perfect. Consider, for instance, the case in which S is a single point and in which V_∞ turns out to be a single point also. This difficulty is circumvented as follows. If V_∞ is totally disconnected and C is the Cantor set, then $V_\infty \times C$ is both totally disconnected and perfect. There is then a homeomorphism $h':C \to V_\infty \times C$, and if we let $\pi:V_\infty \times C \to V$ be the projection mapping, $h\varphi\pi h':C \to S$ is the desired mapping. It only remains to show that V_∞ is totally disconnected.

Consider two points $x = [(u, i), (u, i, j), (u, i, j, k), \ldots]$ and $y = [(v, i'), (v, i', j'), (v, i', j', k'), \ldots]$ of V_∞. Since the points of S_∞ having their nth coordinate in a set $V_{n,i,j,\ldots}$ form an open and closed set in S_∞, if we can show that x and y have coordinates in different sets $V_{n,i,\ldots}$ for some n, it will follow that x and y lie in different components of V_∞. If $u \neq v$, there is an integer n such that no element of $\mathfrak{U}_n$ contains both u and v. In this case, certainly x and y have nth coordinates in different sets $V_{n,i,\ldots}$. If $u = v$, then the only way x and y can be different points is to differ in some nth coordinate, meaning that their nth coordinates lie in different sets $V_{n,i,\ldots}$. Hence V_∞ is totally disconnected. □

An instructive example of a mapping of the interval I^1 onto the unit square I^2 follows from Lemma 3–20. First, there is a continuous mapping f of the Cantor set C onto the unit square. Now let (a, b) be an open interval in $I^1 - C$. In I^2 there is a straight-line interval L joining $f(a)$ to $f(b)$. We map the closed interval $[a, b]$ onto L by a similarity transformation sending a into $f(a)$ and b into $f(b)$. This provides an extension of f to the open interval (a, b). The mapping $f^*:I^1 \to I^2$ obtained by so extending f over all complementary intervals is certainly onto and can easily be shown by the reader to be continuous. We observe that f^* has a derivative everywhere but at a set of measure zero, namely the Cantor set. This shows that the attempt to avoid this sort of pathology cannot be successful even by requiring a mapping to be differentiable almost everywhere.

3–5 The Hahn-Mazurkiewicz theorem. We are now in a position to prove the converse of Theorem 3–24, that every Peano space is a continuous image of the unit interval I^1. Our proof is modeled upon the construction given in the example at the end of the previous section. That is, we will use Theorem 3–28 to obtain a mapping of the Cantor set onto the Peano space and the use of the arcwise connectivity of the Peano space to extend this mapping over the intervals in $I^1 - C$. In the example of Section 3–4, we joined images of adjacent end points of the Cantor set by straight-line segments, and this made continuity of the extended mapping very easy to see. We must do something similar here.

LEMMA 3–29. If P is a Peano space and $\epsilon > 0$ is given, there is a number δ such that if a and b are any two points with $d(a, b) < \delta$, then there is an arc A from a to b of diameter $< \epsilon$.

Proof: By Theorem 3–13, P is uniformly locally connected. Hence given $\epsilon > 0$, there is a $\delta > 0$ such that if $d(a, b) < \delta$, then there is a connected set B of diameter $< \epsilon/2$ containing a and b. About each point x of B there is a connected open set U_x of diameter $< \epsilon/4$. Then $U = \cup U_x$ is a connected open set of diameter $< \epsilon$, and U contains B. In U there is an arc A from a to b, by Theorem 3–16. □

The property established in Lemma 3–29 is usually expressed by saying that a Peano space is *uniformly locally arcwise connected.*

THEOREM 3–30 (Hahn-Mazurkiewicz). For a space P to be compact, connected, locally connected, and metric, it is necessary and sufficient that P be the image of the unit interval under a continuous mapping into a Hausdorff space.

Proof: Theorem 3–24 is proof of the sufficiency part of this theorem, so it remains to establish the necessity of the condition. Let C denote the middle-third Cantor set on I^1, and let the components of $I^1 - C$ be $I_1, I_2, \ldots$. Let the left- and right-hand end points of I_n be denoted by p_n and q_n, respectively. Using Theorem 3–28, let $f:C \to P$ be onto. If $f(p_n) = f(q_n)$ for any n, define f^* on I_n by $f^*(x) = f(p_n)$.

Next let $\epsilon_1, \epsilon_2, \ldots$ be a sequence of positive numbers approaching zero. There is a number $\eta_1 > 0$ such that any two points of P at a distance $< \eta_1$ apart can be joined by an arc of diameter $< \epsilon_1$ by Lemma 3–29, and by uniform continuity there is a number $\delta_1 > 0$ such that if x and y are points of C with $|x - y| < \delta_1$, then $d(f(x), f(y)) < \eta_1$. Thus there is only a finite number of values of n, say $n_1, \ldots, n_{k_1}$, such that if n is one of these, then $d(f(p_n), f(q_n)) \geqq \eta_1$. There are arcs $A_{n_1}, \ldots, A_{n_{k_1}}$ with A_{n_i} joining $f(p_{n_i})$ to $f(q_{n_i})$, but we can make no claim about the diameter of the arcs. We define the desired extension f^* over each closed interval $p_{n_i} \cup I_{n_i} \cup q_{n_i}$ to be a homeomorphism onto A_{n_i}.

Now there is a number $\eta_2 > 0$ such that any two points of P at a distance $< \eta_2$ apart can be joined by an arc of diameter $< \epsilon_2$, and there is a number δ_2 such that any two points of C at a distance $< \delta_2$ apart have images at a distance $< \eta_2$ apart. There is only a finite number of intervals $I_{m_1} \ldots, I_{m_{k_2}}$ such that $\eta_1 > d(f(p_{m_i}), f(q_{m_i})) \geqq \eta_2$. Then $f(p_{m_i})$ and $f(q_{m_i})$ can be joined by an arc A_{m_i} of diameter $< \epsilon_1$. We extend f over each closed interval $p_{m_i} \cup I_{m_i} \cup q_{m_i}$ by a homeomorphism onto A_{m_i}.

There is a number $\eta_3 > 0$ such that any two points of P less than η_3 apart can be joined by an arc of diameter $< \epsilon_3$. Then there is a number $\delta_3 > 0$ such that any two points of C less than δ_3 apart have images less than η_3 apart. There is only a finite number of intervals $I_{l_1}, \ldots, I_{l_{k_3}}$ such that $\eta_2 > d(f(p_{l_i}), f(q_{l_i})) \geqq \eta_3$. Then $f(p_{l_i})$ and $f(q_{l_i})$ can be joined by an arc A_{l_i} of diameter $< \epsilon_2$. Extend f over $p_{l_i} \cup I_{l_i} \cup q_{l_i}$ by a homeomorphism onto A_{l_i}.

Continuing this process indefinitely, we obtain a function $f^*: I^1 \to P$ such that (a) $f^*|C = f$, (b) on each $\overline{I}_n$, f^* is a homeomorphism onto an arc A_n, and (c) the diameters A_n converge to zero. This last makes it easy to show that f^* is continuous, and f^* is onto because f is onto. $\square$

There is a theorem that, had we given it, would have considerably shortened the proof of Theorem 3–30. The result we have in mind here is that every Peano space P has a metric in which it is convex. That is, there is a metric $\rho(x, y)$ for P such that for each two points a and b of P, the set of points x for which

$$\rho(a, x) + \rho(x, b) = \rho(a, b)$$

is isometric to an interval. This theorem, long a conjecture, was proved by Bing [56]. A detailed treatment is given in Hall and Spencer [9]. Note that it is not asserted, nor is it true, that these "convex paths" are unique.

We remark that we could reproduce the situation of the example at the end of the previous section more closely in a proof of Theorem 3–30.

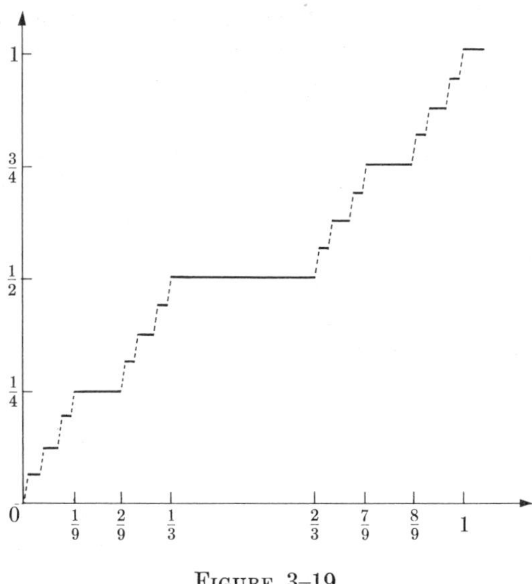

FIGURE 3–19

The technique for doing so is useful in other problems, so we will indicate the method. Let $f:C \to P$ be the mapping with which we started in Theorem 3–30, and let $f^*:I^1 \to P$ be the extension of f given by Theorem 3–30. Let $[a, b]$ be the closure of an interval complementary to C, and let C' be the middle-third Cantor set in $[a, b]$. There is a monotone-increasing continuous function m_{ab} of $[a, b]$ onto $[a, b]$ that is constant on the intervals complementary to C'; this is sometimes called the *Cantor function*, and we show its graph in Fig. 3–19. We define $f^{**}:I^1 \to P$ by setting $f^{**}(x) = f^*(x) = f(x)$ if x is in C. If x is in a complementary interval $[a, b]$, we define $f^{**}(x) = f^*m_{ab}(x)$. The same argument as before shows that f^{**} is continuous. But now having put a middle-third Cantor set in each complementary interval in $I^1 - C$, we have still only a Cantor set in I^1. The function f^{**}, however, is constant on each interval complementary to this new Cantor set.

As a technique for proving facts about Peano spaces, the Hahn-Mazurkiewicz theorem and its proof have not been very successful. Suppose, for example, that we want to use this result to prove the arcwise connectivity theorem (3–15). The difficulty is that, although we know little about the mapping of I^1 onto a Peano space P, we must somehow identify a set on I^1 that will be mapped onto an arc in P joining two given points, and then prove that its image is indeed an arc. This particular problem has been done in an elegant way by J. L. Kelley (see p. 39 of Whyburn [40]).

3–6 Decomposition spaces and continuous transformations. Let S and T be topological spaces, T also being a T_1 space, and let $f:S \to T$ be a continuous mapping. For each point t in T, the set $f^{-1}(t)$ is closed in S, and for two distinct points, t and t', the inverse sets $f^{-1}(t)$ and $f^{-1}(t')$ are disjoint. It follows that the collection of all point-inverses $f^{-1}(t)$, t in T, is a covering of S by disjoint closed sets, a *decomposition* of S into closed sets. Our interest lies in this question: given a decomposition of a space S into closed sets, how can we tell that the decomposition was induced by a continuous mapping? We will restrict our attention to mappings of compact Hausdorff spaces into Hausdorff spaces.

It must be pointed out first that not every decomposition can be induced by a continuous mapping. Suppose, for example, that S is the union of the vertical unit intervals in E^2 with the lower end points on the x-axis at the points $0, 1, \frac{1}{2}, \frac{1}{3}, \ldots$ The collection whose elements are the individual points on the interval over zero and the remaining complete intervals constitutes a decomposition $\mathcal{G}$ of S into closed sets. There is no mapping $f:S \to T$ (T is a Hausdorff space) that induces this decomposition $\mathcal{G}$. For suppose there were such a mapping, say f. Clearly $f(0, 0)$ and $f(0, 1)$ would have to be distinct points of T and hence would lie in disjoint open sets U_o and U_1, respectively. Since $(0, 0)$ is the only limit point of the set of points $(1/n, 0)$ in S, every open set in S that contains $(0, 0)$ also contains all but a finite number of the points $(1/n, 0)$. By continuity, there is an open set V_o containing $(0, 0)$, such that V_o is mapped by f into U_o. Similarly, there is an open set V_1 containing $(0, 1)$ and all but a finite number of points $(1/n, 1)$ that is mapped by f into U_1. But the definition of $\mathcal{G}$ requires that $f(1/n, 0) = f(1/n, 1)$. Hence for sufficiently large values of n, U_o and U_1 both contain $f(1/n, 0)$, contradicting the statement that U and V are disjoint.

EXERCISE 3–15. Show that there is a space Z (not Hausdorff, of course) and a mapping $f:S \to Z$ (S is the example above), such that f induces the above decomposition $\mathcal{G}$ as point-inverses.

In view of the example above, it is evident that we require some condition for a decomposition to be that induced by a continuous mapping. We now define this condition. Let S be a space, and let $\mathcal{G} = \{C_\alpha\}$ be a collection of disjoint compact sets filling up S (covering S). The collection $\mathcal{G}$ is said to be *upper semicontinuous* provided that, for each α, if U is an open set containing C_α, there is an open set V containing C_α and lying in U, such that every element C_β of $\mathcal{G}$ that intersects V lies in U. (The decomposition $\mathcal{G}$ given in the example above is not upper semicontinuous, as the reader will verify easily.)

THEOREM 3–31. If S and T are compact Hausdorff spaces, and if $f:S \to T$ is continuous, then the decomposition $\{f^{-1}(t)\}$ of S, induced by f, is upper semicontinuous.

Proof: Let U be an open set in S containing a set $f^{-1}(t)$. Then $S - U$ is compact, and $f(S - U)$ does not contain t. The set $T - f(S - U)$ is open, so $V = f^{-1}(T - f(S - U))$ is open. Clearly V is a subset of U. The set V is a union of point-inverses, and hence V satisfies the desired conditions. □

The above theorem shows that upper semicontinuity is a necessary condition on a decomposition of a compact T_2 space if it is to be induced by a mapping. We next prove a converse. If we have an upper semicontinuous decomposition $\mathcal{G}$ of a Hausdorff space S, then we define a space T and a mapping $f:S \to T$ such that f induces the decomposition $\mathcal{G}$. As the points T, we take the elements of $\mathcal{G}$. Precisely, the *decomposition space* of $\mathcal{G}$ is the topological space $D(\mathcal{G})$ whose points are the elements of $\mathcal{G}$ and wherein a set U of points of $D(\mathcal{G})$ is open if the union in S of those elements of $\mathcal{G}$ in U is an open set in S.

There is a natural mapping $f:S \to D(\mathcal{G})$, defined by letting $f(x)$ be that element of $\mathcal{G}$ which contains x (in S). We prove the results needed to show that the decomposition space and this natural mapping f provide the converse of Theorem 3–31.

Theorem 3–32. Let S be a Hausdorff space, and let $\mathcal{G}$ be an upper semicontinuous decomposition of S into closed sets. Let U be an open set in S. Then the union of all elements of $\mathcal{G}$ that lie in U is also an open set in S.

Proof: Let Γ denote the union of all elements of $\mathcal{G}$ contained in U. If Γ is empty, the theorem is true. If Γ is not empty and is not open, there is an element X of $\mathcal{G}$ in Γ which contains a limit point of $S - \Gamma$. It follows that every open set that contains X intersects elements of $\mathcal{G}$ that do not lie entirely in U. But this contradicts the definition of upper semicontinuity. □

Exercise 3–16. Show that Theorem 3–32 is false if the phrase "that lie in U" is replaced by "that intersect U."

Theorem 3–33. If S is a compact Hausdorff space, and if $\mathcal{G}$ is an upper semicontinuous decomposition of S, then the decomposition space $D(\mathcal{G})$ is Hausdorff.

Proof: This follows immediately from the normality of S and Theorem 3–32. □

Theorem 3–34. If S is a compact Hausdorff space, and if $\mathcal{G}$ is an upper semicontinuous decomposition of S, then the natural mapping $f:S \to D(\mathcal{G})$ of S onto the decomposition space of $\mathcal{G}$ is continuous.

Proof: This is an immediate consequence of the "two-open-set" definition of continuity (see Section 1–5). □

A real-valued function $y = f(x)$ of a real variable is said to be *upper semicontinuous* in the sense of analysis provided that, for each fixed x_0 in the domain of f,

$$\limsup_{x \to x_0} f(x) \leqq f(x_0).$$

The reason for the topologist's choice of the term *upper semicontinuous collection* is that if $f(x)$ is a nonnegative bounded upper semicontinuous function defined over an interval, then the ordinate sets, defined for each x as being the set of all points (x, y) satisfying $0 \leq y \leq f(x)$, form an upper semicontinuous collection. The proof of this is left as an exercise.

Using the notion of limits of sequences of sets, as in Section 2–16, we can formulate the definition of an upper semicontinuous collection in a metric space in another way.

THEOREM 3–35. Let $\mathcal{G}$ be a collection of disjoint closed sets filling up a compact metric space M. Then a necessary and sufficient condition that $\mathcal{G}$ be upper semicontinuous is that if $\{X_n\}$ is a sequence of elements of $\mathcal{G}$ and if $(\liminf X_n) \cap X$ is not empty, where X is an element of $\mathcal{G}$, then $\limsup X_n$ is contained in X.

Proof: Let $\mathcal{G}$ be upper semicontinuous. Suppose that $\{X_n\}$ is a sequence of elements of $\mathcal{G}$, that $(\liminf X_n) \cap X$ is not empty, where X is an element of $\mathcal{G}$, and that $\limsup X_n$ contains a point p not in X. Now p lies in an open set D whose closure does not meet X. Let $U = S - \overline{D}$, and take the corresponding open set V containing X from the definition of upper semicontinuity. Then every element of $\mathcal{G}$ that intersects V lies in U. In particular, all but a finite number of the elements of $\{X_n\}$ intersect V, so only a finite number of elements can intersect D. Then p is not in $\limsup X_n$, a contradiction.

To prove the sufficiency of the condition, let X be an element of $\mathcal{G}$, and let U be an open set containing X. For each n, let V_n denote the open set of all points p, with $d(p, X) < 1/n$. Suppose that for each n, there is an element X_n of $\mathcal{G}$ intersecting V_n but not lying in U. Let p_n be a point of $X_n \cap V_n$. Then some subsequence $\{p_{n_j}\}$ of $\{p_n\}$ converges to a point of X. The corresponding sets $\{X_{n_j}\}$ have, therefore, a nonempty limit inferior that intersects X. Accordingly, $\limsup X_{n_j}$ is contained in X. But each X_{n_j} contains a point q_i in $S - U$. Every open set containing some point q of $S - U$ then contains infinitely many points q_j by countable compactness. Thus q lies in $\limsup X_{n_j}$, whereas q is not in X, a contradiction. □

We have limited the discussion of upper semicontinuity to compact spaces. The following example gives one reason for this limitation. In the plane, let $f{:}E^2 \to E^1$ be given by $f(x, y) = x$, the projection on the x-axis. The collection of point-inverses is as smooth as one could wish,

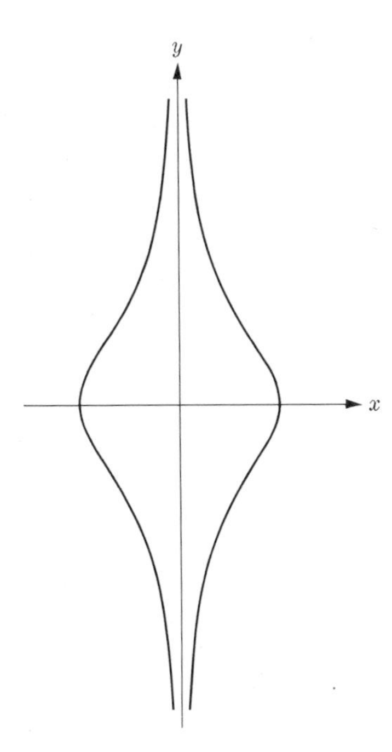

FIGURE 3-20

but it is not upper semicontinuous in the sense of our original definition. For let X be the y-axis, $f^{-1}(0)$, and let U be the set of all points enclosed by the graph of $x^2 = (1 - y^2)^{-2}$, as shown in Fig. 3-20. Then no point-inverse other than X lies in U, so there is certainly no open set V as required. However, if we take the condition of Theorem 3-35 as the *definition* of upper semicontinuity in a metric space, then this collection of point-inverses *is* upper semicontinuous. The definition is not quite perfect, for there exist such collections wherein the decomposition space is not metric [118].

An upper semicontinuous collection $\mathcal{G}$ of disjoint compact sets filling up a Hausdorff space S is said to be *continuous* provided that if C is any element of $\mathcal{G}$, if p and q are points of C, and if U is an open set containing p, then there is an open set V containing q, such that every element of $\mathcal{G}$ that intersects V also intersects U.

THEOREM 3-36. A necessary and sufficient condition that a collection $\mathcal{G}$ of disjoint closed sets filling up a compact metric space be continuous is that if $\{X_n\}$ is a sequence of elements of $\mathcal{G}$, and if X is an element of $\mathcal{G}$ such that $X \cap \liminf X_n$ is not empty, then $X = \lim X_n$.

The proof is left as an exercise.

The term *identification space* is often used for *decomposition space*, but usually with a slight change in emphasis. In this usage, the term implies that most of the elements of the upper semicontinuous collection are points and, frequently, that only a finite number are nondegenerate. Thus if we *identify* the end points of an interval, we get a circle.

EXERCISE 3–17. Show that the space obtained by identifying all the points on the boundary of the n-cube is an n-sphere.

3–7 Monotone and light mappings. We really did not prove the complete equivalence between an upper semicontinuous decomposition and a continuous mapping in the previous section. One further result is required, namely, a theorem to the effect that the decomposition space of the collection $\{f^{-1}(t)\}$ is homeomorphic to the image space $f(S)$ in T. This is the content of the next result.

THEOREM 3–37. Let S and T be compact Hausdorff spaces, and let $f{:}S \to T$ be continuous. Let $\mathcal{G}$ be the collection of all point-inverses $f^{-1}(t)$, t in T, and let $\varphi{:}S \to D(\mathcal{G})$ be the natural mapping of S onto the decomposition space of $\mathcal{G}$. Then there is a homeomorphism $h{:}D(\mathcal{G}) \to T$ such that $f = h\varphi$.

Proof: For each element $f^{-1}(t)$ in $\mathcal{G}$, we define $h(f^{-1}(t)) = t$. Clearly then, if x is any point of $f^{-1}(t)$, we have $\varphi(x) = f^{-1}(t)$ in $D(\mathcal{G})$, and then $h(\varphi(x)) = h(f^{-1}(t)) = t = f(x)$ by definition. It remains to show that h is a homeomorphism of $D(\mathcal{G})$ into T. Since there is an obvious one-to-one correspondence between inverse sets $f^{-1}(t)$ and the point t in T, the transformation h is one-to-one. In view of Exercise 2–43, it only remains to show that h is continuous. To this end, let U be an open set in T. Then $f^{-1}(U)$ is open in S, and by Theorem 3–32 the collection of all inverses $f^{-1}(t)$ in $f^{-1}(U)$ is open in S and hence in $D(\mathcal{G})$. □

If we remove the requirement that S and T be compact, then exactly the same argument also proves the following result.

THEOREM 3–38. Let S and T be Hausdorff spaces, and let $f{:}S \to T$ be continuous with the further property that for each point t in T, $f^{-1}(t)$ is compact. Let $\mathcal{G}$, φ, and $D(\mathcal{G})$ be as defined in Theorem 3–37. Then there is a continuous one-to-one mapping $\psi{:}D(\mathcal{G}) \to T$, such that $f = \psi\varphi$.

To illustrate this result, consider the mapping $f{:}E^2 \to S^1$, defined by setting $f(x, y)$ equal to the point on S^1 with polar coordinates

$$\left(1,\ \frac{2\pi(x^2 + y^2)}{1 + x^2 + y^2}\right).$$

Clearly each circle with center at the origin in E^2 is a point-inverse. The decomposition space is homeomorphic to a closed ray, and the mapping ψ of Theorem 3–38 wraps this ray once around the circle.

We have seen several examples of the factorization of a mapping, that is, the writing of a mapping as the iteration of two or more mappings. The term *factorization* is not intended to suggest the existence of algebraic properties, such as a unique factorization theorem, although we shall see something vaguely related. Our chief example will be the *monotone-light factorization* of a continuous mapping of a compact space.

A mapping $m:S \to T$ is said to be *monotone* provided that, for each point t in T, the inverse $f^{-1}(t)$ is connected. A mapping $l:S \to T$ is said to be *light* provided that, for each t in T, the inverse $l^{-1}(t)$ is totally disconnected, that is, has no component bigger than a point. We will prove that if S is a compact Hausdorff space and if $f:S \to T$ is continuous, then there exist a space M (the *middle space*), a monotone mapping $m:S \to M$, and a light mapping $l:M \to T$, such that $f = lm$. Furthermore, M, m, and l are "unique up to homeomorphisms." (We will define this later.) The method of our proof consists of forming the collection of all components of point-inverses $f^{-1}(t)$. This turns out to be an upper semicontinuous collection, $\mathfrak{M}$. The natural mapping $m:S \to D(\mathfrak{M})$ is monotone, and finding a light mapping $l:D(\mathfrak{M}) \to T$ is easy.

THEOREM 3–39. Let S be a compact Hausdorff space, and let $\mathcal{G}$ be an upper semicontinuous decomposition of S. Let $\mathfrak{M}$ be the collection of all components of elements of $\mathcal{G}$. Then $\mathfrak{M}$ is also upper semicontinuous.

Proof: Let M be an element of $\mathfrak{M}$, and let G be the element of $\mathcal{G}$ having M as a component. Let U be an open set containing M. By Theorem 2–3, there is an open set U' lying in U and containing M, such that $(\overline{U}' - U') \cap G$ is empty. Let $U^* = U' \cup (S - \overline{U}')$, and take the corresponding open set V given by the upper semicontinuity of $\mathcal{G}$. Let $V' = V \cap U'$. If a component M' of some element G' of $\mathcal{G}$ intersects V', then it lies in U^*, since then G' intersects V. Since U' and $S - \overline{U}'$ are disjoint open sets and M' lies in their union, M' can lie only in U'. This shows that V' has the desired property. □

THEOREM 3–40. Let S and T be compact Hausdorff spaces, let $f:S \to T$ be continuous, and let $\mathfrak{M}$ be the upper semicontinuous collection of components of point-inverses $f^{-1}(t)$. Let $m:S \to D(\mathfrak{M})$ be the natural mapping, and define $l:D(\mathfrak{M}) \to T$ by $l(p) = f(m^{-1}(p))$. Then m is monotone, and l is light. Furthermore, if m' and l' are any other mappings, monotone and light respectively, such that $f = l'm'$ and $m'(S) = M'$, then there is a homeomorphism $h:M' \to D(\mathfrak{M})$ such that $m = hm'$ and $l = l'h^{-1}$.

Proof: It is obvious that m is monotone. To see that l is continuous, we note first that if U is an open set in T, then $f^{-1}(U)$ is open in S. The set $f^{-1}(U)$ is the union of elements of $\mathfrak{M}$, so $m(f^{-1}(U))$ is open in $D(\mathfrak{M})$. Now suppose that l is not light. Then for some t_0 in T, there is a component C of $l^{-1}(t_0)$ that is nondegenerate. By the definition of l, each point of C is a component of $f^{-1}(t_0)$. If x and y are two points of C, we can find an open set U^* in S, precisely as was done in the proof of Theorem 3–39, such that U^* contains $f^{-1}(t_0)$, and $U^* = V_x \cup V_y$, where V_x contains $m^{-1}(x)$, and V_y contains $m^{-1}(y)$, and $V_x \cap V_y$ is empty. The set $\mathcal{V}_x$ of all elements of $\mathfrak{M}$ lying in V_x is an open set in $D(\mathfrak{M})$, and so is the set $\mathcal{V}_y$ of all elements of $\mathfrak{M}$ lying in V_y. But C lies in $\mathcal{V}_x \cup \mathcal{V}_y$ and intersects each, contradicting the connectedness of C. Hence l is light. That $f = lm$ is obvious for $lm(p) = f(m^{-1}(m(p))) = f(p)$.

Only the uniqueness part of the theorem remains to be proved. This is a consequence of the statement that the collection of point-inverses $m^{-1}(x)$, x in $D(\mathfrak{M})$, is identical with $\mathfrak{M}$, $\mathfrak{M}$ being the collection of point-inverses $(m')^{-1}(x)$, so that Theorem 3–37 applies. That the collection $\{m^{-1}(x)\} = \mathfrak{M}$ is left as an exercise. □

The power of this result lies in the fact that the middle space $D(\mathfrak{M})$ can often be characterized, or put into a known class of spaces. For example, if the space S is the 2-sphere S^2, the middle space $D(\mathfrak{M})$ is always a *cactoid.* This is a fairly simple type of space, the monotone image of S^2 (see [40]). Then to discover all about mappings of S^2, we need study only light mappings of cactoids. This process is in constant use in the study of Lebesgue area [28]. Unfortunately, the complications (already more severe than we have made them appear) increase rapidly with increased dimension. There is a theorem [84] to the effect that if M is any compact metric space, then there is a monotone mapping m of the unit cube I^3 onto a space S which contains M. The nature of this mapping m can be indicated. In I^3, let C be a Cantor set. There is a continuous mapping $f{:}C \to M$ of C onto M. It is possible to weave disjoint arcs through the sets $f^{-1}(x)$, x in M, in such a way that the resulting collection $\mathcal{G}'$ of arcs is upper semicontinuous. Then add to $\mathcal{G}'$ the collection of all points in I^3 not in any element of $\mathcal{G}'$. The result is an upper semicontinuous collection $\mathcal{G}$ of continua filling up I^3. The space S is $D(\mathcal{G})$, and m is the natural mapping.

Exercise 3–18. Prove that if X is a closed subset of the Hausdorff space S and if $\mathcal{G}'$ is an upper semicontinuous collection filling up X, then the collection $\mathcal{G}$ consisting of the elements of $\mathcal{G}'$ and of the individual points of $S - X$ is also upper semicontinuous.

Exercise 3–19. Apply Exercise 3–18 to the special case of the Cantor set in I^1 to show that every compact metric space can be imbedded in a Peano space.

EXERCISE 3–20. Show that if $f:S \to T$ is an interior mapping of one compact Hausdorff space into another, then (a) the monotone factor need not be interior, but (b) the light factor is always interior.

EXERCISE 3–21. Under the same hypotheses as in Exercise 3–20, show that the union of all sets $f^{-1}(t)$ having at least n components is open in S.

EXERCISE 3–22. Is there an interior mapping of I^2 into I^2 such that each point-inverse consists of exactly two points?

3–8 Indecomposable continua. There are two quite different types of continua, the decomposable and the indecomposable. Although they were originally considered primarily as pathological examples, the indecomposable continua have gained importance in recent years. We give the chief results of this topic here. Our attention is limited to Hausdorff continua, i.e., compact connected Hausdorff spaces.

A continuum is *decomposable* if it is the union of two proper subcontinua; otherwise it is *indecomposable*. We will obtain a few results before giving an example of the latter.

THEOREM 3–41. If a Hausdorff continuum P contains a proper subcontinuum C with interior points, then P is decomposable and conversely.

Proof: If $P - C$ is connected, then $\overline{P - C}$ is not all of P, so $P = C \cup \overline{P - C}$ is a decomposition of P. If $P - C = U \cup V$, where U and V are disjoint open sets, then both $U \cup C$ and $V \cup C$ are proper subcontinua of P and give a decomposition.

Conversely, if P is the union of two proper subcontinua C_1 and C_2, then $C_1 - C_2$ is an open set in P, and hence C_1 has interior points. □

COROLLARY 3–42. Every Peano continuum is decomposable.

Proof: Every point of a Peano continuum lies in arbitrarily small open connected sets because such a space is locally connected. The closure of such an open connected set is a proper subcontinuum with interior points. □

A subset C of a continuum K is a *composant* if, for some point p, C is the set of all points x such that K is not irreducible between p and x. For example, an arc ab has three composants, namely, ab, $ab - b$, and $ab - a$, corresponding to $p \neq a, b$, or $p = a$, or $p = b$. A circle S^1 has just one composant.

THEOREM 3–43. Every decomposable continuum K is a composant for some point.

Proof: Suppose that $K = A \cup B$, where A and B are proper subcontinua of K. Let p be any point in $A \cap B$ (which is not empty since K is connected). Then K is a composant for p. For if x is any point of K, then K contains a proper subcontinuum (either A or B) containing both p and x, and hence K is not irreducible between p and x. □

THEOREM 3–44. Every point of a nondegenerate Hausdorff continuum K is a limit point of any composant C of K.

Proof: We show that every open set intersects C. Let U be open, and let V be an open set whose closure lies in U (K is regular by Theorem 2–1). If the defining point p of the composant C is in $\overline{V}$, then $C \cap U$ is not empty. If the defining point is not in $\overline{V}$, then consider the component of $S - \overline{V}$ that contains p. The closure of this component is a proper subcontinuum of K that contains p and so must lie in C. By Theorem 3–11, this closure meets $\overline{V} - V$, so some point of C is in $\overline{V}$ and hence in U. □

THEOREM 3–45. If K is a metric continuum, then every composant of K is the union of a countable number of proper subcontinua of K.

Proof: Let C be the composant determined by a point p; the open set $K - p$ has a countable basis $\{U_i\}$. For each i, let C_i denote the component of $K - \overline{U}_i$ that contains p. Then $\overline{C}_i$ is a proper subcontinuum of K that contains p and so lies in C. Suppose that x is any point of C. There is a proper subcontinuum K' of K such that K' contains both p and x. Let q be a point in $K - K'$. Then there is an integer j such that q is in U_j and $\overline{U}_j$ lies in $K - K'$. Thus K' is a subset of C_j, so the point x is in C_j. Hence $C = \cup\overline{C}_j$. □

THEOREM 3–46. If K is a metric continuum which is indecomposable, then K has uncountably many composants.

Proof: Suppose that K contains only countably many composants. By Theorem 3–45, every composant of K is a union of countably many proper subcontinua of K. This implies that K is a union of countably many of its subcontinua. But no proper subcontinuum of K can contain an interior point, or else K would be decomposable, by Theorem 3–41. Hence we have a contradiction of Theorem 2–79. □

THEOREM 3–47. If K is an indecomposable continuum, then the composants of K are disjoint.

Proof: Let C_1 and C_2 be composants of K, and suppose that there is a point x in $C_1 \cap C_2$. Let p_1 and p_2 be the defining points of C_1 and C_2, respectively, and let y be any point in C_2. There is a continuum K_1 in C_1 containing p_1 and x and a continuum K_2 in C_2 containing p_2 and x. Also there is a continuum K_3 in C_2 containing p_2 and y. Now $K_1 \cup K_2$ is a proper subcontinuum of K, or else $K = K_1 \cup K_2$ is decomposable. Similarly $(K_1 \cup K_2) \cup K_3$ is a proper subcontinuum of K. But $K_1 \cup$

$K_2 \cup K_3$ contains both p_2 and y; hence y is a point of C_1. Therefore C_2 lies in C_1 and, similarly, C_1 lies in C_2. □

COROLLARY 3–48. *Every indecomposable metric continuum is irreducible between each two points of an uncountable set.*

Proof: Take one point from each composant of such a continuum, and apply Theorems 3–46 and 3–47. □

COROLLARY 3–49. *If a metric continuum is not irreducible, then it is not indecomposable.*

THEOREM 3–50. *No decomposable continuum is irreducible between each two of three points.*

Proof: Let $K = A \cup B$ be the union of two proper subcontinua A and B, and let a, b, and c be three points of K. At least two of these points lie in A (or two lie in B). Then K is not irreducible between these two. □

Combining Corollary 3–49 and Theorem 3–50 yields a necessary and sufficient condition for indecomposability.

THEOREM 3–51. *A necessary and sufficient condition that the metric continuum K be indecomposable is that there exist three points of K such that K is irreducible between each two of these three points.*

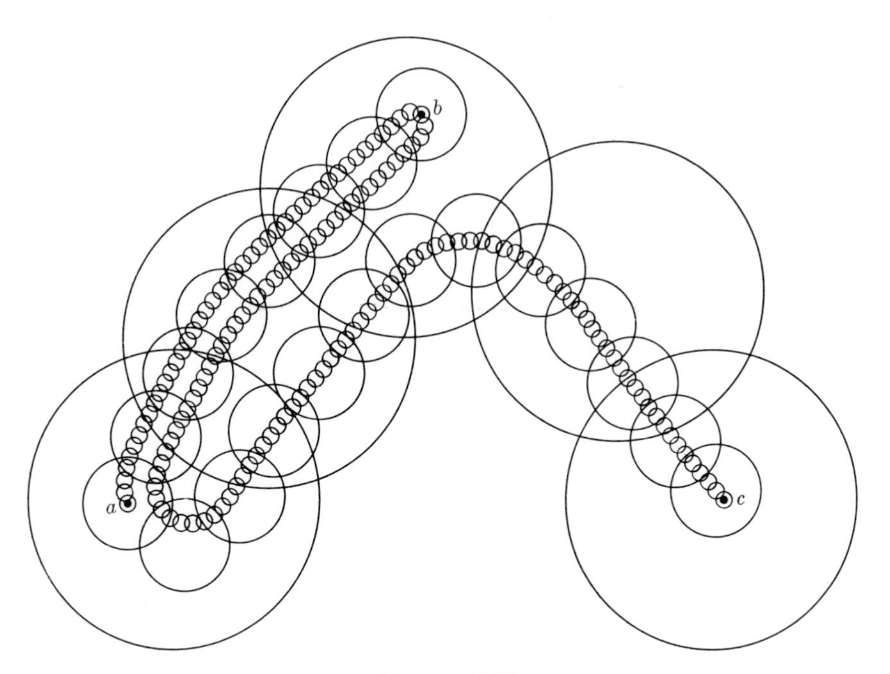

FIGURE 3–21

This result suggests a construction of the "easiest" indecomposable continuum. Let a, b, and c be three points of E^2. Consider simple chains of connected open sets as follows. Let C_1 be a simple chain from a to c through b, that is, one set in C_1 contains b; then let C_2 be a simple chain from b to c through a and such that C_2 lies in C_1; let C_3 be a simple chain from a to b through c and lying in C_2. In general, C_{3n+1} is a simple chain from a to c through b, C_{3n+2} is a simple chain from b to c through a, and C_{3n+3} is a simple chain from a to b through c. And for any integer k, C_k lies in C_{k+1}. The intersection $\cap C_{3n+1}$ is a continuum irreducible between a and c, $\cap C_{3n+2}$ is a continuum irreducible between b and c, and $\cap C_{3n}$ is a continuum irreducible between a and b. But these intersections are all the same, and by Theorem 3–41 constitute an indecomposable continuum. Figure 3–21 gives the first three stages of this construction.

A famous example of an indecomposable continuum is the *pseudo-arc*. This set was first described by Knaster [91] in a different context. Moïse [103] named the set and first investigated its properties. In Fig. 3–22 the first three stages of Moïse's construction are given. We have five open sets $U_{1,1}, \ldots, U_{1,5}$ with $\overline{U}_{1,1}$ meeting only $\overline{U}_{1,2}$, $\overline{U}_{1,2}$ meeting $\overline{U}_{1,3}$, etc., and a point a in $U_{1,1}$ and a point b in $U_{1,5}$. Next there are forty-five open sets $U_{2,1}, \ldots, U_{2,45}$ as pictured, etc. The pseudo-arc is the intersection $\cap_{i=1}^{\infty} \overline{U}_i$, where $U_i = \cup_j U_{i,j}$. There is nothing special about the number 5 here. We could use any integer larger than 4 and obtain a pseudo-arc.

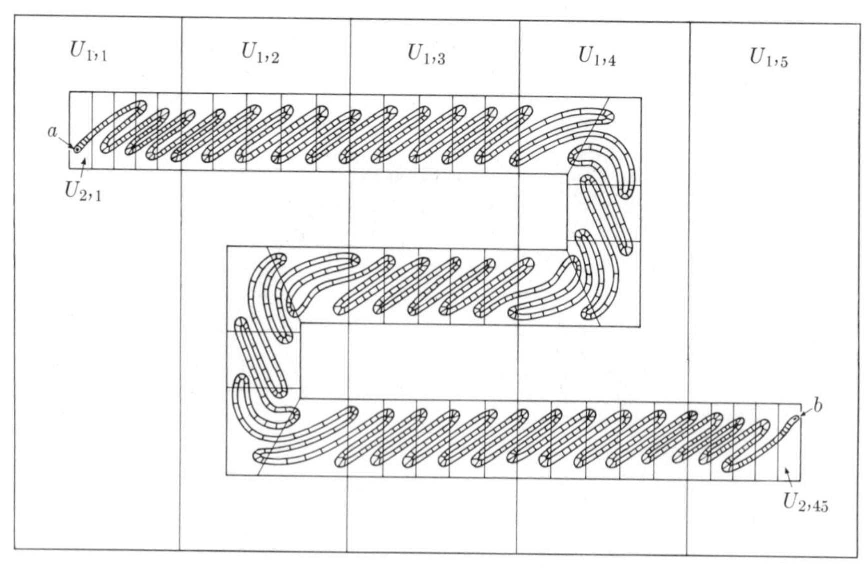

FIG. 3–22. Three steps in constructing the pseudo-arc.

Moïse proved that every two pseudo-arcs are homeomorphic and that the pseudo-arc is indecomposable. His surprising result is this: if N is a nondegenerate subcontinuum of a pseudo-arc M, then N is homeomorphic to N. That is, the pseudo-arc is *hereditarily indecomposable.* Bing [55] showed that the pseudo-arc is homogeneous. [A space S is *homogeneous* provided that for every pair of points a, b in S there is a homeomorphism h of S onto itself such that $h(a) = b$ and $h(b) = a$.] It was known previously that the simple closed curve is the only homogeneous nondegenerate bounded *locally connected* plane continuum (see Mazurkiewicz [101]). In a subsequent paper [59] Bing also showed that "almost every continuum is a pseudo-arc" in the following sense. If the collection of all continua in a Euclidean space or in Hilbert space is topologized by means of the Hausdorff metric, then the pseudo-arcs constitute a dense G_δ-set.

EXERCISE 3–23. Use the properties of the pseudo-arc to show that the plane E^2 contains uncountably many disjoint nondegenerate continua, no one of which contains an arc. (Also see R. L. Moore [108], Roberts [117], and Anderson [49].)

Another interesting example of an indecomposable continuum is known as the *Lakes of Wada* (see Yoneyama [133]). We construct a modification of this example by considering a double annulus, as shown in Fig. 3–23(a). To preserve the poetic flavor of the original, we take this to be an island in the ocean with two lakes, one having blue water and the other green. At time $t = 0$, we dig a canal from the ocean, which brings salt water to within a distance of 1 unit of every point of land. At time $t = \frac{1}{2}$, we dig a canal from the blue lake, which brings blue water to within a distance $\frac{1}{2}$ of every point of land. At time $t = \frac{3}{4}$, we dig a canal to bring green water to within a distance of $\frac{1}{3}$ of every point of land. At time $t = \frac{7}{8}$, we dig a canal from the end of the first canal to bring salt water to within a distance $\frac{1}{4}$ of every point of land, and so forth. If we think of these canals as open sets, at time $t = 1$ the "land" remaining is a plane continuum which bounds *three* open domains in the plane.

If any *plane* continuum is the common boundary of three open sets, then it is either indecomposable or is a union of two indecomposable continua (see Kuratowski [93]). In E^3, this last statement is not true. Indeed, there is an absolute neighborhood retract in E^3 which is the common boundary of three open sets (see [99]).

The Lakes of Wada raise an interesting question about double integrals. Making the construction in the unit square I^2, we obtain three open sets U_1, U_2, and U_3, which are disjoint and have a common boundary. Furthermore, each open set U_i is dense in I^2, and by making each successive canal very narrow, we can adjust the areas of the canals so that the area of

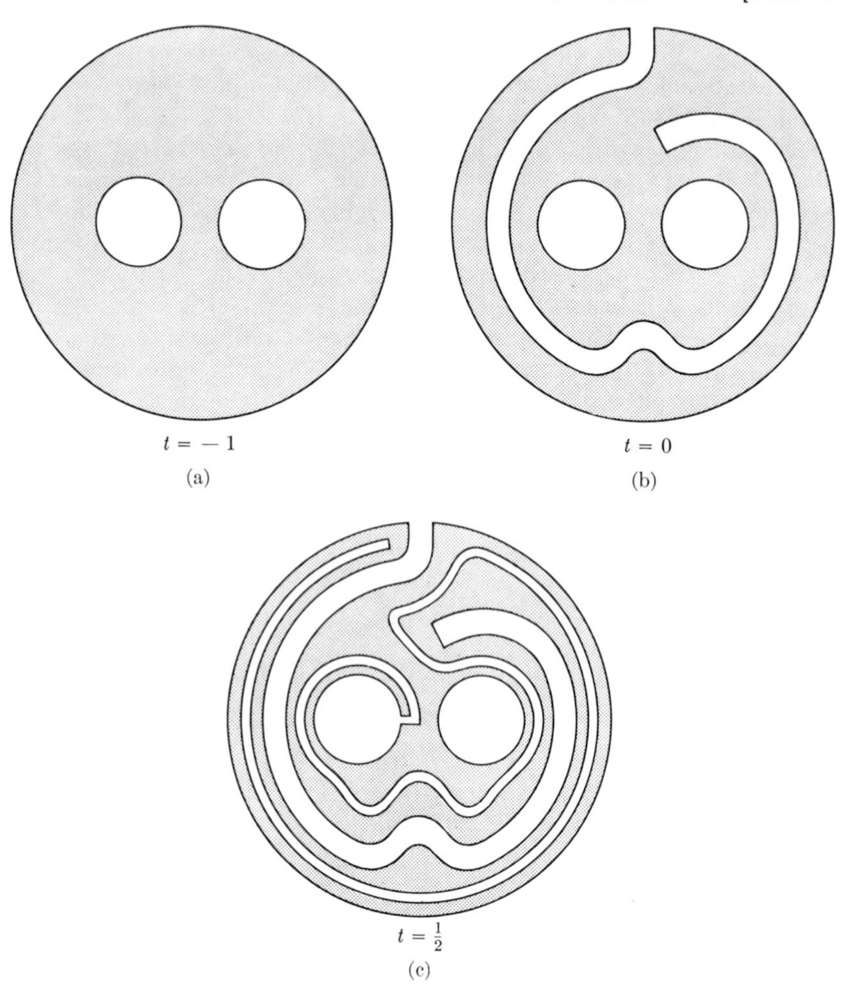

FIG. 3–23. Three stages in constructing the Lakes of Wada.

each U_i is $\frac{1}{10}$. Now given any real-valued integrable function $f(x, y)$ on I^2, can we say that

$$\iint_{I^2} f = \iint_{U_1} f + \iint_{U_2} f + \iint_{U_3} f \text{ ?}$$

No! For if $f \equiv 1$ on I^2, then

$$\iint_{I^2} f = 1 \qquad \text{while} \qquad \iint_{U_i} f = \frac{1}{10}, \qquad \text{and} \qquad 1 \neq \frac{3}{10}.$$

On the other hand, the boundary of each U_i has measure $\frac{7}{10}$. Thus

$$\iint_{I^2} f \neq \iint_{\overline{U}_1} f + \iint_{\overline{U}_2} f + \iint_{\overline{U}_3} f.$$

For, again taking $f \equiv 1$, we obtain $1 \neq \frac{7}{10} + \frac{7}{10} + \frac{7}{10}$!

An often-quoted example of an indecomposable continuum is the *solenoid*. Let $K_n = S^1$, the unit circle, for each positive integer n, and let h_m be the mapping of K_{n+1} onto K_m given by the complex function $w = z^2$. The inverse limit space K of the inverse limit sequence $\{K_m, h_m\}$ is the solenoid. Further, each space $K_m = S^1$ is a topological group, and h_m is a continuous homomorphism. Therefore K is also a compact topological group, the *solenoid group*.

Remark. There exist indecomposable continua of any positive dimension (See Bing [60].)

Remark. A *clan* is a continuum on which there is defined a continuous multiplication with a two-sided identity element. If the continuum is indecomposable, then the clan is a group! For a discussion of this and similar results about topological semigroups, see A. D. Wallace [130].

EXERCISE 3–24. Show that if p is a point of an indecomposable continuum K, the set of all points x of K such that K is irreducible between p and x is dense in K.

EXERCISE 3–25. Show that the union of a countable number of proper subcontinua of an indecomposable metric continuum K cannot separate K.

EXERCISE 3–26. Show that if p is a point of a decomposable continuum K, then the set of all points x of K such that K is irreducible between p and x does not have p as a limit point.

EXERCISE 3–27. Let A be an indecomposable continuum and ab be an arc having only the point b in common with A. Let $K = A \cup ab$. Show that the set of points x of K such that K is irreducible from a to x is not closed.

EXERCISE 3–28. A continuum C is *unicoherent* provided that if $C = H \cup K$, H and K subcontinua, then $H \cap K$ is connected. A continuum is *hereditarily unicoherent* if every subcontinuum is unicoherent. Show that if a continuum C is hereditarily unicoherent and contains a subset R that is the continuous image of a straight-line ray and which is such that every point of C is a limit point of R, then C is indecomposable.

3–9 Dimension theory. The study of dimension theory is extensive, and only a brief introduction to the topic can be given here. Once again our purpose is to provide merely an indication of an important unifying concept in topology that is aside from the major interests of this book. Hurewicz and Wallman's *Dimension Theory* [15] is an excellent reference for the interested reader.

Assigning an integer to every space in such a way as to satisfy our intuitive geometric idea of dimension is far from being a trivial problem. Indeed, it does not seem to have a solution. By this, we mean that while the inductive *definition* given below applies to any space, a satisfactory *theory* of dimension has not been developed for arbitrary spaces. In this section, *all spaces are assumed to be separable metric spaces.*

A space X has *dimension zero at a point* p (dim (X at p) $= 0$) if there are arbitrarily small open sets with empty boundaries containing the point p. Then X has *dimension zero* (dim $X = 0$) if dim X at $p = 0$ for all points p in X.

As an example of a 0-dimensional space, consider the rational numbers F as a subspace of E^1. Each rational number r lies in an arbitrarily small interval I in E^1 with irrational end points. The relative open set $F \cap I$ contains r and has an empty boundary. Thus dim $F = 0$. Indeed, *any countable* (separable metric) *space is* 0-*dimensional.* Furthermore, a similar argument shows that the set R of irrational numbers is a 0-dimensional subspace of E^1. Thus the 1-dimensional space E^1 is the union of two 0-dimensional subspaces. This is a special case of Theorem 3–57.

Consider the following three subsets of E^2. Let F^2 be the set of all points in E^2 both of whose coordinates are rational, let R^2 be the set of all points both of whose coordinates are irrational, and let $X = E^2 - (F^2 \cup R^2)$. All three of these sets are 0-dimensional. Since F^2 is countable, it is 0-dimensional. Let p be any point in R^2. Then there is an arbitrarily small rectangle I in E^2 containing p and bounded by lines $x = f_1$, $x = f_2$, $y = f_3$, $y = f_4$, where each f_i is rational. No such line meets R^2, and hence the relative open set $R^2 \cap I$ has an empty boundary, and R^2 is 0-dimensional. Finally, let q be any point of X. Then there is an arbitrarily small rectangle I in E^2 containing q and bounded by lines $y = x + f_1$, $y = x + f_2$, $y = -x + f_3$, $y = -x + f_4$, where each f_i is rational. Any point on such a line either has both coordinates rational or has both coordinates irrational. Since X consists of all points having just one coordinate rational, it follows that no such line meets X. Hence $X \cap I$ has an empty boundary, and X is 0-dimensional. This proves that E^2 is the union of three 0-dimensional subspaces. (Again see Theorem 3–57 below.)

Other examples of 0-dimensional spaces are the Cantor set and the set F_1^ω of all points in the Hilbert cube I^ω all of whose coordinates are rational. Oddly enough, the set F^ω of all points in Hilbert space H all of whose coordinates are rational is *not* 0-dimensional but 1-dimensional [76].

The following result is easily proved and explicitly states one of our intuitive ideas of the properties that "dimension" should have.

LEMMA 3–52. A nonempty subset of a 0-dimensional space is 0-dimensional.

THEOREM 3–53. A space that is a countable union of *closed* 0-dimensional subsets is 0-dimensional.

THEOREM 3–54. Among compact spaces, the 0-dimensional spaces and the totally disconnected spaces are identical.

For proofs of Theorems 3–53 and 3–54, see Chapter I of Hurewicz and Wallman [15].

The following inductive definition is due essentially to Menger. The empty set $\emptyset$ and only this set has dimension -1. A space X *has dimension* $\leq n$ $(n \geq 0)$ *at a point* p (dim (X at p) $\leq n$) if p lies in arbitrarily small open sets whose boundaries have dimension $\leq (n - 1)$. Then X *has dimension* n *at* p (dim (X at p) $= n$) if dim (X at p) $\leq n$ but dim (X at p) $= (n - 1)$ is false. The space X *has dimension* $\leq n$ (dim $X \leq n$) if dim (X at p) $\leq n$ for all points p in X; and X *has dimension* n (dim $X = n$) if dim $X \leq n$ but dim $X \leq (n - 1)$ is false.

A space may be n-dimensional, without having dim (X at p) $= n$ at each point. For example, the union of an arc and a disk with one point in common has dimension 2, but is 1-dimensional at some points.

It is easy to show that E^1 and I^1 have dimension 1. Also any polygon has dimension 1. An inductive argument showing that Euclidean n-space has dimension $\leq n$ is left as an easy exercise. Also an inductive proof of the following "desirable" property is quite easy to construct.

LEMMA 3–55. A subspace of a space of dimension $\leq n$ has dimension $\leq n$.

As a generalization of Theorem 3–53, we have the following result.

THEOREM 3–56. A space that is a countable union of *closed* subsets of dimension $\leq n$ has dimension $\leq n$.

Generalizing the examples wherein E^2 and E^3 were decomposed into a union of 0-dimensional (nonclosed) subsets, we have the next result.

THEOREM 3–57. A space X has dimension $\leq n$, n finite, if and only if X is a union of $n + 1$ subspaces of dimension zero.

Again the reader is referred to Hurewicz and Wallman, Chapter II [15], for proofs of these theorems. Another interesting situation arises from the next result.

THEOREM 3–58. If one of the two spaces X and Y is nonempty, then dim $(X \times Y) \leqq$ dim X + dim Y.

One might expect equality to hold in the relation dim $(X \times Y) \leqq$ dim X + dim Y. Indeed, this is such an intuitively appealing property that one is tempted to require it for "dimension." Unfortunately, equality need not hold. As was mentioned above, the subset F^ω of Hilbert space has dimension 1. It is easy to prove that $F^\omega \times F^\omega$ is homeomorphic to F^ω and hence $\dim (F^\omega \times F^\omega) = 1$! Pontrjagin [114] has given an example of two compact 2-dimensional spaces whose product is only 3-dimensional.

Perhaps the three most important results concerning dimension of Euclidean spaces are the following. The first of these is by way of being a justification for the definition of dimension.

THEOREM 3–59. E^n has dimension n (Brouwer [70]).

THEOREM 3–60. Every n-dimensional subset of E^n has interior points.

THEOREM 3–61. E^n cannot be separated by a subset of dimension $\leqq (n - 2)$.

Two last theorems are of conceptual interest.

THEOREM 3–62. Let X be an arbitrary separable metric space of dimension $\leqq n$, when n is finite. Then X is homeomorphic to a subset of I^{2n+1}. (See Menger [102] and Nöbeling [109].)

THEOREM 3–63. Let X be an arbitrary separable metric space. Then X is homeomorphic to a subset of the Hilbert cube I^ω.

A special case of Theorem 3–62 is proved in Section 5–8. Also, we may point out that some results to be found in Section 6–17 are intimately connected with dimension theory, as will be explicitly stated.

A systematic study of dimension theory embodies many important concepts in topology and will well reward the reader whose interests are primarily in topology itself. It would be difficult to recommend a better source than the Hurewicz and Wallman book, which certainly should be in every topologist's personal library.

CHAPTER 4

THE ELEMENTS OF HOMOTOPY THEORY

4–1 Introduction. In this chapter we strive for two goals, the presentation of the basic concept of *homotopy* and an introduction to the extremely broad topic called *algebraic topology.* Since the usefulness of homotopy will become apparent in this and succeeding chapters, we will say a few words only about the latter goal. A typical process in algebraic topology is to associate certain algebraic groups with a given topological space. These associated groups turn out to be topological invariants in the sense that to homeomorphic spaces our processes always associate isomorphic groups. Then, in some way, the structure of these groups yields information about the structure of the space with which they are associated.

Our reason for approaching algebraic topology via homotopy is found in the strongly geometric flavor of the theory of homotopic mappings. Speaking intuitively, two mappings are homotopic if one can be deformed continuously into the other. Or we may view homotopic mappings as being members of a one-parameter family of mappings with a continuous parameter. Since precision lies in this direction, we may begin with a general definition, which will be specialized to give the concept we desire.

A *parametrized family of mappings* of a space X into a space Y is a continuous function $h:X \times C \rightarrow Y$, where C is any space and is called the *parameter space.* Given any fixed point p in C, the subset $X \times p$ of $X \times C$ consisting of all pairs (x, p), x in X, is a *cross section* of the product space $X \times C$. Then for each point p in C, the mapping $h|X \times p$ (h restricted to the cross section $X \times p$) is a *member of the parametrized family.*

The generality of the above definition incorporates many situations, and we will not attempt a theory covering all of them. Rather, we mention a few examples and go on to homotopy. First, let C be the positive integers with the discrete topology. In this case, our parametrized family is simply a sequence of mappings of X into Y. Or take the parameter space C to be a linear interval $[a, b]$. The resulting family is that which one usually calls a 1-parameter family. Of course, with C taken as a parallelotope in E^n, we have an n-parameter family, etc. One finds the spheres S^n being used as parameter spaces, and many other examples. As we remarked, a unified theory covering all such cases cannot be developed here. We might mention that mapping theorems on product spaces, fibre spaces, and fibre bundles all incorporate similar considerations. (See Steenrod [35].)

4–2 Homotopic mappings. For our purposes, the most important instance of a parametrized family of mappings (continuous understood) is obtained by taking the parameter space to be the closed unit interval I^1. As a first example, let X be the unit circle S^1, and let Y be the Euclidean plane E^2. Then any mapping $h:S^1 \times I^1 \to E^2$ is such a family. Each member $h|S^1 \times t$, $0 \leq t \leq 1$, may be considered as a mapping of S^1 into E^2 and, in particular, the two members $h|S^1 \times 0$ and $h|S^1 \times 1$ may be viewed as continuous deformations of each other. Figure 4–1 is a simple example of such a family.

To be precise, two mappings f and g of a space X into a space Y are *homotopic* (and we write $f \simeq g$) if there is a mapping $h:X \times I^1 \to Y$ such that for each point x in X,

$$h(x, 0) = f(x) \qquad \text{and} \qquad h(x, 1) = g(x).$$

This is just another way of saying that $h|X \times 0 = f$ and $h|X \times 1 = g$, and hence we have the connection with 1-parameter families. The mapping h is called a *homotopy between f and g* and the product space $X \times I^1$ is the *homotopy cylinder.*

In these terms, the mappings $h|X^1 \times 0$ and $h|X^1 \times 1$ shown in Fig. 4–1 are homotopic mappings of S^1 into E^2. As we shall see later, any mapping of S^1 into E^2 is homotopic to any other such mapping, so our example is rather trivial. Such a statement is not true for every space Y, of course. For instance, let Y be the punctured plane $E^2 - (0, 0)$. Then a constant mapping c of S^1 onto a single point p of Y cannot be homotopic to a mapping of f of S^1 onto a simple closed curve J passing around the (missing) origin (see Fig. 4–2). Intuitively, it is impossible to deform J continuously onto the point p *while remaining in the space* Y.

The question of the existence of a homotopy between two mappings $f, g:X \to Y$ can be very difficult. The answer depends upon f and g, certainly, and also upon the structure of the spaces X and Y. It is evident that this question is one of extending a given mapping. For if f and g are

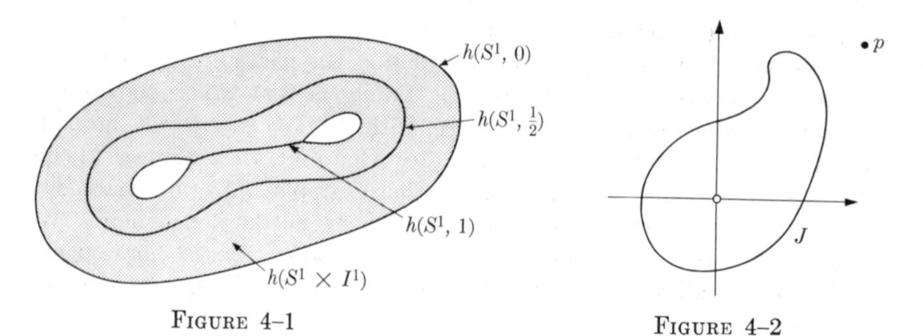

FIGURE 4–1 FIGURE 4–2

two mappings of X into Y, then we have a mapping h' on the closed subset $(X \times 0) \cup (X \times 1)$ of $X \times I^1$ given by $h'(x, 0) = f(x)$ and $h'(x, 1) = g(x)$. Then f and g are homotopic if and only if h' can be extended to a mapping h of the entire product space $X \times I^1$ into Y. Thus it would seem that theorems about homotopy are but special cases of more general theorems on the extension of mappings. Indeed such is the case, but the general extension problem is far from being solved, and also the special case of homotopy plays an important role in the more general problem.

It might appear at first glance that too much importance is being attached to the interval I^1 in the definition of homotopy. Why not use a 2-sphere, for example, in place of I^1? Part of the reason is this: *if A is an arcwise-connected space and f_1 and f_2 are mappings of a space X into a space Y, and if f_1 and f_2 are members of a parametrized family with parameter space A, then f_1 and f_2 are homotopic.* To see this, let $h:X \times A \to Y$ be the parametrized family, and let a_1, a_2 be the points of A such that $h(x, a_1) = f_1(x)$ and $h(x, a_2) = f_2(x)$. There is an arc a_1a_2 in A. The mapping $h|X \times a_1a_2$ is not quite a homotopy between f_1 and f_2, but only because of the use of I^1 in the definition. A "coordinatization" of the arc a_1a_2 gives the homotopy. □

EXERCISE 4–1. Use the fact that I^1 is an absolute retract to show that the converse of this result is true.

If C is a connected space that is not arcwise-connected, and if X and Y are spaces, then there may be no relationship between "homotopy" and "being in a parametrized family with parameter space C." For example, if X is a Peano space, and Y is a continuum that contains no arc (e.g., the pseudo-arc, Section 3–8), then the only continuous mappings of X into Y are constants. Two such mappings would be homotopic only if they were identical. But if we take $C = Y$, any two such mappings lie in the same parametrized family. We can say this: *if X and C are metric continua and Y is an absolute neighborhood retract, then if f and g are mappings of X into Y that lie in a family parametrized by C, then f and g are homotopic.* For we can imbed C in a Peano space P (Exercise 3–19). The mapping $h:X \times C \to Y$ can be extended to a neighborhood U of $X \times C$ in $X \times P$. For each point c in C, there is an open set U_c in P containing c and such that $X \times U_c$ lies in U. This last is easily seen from compactness. Now let c_1, c_2 be the points of C such that $h(x, c_1) = f(x)$ and $h(x, c_2) = g(x)$. The component of the union $\cup U_c$ that contains C also contains an arc c_1c_2, by Theorem 3–16. The set $X \times c_1c_2$ lies in $X \times \cup U_c$ and so lies in U. Thus the extension of h to U can be restricted to $X \times c_1c_2$ to give a homotopy between f and g. □

We saw in Section 1–11 that the collection of all continuous mappings of a space X into a space Y can be topologized (in several ways) so as to obtain a function space Y^X. In our work here, we will assume that the compact-open topology has been assigned to Y^X.

THEOREM 4–1. *The homotopy relation between mappings of a space X into a space Y is an equivalence relation on Y^X. That is, the relation "$\simeq$" satisfies*

(1) $f \simeq f$ for each mapping f in Y^X (reflexive law),

(2) $f \simeq g$ implies $g \simeq f$ (symmetry law),

and

(3) $f \simeq g$ and $g \simeq k$ implies $f \simeq k$ (transitive law).

Proof: (1) For any mapping f in Y^X, define $h:W \times I^1 \to Y$ by

$$h(x, t) = f(x) \qquad (0 \leq t \leq 1).$$

It is evident that h is continuous and that $h(x, 0) = f(x) = h(x, 1)$ for all points x in X.

(2) If $f \simeq g$, then there is a homotopy $h:X \times I^1 \to Y$ such that $h(x, 0) = f(x)$ and $h(x, 1) = g(x)$ for all points x in X. We define

$$\overline{h}(x, t) = h(x, 1 - t).$$

Again $\overline{h}$ is obviously continuous and $\overline{h}(x, 0) = g(x)$, while $\overline{h}(x, 1) = f(x)$. Thus $g \simeq f$.

(3) If $f \simeq g$ and $g \simeq k$, then there are homotopies h_1 and h_2, with $h_1(x, 0) = f(x)$, $h_1(x, 1) = g(x)$, $h_2(x, 0) = g(x)$, and $h_2(x, 1) = k(x)$. We define a homotopy h between f and k by setting

$$\begin{aligned} h(x, t) &= h_1(x, 2t) && (0 \leq t \leq \tfrac{1}{2}) \\ &= h_2(x, 2t - 1) && (\tfrac{1}{2} \leq t \leq 1). \end{aligned}$$

Then $h(x, \frac{1}{2}) = g(x)$ by both definitions, so h is well-defined and continuous on $X \times I^1$. Clearly $h(x, 0) = h_1(x, 0) = f(x)$, while $h(x, 1) = h_2(x, 1) = k(x)$. Thus $f \simeq k$. □

The following result should be familiar, and is quoted without proof.

THEOREM 4–2. *Let A be any set, and let R be an equivalence relation on A. Then A is decomposed by R into disjoint subsets called equivalence classes.*

In view of Theorems 4–1 and 4–2, the function space Y^X is decomposed by the homotopy relation into disjoint *homotopy classes*. Although one does not attempt to visualize these homotopy classes, they are easily characterized.

THEOREM 4–3. The homotopy classes of Y^X are precisely the arcwise-connected components of Y^X.

Proof: This is merely a matter of checking definitions. For if $f \simeq g$, then the homotopy $h(x, t)$ between f and g defines a mapping $F:I^1 \to Y^X$ given by

$$F(t) = f_t(x) = h(x, t).$$

Then $F(I^1)$ is a Peano continuum in Y^X, and as such contains an arc between f and g. Conversely, an "arc" of mappings between f and g provides a homotopy between the two. □

We will close this section with an interesting result, due to Borsuk [69], which associates homotopy and the extension of mappings.

THEOREM 4–4. Let A be a closed subset of a separable metric M, and let f' and g' be homotopic mappings of A into the n-sphere S^n. If there exists an extension f of f' to all of M, then there also exists an extension g of g' to all of M, and the extensions f and g may be chosen to be homotopic also.

Proof (we follow Dowker [74]): Let $h':A \times I^1 \to S^n$ be the assumed homotopy between f' and g', and let f be the given extension of f' to all of M. Let D be the set in $M \times I^1$ given by

$$D = (A \times I^1) \cup (M \times 0).$$

Clearly D is a closed subset of $M \times I^1$, and on D we may define the mapping $F':D \to S^n$ given by

$$F'(x, 0) = f(x) \qquad \text{for all } x \text{ in } M,$$

and

$$F'(x, t) = h'(x, t) \qquad \text{for all } x \text{ in } A \text{ and } 0 \leq t \leq 1.$$

Since $h'(x, 0) = f'(x) = f(x)$ for all points x in A, this mapping F' is well-defined and continuous.

Theorem 2–35 states that there is an open set U in $M \times I^1$ such that U contains D and such that F' can be extended to a mapping F on U. It is easy to show that there is an open set V in M such that V contains A and such that $V \times I^1$ lies in U (see Exercise 1–31). Therefore the mapping F is defined on the subset $(V \times I^1) \cup (M \times 0)$. Theorem 2–37 now applies to give a mapping $H(x, t)$ which agrees with F on $(A \times I^1) \cup (M \times 0)$. Defining $g(x)$ to be $H(x, 1)$ gives the desired extension of g'. The details here are easily checked. □

An important feature of this result is that not only can the mapping g' be extended (if f' can), but also the connecting homotopy can be extended.

4–3 Essential and inessential mappings. A mapping $f:X \to Y$ of a space X into a space Y is said to be *inessential* if f is homotopic to a constant mapping $c(X) = y_0$, a single point of Y; otherwise, f is *essential.* Our results deal with inessential mappings because the very existence of an essential mapping of X into Y may be very difficult to prove. For instance, given $n > m$, is there an essential mapping of S^n into S^m? (More on this later.)

As a corollary to the Borsuk theorem (4–4), one easily proves the following result.

THEOREM 4–5. Any inessential mapping of a closed subset of a separable metric space M into S^n can always be extended over all of M in such a way that the resulting extension is also inessential.

Proof: A constant mapping can always be extended. □

We may characterize inessential mappings if we introduce a new definition. Given a space X and a point p not in X, we form the *join* pX of X and p as follows. Consider the (disjoint) union $p \cup (X \times I^1)$ of the point p and the product space $X \times I^1$. Define the *identification mapping* π on $p \cup (X \times I^1)$ by

$$\begin{aligned} \pi(p) &= p, \\ \pi(x, 1) &= p \qquad \text{for all } x \text{ in } X, \\ \pi(x, t) &= (x, t) \qquad \text{for all } x \text{ in } X \text{ and } 0 \leq t < 1. \end{aligned}$$

The image of $p \cup (X \times I^1)$ under π is the join pX. The *identification topology* is used in pX, which means that a set U in pX is open if and only if $\pi^{-1}(U)$ is open in $X \times I^1$. Essentially then, pX is obtained by assigning a new topology to $X \times I^1$ in which any open set that meets $X \times 1$ actually contains $X \times 1$. If X is imbedded in a linear subspace L of Hilbert space or E^n, and p is not in L, the join pX can be geometrically realized as the union of all intervals px, x in X.

THEOREM 4–6. A mapping $f:X \to Y$ is inessential if and only if f may be extended to all of a join pX.

Proof: Suppose first that $\bar{f}:pX \to Y$ is an extension of f. Define the mapping g on $X \times I^1$ by

$$g(x, t) = \bar{f}(x, t) \qquad \text{for all } x \text{ in } X \text{ and } 0 \leq t < 1$$

and

$$g(x, 1) = \bar{f}(p).$$

Since every open set in pX is open in $X \times I^1$, it follows that if U is an open set in Y, then $\bar{f}^{-1}(U)$ is open in pX, and hence $g^{-1}(U)$ is open in

$X \times I^1$. Therefore g is continuous. But now $g(x, 0) = \bar{f}(x, 0) = f(x)$, by definition, while $g(x, 1) = \bar{f}(p)$ is a constant mapping. That is, g is a homotopy between f and a constant mapping.

Conversely, let f be homotopic to a constant mapping $c(X) = y_0$, where y_0 is some fixed point of Y. By definition, there is a homotopy $h:X \times I^1 \to Y$ such that $h(x, 0) = f(x)$ and $h(x, 1) = y_0$ for all points x in X. Define the mapping $\bar{f}$ on pX by setting

$$\bar{f}(x, t) = h(x, t) \qquad \text{for all } x \text{ in } X \text{ and } 0 \leq t < 1$$

and

$$\bar{f}(x, 1) = \bar{f}(p) = y_0.$$

Now $\bar{f}$ is continuous, for if U is an open set in Y, then $\bar{f}^{-1}(U)$ will be open in $X \times I^1$, and if $\bar{f}^{-1}(U)$ meets $X \times 1$, then it contains $X \times 1$; in short, $\bar{f}^{-1}(U)$ is open in pX. Thus $\bar{f}$ is the desired extension of f. □

As we said earlier, the existence of a homotopy between two mappings f and g of X into Y depends upon the spaces X and Y as well as on the mappings themselves. For certain spaces Y, *all* mappings $f:X \to Y$ are homotopic. A space Y is said to be *contractible to a point* p in Y, or simply *contractible*, if the identity mapping $i(y) = y$ of Y onto itself is homotopic to the constant mapping $c(Y) = p$.

THEOREM 4–7. If Y is contractible to a point, then every mapping f of a space X into Y is inessential. (Hence all mappings $f:X \to Y$ are homotopic.)

Proof: Given $f:X \to Y$, the composite mapping if, where i is the identity mapping of Y onto itself, certainly coincides with f. If, as assumed, i is homotopic to a constant mapping $c(Y) = p$, then the composite mapping cf carries X onto the point p. By definition, there is a homotopy $h':Y \times I^1 \to Y$ such that $h'(y, 0) = y$ and $h'(y, 1) = p$ for all points y in Y. Define the mapping $h:X \times I^1 \to Y$ given by

$$h(x, t) = h'[f(x), t].$$

Certainly h is continuous, and we have

$$h(x, 0) = h'[f(x), 0] = f(x)$$

and

$$h(x, 1) = h'[f(x), 1] = p.$$

Therefore h is a homotopy between f and a constant mapping. □

To obtain some examples of contractible spaces, consider the following definition. A metric space M with metric d is *starlike* in that metric if there is a point p in M such that each other point x in M can be joined to p by a unique arc congruent in the metric of M to a line segment.

THEOREM 4–8. If M is a metric space and has a metric in which M is starlike, then M is contractible.

Proof: Let p be the point and d the metric on M such that from each point x in M there is a unique arc px congruent to an interval. Define a mapping $h:M \times I^1 \to M$ by taking $h(x, t)$ to be the unique point y on px such that $d(p, y) = t \cdot d(p, x)$. Then $h(x, 0) = p$ for all points x, and $h(x, 1) = x$ for all x. Thus if h is continuous, then h is a homotopy between the identity and a constant mapping. A proof that h is continuous may easily be given if the reader uses Theorem 1–37. □

From Theorem 4–8 it follows that *any Euclidean cube I^n, and the Hilbert cube I^ω, is contractible.* Hence as a corollary to Theorems 4–7 and 4–8, we have the following result.

COROLLARY 4–9. Any mapping of a space X into I^n or I^ω is inessential.

COROLLARY 4–10. Any mapping of a compact space into E^n or Hilbert space H is inessential.

Proof: Since the continuous image of any compact space in E^n or in H will be compact, it may be taken to lie in some sufficiently large cube in $E^n(H)$, and such a cube is contractible. □

This gives a proof of the statement made at the beginning of Section 4–2 to the effect that every pair of mappings of S^1 into E^2 are homotopic.

We may use the next result to obtain other contractible spaces.

THEOREM 4–11. Any retract of a contractible space is contractible.

Proof: Let X be a contractible space, and let $r:X \to A$ be a retraction of X onto a subset A. By definition, the identity mapping $i:X \to X$ is homotopic to a constant mapping $c(X) = x_0$ via a homotopy h'. Define the mapping $h:A \times I^1 \to A$, given by

$$h(x, t) = r[h'(x, t)].$$

Then h is certainly continuous on $A \times I^1$, and

$$h(x, 0) = r[h'(x, 0)] = r(x) = x$$

and

$$h(x, 1) = r[h'(x, 1)] = r(x_0) \qquad \text{for each point } x \text{ in } A.$$

Thus h is a homotopy between the identity mapping $r|A$ (r restricted to A) of A onto itself and the constant mapping $c'(A) = r(x_0)$. □

THEOREM 4–12. Any compact metric absolute retract is contractible.

Proof: We stated, in Theorem 3-63, that every separable metric space can be imbedded in the Hilbert cube I^ω. Thus if A is a compact metric absolute retract, it is homeomorphic to a subset A' of I^ω. The subset A' is a retract of I^ω, by the definition of absolute retract, and so by Theorem 4-11, A' is contractible. Since contractibility is a topological property, A is also contractible. □

The converse of Theorem 4-12 is false. For example, the join of a Cantor set and a point, the *Cantor star,* is compact metric and contractible but is not an absolute retract.

The next result, which will be cited in Section 6-14, may be proved as an exercise.

THEOREM 4-13. If $f:X \to S^n$ is a mapping of a space X into the n-sphere such that $S^n - f(X)$ is not empty, then f is inessential.

4-4 Homotopically equivalent spaces. This brief section introduces a concept that becomes important in our later discussions. Two spaces X and Y *are of the same homotopy type* (are *homotopically equivalent*) if there exist mappings $f:X \to Y$ and $g:Y \to X$ such that the composite mappings $fg:Y \to Y$ and $gf:X \to X$ are homotopic, respectively, to the identity mappings $i:Y \to Y$ and $i:X \to X$. All the forthcoming algebraic groups to be associated with a space fail to distinguish between two homotopically equivalent spaces. It is obvious that homeomorphic spaces are of the same homotopy type, but the converse is not necessarily true. To give an example of a general procedure for obtaining two homotopically equivalent spaces that are not homeomorphic, we prove a theorem.

Let $f:X \to Y$ be continuous. In the (disjoint) union $(X \times I^1) \cup Y$, identify each point $(x, 1)$ with the point $f(x)$ in Y. Using the identification topology, the resulting space $Y_{f(X)}$ is called the *mapping cylinder* of f. As a special case, if $c:X \to p$ is a constant mapping of X onto a space with only one point p, then the mapping cylinder of c is homeomorphic to the join pX.

THEOREM 4-14. Let $f:X \to Y$ be any continuous mapping of a space X into a space Y. Then the mapping cylinder $Y_{f(X)}$ is homotopically equivalent to Y.

Proof: Define a mapping $g:Y_{f(X)} \to Y$ by setting

$$g(x, t) = f(x) \qquad \text{for } (x, t) \text{ in } X \times I^1$$

and

$$g(y) = y \qquad \text{for } y \text{ in } Y.$$

This mapping is well-defined and continuous on $Y_{f(X)}$ because it is continuous on each of two closed subsets of $Y_{f(X)}$ and agrees on the intersection of these subsets. Next let $h: Y \to Y_{f(X)}$ be the identity injection $h(y) = y$. Clearly we have

$$gh(y) = g(y) = y,$$

so the composite mapping $gh: Y \to Y$ *is* the identity mapping.

Considering the composite mapping hg of $Y_{f(X)}$ into itself, we have

$$hg(y) = h(y) = y \qquad \text{for all points } y \text{ in } Y$$

and

$$hg(x, t) = h(f(x)) = f(x) \qquad \text{for all points } (x, t) \text{ in } X \times I^1.$$

We define a mapping $H: Y_{f(X)} \times I^1 \to Y_{f(X)}$ by setting

$$H(y, s) = y \qquad \text{for all } y \text{ in } Y \text{ and } 0 \leq s \leq 1$$
$$H((x, t), s) = (x, (1 - s)t + s) \qquad \text{for } (x, t) \text{ in } X \times I^1 \text{ and } 0 \leq s \leq 1.$$

When $t = 1$, we have

$$H((x, 1), s) = (x, 1) = f(x) = H(f(x), s) \qquad (0 \leq s \leq 1),$$

so the two definitions agree on those points identified in $Y_{f(X)}$. Hence H is well-defined and continuous. But now

$$H(y, 0) = y,$$
$$H((x, t), 0) = (x, t),$$

or $H(z, 0)$ is the identity mapping on $Y_{f(X)}$, while

$$H(y, 1) = y$$

and

$$H((x, t), 1) = (x, 1) = f(x),$$

so $H(z, 1) = hg(z)$ for all points z in $Y_{f(X)}$. Therefore H is a homotopy between the identity mapping on $Y_{f(X)}$ and the composite mapping hg. □

We can state a corollary to Theorem 4–14 by giving another definition. A subset D of a space X is a *deformation retract* of X if there is a retraction r of X onto D which is homotopic to the identity mapping of X onto itself under a homotopy that leaves D fixed. That is, there is a homotopy $h: X \times I^1 \to X$ such that

$$h(x, 0) = x \qquad \text{for all } x \text{ in } X,$$
$$h(x, 1) = r(x) \qquad \text{for all } x \text{ in } X,$$

and

$$h(x, t) = x \qquad \text{for all } x \text{ in } D \text{ and } 0 \leq t \leq 1.$$

COROLLARY 4–15. The space Y is a deformation retract of the mapping cylinder $Y_{f(X)}$.

Proof: Consider the mapping $g:Y_{f(X)} \to Y$ given in the proof of Theorem 4–14. Clearly $g(y) = y$ for each point y in Y, so g is a retraction of $Y_{f(X)}$ onto Y. The homotopy $H(z, s)$ given in Theorem 4–14 has the property that

$$H(z, 0) = z$$

and

$$H(z, 1) = g(z).$$

Thus H is a homotopy between the identity mapping on $Y_{f(X)}$ and the mapping g. Finally, for any point y in Y, we have

$$H(y, s) = y,$$

by definition. □

EXERCISE 4–2. Show directly and by Theorem 4–14 that the circle S^1 is of the same homotopy type as the cylinder $S^1 \times I^1$.

EXERCISE 4–3. Assume that S^n is not contractible, and show that it is not a mapping cylinder.

EXERCISE 4–4. Show that there are two continua, one locally connected and the other not, that are of the same homotopy type.

4–5 The fundamental group. Here for the first time we construct an algebraic group that is a topological invariant of the space Y to which it is associated. This so-called *fundamental group*, a conception of H. Poincaré, was possibly suggested to him by a study of plane regions as used in the theory of functions. In that study the concept of simply-connected and multiply-connected regions plays an important role in complex integration. The very definition of a simply-connected region in terms of "shrinking" simple closed curves should strongly suggest homotopy to the now-sophisticated reader. We will clarify this matter in a subsequent remark.

Let Y be a topological space, and let y_0 be a point in Y. Then *the y_0-neighborhood of curves in Y, $C(Y, y_0)$*, is the collection of all continuous mappings $f:I^1 \to Y$ of the unit interval into Y such that $f(0) = y_0 = f(1)$. Note that $C(Y, y_0)$ is a subspace of the function space Y^{I^1} and is not a neighborhood in Y in the usual sense.

Let f and g be two mappings in $C(Y, y_0)$. Then *f is homotopic to g modulo y_0* (abbreviated $f \underset{y_0}{\simeq} g$) if there exists a homotopy $h:I^1 \times I^1 \to Y$ such that

$$h(x, 0) = f(x) \text{ and } h(x, 1) = g(x) \qquad \text{for all } x \text{ in } I^1$$

and

$$h(0, t) = y_0 = h(1, t) \qquad \text{for all } t \text{ in } I^1.$$

This is illustrated by Fig. 4–3.

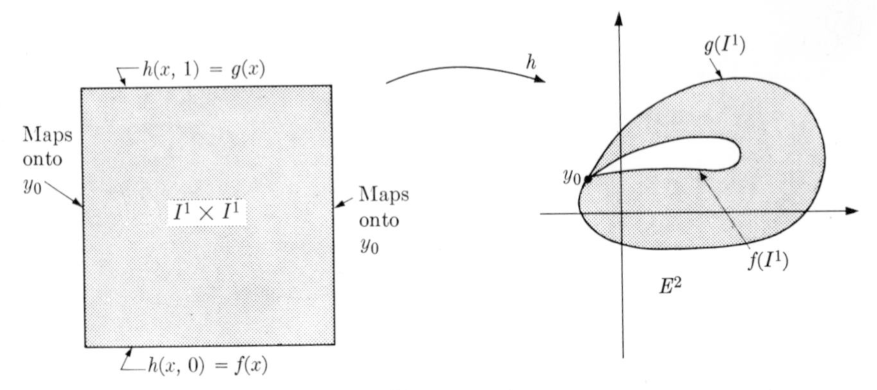

FIGURE 4–3

LEMMA 4–16. Homotopy modulo y_0 is an equivalence relation on $C(Y, y_0)$.

It has to be shown that homotopy modulo y_0 is reflexive, symmetric, and transitive. Since the details are so very similar to the proof of Theorem 4–1, we leave the proof of this lemma as an exercise.

Now, in accord with Theorems 4–2 and 4–3, Lemma 4–16 tells us that $C(Y, y_0)$ is decomposed by the relation $\underset{y_0}{\simeq}$ into disjoint equivalence classes, which are the arcwise-connected components of $C(Y, y_0)$. We let $\pi_1(Y, y_0)$ denote the collection of these equivalence classes. By introducing a suitable group operation, this collection becomes the *fundamental group* of Y modulo y_0 (or the *Poincaré group* of Y or the *first homotopy group* of Y modulo y_0).

Let $[f]$ denote the homotopy class of which the mapping f in $C(Y, y_0)$ is a representative. That is, $[f]$ denotes the collection of all elements g of $C(Y, y_0)$ such that $f \underset{y_0}{\simeq} g$. We will define a "multiplication" $[f] \circ [g]$ of two such elements of $\pi_1(Y, y_0)$. This operation will yield another element of $\pi_1(Y, y_0)$ and will satisfy the group axioms. Let f and g be two mappings in $C(Y, y_0)$. The *juxtaposition* $f{*}g$ of f and g is a new element of $C(Y, y_0)$ given by

$$(f{*}g)(x) = f(2x) \qquad \text{for } 0 \leq x \leq \tfrac{1}{2},$$
$$(f{*}g)(x) = g(2x - 1) \qquad \text{for } \tfrac{1}{2} \leq x \leq 1.$$

Since $(f{*}g)(\frac{1}{2}) = f(1) = g(0) = y_0$, the mapping $f{*}g$ is a well-defined element of $C(Y, y_0)$. Then if $[f]$ and $[g]$ are two elements of $\pi_1(Y, y_0)$, we define their product by means of the formula

$$[f] \circ [g] = [f{*}g].$$

Our first task is to show that the operation "$\circ$" is well-defined. That is, we must show that we obtain the same equivalence class $[f{*}g]$ regardless of what representatives f and g of $[f]$ and $[g]$ are used. This is done by proving that if $f_1 \underset{y_0}{\simeq} f_2$ and $g_1 \underset{y_0}{\simeq} g_2$, then $f_1 * g_1 \underset{y_0}{\simeq} f_2 * g_2$, and is purely a matter of calculation. By definition, there are homotopies h_1 and h_2 such that

$$h_1(x, 0) = f_1(x), \quad h_1(x, 1) = f_2(x), \quad h_1(0, t) = y_0 = h_1(1, t);$$
$$h_2(x, 0) = g_1(x), \quad h_2(x, 1) = g_2(x), \quad h_2(0, t) = y_0 = h_2(1, t).$$

Define a homotopy h between $f_1 * g_1$ and $f_2 * g_2$ as follows:

$$\begin{aligned} h(x, t) &= h_1(2x, t) && \text{for } 0 \leq x \leq \tfrac{1}{2} \\ &= h_2(2x - 1, t) && \text{for } \tfrac{1}{2} \leq x \leq 1. \end{aligned}$$

Since at $x = \frac{1}{2}$, $h(\frac{1}{2}, t) = h_1(1, t) = y_0$, or $h(\frac{1}{2}, t) = h_2(0, t) = y_0$, the mapping h is well-defined and continuous. Also

$$\left.\begin{aligned} h(x, 0) &= h_1(2x, 0) = f_1(2x) && (0 \leq x \leq \tfrac{1}{2}) \\ &= h_2(2x - 1, 0) = g_1(2x - 1) && (\tfrac{1}{2} \leq x \leq 1) \end{aligned}\right\} = (f_1 * g_1)(x)$$

and

$$\left.\begin{aligned} h(x, 1) &= h_1(2x, 1) = f_2(2x) && (0 \leq x \leq \tfrac{1}{2}) \\ &= h_2(2x - 1, 1) = g_2(2x - 1) && (\tfrac{1}{2} \leq x \leq 1) \end{aligned}\right\} = (f_2 * g_2)(x).$$

This proves that the operation "$\circ$" is well-defined and single-valued. Certainly $\pi_1(Y, y_0)$ is closed under this operation. We now set out to prove that this operation satisfies the axioms for a group. This requires some manipulation, and we have included several diagrams to assist in the necessary careful study.

It should be apparent that the desired associative law for the operation "$\circ$" follows at once if we can show that

$$(f_1 * f_2) * f_3 \underset{y_0}{\simeq} f_1 * (f_2 * f_3).$$

Let us analyze these juxtapositions. It is not difficult to see that

$$\begin{aligned} [(f_1 * f_2) * f_3](x) &= f_1(4x) && (0 \leq x \leq \tfrac{1}{4}) \\ &= f_2(4x - 1) && (\tfrac{1}{4} \leq x \leq \tfrac{1}{2}) \\ &= f_3(2x - 1) && (\tfrac{1}{2} \leq x \leq 1) \end{aligned}$$

and that

$$\begin{aligned} [f_1 * (f_2 * f_3)](x) &= f_1(2x) && (0 \leq x \leq \tfrac{1}{2}) \\ &= f_2(4x - 2) && (\tfrac{1}{2} \leq x \leq \tfrac{3}{4}) \\ &= f_3(4x - 3) && (\tfrac{3}{4} \leq x \leq 1). \end{aligned}$$

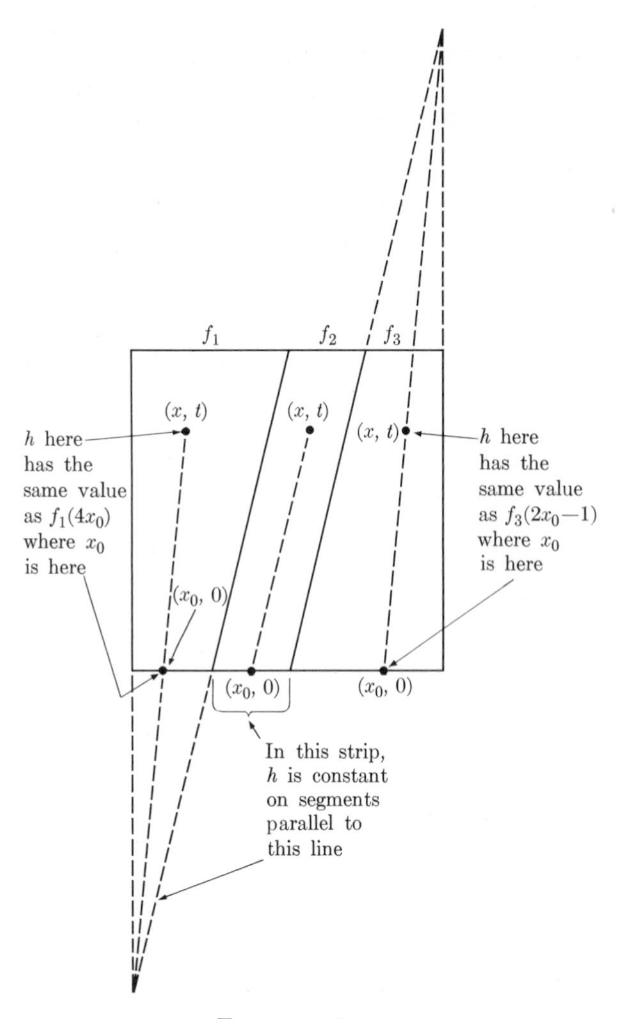

FIGURE 4–4

A homotopy h between these two mappings may be given as follows:

$$
\begin{aligned}
h(x, t) &= f_1\left(\frac{4x}{t+1}\right) && \text{for pairs } (x, t) \text{ with } t \geqq 4x - 1 \\
&= f_2(4x - t - 1) && \text{for pairs } (x, t) \text{ with } 4x - 1 \geqq t \geqq 4x - 2 \\
&= f_3\left(\frac{4x - t - 2}{2 - t}\right) && \text{for pairs } (x, t) \text{ with } 4x - 2 \geq t.
\end{aligned}
$$

Elementary analytic geometry applied to Fig. 4–4 will show how these expressions were obtained.

It is a simple matter to check that h is the desired homotopy modulo y_0. For

$$\left.\begin{aligned} h(x, 0) &= f_1(4x) && \text{for } 0 \geq 4x - 1 \\ & && \text{or } 0 \leq x \leq \tfrac{1}{4} \\ &= f_2(4x - 1) && \text{for } 4x - 1 \geq 0 \geq 4x - 2 \\ & && \text{or } \tfrac{1}{4} \leq x \leq \tfrac{1}{2} \\ &= f_3(2x - 1) && \text{for } 4x - 2 \geq 0 \\ & && \text{or } \tfrac{1}{2} \leq x \leq 1 \end{aligned}\right\} = (f_1 * f_2) * f_3$$

while

$$\left.\begin{aligned} h(x, 1) &= f_1(2x) && \text{for } 1 \geqq 4x - 1 \\ & && \text{or } 0 \leq x \leq \tfrac{1}{2} \\ &= f_2(4x - 2) && \text{for } 4x - 1 \geqq 1 \geqq 4x - 2 \\ & && \text{or } \tfrac{1}{2} \leq x \leq \tfrac{3}{4} \\ &= f_3(4x - 3) && \text{for } 4x - 2 \geqq 1 \\ & && \text{or } \tfrac{3}{4} \leq x \leq 1 \end{aligned}\right\} = f_1 * (f_2 * f_3).$$

Since for $t = 4x - 1$, we have $h(x, t) = f_1(x)$, etc., the continuity of h is assured and the associative law has been proved.

Next, let j denote the constant mapping $j(x) = y_0$ for each point x in I^1. We claim that the equivalence class $[j]$ is the identity element of $\pi_1(Y, y_0)$. To prove this, it will suffice to show that $f*j \underset{y_0}{\simeq} f$ for any function f in $C(Y, y_0)$. This is done by constructing the homotopy

$$\begin{aligned} h(x, t) &= f\left(\frac{2x}{1 + t}\right) && \text{for pairs } (x, t) \text{ with } t \geqq 2x - 1 \\ &= y_0 && \text{for pairs } (x, t) \text{ with } t \leqq 2x - 1. \end{aligned}$$

(To see where we got this, examine Fig. 4-5.) The continuity of h is only in question where $t = 2x - 1$, but for any such point, $h(x, t) = y_0$, so h is continuous as required. A check of the boundary conditions shows that

$$\left.\begin{aligned} h(x, 0) &= f(2x) && \text{for } 0 > 2x - 1 \text{ or } 0 \leq x \leq \tfrac{1}{2} \\ &= y_0 && \text{for } 0 \leq 2x - 1 \text{ or } \tfrac{1}{2} \leq x \leq 1 \end{aligned}\right\} = f*j$$

and

$$h(x, 1) = f(x) \qquad \text{for } 1 > 2x - 1 \text{ or } 0 \leq x \leq 1.$$

The other boundary conditions are obvious, and we know that $[j]$ is the identity element of $\pi_1(Y, y_0)$.

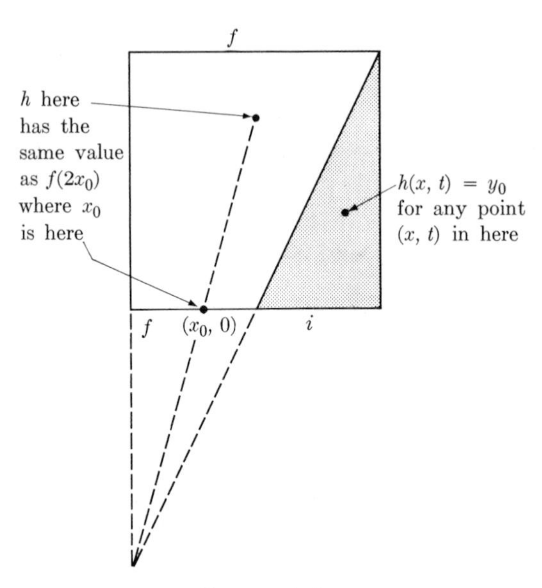

FIGURE 4–5

Finally we must show the existence of inverses. To do so, let f be any mapping in $C(Y, y_0)$, and define a new mapping f by setting

$$\bar{f}(x) = f(1 - x).$$

Clearly, $\bar{f}(0) = f(1) = y_0 = \bar{f}(1) = f(0)$, so $\bar{f}$ is an element of $C(Y, y_0)$. We show that $f * \bar{f} \underset{y_0}{\simeq} j$. By definition,

$$\begin{aligned}(f*\bar{f})(x) &= f(2x) && (0 \leq x \leq \tfrac{1}{2})\\ &= \bar{f}(2x - 1) = f(2 - 2x) && (\tfrac{1}{2} \leq x \leq 1).\end{aligned}$$

We may construct a homotopy between $f*\bar{f}$ and j by setting

$$\begin{aligned}h(x, t) &= f\left(\frac{2x}{1 - t}\right) && \text{for } t \leqq 1 - 2x,\ 0 \leq x \leq \tfrac{1}{2}\\ &= y_0 && \text{for } t \geqq 1 - 2x,\ 0 \leqq x \leqq \tfrac{1}{2};\\ & && t \geqq 2x - 1,\ \tfrac{1}{2} \leqq x \leqq 1\\ &= f\left(\frac{2x - 2}{t - 1}\right) && \text{for } t \leqq 2x - 1,\ \tfrac{1}{2} \leqq x \leqq 1.\end{aligned}$$

Notice that at $t = 1 - 2x$ we have

$$h(x, t) = f\left(\frac{2x}{1 - (1 - 2x)}\right) = f(1) = y_0$$

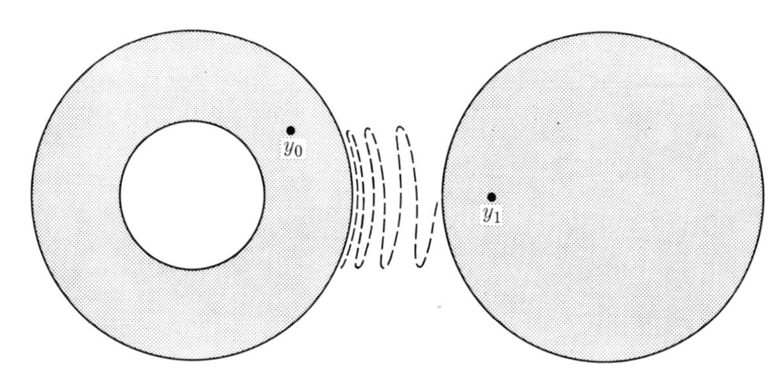

FIGURE 4-6

and that for $t = 2x - 1$,

$$h(x, t) = f\left(\frac{2x - 2}{2x - 1 - 1}\right) = f(1) = y_0.$$

Thus h has the necessary continuity. The only question here concerns continuity at $t = 1$, but we need only insert the limiting values of the arguments to complete the argument. Checking the boundary conditions, we see that

$$\left.\begin{aligned} h(x, 0) &= f(2x) && \text{for } 0 \leq 1 - 2x \text{ or } 0 \leq x \leq \tfrac{1}{2} \\ &= y_0 && \text{for } 2x - 1 \leq 0 \leq 1 - 2x \text{ or } x = \tfrac{1}{2} \\ &= f\left(\frac{2x - 2}{-1}\right) = f(2 - 2x) && \text{for } 0 \leq 2x - 1 \text{ or } \tfrac{1}{2} \leq x \leq 1 \end{aligned}\right\} = f * \bar{f}$$

while

$$h(x, 1) = y_0 \qquad \text{for all } x \text{ satisfying the inequalities,}$$

and we note that the various limiting values agree. This suffices to show that the class $[\bar{f}]$ is the inverse under the operation "$\circ$" of the class $[f]$ and completes the proof that $\pi_1(Y, y_0)$ is a group. □

We notice that the fundamental group as defined seems to depend upon the *base point* y_0 in Y, and in general this is true. If, for instance, Y is the union of an annular region in E^2 and a disjoint disc in E^2 (see Fig. 4-6), then for y_0 (any point in the annular region), $\pi_1(Y, y_0)$ is infinite cyclic, whereas if y_1 is any point in the disc, $\pi_1(Y, y_1)$ consists only of the identity element. One notes that this example fails to be connected and might conjecture that for a connected space, the groups $\pi_1(Y, y_0)$ and $\pi_1(Y, y_1)$, $y_0 \neq y_1$, would necessarily be isomorphic. It is easy to modify the above example by simply adding a $\sin(1/x)$ curve as the broken line in Fig. 4-6, and so disprove this conjecture.

We do have the desired isomorphism in the case of an arcwise-connected space.

THEOREM 4–17. If Y is an arcwise-connected space, then for any pair of points y_0 and y_1 in Y, the groups $\pi_1(Y, y_0)$ and $\pi_1(Y, y_1)$ are isomorphic.

Proof: We give a brief sketch of the proof, leaving the details to be checked by the reader. Because Y is arcwise-connected, there is a homeomorphism p of I^1 into Y such that $p(0) = y_0$ and $p(1) = y_1$. Define $\overline{p}(x) = p(1 - x)$. It is easily shown that

$$[p*\overline{p}] = [j_0],$$

where $[j_0]$ is the identity element in $\pi_1(Y, y_0)$, and that

$$[\overline{p}*p] = [j_1],$$

where $[j_1]$ is the identity in $\pi_1(Y, y_1)$.

Now consider any element $[f]$ in $\pi_1(Y, y_0)$. Define the algebraic transformation $\lambda:\pi_1(Y, y_0) \to \pi_1(Y, y_1)$ given by

$$\lambda([f]) = [\overline{p}*f*p].$$

One easily sees that $\overline{p}*f*p$ is an element of the y_1-neighborhood of curves, $C(Y, y_1)$. To complete the proof, it must be shown that

(1) λ *is single-valued,* which entails proving that if $f \underset{y_0}{\simeq} g$, then $\overline{p}*f*p \underset{y_0}{\simeq} \overline{p}*g*p$;

(2) λ *is one-to-one,* which is shown by proving that if $\overline{p}*f*p \underset{y_1}{\simeq} \overline{p}*g*p$, then $f \underset{y_0}{\simeq} g$;

(3) λ *is onto,* which can be done by showing that any element $[f]$ of $\pi_1(Y, y_1)$ has a representative of the form $\lambda(p*f*\overline{p})$; and finally,

(4) λ *is a homomorphism,* which means that

$$\lambda([f] \circ [g]) = \lambda([f]) \circ \lambda([g]),$$

which is merely an exercise in the use of the definitions.

The details of the proof are largely routine, but they should be completed as a valuable exercise. □

In view of this result, we may suppress the role of the base point y_0 in discussing arcwise-connected spaces Y and simply refer to the fundamental group $\pi_1(Y)$. In general, this is not the case, and we have an entire system of groups $\pi_1(Y, y)$, one for each point y in Y. For now we only note that if C_{y_0} is the arcwise-connected component, of space Y, which contains y_0, then our proof of Theorem 4–17 shows that $\pi_1(Y, y_0)$ is $\pi_1(C_{y_0})$.

Although Theorem 4–17 states that $\pi_1(Y, y_0)$ and $\pi_1(Y, y_1)$ are isomorphic if Y is arcwise-connected, there is no canonical (uniquely defined) isomorphism between the two in the general case. It can be shown that there is such a canonical isomorphism if the fundamental group of the space is abelian (we will not do this). An arcwise-connected space whose fundamental group is abelian is called 1-*simple*. We will give examples of this concept shortly, but one important class of 1-simple spaces comprises the arcwise-connected topological groups. This is a corollary of our next result.

A space X is called a *Hopf space* if there exists a mapping $\varphi: X \times X \to X$ and a point p of X such that $\varphi(p, p) = p$ and such that both $\varphi(p, x): X \to X$ and $\varphi(x, p): X \to X$ are homotopic to the identity mapping, the homotopy leaving p fixed.

THEOREM 4–18. An arcwise-connected Hopf space has an abelian fundamental group.

Proof: Let $[f]$ and $[g]$ be two elements of $\pi_1(X, p)$. If we show that $f*g$ is homotopic to $g*f$, the homotopy leaving end points fixed at p, the theorem will be established. Consider the cube I^3. Define a mapping F on the base and sides of I^3 into $X \times X$ as follows. On the base, let

$$F(x, y, 0) = \varphi(f(x), g(y)).$$

Then on the four edges of the base, we have mappings as indicated in Fig. 4–7.

Next let h_1 and h_2 be the homotopies between $\varphi(x, p)$ and $\varphi(p, x)$ and the identity. That is,

$$h_1(x, 0) = \varphi(x, p), \qquad h_1(x, 1) = x,$$
$$h_2(x, 0) = \varphi(p, x), \qquad h_2(x, 1) = x;$$

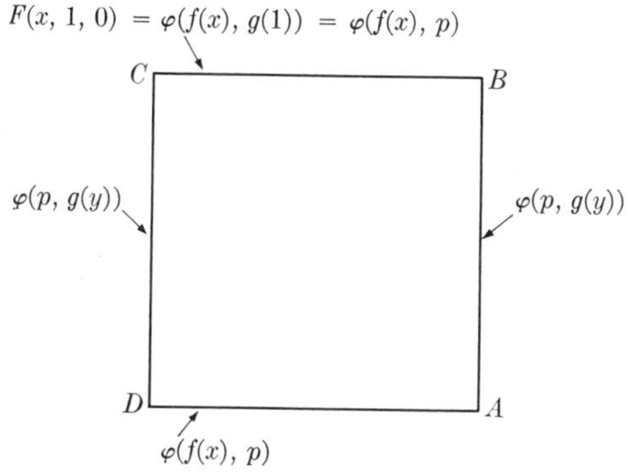

FIGURE 4–7

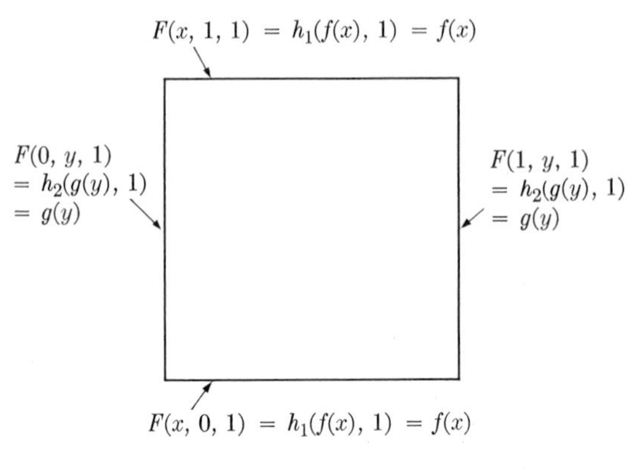

FIGURE 4–8

and
$$h_1(p, t) = h_2(p, t) = p \qquad \text{for all } t.$$

Then we can define F on the sides of I^3 by setting

$$F(x, 0, z) = h_1(f(x), z),$$
$$F(x, 1, z) = h_1(f(x), z),$$
$$F(0, y, z) = h_2(g(y), z),$$

and

$$F(1, y, z) = h_2(g(y), z).$$

It is clear that these agree with $F(x, y, 0)$ on the bottom edges of I^3 and that all are constant and equal to p on the vertical edges. Thus F is well-defined and continuous on the base and sides of I^3. Now the base and sides of I^3 obviously constitute a closed subset of I^3 which is homeomorphic to I^2. Since I^2 is an absolute retract, the base and sides of I^3 constitute a retract of I^3. Hence by Theorem 2–38, the mapping F can be extended to all of I^3, in particular, to the top of I^3. Let this extension still be denoted by F.

On the top edges of I^3, we have mappings as shown in Fig. 4–8. Now we want a mapping $H(x, t)$ on $I^1 \times I^1$ such that

$$H(x, 0) = f(2x) \qquad (0 \leq x \leq \tfrac{1}{2})$$
$$= g(2x - 1) \qquad (\tfrac{1}{2} \leq x \leq 1)$$

and

$$H(x, 1) = g(2x) \qquad (0 \leq x \leq \tfrac{1}{2})$$
$$= f(2x - 1) \qquad (\tfrac{1}{2} \leq x \leq 1).$$

The above square suggests that H might be obtained as follows. The de-

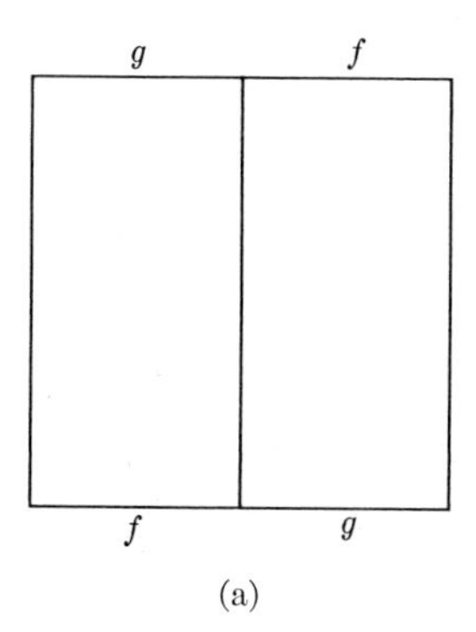

(a)

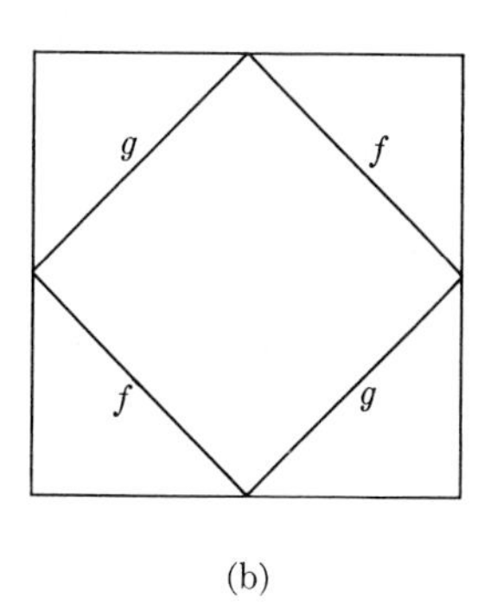

(b)

FIGURE 4–9

sired homotopy square in Fig. 4–9(a) can be squeezed down to form that in Fig. 4–9(b), which is the square we have above. A mapping that carries this out may be given by

$$\begin{aligned} H(x, t) &= F(2xt, 2x(1 - t), 1) && \text{for } 0 \leq x \leq \tfrac{1}{2} \\ &= F(x + (2t - 1)(1 - x), 1 - 2t(1 - x), 1) && \text{for } \tfrac{1}{2} \leq x \leq 1. \end{aligned}$$

It is easy to see that with x in the appropriate interval, the arguments of F are in the correct range. Also

$$F(2xt, 2x(1 - t), 1) = F(t, 1 - t, 1) \qquad \text{when } x = \tfrac{1}{2},$$

and

$$F(x + (2t - 1)(1 - x), 1 - 2t(1 - x) = F(t, 1 - t, 1) \qquad \text{when } x = \tfrac{1}{2}.$$

Thus H is well-defined and continuous. Finally,

$$\begin{aligned} H(x, 0) &= F(0, 2x, 1) \\ &= h_2(g(2x), 1) = g(2x) && (0 \leq x \leq \tfrac{1}{2}) \\ &= F(2x - 1, 1, 1) \\ &= h_1(f(2x - 1, 1) = f(2x - 1) && (\tfrac{1}{2} \leq x \leq 1) \end{aligned}$$

and

$$\begin{aligned} H(x, 1) &= F(2x, 0, 1) \\ &= h_1(f(2x), 1) = f(2x) && (0 \leq x \leq \tfrac{1}{2}) \\ &= F(1, 2x - 1, 1) \\ &= h_2(g(2x - 1), 1) = g(2x - 1) && (\tfrac{1}{2} \leq x \leq 1). \end{aligned}$$

This completes the proof. □

COROLLARY 4–19. Every arcwise-connected topological group has an abelian fundamental group.

Proof: We show that every arcwise-connected topological group G is a Hopf space. Let e be the identity element in G and define $\varphi: G \times G \to G$ by $\varphi(x, y) = x \cdot y$. Then $\varphi(e, e) = e$, and $\varphi(e, x) = e \cdot x = x$, and $\varphi(x, e) = x \cdot e = x$. Since the group multiplication is continuous, φ is continuous and G is a Hopf space. □

We claimed earlier that the fundamental group somehow reflects the connectivity structure of the underlying space. It will be difficult to envision this concept without a number of examples. We give examples without proof.

EXAMPLE 1. Spaces for which π_1 is trivial, i.e., consists only of the identity element:

(a) any contractible space; hence E^n, I^n, and any compact metric absolute retract,
(b) any sphere S^n, $n > 1$, and
(c) $E^3 - p$ (punctured 3-space).

EXAMPLE 2. Spaces for which π_1 is infinite cyclic:

(a) S^1,
(b) $E^2 - p$ (punctured plane),
(c) any annular region in E^2, and
(d) E^3 with a line removed.

EXAMPLE 3. Spaces whose fundamental group has two generators:

(a) The torus T. $\pi_1(T)$ is the direct sum of two infinite cyclic groups (see Theorem 4–23).

(b) The figure-8 curve and the doubly punctured plane. Each of these has the same fundamental group $\pi_1(X)$ which is a "free group" on two generators in the algebraist's meaning of the term "free group." To the algebraist a free group on the two generators a and b consists of all "words" of the form

$$a^{m_1}b^{n_1}a^{m_2}b^{n_2}\ldots a^{m_k}b^{n_k}$$

where m_i and n_i are integers and k is a natural number. Note that the only free group which is abelian is that on *one* generator. (In Section 6–5 we use another and different definition of the term "free group.")

It begins to seem that the number of "holes" in the space has some bearing upon the structure of its fundamental group. But note that the punctured 3-space does not have a hole as far as the fundamental group can determine! We might note here that *a simply-connected domain R, as used in analysis, is precisely a domain whose fundamental group is trivial.*

The above examples [except 3(a)] are all of 1-simple spaces. As another example of a space that is not 1-simple, having a nonabelian fundamental group, let J be a simple closed curve in E^3 which has been tied in an overhand knot (see Fig. 4-10, in Section 4-6). Consider the space $E^3 - J$. As an exercise, the reader may compute the generators of $\pi_1(E^3 - J)$ and find relations between those which imply that this group is not abelian.

There is just one other algebraic group used in topology which is not always abelian. This is a relative homotopy group, and we will consider it briefly in Section 7-8.

We arrive next at a concept of immense importance throughout algebraic topology. It is the idea that a homomorphism on groups of a space can be induced by a continuous mapping of a space. This will arise time and again as we progress. Thus the following development is an introduction to an extensive area of study.

If A is a closed subset of a space X, then we speak of the *pair* of spaces (X, A). By a *mapping* $f:(X, A) \to (Y, B)$ *of the pair* (X, A) into the pair (Y, B), we mean a mapping $f:X \to Y$ such that $f(A)$ is contained in B.

THEOREM 4-20. A mapping $h:(X, x_0) \to (Y, y_0)$ induces a homomorphism $h_*:\pi_1(X, x_0) \to \pi_1(Y, y_0)$.

Proof: Define a mapping $h_\#$ of the x_0-neighborhood of curves $C(X, x_0)$ into $C(Y, y_0)$ which takes each f in $C(X, x_0)$ into an element $h_\# f$ in $C(Y, y_0)$ given by

$$(h_\# f)(t) = h(f(t)).$$

To prove that $h_\#$ is continuous on the function space $C(W, x_0)$, let f_0 be any element of $C(X, x_0)$, and let U be any basis element in the compact-open topology of $C(Y, y_0)$ which contains $h_\# f_0$. Now by definition, U is the collection of all functions in $C(Y, y_0)$ that carry a compact set K into an open set O. So consider the basis element U^{-1} of $C(X, x_0)$ consisting of all functions carrying K into $h^{-1}(O)$. Now f_0 lies in this basis element since $[h_\# f](K)$ lies in O, and so $h(f(K))$ lies in O and $f(K)$ lies in $h^{-1}(O)$. On the other hand, if g lies in U^{-1}, then $g(K)$ lies in $h^{-1}(O)$ and $[h_\# g](K) = h(g(K))$ lies in O, so $h_\# g$ is an element of U. Thus $h_\#$ is continuous.

We define the induced homomorphism h_* by

$$h_*([f]) = [h_\# f].$$

Since $h_\#$ is continuous, it certainly carries arcwise-connected components of $C(X, x_0)$ into arcwise-connected components of $C(Y, y_0)$, so h_* is well-defined.

To prove that h_* is a homomorphism, we need only to show that

$$h_*([f] \circ [g]) = h_*([f]) \circ h_*([g]),$$

and this will follow if we show that

$$h_{\#}(f*g) = h_{\#}f*h_{\#}g.$$

But this is immediate, for

$$\begin{aligned}[h_{\#}(f*g)](x) &= h(f(2x)) = [h_{\#}f](2x) && (0 \le x \le \tfrac{1}{2}) \\ &= h(g(2x - 1)) = [h_{\#}g](2x - 1) && (\tfrac{1}{2} \le x \le 1) \\ &= [h_{\#}f*h_{\#}g](x).\end{aligned}$$

Passing to equivalence classes yields the homomorphism. □

Among the important properties of the induced homomorphism, we have those stated in the next result.

THEOREM 4–21. If f and g are homotopic mappings of (X, x_0) into (Y, y_0), then the induced homomorphisms coincide. If $f:(X, x_0) \to (Y, y_0)$ and $g:(Y, y_0) \to (Z, z_0)$, then $(gf)_* = g_*f_*$.

Proof: To show that the induced homomorphism depends only upon the homotopy class of the mapping, we need only point out that if f and g are homotopic, leaving the point x_0 fixed, then the mappings $f(\varphi(t))$ and $g(\varphi(t))$ are also homotopic leaving the point y_0 fixed. Thus $f_{\#}\varphi$ and $g_{\#}\varphi$ are homotopic and, by definition, $f_* = g_*$.

To prove the composition rule, let φ be any element of $C(X, x_0)$. Then

$$\begin{aligned}[(gf)_{\#}\varphi](t) &= (gf)(\varphi(t)) = g[f(\varphi(t))] \\ &= g_{\#}[f(\varphi(t))] = g_{\#}[f_{\#}(\varphi(t))] \\ &= [(g_{\#}f_{\#})\varphi](t).\end{aligned}$$

This obviously implies that $(gf)_* = g_*f_*$. □

COROLLARY 4–22. If (X, x_0) and (Y, y_0) are homotopically equivalent, then $\pi_1(X, x_0)$ and $\pi_1(Y, y_0)$ are isomorphic.

Proof: By definition, there exist mappings $f:X \to Y$ and $g:Y \to X$ such that both fg and gf are homotopic to the identity mappings $i:Y \to Y$ and $i:X \to X$, respectively. Hence both $(fg)_* = f_*g_*$ and $(gf)_* = g_*f_*$ are isomorphisms onto. Consider f_*. Since f_*g_* is onto, f_* must be onto, and since g_*f_* is an isomorphism, f_* must be an isomorphism. Therefore f_* is an isomorphism of $\pi_1(X, x_0)$ onto $\pi_1(Y, y_0)$. □

This corollary has a corollary, too. Namely, it is obvious from Corollary 4–22 that *two homeomorphic spaces have isomorphic fundamental groups.* Here we have our first example of an algebraic group associated with a space which is a topological invariant of that space. We must add however that Corollary 4–22 also proves that the fundamental group cannot characterize a topological space. By this we mean that two nonhomeomorphic spaces may well have isomorphic fundamental groups. This we have already seen by example, but it is worth pointing out here because we will see similar statements again in more complicated situations.

We quote one last result to be used as a comparison later on.

THEOREM 4–23. Let (X, x_0) and (Y, y_0) be pairs. Then the fundamental group $\pi_1(X \times Y, x_0 \times y_0)$ is isomorphic to the direct product $\pi_1(X, x_0) \otimes \pi_1(Y, y_0)$.

Proof: Let π_X and π_Y denote the projections of $X \times Y$ onto X and Y respectively. Then for any mapping f in $C(X \times Y, x_0 \times y_0)$, the mappings $\pi_X f$ and $\pi_Y f$ are in $C(X, x_0)$ and $C(Y, y_0)$ respectively. We define a transformation T of $\pi_1(X \times Y, x_0 \times y_0)$ into the direct product $\pi_1(X, x_0) \otimes \pi_1(Y, y_0)$ by setting

$$T([f]) = ([\pi_X f], [\pi_Y f]).$$

We show that T is the desired isomorphism as follows.

(1) T is well-defined. For suppose that $f_0 \underset{x_0 \times y_0}{\simeq} f_1$. Then by definition, there is a homotopy $H: I^1 \times I^1 \to X \times Y$ such that $H(t, 0) = f_0(t)$, $H(t, 1) = f_1(t)$, and $H(0, s) = x_0 \times y_0 = H(1, s)$, $0 \leq s \leq 1$. Consider the mappings $\pi_X f_0$ and $\pi_X f_1$ and the mapping $\pi_X H: I^1 \times I^1 \to X$. It is clear that we have

$$\pi_X H(t, 0) = \pi_X f_0(t), \qquad \pi_X H(t, 1) = \pi_X f_1(t),$$

and

$$\pi_X H(0, s) = \pi_X H(1, s) = \pi_X(x_0 \times y_0) = x_0 \qquad (0 \leq x \leq 1).$$

Thus $\pi_X H$ is a homotopy modulo x_0 between $\pi_X f_0$ and $\pi_X f_1$, and hence the class $[\pi_X f]$ is well-defined. A similar argument holds for $[\pi_Y f]$.

(2) T is onto. For if (g, h) is any pair in $C(X, x_0) \times C(Y, y_0)$, then the element f of $C(X \times Y, x_0 \times y_0)$ defined by

$$\begin{aligned} f(t) &= (g(2t), y_0) && (0 \leq t \leq \tfrac{1}{2}) \\ &= (x_0, h(2t - 1)) && (\tfrac{1}{2} \leq t \leq 1) \end{aligned}$$

clearly has the property $\pi_X f \underset{x_0}{\simeq} g$ and $\pi_Y f \underset{y_0}{\simeq} h$. That f is well-defined and continuous follows from the fact that $f(\frac{1}{2}) = (x_0, y_0)$ by each definition.

(3) T is one-to-one. For if $\pi_X f_0 \underset{y_0}{\simeq} \pi_X f_1$ and $\pi_Y f_0 \underset{y_0}{\simeq} \pi_Y f_1$ via homotopies $h_1(t, s)$ and $h_2(t, s)$, then we may define a homotopy $H: I^1 \times I^1 \to X \times Y$ by

$$H(t, s) = (h_1(t, s), h_2(t, s)).$$

Certainly H is continuous, being continuous into each factor of $X \times Y$. Also we see that

$$H(t, 0) = (\pi_X f_0(t), \pi_Y f_0(t)) = f_0(t)$$

and

$$H(t, 1) = (\pi_X f_1(t), \pi_Y f_1(t)) = f_1(t),$$

while

$$H(0, s) = (h_1(0, s), h_2(0, s)) = (x_0, y_0)$$

and

$$H(1, s) = (h_1(1, s), h_2(1, s)) = (x_0, y_0).$$

(4) T is a homomorphism. For if $[f]$ and $[g]$ are elements of $\pi_1(X \times Y, x_0 \times y_0)$ then

$$\begin{aligned} T([f] \circ [g]) &= T([f*g]) = ([\pi_X(f*g)], [\pi_Y(f*g)]) \\ &= ([\pi_X f*\pi_X g], [\pi_Y f*\pi_Y g]) \\ &= ([\pi_X f] \circ [\pi_X g], [\pi_Y f] \circ [\pi_Y g]). \square \end{aligned}$$

4–6 Knots and related imbedding problems. Two simple closed curves in E^3 may be said to be equivalent if there is an orientation-preserving homeomorphism of E^3 onto itself which throws one curve onto the other. Then a simple closed curve J is *unknotted* if it is equivalent to the plane circle in E^3 with equation $x_1^2 + x_2^2 = 1$, $x_3 = 0$; otherwise J is *knotted* or is a *knot*. These definitions lead to equivalence classes of knots in the obvious way, and the chief problem of knot theory is to find topological properties that will serve to classify these equivalence classes.

Since each knot, as a subspace of E^3, is a simple closed curve, we see that the knots themselves are all homeomorphic. They only differ in the manner in which they are imbedded in E^3. This observation leads one to a study of the complement of a knot. Let J be a knot in E^3. The fundamental group $\pi_1(E^3 - J)$ of the complement of J is called *the group of the knot J*. Will this group serve to classify knots? The answer must be negative. For consider the knots J_1 and J_2 in Fig. 4–10. It should be

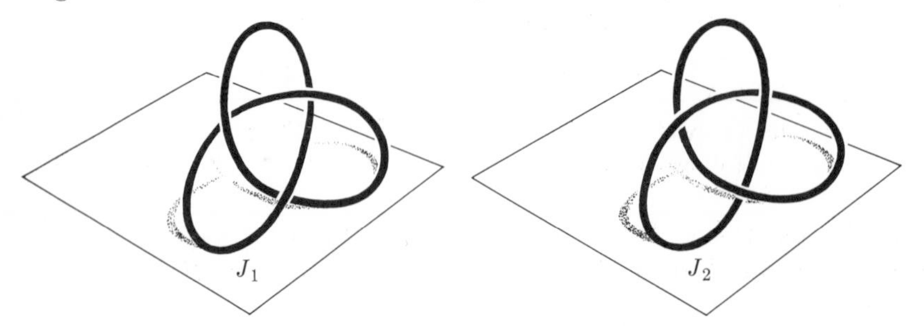

FIG. 4–10. Right- and left-hand trefoil knots.

obvious that their groups are isomorphic, and yet there is no *orientation-preserving* homeomorphism of E^3 onto itself which carries J_1 onto J_2.

One is tempted by this example to eliminate the "orientation-preserving" requirement and try again. But this does not help either. There exist nonequivalent knots having isomorphic groups even without the orientation requirement. This means that the group of a knot cannot fully characterize the equivalence class of that knot. In practice, a knot theorist uses topology, combinatorial analysis, differential geometry, and anything else he finds applicable. Such attacks have produced a wealth of information but no complete solution. The reader is referred to Reidemeister [30], whose book *Knotentheorie* contains the basic work. Then recent surveys by Fox [79] and Seifert and Threlfall [119] will carry the interested reader up to the point of studying the current literature.

When first confronted with the problem of knots in E^3, one rarely sees its significance and may tend to dismiss the topic as being of limited interest. We give the ensuing discussion to place knot theory in its properly important place. In essence, we are faced here with the problem of extending a given mapping. For if J_1 and J_2 are two knots in E^3, there is a homeomorphism $h: J_1 \to J_2$. Indeed there are many such homeomorphisms. The question of the equivalence of J_1 and J_2 is then, does there exist a homeomorphism $\tilde{h}$ of E^3 onto itself such that $\tilde{h}|J_1 = h$? Viewed in this light, the general problem (of which knot theory is a part) may be considered to have been initiated by Schoenflies [32].

In 1908, Schoenflies proved the following result (which is paraphrased here): *let J be a simple closed curve in the plane E^2, and let h be a homeomorphism of J onto the unit circle S^1 in E^2. Then h may be extended to a homeomorphism $\tilde{h}$ of E^2 onto itself.* In other words, there are no knots in the plane. The very existence of knots in E^3 constitutes a major hurdle in generalizing any result from E^2 to E^3. For instance, the Schoenflies theorem above cannot be generalized by replacing E^2 by E^3.

There is another natural way in which we might try to generalize the Schoenflies theorem, and this attempt leads to further problems. Let S be a simple closed surface in E^3, that is, S is a homeomorph of S^2, and let h be a homeomorphism of S onto the unit sphere S^2 in E^3. Is there an extension $\tilde{h}$ of h such that $\tilde{h}$ is a homeomorphism of E^3 onto itself? In the special case that S is a finite polytope (see Chapter 5) in E^3, Alexander [46] was able to give an affirmative answer to this question. At the same time, however, he gave a famous example, the *Alexander horned sphere*, showing that the answer must be "no" in the general case. This example is pictured in Fig. 4–11. We can see from the picture alone that it is quite obvious that the complement of the horned sphere is not simply connected. Since the complement of S^2 in E^3 *is* simply connected, it follows that no homeomorphism of E^3 onto itself will throw the horned sphere

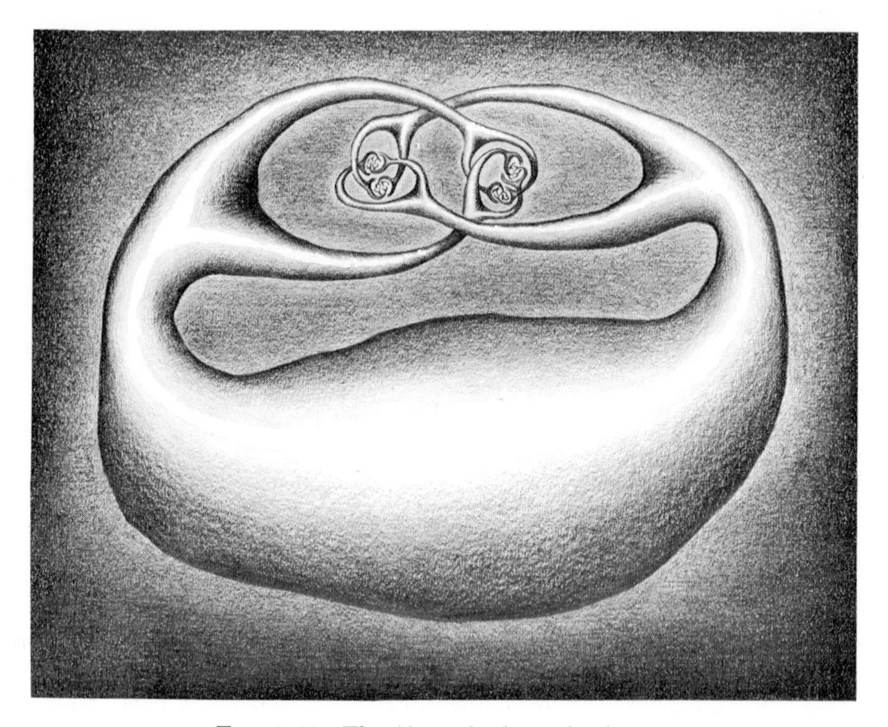

FIG. 4–11. The Alexander horned sphere.

onto S^2. Note that there is a Cantor set of "bad" points on the horned sphere.

Alexander's work was published in 1924. The problem has been revived recently, and further results have been obtained. In 1948, Artin and Fox [50] were led to the following definition. Let P be a homeomorph of a finite polytope P', both imbedded in E^n. If there is a homeomorphism of E^n onto itself which carries P onto P', then P is said to be *tamely imbedded* (or *tame*) in E^n; otherwise P is *wildly imbedded* (or *wild*) in E^n. In these terms, the Schoenflies theorem may be paraphrased as, "Every homeomorph of a polytope in E^2 is tame." And, of course, the Alexander horned sphere is wild in E^3.

Artin and Fox then proceeded to give a number of surprising examples. In Fig. 4–12 we picture one of these, a *wild arc* in E^3. This shows that even the most simple polytope, the closed interval, may be wildly imbedded in E^3! We remark that by "swelling" the arc in Fig. 4–12 into a tube tapering to the two points p and q, we obtain another wild sphere, this example having only two bad points.

The examples of Artin and Fox inspired a renewed attack upon the difficult problem of extending the Schoenflies theorem. Recent papers by

FIG. 4–12. A simple arc in E^3 whose complement is not simply connected.

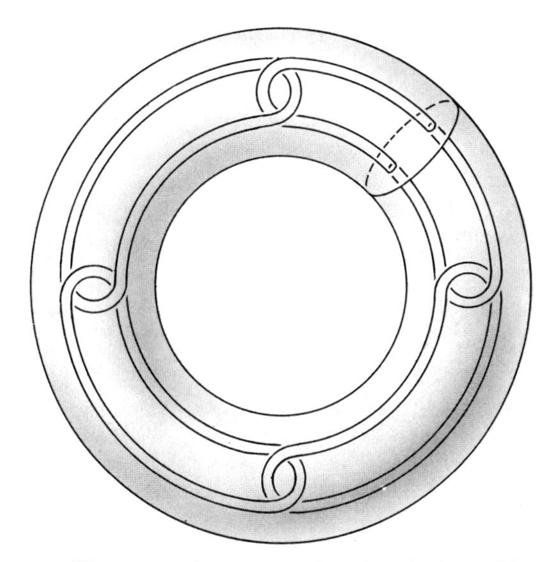

FIG. 4–13. First stage in constructing Antoine's necklace.

Bing [62], Harrold [82], Moise [104], Mazur [100], Brown [72a], and many others have contributed much new knowledge, particularly regarding wild sets in E^3. A theorem due to Klee [90], for instance, may be used to prove that any simple closed curve in E^3 is tame in E^5.

Another wild set, Antoine's necklace, deserves mention before we leave this subject if only for historical interest. Let T be a solid torus, and let T_1, T_2, T_3, and T_4 be four solid tori imbedded in T and linked as shown in Fig. 4–13. In each T_i, let T_{i1}, T_{i2}, T_{i3}, and T_{i4} be four solid tori imbedded and linked in T_i as the T_i are imbedded and linked in T. This imbedding of tori is done for each positive integer k. At the kth step, we will have 4^k tori, whose union we denote by A_k. *Antoine's necklace* is the intersection $\bigcap_{k=1}^{\infty} A_k$ of all the sets A_k. Since each A_k is compact and A_k contains A_{k+1} for each k, these sets satisfy the finite intersection hypothesis and their intersection is nonempty. By construction, it is obvious that the components of the necklace are single points. In fact, it is not difficult to prove that this set is a totally disconnected, compact, perfect metric space. Hence by Corollary 2–98, Antoine's necklace is homeomorphic to the Cantor set. One easily sees that the complement of the necklace in E^3 is not simply connected. For such a simple closed curve as J in Fig. 4–13 cannot be deformed to a point in the complement. And there are infinitely many simple closed curves in the complement which are not deformable into each other [65]!

We leave this topic with the observation that it is among the most active of the current research problems in topology. In particular, the study of dimensions greater than three is practically untouched.

4–7 The higher homotopy groups. The consistent use of the symbol π_1 for the fundamental group should have suggested that $\pi_2, \pi_3, \ldots,$ must be defined. These so-called *higher homotopy groups* were invented by Hurewicz [85] in 1935. We give a brief introduction to this concept here and return to it again in Section 7–8.

Again we consider a space Y and a particular base point y_0. In a generalization of the y_0-neighborhood of curves $C(Y, y_0)$, we consider mappings $f:I^n \to Y$ of the n-cube into Y such that f throws the boundary of I^n onto the point y_0. We recall that I^n is taken to be the collection of all n-tuples $(x_1, x_2, \ldots, x_n)$ of real numbers such that $0 \leq x_i \leq 1$, $i = 1, 2, \ldots, n$. The boundary βI^n of I^n consists of all such n-tuples such that the product $\Pi_{i=1}^n x_i(1 - x_i) = 0$. This simply says that at least one coordinate in the n-tuple equals either zero or one. Thus we consider the collection $C_n(Y, y_0)$ of all mappings $f:I^n \to Y$ such that $f(\beta I^n) = y_0$. Clearly, $C_n(Y, y_0)$ is a subset of Y^{I^n}, and we may topologize it with the compact-open topology.

To define a homotopy relation in $C_n(Y, y_0)$, we say that f and g are homotopic modulo y_0, $f \underset{y_0}{\simeq} g$, provided there is a continuous mapping $h:I^n \times I^1 \to Y$ such that

$$h(x, 0) = f(x) \qquad \text{for all } x \text{ in } I^n,$$
$$h(x, 1) = g(x) \qquad \text{for all } x \text{ in } I^n,$$

and

$$h(\beta I^n, t) = y_0 \qquad \text{for } 0 \leq t \leq 1.$$

It is easily shown that this is an equivalence relation on $C_n(Y, y_0)$. Details of this proof are very similar to that of Lemma 4–16 and may be carried out as an exercise. It follows that $C_n(Y, y_0)$ is decomposed into disjoint equivalence classes, which are the arcwise-connected components of $C_n(Y, y_0)$.

The *juxtaposition* of two mappings f and g in $C_n(Y, y_0)$ is similar to that in Section 4–5. We define

$$\begin{aligned}(f*g)(x_1, x_2, \ldots, x_n) &= f(2x_1, x_2, \ldots, x_n) && (0 \leq x_1 \leq \tfrac{1}{2})\\ &= g(2x_1 - 1, x_2, \ldots, x_n) && (\tfrac{1}{2} \leq x_1 \leq 1).\end{aligned}$$

Since at $x_1 = \frac{1}{2}$, we have $f(1, x_2, \ldots, x_n) = y_0 = g(0, x_2, \ldots, x_n)$, the mapping $f{*}g$ is a well-defined element of $C_n(Y, y_0)$.

The *nth homotopy group of Y at the point* y_0, $\pi_n(Y, y_0)$, is defined as having elements that are the arcwise-connected components of $C_n(Y, y_0)$ and having the group operation given by

$$[f] \circ [g] = [f{*}g],$$

the heavy brackets again denoting equivalence classes. Of course, it is necessary to prove that this operation is well-defined, the result depending only upon the equivalence classes and not upon the representatives used, and that this operation satisfies the axioms for a group.

We show only that if $f \underset{y_0}{\simeq} f_1$ and $g \underset{y_0}{\simeq} g_1$, then $f{*}g \underset{y_0}{\simeq} f_1{*}g_1$. This will prove that the operation "$\circ$" is well-defined. By definition, there are homotopies h_1 and h_2 such that

$$\begin{aligned} h_1(x, 0) &= f(x), & h_2(x, 0) &= g(x), \\ h_1(x, 1) &= f_1(x), & h_2(x, 1) &= g_1(x), \\ h_1(\beta I^n, t) &= y_0, & h_2(\beta I^n, t) &= y_0. \end{aligned}$$

Define the mapping

$$\begin{aligned} h(x_1, x_2, \ldots, x_n, t) &= h_1(2x_1, x_2, \ldots, x_n, t) & (0 \le x_1 \le \tfrac{1}{2}) \\ &= h_2(2x_1 - 1, x_2, \ldots, x_n, t) & (\tfrac{1}{2} \le x_1 \le 1). \end{aligned}$$

At $x = \frac{1}{2}$, we have $h_1(1, x_2, \ldots, x_n, t) = y_0 = h_2(0, x_2, \ldots, x_n, t)$, so h is well-defined and continuous. Furthermore,

$$\begin{aligned} h(x_1, x_2, \ldots, x_n, 0) &= h_1(2x_1, x_2, \ldots, x_n, 0) & \\ &= f(2x_1, x_2, \ldots, x_n) & (0 \le x_1 \le \tfrac{1}{2}) \\ &= h_2(2x_1 - 1, x_2, \ldots, x_n, 0) & \\ &= g(2x_1 - 1, x_2, \ldots, x_n) & (\tfrac{1}{2} \le x_1 \le 1), \end{aligned}$$

which is the definition of $f{*}g$, and

$$\begin{aligned} h(x_1, x_2, \ldots, x_n, 1) &= h_1(2x_1, x_2, \ldots, x_n, 1) & \\ &= f_1(2x_1, x_2, \ldots, x_n) & (0 \le x_1 \le \tfrac{1}{2}) \\ &= h_2(2x_1 - 1, x_2, \ldots, x_n, 1) & \\ &= g_1(2x_1 - 1, x_2, \ldots, x_n) & (\tfrac{1}{2} \le x_1 \le 1), \end{aligned}$$

which is $f_1{*}g_1$. Thus h is a homotopy between $f{*}g$ and $f_1{*}g_1$.

The remainder of the group axioms are established in the same way as was done for $\pi_1(Y, y_0)$. The associative law will hold if it is shown that (1)

$$(f_1 * f_2) * f_3 \underset{y_0}{\simeq} f_1 * (f_2 * f_3).$$

The constant mapping $c(I^n) = y_0$ is proved to represent the identity element by showing that (2) $f * c \underset{y_0}{\simeq} f$. The mapping $\bar{f}(x_1, x_2, \ldots, x_n) = f(1 - x_1, x_2, \ldots, x_n)$ represents the inverse of $[f]$, and this is proved by showing that (3) $f * \bar{f} \underset{y_0}{\simeq} c$. Details are again left as an exercise.

There is one further property of $\pi_n(Y, y_0)$, $n > 1$, not necessarily shared by $\pi_1(Y, y_0)$.

THEOREM 4–24. $\pi_n(Y, y_0)$, $n > 1$, is an abelian group.

Proof: This may be established by showing that for any pair f and g in $C_n(Y, y_0)$, $n > 1$, we have

$$f * g \underset{y_0}{\simeq} g * f.$$

We indicate a proof as follows. Consider the x_1x_2-face of I^n. The mappings f and g in $f * g$ are arranged as in Fig. 4–14. Of course $(f * g)(x_1, x_2) = y_0$ for every point on the four edges and on the vertical center line of this face.

Consider the mapping

$$
\begin{aligned}
&h_1(x_1, x_2, \ldots, x_n, t) \\
&\quad = f\left(2x_1, \frac{2x_2}{2 - t}, x_3, \ldots, x_n\right) && (0 \le x_1 \le \tfrac{1}{2}) && (0 \le x_2 \le 1 - \tfrac{1}{2}t) \\
&\quad = y_0 && (0 \le x_1 \le \tfrac{1}{2}) && (1 - \tfrac{1}{2}t \le x_2 \le 1) \\
&\quad = g\left(2x_1 - 1, \frac{2x_2 - t}{2 - t}, x_3, \ldots, x_n\right) \\
&&& (\tfrac{1}{2} \le x_1 \le 1) && (\tfrac{1}{2}t \le x_2 \le 1) \\
&\quad = y_0 && (\tfrac{1}{2} \le x_1 \le 1) && (0 \le x_2 \le \tfrac{1}{2}t).
\end{aligned}
$$

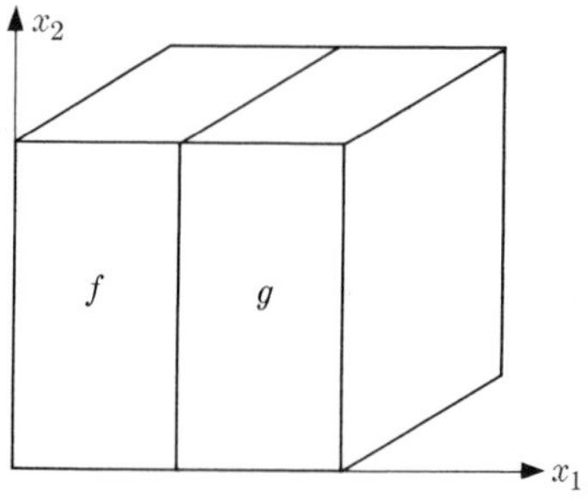

FIGURE 4–14

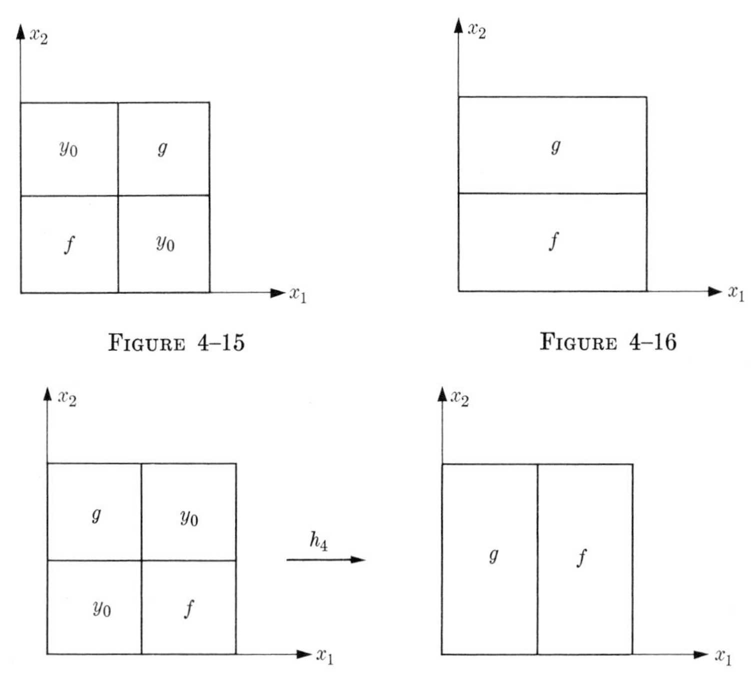

FIGURE 4-15

FIGURE 4-16

FIGURE 4-17

It is readily seen that h_1 is well-defined and continuous. Also

$$\begin{aligned} h_1(x_1, x_2, \ldots, x_n, 0) & \\ = f(2x_1, x_2, \ldots, x_n) & \quad (0 \leq x_1 \leq \tfrac{1}{2}) \quad (0 \leq x_2 \leq 1) \\ = g(2x_1 - 1, x_2, \ldots, x_n) & \quad (\tfrac{1}{2} \leq x_1 \leq 1) \quad (0 \leq x_2 \leq 1), \end{aligned}$$

which is $f*g$, and

$$\begin{aligned} h_1(x_1, x_2, \ldots, x_n, 1) & \\ = f(2x_1, 2x_2, x_3, \ldots, x_n) & \quad (0 \leq x_1 \leq \tfrac{1}{2}) \quad (0 \leq x_2 \leq \tfrac{1}{2}) \\ = y_0 & \quad (0 \leq x_1 \leq \tfrac{1}{2}) \quad (\tfrac{1}{2} \leq x_2 \leq 1) \\ = g(2x_1 - 1, 2x_2 - 1, x_3, \ldots, x_n) & \quad (\tfrac{1}{2} \leq x_1 \leq 1) \quad (\tfrac{1}{2} \leq x_2 \leq 1) \\ = y_0 & \quad (\tfrac{1}{2} \leq x_1 \leq 1) \quad (0 \leq x_2 \leq 1). \end{aligned}$$

Therefore h_1 is a homotopy between $f*g$ and a new mapping, which on the x_1x_2-face of I^n is as in Fig. 4-15. Working now on the x_1-coordinates only, it should be intuitively clear that we can construct a homotopy h_2 to yield the picture in Fig. 4-16. Then there are similar homotopies h_3 and h_4, which yield the diagrams in Fig. 4-17. It is a simple application of analytic geometry (albeit tedious) to construct those homotopies, and the reader should do so. We give a more sophisticated proof of this result shortly. □

We now have defined the higher homotopy groups, but the problem of determining these groups for a given space has not been touched here. In fact, this determination of homotopy groups is very difficult. For instance, it is only very recently that many of the groups $\pi_n(S^k, y_0)$, $n > k$, of the k-sphere have been determined. We will not attempt the calculation of homotopy groups, but will give some examples of known groups later.

The y_0-neighborhood of curves $C(Y, y_0)$, in the compact-open topology, is a space in its own right, the *space of loops* at y_0. In general, it will not be connected. If Y is a torus, for example, each homotopy class in $C(Y, y_0)$ will be a component of the loop space. These homotopy classes are arcwise connected (why?), and in fact are sets in $C(Y, y_0)$ maximal with respect to the property of being arcwise connected.

Let $c(I^1) = y_0$ be the constant mapping in $C(Y, y_0)$, and consider the pair $(C(Y, y_0), c)$. Again we may consider the space of curves $C(C(Y, y_0), c)$ consisting of all mappings $\varphi: I^1 \to C(Y, y_0)$ such that $\varphi(0) = \varphi(1) = c$, the constant mapping. Given any element φ in $C(C(Y, y_0), c)$, each functional value $\varphi(t)$ is itself a mapping $\varphi(t): I^1 \to Y$ such that $[\varphi(t)](0) = y_0$ and $[\varphi(t)](1) = y_0$.

On I^2, we may consider the mapping defined for each point (x_1, x_2) as $[\varphi(x_1)](x_2)$. Clearly, $[\varphi(x_1)](x_2)$ carries I^2 into Y and has the property that

$$[\varphi(0)](x_2) = c(x_2) = y_0,$$

$$[\varphi(1)](x_2) = c(x_2) = y_0,$$

$$[\varphi(x_1)](0) = y_0,$$

and

$$[\varphi(x_1)](1) = y_0.$$

It is easy to see that $[\varphi(x_1)](x_2)$ is continuous on I^2. Hence to each element of $C(C(Y, y_0), c)$, there corresponds an element of $C_2(Y, y_0)$. It should not be too surprising that we have the following result.

THEOREM 4–25. $\pi_2(Y, y_0)$ is isomorphic to $\pi_1(C(Y, y_0), c)$.

Proof: Consider the transformation λ which carries an element $[\varphi]$ of $\pi_1(C(Y, y_0), c)$ onto the element of $\pi_2(Y, y_0)$ given by

$$\lambda([\varphi]) = [[\varphi(x_1)](x_2)].$$

To show that λ is well-defined, let $\varphi_1 \underset{c}{\simeq} \varphi_2$ in $C(C(Y, y_0), c)$. This means that there is a homotopy $h(x, t): I^1 \times I^1 \to C(Y, y_0)$ such that

$$h(x, 0) = \varphi_1(x),$$

$$h(x, 1) = \varphi_2(x),$$

and

$$h(0, t) = h(1, t) = c.$$

Define the mapping

$$h'(x, y, t) = [h(x, t)](y).$$

Then one checks that

$$h'(x, y, 0) = [h(x, 0)](y) = [\varphi_1(x)](y)$$

and

$$h'(x, y, 1) = [h(x, 1)](y) = [\varphi_2(x)](y)$$

and that

$$h'(0, y, r) = [h(0, t)](y) = c(y) = y_0,$$
$$h'(1, y, r) = [h(1, t)](y) = c(y) = y_0,$$
$$h'(x, 0, t) = [h(x, t)](0) = y_0,$$

and

$$h'(x, 1, t) = [h(x, t)](1) = y_0.$$

Therefore we have $[\varphi_1(x)](y) \underset{y_0}{\simeq} [\varphi_2(x)](y)$, which shows that λ *is well-defined.*

A reversal of the above construction will prove that λ *is one-to-one.* Thus if $[\varphi_1(x)](y) \underset{y_0}{\simeq} [\varphi_2(x)](y)$, there is a homotopy $h'(x, y, t)$ such that

$$h'(x, y, 0) = [\varphi_1(x)](y),$$
$$h'(x, y, 1) = [\varphi_2(x)](y),$$

and

$$h'(\beta I^2, t) = y_0.$$

We define the mapping

$$h(x, t) = h'(x, y, t) \qquad (0 \leq y \leq 1).$$

For every pair (x, t), this is a mapping of I^1 into Y. Then

$$h(x, 0) = h'(x, y, 0) = [\varphi_1(x)](y),$$
$$h(x, 1) = h'(x, y, 1) = [\varphi_2(x)](y),$$

and also

$$h(0, t) = h'(0, y, t) = y_0 = h'(1, y, t) = h(1, t).$$

That is, $h(0, t)$ and $h(1, t)$ are constant mappings. Thus h is a homotopy between φ_1 and φ_2 in $C(Y, y_0)$.

Given $F(x, y)$ in $C_2(Y, y_0)$, we simply define

$$\varphi(x) = f(x, y) \qquad (0 \leq y \leq 1).$$

This is certainly a mapping of $x \times I^1$ for each x so that φ is a mapping of I^1 into $C(Y, y_0)$, and clearly $[\varphi(x)](y) = f(x, y)$. Thus λ *is onto.*

Finally, consider $\varphi_1 * \varphi_2$. By definition,

$$(\varphi_1 * \varphi_2)(x) = \varphi_1(2x) \qquad (0 \leq x \leq \tfrac{1}{2})$$
$$= \varphi_2(2x - 1) \qquad (\tfrac{1}{2} \leq x \leq 1).$$

Therefore

$$[(\varphi_1 * \varphi_2)(x)](y) = [\varphi_1(2x)](y) \qquad (0 \leq x \leq \tfrac{1}{2})$$
$$= [\varphi_2(2x - 1)](y) \qquad (\tfrac{1}{2} \leq x \leq 1)$$
$$= [\varphi_1(x)](y) * [\varphi_2(x)](y),$$

and we have shown that λ is an isomorphism. □

COROLLARY to Theorem 4–25. The group $\pi_n(Y, y_0)$ is isomorphic to $\pi_{n-1}(C(Y, y_0), c)$, $n > 1$.

We leave the proof as an exercise; it is very much like that for Theorem 4–25.

This suggests that we can give an alternative definition of the higher homotopy groups. We begin with the pair (Y, y_0), let Ω_1 be the space of loops $C(Y, y_0)$, Ω_2 be the space $C(C(Y), c)$, Ω_3 be the loop space over Ω_2, etc. Then our corollary implies that $\pi_n(Y, y_0)$ is isomorphic to $\pi_1(\Omega_{n-1})$. We could have taken this as our definition of π_n, which, historically, is what Hurewicz did in his original papers [85]. There are advantages in this approach, as the next two proofs show. Still another approach to the definition, due to Serre [120], will be discussed in Section 4–8.

THEOREM 4–26. For any pair (Y, y_0), $C(Y, y_0)$ is a Hopf space.

Proof: We consider the product space $C(Y, y_0) \times C(Y, y_0)$ and the constant mapping c. Define a mapping φ of the product space onto $C(Y, y_0)$ by setting

$$\varphi(f, g) = f * g.$$

Clearly,

$$\varphi(c, c) = c * c = c,$$
$$\varphi(c, g) = c * g \underset{y_0}{\simeq} g,$$

and

$$\varphi(f, c) = f * c \underset{y_0}{\simeq} f.$$

Therefore $C(Y, y_0)$ will be shown to be a Hopf space if we show that φ is continuous. To do this, let U be a member of the basis in $C(Y, y_0)$. By definition, U is the collection of all mappings in $C(Y, y_0)$ that carry a compact set K in I^1 into an open set O in Y. Thus $\varphi^{-1}(U)$ is the set of all pairs f, g such that $f * g$ carries K into O. But if $f * g$ carries K into O, then either f or g carries K into O, and conversely, if either f or g carries K into O, so does $f * g$. Therefore $\varphi^{-1}(U) = U \times C(Y, y_0) \cup C(Y, y_0) \times U$, and this is a basis element in the product space $C(Y, y_0) \times C(Y, y_0)$. □

COROLLARY 4–27. For $n > 1$, $\pi_n(Y, y_0)$ is abelian.

Proof: Using the new loop-space definition of the higher homotopy groups, for $n > 1$, $\pi_n(Y, y_0)$ is the fundamental group of a Hopf space and Theorem 4–18 applies. □

We state the following important theorem without proof. It may be proved by techniques quite similar to those used in Theorems 4–20 and 4–21.

THEOREM 4–28. *Let $h:(X_1, x_0) \to (Y, y_0)$ be continuous. Then h induces a homomorphism $h_*:\pi_n(X, x_0) \to \pi_n(Y, y_0)$ such that (1) if h is the identity mapping i, then i_* is the identity isomorphism, (2) if h and h' are homotopic mappings, then $h_* = h'_*$, and (3) if $h:(X, x_0) \to (Y, y_0)$ and $h':(Y, y_0) \to (Z, z_0)$, then $(h'h)_* = h'_*h_*$.*

As we will note in Section 7–8, this result tells us that homotopy theory satisfies some of the Eilenberg-Steenrod axioms for homology. In analogy to Corollary 4–22, we have the immediate corollary below.

COROLLARY 4–29. *If (X, x_0) and (Y, y_0) are homotopically equivalent, then $\pi_n(X, x_0)$ and $\pi_n(Y, y_0)$ are isomorphic for each $n \geq 1$.*

Of course it follows that homeomorphic pairs have isomorphic homotopy groups and hence that *the homotopy groups are topological invariants.* One further remark is in order before examining a few examples. In defining the fundamental group, we considered mapping $f:I^1 \to Y$ such that $f(0) = f(1) = y_0$. It is easy to see that this is equivalent to studying mappings $f:S^1 \to Y$ such that a fixed point s_0 of S^1 always maps onto y_0. That is, we could identify the (two) points in the boundary of I^1 first, thus obtaining S^1, and then map into the pair (Y, y_0). In the general case, too, mappings of the pair $(I^n, \beta(I^n))$ into (Y, y_0) are equivalent to mappings of (S^n, s_0) into (Y, y_0), where s_0 is some fixed point of S^n. For, identifying the points of $\beta(I^n)$ to a single point yields a space homeomorphic to S^n (see Section 3–6). Such a formulation of the homotopy groups is sometimes more convenient than that which we have given. It may be found in detail in a paper by Eilenberg [44].

Let us now examine a few examples. Theorem 4–7 clearly applies to give us the fact that *for any contractible space Y, all homotopy groups $\pi_n(Y, y_0)$* are trivial, i.e., consist of the identity element only. This applies to Euclidean cubes I^n, the Hilbert cube I^ω, and in view of Theorem 4–11, to any retract of a contractible space.

It is fairly easy to compute the homotopy groups $\pi_k(S^n, s_0)$, $k \leq n$. In particular, for $k < n$, $\pi_k(S^n, s_0)$ is trivial, while $\pi_n(S^n, s_0)$ is infinite cyclic. These facts will be established in Section 6–14. On the other hand, it has been a difficult and important problem in homotopy theory to determine the groups $\pi_k(S^n, s_0)$ for $k > n$. (Henceforth, we suppress the base point s_0 in our symbol.) By definition (see Section 6–4), the higher *homology* groups of S^n are trivial, and it is natural to ask if this might not be the case for $\pi_k(S^n)$, $k > n$. Equivalently, is every mapping of S^k into S^n inessential for $k > n$? The following example, due to H. Hopf [83], provides a negative answer to this question.

Let S^3 be the unit sphere in E^4 referred to rectangular coordinates, and let S^2 be the unit sphere in E^3 referred to spherical coordinates. For each point $(1, \alpha, \beta)$, $\beta \neq 0$, of S^2, there is a unique 2-plane in E^4 having the equations

$$x_3 = 2x_1 \cdot \cos \alpha \cdot \cot \beta/2 - 2x_2 \cdot \sin \alpha \cdot \cot \beta/2$$

and

$$x_4 = 2x_1 \cdot \sin \alpha \cdot \cot \beta/2 + 2x_2 \cdot \cos \alpha \cdot \cot \beta/2.$$

The "north pole" $(1, \alpha, 0)$ of S^2 corresponds to the plane $x_1 = 0 = x_2$.

Each of these 2-planes intersects S^3 in a circle $S^1(\alpha, \beta)$. These circles are disjoint, for if either $\alpha_1 \neq \alpha_2$ or $\beta_1 \neq \beta_2$, then the 2-planes corresponding to the points $(1, \alpha_1, \beta_1)$ and $(1, \alpha_2, \beta_2)$ intersect only at the origin in E^4. Indeed, it is possible to show that these circles constitute an upper semicontinuous collection of continua filling up S^3. We define the (monotone) mapping $f[S^1(\alpha, \beta)] = (1, \alpha, \beta)$. This is an *essential mapping* of S^3 onto S^2.

For suppose that f is homotopic to a constant mapping $c(S^3) = (1, \alpha_0, \beta_0)$ via a homotopy $h: S^3 \times I^1 \to S^2$. Given a point $(1, \alpha, \beta)$ of S^2, we choose the point of $S^1(\alpha, \beta)$ in which this circle intersects the 3-dimensional half-space $x_1 = x_2 \geq 0$. In particular, this point has coordinates

$$x_1 = x_2 = \left(\frac{1 + \cos \beta}{2(5 - 3 \cos \beta)}\right)^{1/2},$$

$$x_3 = 2 (\cos \alpha - \sin \alpha) \left(\frac{1 - \cos \beta}{2(5 - 3 \cos \beta)}\right)^{1/2},$$

and

$$x_4 = 2 (\sin \alpha + \cos \alpha) \left(\frac{1 - \cos \beta}{2(5 - 3 \cos \beta)}\right)^{1/2}.$$

Let this point be denoted by $y(\alpha, \beta)$. We define a mapping of $S^2 \times I^1$ onto S^2 by setting

$$\lambda[(1, \alpha, \beta), t] = h[y(\alpha, \beta), t].$$

It is easy to show that λ is continuous [one need only show that $y(\alpha, \beta)$ is a continuous function of the point $(1, \alpha, \beta)$]. Then we have

$$\lambda[(1, \alpha, \beta), 0] = h[y(\alpha, \beta), 0] = f[y(\alpha, \beta)] = (1, \alpha, \beta)$$

and

$$\lambda[(1, \alpha, \beta), 1] = h[y(\alpha, \beta), 1] = c[y(\alpha, \beta)] = (1, \alpha_0, \beta_0).$$

That is, λ is a homotopy between the identity mapping on S^2 and a constant mapping. This means that S^2 is contractible, which is false. Thus h cannot exist, and f is not inessential.

For some years the bulk of the information concerning the higher homotopy groups of spheres came from the application of the *Freudenthal suspension* homomorphism. We will not use this operation, but we will describe it and quote two results. The description will be somewhat simplified if we adopt the following conventions. In E^{n+2}, let S^{n+1} denote the set of points $(x_1, \ldots, x_{n+2})$ such that $\sum_{i=1}^{n+2} x_i^2 = 1$, and let S^n be the subset of S^{n+1} for which $x_{n+2} = 0$. Let H_+^{n+1} and H_-^{n+1} be the subsets of S^{n+1} for which $x_{n+2} \geqq 0$ and $x_{n+2} \leqq 0$, respectively. Each of these "hemispheres" is an $(n+1)$-cell and may be taken to be a join over S^n, H_+^{n+1} having vertex $(0, \ldots, 0, 1)$ and H_-^{n+2} having vertex $(0, \ldots, 0, -1)$. Clearly, $S^{n+1} = H_+^{n+1} \cup H_-^{n+1}$, while $S^n = H_+^{n+1} \cap H_-^{n+1}$.

Given any mapping $f:S^n \to S^m$, we may extend f to a mapping $f_+:H_+^{n+1} \to H_+^{m+1}$ by mapping the vertex of H_+^{n+1} onto that of H_+^{m+1} and extending radially. Similarly we obtain $f_-:H_-^{n+1} \to H_-^{m+1}$. In this way we can associate with f its *suspension* $E(f):S^{n+1} \to S^{m+1}$. If f and g are homotopic mappings of S^n onto S^m, then the connecting homotopy can also be suspended to provide a homotopy between $E(f)$ and $E(g)$. Thus with each element $[f]$ of $\pi_n(S^m)$, we have associated a unique element $[E(f)]$ of $\pi_{n+1}(S^{m+1})$, and hence have a well-defined transformation of $\pi_n(S^m)$ into $\pi_{n+1}(S^{m+1})$ given by $E([f]) = [E(f)]$. For proofs of the following results, see Freudenthal [80].

THEOREM 4–30. E is a homomorphism of $\pi_n(S^m)$ into $\pi_{n+1}(S^{m+1})$.

THEOREM 4–31. For $n < 2m$, E is a homomorphism onto, and for $n < 2m - 1$, E is an isomorphism onto.

More than this is known about the Freudenthal suspension homomorphism E, but we have not yet developed the machinery needed to describe all its properties. The homomorphism E, together with certain specialized constructions which are too involved to be duplicated here, accounted for most of our knowledge of the groups $\pi_n(S^m)$, $n > m$, until recently. For a listing of this information, the reader may consult Section 21 of Steenrod's *The Topology of Fibre Bundles* [35]. In 1951, Serre [120], utilizing newly developed methods, gave a method whereby $\pi_n(S^m)$ can be calculated for many values of $n > m$. His methods are beyond the scope of this book but are currently being used extensively. We will mention the problem of homotopy groups of spheres again, giving examples when we have the necessary developments to do so.

4–8 Covering spaces. Let X be an arcwise and locally arcwise-connected space. A mapping $p:B \to X$ of a space B *onto* X is a *covering mapping* if for each point x in X there is an arcwise-connected open set U containing x such that each component of $p^{-1}(U)$ is open in B and is mapped homeomorphically onto U by p. The space B is called a *covering space* of X.

As an example, consider the mapping $p:E^1 \to S^1$ defined by $p(t) = (\cos 2\pi t, \sin 2\pi t)$. Given any point (x, y) on S^1, its antipodal point is $(-x, -y)$. Let $U = S^1 - (-x, -y)$. It is readily seen that $p^{-1}(U)$ consists of the union of all open intervals of unit length centered at the points $1/2\pi \arccos x$. Also each such interval maps homeomorphically onto U under p. Therefore E^1 is a covering space of S^1.

Now let $\{U_\alpha\}$ be a covering of X by open sets satisfying the conditions of the above definition. For any point b in $p^{-1}(U_\alpha)$, let $U_\alpha(b)$ denote that component of $p^{-1}(U_\alpha)$ containing b. Suppose that we have a path in U_α from a point x_0 to a point x_1. That is, we have a mapping $f:I^1 \to U_\alpha$ such that $f(0) = x_0, f(1) = x_1$. Let b_0 be any (fixed) point in $p^{-1}(x_0)$. Applying the homeomorphism $[p|U_\alpha(b_0)]^{-1}$, we have the path in $U_\alpha(b_0)$ given by the mapping $[p|U_\alpha(b_0)]^{-1}f:I^1 \to U_\alpha(b_0)$. It is obvious that this path covers the path in U_α. Since $U_\alpha(b_0)$ is open in B, it follows that this path is the *only* one in B which covers the given path and *emanates from the point* b_0.

It is now an easy matter to give greater generality to the last statement. For let $P = f(I^1)$ be any path in X from a point x_0 to a point x_1. Since I^1 is compact, we may subdivide I^1 into a finite number of closed intervals, $I^1 = I_1 \cup \cdots \cup I_k$, such that each $f(I_j)$ lies entirely in some open set of the covering $\{U_\alpha\}$. Then if b_0 is any point in $p^{-1}(x_0)$, a step-by-step construction as above provides a *unique* path P' in B such that P' covers P and emanates from b_0.

Suppose next that P_1 and P_2 are two paths in X from x_0 to x_1 given by mappings $f_1, f_2:I^1 \to X$. If f_1 and f_2 are homotopic modulo the set $x_0 \cup x_1$, then there is a homotopy $h:I^1 \times I^1 \to X$ with $h(t, 0) = f_1(t)$, $h(t, 1) = f_2(t)$, $h(0, u) = x_0$, and $h(1, u) = x_1$. Since the unit square $I^1 \times I^1$ is compact, there exists an integer N sufficiently large so that each square $i/N \leqq t \leqq (i + 1)/N$, $j/N \leqq u \leqq (j + 1)/N$, $i,j = 0, \ldots, N - 1$, is mapped by h into an open set in the covering $\{U_\alpha\}$ of X. If we again apply the local homeomorphisms $[p|U_\alpha(b)]^{-1}$ one at a time, the homotopy h can be "lifted" into the space B. Filling in the details of this construction provides a proof of the following result.

THEOREM 4–32. Let $p:B \to X$ be a covering mapping onto the arcwise and locally arcwise-connected space X. Let P_1 and P_2 be homotopic paths from a point x_0 to a point x_1 in X. Then for each point b in $p^{-1}(x_0)$, there exist unique paths P_1' and P_2' in B covering P_1 and P_2, respectively, and emanating from the point b. Furthermore, the paths P_1' and P_2' are homotopic in B.

As an application of Theorem 4–32, we may prove a result that affords one means of obtaining precise information about the fundamental group.

THEOREM 4–33. Let $p: B \to X$ be a covering mapping onto the arcwise and locally arcwise-connected space X, let b be any point in B, and set $p(b) = x$. Then the induced homomorphism $p_*: \pi_1(B, b) \to \pi_1(X, x)$ is an isomorphism into.

Proof: The fact that p_* is an isomorphism into follows immediately from Theorem 4–32. For p_* is defined by setting $p_*([f]) = [pf]$, and if the two paths pf_1 and pf_2 are homotopic modulo the base point x, then Theorem 4–32 says that we may construct a covering homotopy between f_1 and f_2 in B. Therefore $p_*([f])$ is the identity element of $\pi_1(X, x)$ if and only if f is homotopic to a constant, that is, $[f]$ is the identity in $\pi_1(B, b)$. □

Utilizing similar procedures, we can also prove the following generalizations of Theorem 4–33 to higher dimensions.

THEOREM 4–34. If B is a covering space of X, and if $p(b) = x$, then the induced homeomorphism $p_*: \pi_n(B, b) \to \pi_n(X, x)$, $n \geqq 2$, is an isomorphism onto.

From the example at the beginning of this section and Theorem 4–34, one easily sees the fact that $\pi_n(S^1)$ is trivial for all $n > 1$. Another example is obtained by recalling the mapping $p: S^2 \to P$ of the 2-sphere onto the projective plane where p identifies antipodal points of S^2. It is not hard to show that this is a covering mapping. Hence, from Theorem 4–34, it follows that *the higher homotopy groups $\pi_n(P)$, $n > 1$, of the projective plane P are isomorphic to those of the 2-sphere.*

One further concept may be developed in this setting. Let x_0 be a fixed point of the arcwise and locally arcwise-connected space X. For each point x of X and each path $f: I^1 \to X$ from x_0 to x, we have a pair (x, f). Two such pairs (x, f) and (x', f') are equivalent if and only if $x = x'$ and f is homotopic to f' modulo $x_0 \cup x$. The corresponding equivalence classes $[(x, f)]$ constitute the points of a new space, $R(X)$. A topology is assigned to $R(X)$ as follows. Each point x of X lies in an arcwise-connected open set U. For any point x' in U, there is an arc $g: I^1 \to U$ from x to x'. Consider the equivalence class $[(x', f*g)]$, where f is a path from x_0 to x and $f*g$ is the juxtaposition of f and g. The union over U of all such equivalence classes is a set Q_x in $R(X)$. The collection of all such sets $\{Q_x\}$ is taken to be a basis for a topology in $R(X)$. Then the space $R(X)$ is the *universal covering space* of X. The meaning of the word *universal* is explained by the following lemma.

LEMMA 4–35. If B is any covering space of X and X is locally simply connected, then $R(X)$ is a covering space of B.

However, unless X is locally simply connected, the universal covering space of our definition may fail to be a covering space. The natural map of $R(X)$ onto X is locally one-to-one, but may not be a local homeomorphism. The reader should find an example of one such space X. For this reason, it is frequently required in the definition that X be locally simply connected.

We quote three results here that are of interest. The first of these may be proved by the reader as an exercise.

THEOREM 4–36. $R(X)$ is simply connected.

THEOREM 4–37. $R(S^1)$ is the real line E^1.

THEOREM 4–38. S^2 is the universal covering space of itself and of the projective plane.

Theorems 4–33, 4–34, and 4–35 provide the motivation for the definition of the higher homotopy groups given by Serre [120]. Begin with the pair (X, x_0). We define $\pi_1(X, x_0)$ as usual. Let T_1 denote the universal covering space of $C(X, x_0)$, and define $\pi_2(X, x_0)$ to be $\pi_1(T_1, t)$, t a point in T_1 mapped onto the constant loop c. It should be clear how to proceed.

4–9 Homotopy connectedness and homotopy local connectedness. If we examine the property of arcwise connectedness (see Section 3–2) in the light of our knowledge of homotopy, it becomes apparent that the definition may be rephrased as follows. A space Y is arcwise connected if every mapping $f{:}S^0 \to Y$ of the 0-sphere into Y is homotopic to a constant. To see that this is equivalent to the original definition, we note that S^0 consists of the two points ± 1 in E^1. Hence $S^0 \times I^1$ is a pair of line segments. Any mapping which is constant on $S^0 \times 1$ identifies these upper end points in Y. Hence a homotopy between a mapping f of S^0 and the constant mapping of S^0 is equivalent to a mapping of the interval $[-1, +1]$ into Y. That the resulting Peano continuum is arcwise connected (Theorem 3–16) then shows that the new definition implies the original. A proof of the implication in the other direction is even easier, but the reader should write out the details.

This new point of view leads to an immediate generalization of arcwise connectedness. A space Y is said to be *connected in dimension n in the sense of homotopy* (abbreviated "*n-C*") if every mapping of the n-sphere S^n into Y is homotopic to a constant. This means that the nth homotopy group $\pi_n(Y, y_0)$ is trivial for any base point y_0 in Y. A space which is

k-C for all $k \leq n$ will be called a C^n-*space*, and if it is k-C for all k, the space is a C^ω-space. It is easy to see that 0-$C = C^0 =$ arcwise connected.

Our first theorem is an immediate consequence of Theorem 4–7.

THEOREM 4–39. Any contractible space is a C^ω-space.

By Theorem 4–12 then, any compact metric absolute retract is a C^ω-space.

We apply the standard procedure for localizing a topological property to obtain the following definition. A space Y is *locally connected at the point y in dimension n in the sense of homotopy* (abbreviated "n-LC at y") if every open set U containing y contains an open set V containing y such that every mapping of S^n into V is homotopic to a constant mapping with the image of the homotopy cylinder contained in U. (This is a *homotopy over* U.) The space is n-LC if it is n-LC at every point, and it is LC^n (or LC^ω) if it is k-LC for all $k \leqq n$ (or for all k).

A space is *locally contractible at a point x* if every open set U containing x contains an open set V containing x such that V is contractible over U to a point y in U. The space is *locally contractible* if it has this property at every point.

A simple application of Theorem 4–7 also proves the next result.

THEOREM 4–40. A locally contractible space is LC^ω.

THEOREM 4–41. Convex subsets of a Euclidean cube I^n or the Hilbert cube I^ω are both contractible and locally contractible. Hence such sets are both C^ω and LC^ω.

Proof: A subset of I^n or I^ω is *convex* if every pair of points in the subset are end points of a line segment that lies entirely within the subset. This immediately implies that such a convex set is starlike and hence Theorem 4–8 applies to give contractibility. Furthermore, any spherical neighborhood in I^n or I^ω is obviously convex, and the intersection of convex sets is convex, so every point of a convex subset of I^n or I^ω lies in arbitrarily small convex open sets. This implies local contractibility, and the present theorem follows from Theorems 4–39 and 4–40. □

THEOREM 4–42. Any neighborhood retract of a locally contractible space is itself locally contractible. Hence a compact metric absolute neighborhood retract is locally contractible.

Proof: Let X be locally contractible, and let A be a neighborhood retract of X. Then there exists an open set W in X such that W contains A and there is a retraction $r{:}W \to A$. Now let x be any point of A, and let U be an open set in A containing x. By definition, there is an open set U' in X, and we might as well say W, such that U' contains x and $U' \cap A = U$. Since X is locally contractible, there is a second open set V' in W such that V' contains x and is contractible over U' to x. Let f be the mapping which does the contraction so that $f(V') = y$, a point of U'. Then the set $V = V' \cap R$ is an open set in A containing x and the mapping $rf|A$ contracts V over U into the point $r(y)$. Thus A is locally contractible. The remainder of the theorem is proved just as was Theorem 3–8. □

COROLLARY 4–43. *Every compact metric absolute neighborhood retract is* LC^{ω}.

We state the last result of this section without proof. For a proof and for a development of the ideas which are embodied in this section, the reader is referred to Lefschetz's *Topics in Topology* [21].

THEOREM 4–44. *Every finite polytope (see Section 5–4) is locally contractible and, indeed, is an absolute neighborhood retract.*

CHAPTER 5

POLYTOPES AND TRIANGULATED SPACES

5–1 Introduction. The word *polytope* has become a generic term used to denote those subsets of a Euclidean space, such as polygons, polyhedra, etc., which are constructed with rectilinear elements. The reader will recognize that many of our examples have been spaces which are homeomorphic to some polytope. We refer here to such things as arcs, spheres, tori, and so on. In the succeeding chapters, we develop algebraic mechanisms (homology and cohomology theory) to aid in our study of these important spaces. In this chapter, we will study the basic geometry of polytopes.

5–2 Vector spaces. Throughout this chapter, we will use the algebraic properties of vector spaces to prove geometric theorems. This implies that the fundamental properties of vector spaces should be familiar, so we state these properties in this section, largely without proofs. Insofar as the statement of theorems is concerned, this section is self-contained. However, the reader who lacks preparation is strongly recommended to consult either Halmos [10] or Thrall and Tornheim [36].

A *vector space V over a field F* is an abelian (additively written) group for which a multiplication on the left by members of F has been defined with the usual associative and distributive properties. The additive identities of V and F will be denoted by $\overline{0}$ and 0, respectively.

A finite collection $v_1, v_2, \ldots, v_k$ of vectors (i.e., elements of V) is said to be *linearly independent* provided that if

$$\sum_{i=1}^{k} f_i \cdot v_i = \overline{0}, \qquad f_i \text{ in } F,$$

then, for each i, $f_i = 0$.

LEMMA 5–1. A finite set of vectors in V is linearly independent if and only if every subset of this finite set is linearly independent.

An arbitrary set K of vectors is said to be *linearly independent* if every finite subset of K is linearly independent.

A subset B of a vector space V is a *basis for* V if (1) B is linearly independent and (2) for every vector u in $V - B$, the set $B \cup \{u\}$ is not linearly independent.

The following is an existence theorem.

THEOREM 5–2. Every vector space over a field has a basis.

THEOREM 5–3. If $b_1, b_2, \ldots, b_k$ and $b'_1, b'_2, \ldots, b'_n$ are two bases for the same vector space V, than $n = k$.

If the vector space V has a basis of n elements, then we say that the *dimension of* V is n. Theorem 5–3 implies that dimension does not depend upon a particular basis. Note also that Theorem 5–2 says nothing about the cardinality of the basis. It is a fact that given any cardinal number $\aleph$, there is a vector space with a basis of cardinality $\aleph$.

If $B = \{b_1, b_2, \ldots, b_n\}$ is a basis for a vector space V, and if v is a vector in $V - B$, then since $B \cup \{v\}$ is not linearly independent, there exist elements $f_0, f_1, \ldots, f_n$, not all 0, in F such that

$$f_0 \cdot v + f_i b_1 + \cdots + f_n b_n = \bar{0}.$$

Now f_0 cannot be zero, for this would contradict the linear independence of B. Thus we may write

$$v = -\sum_{i=1}^{n} f_0^{-1} f_i \cdot b_i.$$

This implies the following.

THEOREM 5–4. If V is a vector space of dimension n over a field F, and if $B = \{b_1, \ldots, b_n\}$ is a basis for V, then for any element v of V there exist unique elements $f_1, \ldots, f_n$ of F such that

$$v = \sum_{i=1}^{n} f_i \cdot b_i.$$

The uniqueness claimed in Theorem 5–4 is easy to prove, for if v were also expressed as

$$v = \sum_{i=1}^{n} g_i \cdot b_i, \qquad g_i \text{ in } F,$$

then

$$\bar{0} = v - v = \sum_{i=1}^{n} f_i \cdot b_i - \sum_{i=1}^{n} g_i \cdot b_i = \sum_{i=1}^{n} (f_i - g_i) \cdot b_i.$$

The independence of B then implies that $f_i - g_i = 0$ for each i.

The dimension of a vector space characterizes the vector spaces. More precisely, we have the following result.

THEOREM 5–5. Two vector spaces over the same field are isomorphic if and only if they have the same dimension.

5–3 E^n as a vector space over E^1. Barycentric coordinates. Perhaps the most common example of a vector space is that obtained from Euclidean n-space. We defined E^n as the set of all ordered n-tuples of real numbers (with the usual metric topology). To consider E^n as a vector space, we must give an *addition of vectors* and a *scalar multiplication.* This is done by setting

$$(a_1, a_2, \ldots, a_n) + (b_1, b_2, \ldots, b_n) = (a_1 + b_1, a_2 + b_2, \ldots, a_n + b_n)$$

and

$$c \cdot (a_1, a_2, \ldots, a_n) = (ca_1, ca_2, \ldots, ca_n),$$

where the a_i's, the b_i's, and c are real numbers. It is easily verified that with these definitions, E^n becomes a vector space over the field of real numbers E^1. It has a basis of the form $(1, 0, \ldots, 0)$, $(0, 1, 0, \ldots, 0)$, $\ldots$, $(0, 0, \ldots, 0, 1)$. More briefly, if δ_{ij} is the Kronecker delta, given by $\delta_{jj} = 1$ and $\delta_{ij} = 0$, $i \neq j$, this basis is $(\delta_{1j}, \delta_{2j}, \ldots, \delta_{nj})$, $j = 1, 2, \ldots, n$. Hence as a vector space, E^n has dimension n.

A set of points H^k in E^n is a *k-dimensional hyperplane* if there is a linearly independent set of vectors (points) $\{a_i\}$, $i = 1, 2, \ldots, k \leqq n$, and a vector a_0 such that H^k is exactly the set of all points h which may be expressed as

$$h = a_0 + \sum_{i=1}^{k} t_i \cdot a_i, \qquad t_i \text{ real numbers.}$$

We remark that if $a_0 = \overline{0} = (0, 0, \ldots, 0)$, then H^k is a *k-dimensional vector subspace of E^n*, so in general each hyperplane is a "translation" of some vector subspace. The reader should see that this definition reduces to that of a line in E^2 ($k = 1, n = 2$), to a line in E^3 ($k = 1, n = 3$), and to a plane in E^3 ($k = 2, n = 3$).

In geometry, one says that a set of $k + 1$ points in E^n is *geometrically independent* if no $(k - 1)$-dimensional hyperplane contains all the points. The algebraic equivalent of this condition is as follows. A set $\{a_0, a_1, \ldots, a_k\}$ of vectors in E^n is *pointwise independent* provided that the k vectors $a_1 - a_0, a_2 - a_0, \ldots, a_k - a_0$ are linearly independent.

THEOREM 5–6. The set $A = \{a_0, a_1, \ldots, a_k\}$ in E^n is pointwise independent if and only if the two conditions (1) $\sum_{i=0}^{k} g_i \cdot a_i = \overline{0}$ and (2) $\sum_{i=0}^{k} g_i = 0$ imply that (3) $g_i = 0$ for all $i = 0, 1, \ldots, k$.

Proof: Suppose that A is pointwise independent and that conditions (1) and (2) hold. Then

$$\sum_{i=0}^{k} g_i(a_i - a_j) = \sum_{i=0}^{k} g_i \cdot a_i - \left(\sum_{i=0}^{k} g_i\right) a_j = \overline{0}.$$

Since for j fixed, the vectors $a_i - a_j$ are linearly independent, condition (3) follows.

On the other hand, if conditions (1) and (2) imply (3), and if there exist real numbers $g_1, g_2, \ldots, g_k$ such that

$$\sum_{i=1}^{k} g_i \cdot (a_i - a_0) = \bar{0},$$

then

$$\sum_{i=1}^{k} g_i \cdot a_i = \left(\sum_{i=1}^{k} g_i\right) \cdot a_0.$$

Letting

$$g_0 = -\sum_{i=1}^{k} g_i,$$

we have

$$\sum_{i=0}^{k} g_i \cdot a_i + \left(\sum_{i=0}^{k} g_i\right) \cdot a_0 = \bar{0}.$$

Clearly, (1) and (2) are satisfied. Hence each $g_i = 0$ by (3), and this implies that the vectors $\{a_i - a_0\}$ are linearly independent. □

THEOREM 5–7. If $A = \{a_0, a_1, \ldots, a_k\}$, $k \leqq n$, is a pointwise independent set in E^n, then there exists a unique k-dimensional hyperplane H^k containing A and having the property that a vector h is in H^k if and only if $h = a_0 + \sum_{i=1}^{k} g_i(a_i - a_0)$, the g_i being unique if $h \neq \bar{0}$.

Proof: Let H^k be the set of vectors of the form

$$h = a_0 + \sum_{i=1}^{k} g_i(a_i - a_0).$$

Then by definition, H^k is a hyperplane. That H^k contains the set A follows from the equations

$$a_j = a_0 + \sum_{i=1}^{k} \delta_{ij}(a_i - a_0),$$

where δ_{ij} is the Kronecker delta. The uniqueness of the numbers g_i follows from Theorem 5–4, since the set of all vectors $\{h - a_0\}$, h in H^k, is a k-dimensional vector subspace of E^n with basis $\{a_i - a_0\}$.

It only remains to show that the hyperspace H^k is unique. Suppose that there exists another k-dimensional hyperplane F^k containing A. By definition, there must be a linearly independent set $B = \{b_1, \ldots, b_k\}$ and a vector b_0 such that p lies in F^k if and only if $p = b_0 + \sum_{i=1}^{k} f_i \cdot b_i$.

Since F^k contains A, for each a_j there are coefficients f_{ij} such that

$$a_j = b_0 + \sum_{i=1}^{k} f_{ij} b_i \qquad (j = 0, 1, \ldots, k).$$

In particular, $a_0 = b_0 + \sum_{i=1}^{k} f_{i0} \cdot b_i$, so we have

$$a_j - a_0 = \sum_{i=1}^{k} (f_{ij} - f_{i0}) \cdot b_i.$$

Since B and the set of vectors $\{a_i - a_0\}$, $i = 1, \ldots, k$, are both assumed to be linearly independent sets, there are *unique* solutions

$$b_i = \sum_{j=1}^{k} g_{ij}(a_j - a_0) \qquad (i = 1, 2, \ldots, k).$$

Substituting these solutions into the characterizing equation for elements of F^k, we see that $F^k = H^k$. Thus H^k is unique. □

We have already observed that any hyperplane H^k is a "translation" of a k-dimensional vector space p^k imbedded in E^n. Since such a translation is an *isometry* (i.e., a one-to-one, distance-preserving mapping onto), we expect H^k to be homeomorphic to p^k. Now by Theorem 5–5, the subspace p^k is isomorphic to E^k. By some means, then, a vector of H^k should be determined uniquely by a vector in E^k, that is, by an ordered k-tuple of real numbers. The preceding theorem showed that each vector in H^k has a unique expression as a linear combination of $k + 1$ pointwise independent vectors. If we can find a dependence among the $k + 1$ coefficients in this combination, then giving a k-tuple of real numbers, i.e., a vector in E^k, would prescribe the $k + 1$ coefficients determining a vector in H^k. In this way, a specific mapping of E^k onto H^k could be defined.

THEOREM 5–8. Let $A = \{a_0, a_1, \ldots, a_k\}$ be a pointwise independent set in E^n. Then the k-dimensional hyperplane H^k containing A is characterized by the condition

(1) h is in H^k if and only if

$$\text{(i)} \quad h = \sum_{i=0}^{k} f_i a_i \qquad \text{and} \qquad \text{(ii)} \quad \sum_{i=0}^{k} f_i = 1,$$

where for each h in H^k, the coefficients f_i are unique.

Proof: By Theorem 5–7 we know that h is in H^k if and only if

$$h = a_0 + \sum_{i=1}^{k} g_i(a_i - a_0) = \sum_{i=1}^{k} g_i a_i + \left(1 - \sum_{i=1}^{k} g_i\right) a_0.$$

If we set $g_0 = 1 - \sum_{i=1}^{k} g_i$, then h is of the form (i), and (ii) is obviously satisfied. The field elements (real numbers) g_i are unique for each h so that the f_i are also.

Moreover, suppose that a vector v satisfies (i) and (ii). Then we have

$$v = \sum_{i=0}^{k} f_i a_i = \sum_{i=1}^{k} f_i(a_i - a_0) + \left(\sum_{i=0}^{k} f_i\right) a_0 = a_0 + \sum_{i=1}^{k} f_i(a_i - a_0),$$

and hence v belongs to H^k.

In the other direction, we have already shown above that conditions (i) and (ii) are equivalent to condition (1) and then Theorem 5–7 applies to show that H^k is the k-dimensional hyperplane containing A. □

This result seems to be merely a slightly different restatement of Theorem 5–7, and so it is. But the difference is significant. Using the dependence (ii) among the coefficients f_i, we have, for a given pointwise independent set $A = \{a_0, a_1, \ldots, a_k\}$ and the hyperplane H^k containing A, the following fact. For each point $p = (p_1, \ldots, p_k)$ of E^k, there is one and only one vector h in H^k such that $h = \sum_{i=0}^{k} p_i a_i$, where we take $p_0 = 1 - \sum_{k=1}^{k} p_i$. Our results above show that this gives us a one-to-one transformation of E^k onto H^k. By means of lengthy, but direct, arguments the reader can prove that this transformation preserves both

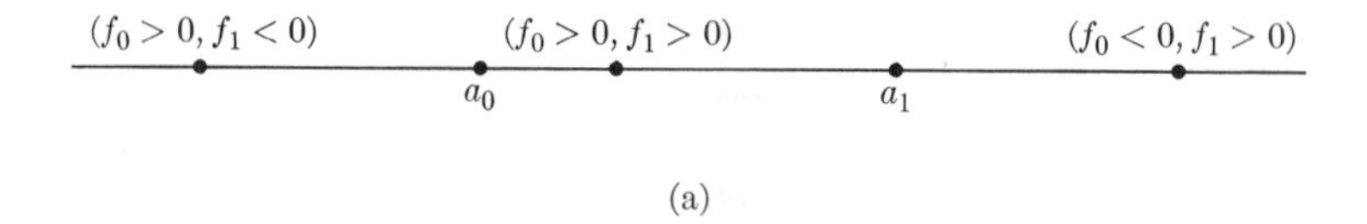

(a)

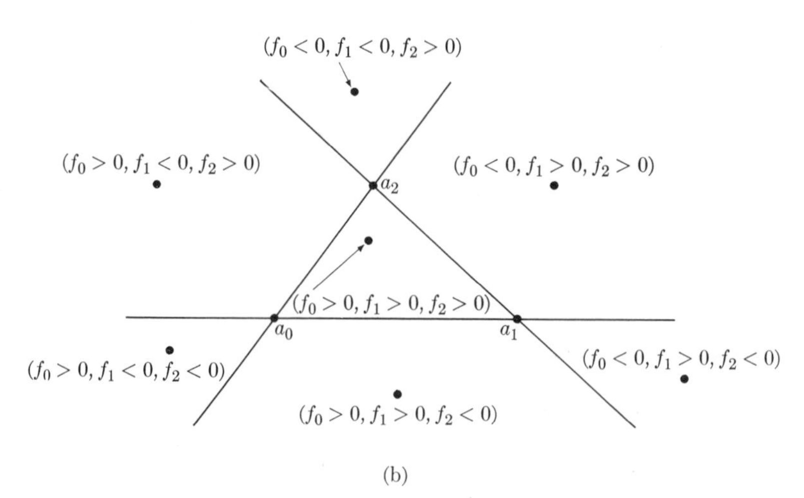

(b)

FIG. 5–1. Barycentric coordinates. (a) One-dimensional. (b) Two-dimensional.

linear and pointwise independence. It can also be shown to preserve distance so that E^k and H^k are homeomorphic. That is, we may say that the transformation $b_A : E^k \to H^k$ defined for a vector $v = (v_1, v_2, \ldots, v_k)$ in E^k by

$$b_A(v) = \left(1 - \sum_{i=1}^{k} v_i\right) a_0 + \sum_{i=1}^{k} v_i a_i$$

is one-to-one, onto, and preserves all linear relations.

Let $A = \{a_0, a_1, \ldots, a_k\}$ be a pointwise independent set of $k + 1$ vectors. Then the real numbers $f_0, f_1, \ldots, f_k$ are the *barycentric coordinates of a vector h with respect to A* if and only if

$$\text{(i)}\quad h = \sum_{i=0}^{k} f_i a_i \qquad \text{and} \qquad \text{(ii)}\quad \sum_{i=0}^{k} f_i = 1.$$

Thus the totality of vectors in E^n having barycentric coordinates with respect to A is the unique k-dimensional hyperplane H^k containing A.

There is a more intuitive approach to barycentric coordinates. The real numbers f_i can be considered as weights (both positive and negative weights being permitted) which are assigned to the points of A. The resulting system of $k + 1$ particles has a centroid which is precisely the point h for which the numbers f_i are the barycentric coordinates with respect to A. For examples, look at Fig. 5–1, which is self-explanatory.

5–4 Geometric complexes and polytopes. Let $A = \{a_0, a_1, \ldots, a_k\}$ be a set of $k + 1$ pointwise independent points in E^n. The *geometric k-simplex in E^n* determined by A is the set of all points of the hyperplane H^k containing A for which the barycentric coordinates with respect to A are all nonnegative. It is quite easy to see that a geometric 0-simplex is a single point, a geometric 1-simplex is a closed line segment, a geometric 2-simplex is a closed triangular plane region, a geometric 3-simplex is a solid tetrahedron, and so on. At times, it is convenient to use an *open geometric k-simplex* which is the set of points whose barycentric coordinates are all positive. Here again the set is simple, a point, an open line segment, etc. If the $k + 1$ points $p_0, p_1, \ldots, p_k$ determine a geometric k-simplex, then we will denote that simplex by the symbol $\langle p_0 p_1 \cdots p_k \rangle$ and call the points p_i the *vertices* of the simplex. When we wish to speak about k-simplexes in general, we will use the generic symbol s^k; that is, s^k will denote any geometric k-simplex, whereas $\langle p_0 p_1 \cdots p_k \rangle$ denotes the particular simplex with vertices $p_0, \ldots, p_k$.

Another geometric concept is of value in dealing with simplexes. A subset B of E^n is said to be *convex* if, given any two points x and y of B, the line segment joining x and y is entirely contained in the set B. It is easily seen that *the intersection of any number of convex sets is again a convex set.* Given any subset A of E^n, the *convex hull* of A is the intersection of all convex subsets containing A. By the remark above, the convex hull of any subset A of E^n is convex.

LEMMA 5–9. The geometric k-simplex $\langle p_0 p_1 \cdots p_k \rangle$ determined by a set $A = \{p_0, p_1, \ldots, p_k\}$ of $k + 1$ pointwise independent points of E^n is the convex hull of the set A.

Proof: The hyperplane H^k containing A is a convex set, and the k-simplex $\langle p_0 \cdots p_k \rangle$ clearly lies in H^k. Each half-plane of H^k determined by taking the barycentric coordinate f_i to be nonnegative is also a convex subset of E^n. The intersection of these $k + 1$ half-planes of H^k is precisely the simplex $\langle p_0 \cdots p_k \rangle$, which is therefore a convex set.

Next let $0 \leq r < k$. For each point x in $\langle p_0 \cdots p_k \rangle$, there are points x' in $\langle p_0 \cdots p_r \rangle$, x'' in $\langle p_{r+1} \cdots p_k \rangle$, such that x lies on the segment $x'x''$. To see this, let x be written in vector notation as $x = \sum_{i=0}^{k} x_i p_i$, where $x_i \geqq 0$ and $\sum_{i=0}^{k} x_i = 1$. Set $a' = \sum_{i=1}^{r} x_i$ and $a'' = \sum_{i=r+1}^{k} x_i$. If either $a' = 0$ or $a'' = 0$, the statement is obviously true. If both fail to be zero, then we set

$$x' = \sum_{i=0}^{r} \left(\frac{x_i}{a'}\right) p_i \qquad \text{and} \qquad x'' = \sum_{i=r+1}^{k} \left(\frac{x_i}{a''}\right) p_i.$$

These are obviously points of $\langle p_0 \cdots p_r \rangle$ and $\langle p_{r+1} \cdots p_k \rangle$, respectively, and

$$x = a'x' + a''x'',$$

with $a' + a'' = 1$.

To finish the proof, let B be any convex set containing A. We use induction to prove that $\langle p_0 \cdots p_k \rangle$ also lies in B. This is easily seen for $k = 0$. Suppose that it is true for $k - 1$, and let x be a point in $\langle p_0 \cdots p_k \rangle$. Then x is on a line segment from p_0 to a point x'' in $\langle p_1 \cdots p_k \rangle$. By the induction hypothesis, x'' lies in B. Since p_0 and x'' lie in B, and since B is convex, it follows that x lies in B. Hence B contains $\langle p_0 \cdots p_k \rangle$, and the lemma is immediate. □

The geometric simplexes are the basic building blocks from which we will construct spaces. As a simple instance, we may glue four 2-simplexes together along their edges so as to form a tetrahedral surface (a homeomorph of the 2-sphere). For reasons that will become clear later, we have some rules about the way in which simplexes can be joined together. Roughly speaking, we cannot be haphazard about placing our bricks; we must be expert bricklayers and "line up the edges." To be precise, we say

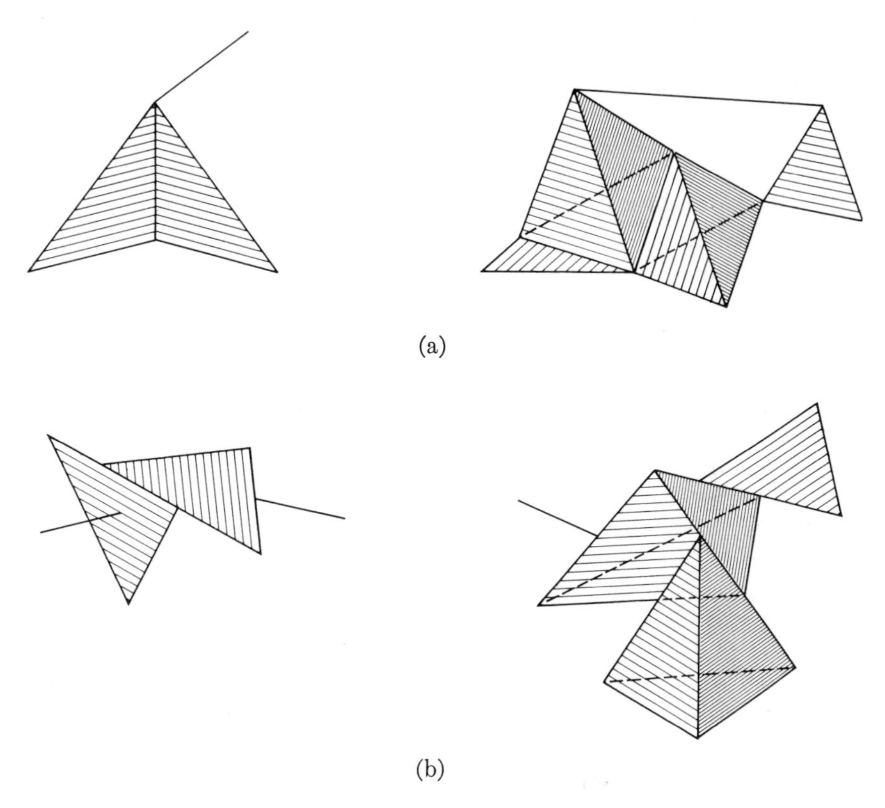

FIG. 5–2. (a) Proper joining. (b) Improper joining.

that two geometric simplexes, s^m and s^n, $m \leqq n$, are *properly joined* if either

$$(1) \quad s^m \cap s^n = \emptyset \qquad \text{(the empty set)}$$

or

$$(2) \quad s^m \cap s^n = s^k, \qquad k \leqq m,$$

where s^k is a subsimplex of both s^m and s^n. In Fig. 5–2, we show examples of both proper and improper joining of simplexes.

This joining can be more easily expressed if we introduce a natural concept. Let $\langle p_0 \cdots p_n \rangle$ be a geometric n-simplex. By Lemma 5–1 and the definition of pointwise independent vectors, it follows that any subset of the vertices $p_0, \ldots, p_n$ is itself the set of vertices of a geometric simplex. Each such subsimplex is called a *face* of $\langle p_0 \cdots p_n \rangle$. In particular, we will use the simplex $\langle p_0 \cdots \hat{p}_j \cdots p_n \rangle$ to denote that face of $\langle p_0 \cdots p_n \rangle$ obtained by deleting the vertex p_j from the collection of vertices $p_0, \ldots, p_n$.

It is clear that if $\langle p_0 \cdots p_n \rangle$ is in E^n, then $\langle p_0 \cdots \hat{p}_j \cdots p_n \rangle$ is a closed geometric $(n-1)$-simplex in the point-set boundary of $\langle p_0 \cdots p_n \rangle$ relative to E^n.

We may now say that two geometric simplexes are properly joined if they do not meet at all, or if their intersection is a face of each of them. Note that a simplex is a face of itself.

This leads to the chief concept of this chapter, the *geometric complex.* What we would like to say is that a geometric complex K is a (countable) collection of properly joined geometric simplexes with the property that if s^n is any simplex of K, then every face of s^n also belongs to K. This is the customary definition, but we will have to say more for reasons to be explained.

Since a simplex is defined as a certain subset of some Euclidean space E^n, two simplexes cannot be properly joined unless they lie in the same Euclidean space. The "components" of a complex, then, would all have to lie in the same Euclidean space. This would mean that a configuration consisting of a 1-simplex having a vertex in common with a 2-simplex, which has a vertex in common with a 3-simplex, which has a vertex in common with a 4-simplex, etc., could not be in a complex. But we do not want to bar this possibility. One way out of this difficulty would be to re-do the several preceding sections in terms of finite-dimensional linear subsets of Hilbert space (which is an infinite-dimensional vector space). This would imply, however, that we could not consider a complex with more than c simplexes (c being the cardinality of the real numbers). In this book, we do nothing with a geometric complex that would force us out of Hilbert space or countable complexes. But we do want to make the definition sufficiently general to permit easy extension to such cases.

We want to arrive at the idea of considering simplexes from different Euclidean spaces, taking them away somewhere and joining them together. We define now a *topological geometric simplex,* a term we will abandon later. A topological geometric n-simplex σ^n is a pair (A, h) consisting of a topological space A and a definite homeomorphism h between A and some geometric n-simplex s^n_A. The space A is said to be the *carrier* of this simplex. The topological geometric simplex $\sigma^m = (B, h')$ is a *face* of σ^n if B is a subset of A and $h' = h|B$ (h restricted to B). Two topological geometric simplexes (A, h) and (B, h') are *properly joined* (1) if $A \cap B$ is a face of each simplex and, (2) if s_1 is the face of $s_A = h(A)$ corresponding to $A \cap B$ under the homeomorphism h and if s_2 is the face of $s_B = h'(B)$ corresponding to $A \cap B$ under h', then there is a linear mapping $l{:}s_1 \to s_2$ such that $h^{-1}|A \cap B = (h'^{-1}|A \cap B)l$. A *topological geometric complex* K is a (countable) collection of properly joined topological geometric simplexes with the property that every face of a simplex in K is a simplex in K.

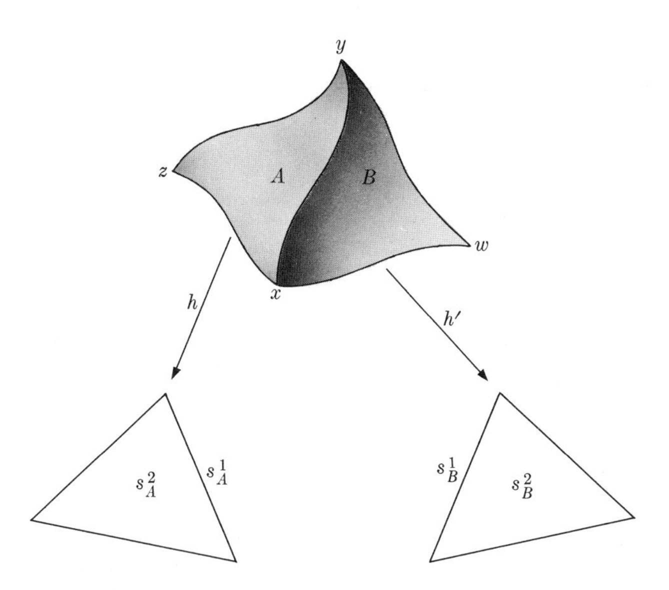

FIGURE 5-3

To illustrate this rather complicated definition, consider the two sets A and B in Fig. 5-3. These are closures of plane regions. There are homeomorphisms $h:A \rightarrow s_A^2$, $h':B \rightarrow s_B^2$ such that the pairs $(xy, h|xy)$, $(yz, h|yz)$, and $(xz, h|xz)$ are faces of A and $(xy, h'|xy)$, $(xw, h'|xw)$, and $(yw, h'|yw)$ are faces of B. To show that (A, h) and (B, h') are properly joined, we must show that there is a linear mapping l from the face s_A^1 of s_A^2 corresponding to xy onto the face s_B^1 of s_B^2 corresponding to xy such that $(h|s_A^1)^{-1} = (h'|s_B^1)^{-1}l$. There are only two possible choices for l, so there is very little room for flexibility in picking the homeomorphisms h and h'. The slightest variation in one or the other, under our definitions, would change the simplexes from properly joined to improperly joined simplexes. It would be possible to include still more machinery in our definitions and give more flexibility here. But the difficulty is not really a practical one. It is usually quite clear that the desired mappings exist.

Now let K be a topological geometric complex, and consider the set S that is the union of all of its simplexes. It may happen that the sets carrying the simplexes of K all lie in some topological space T. In such a case, S is a subspace of T and so has a topology. This topology may or may not be a "natural" one. For example, let K be the infinite complex composed of all closed intervals $[n, n + 1]$, where n is a nonnegative integer, and their vertices. In the union of the simplexes of K, that is, the nonnegative real numbers, only sets intersecting $[0, 1]$ in an infinite set can have 0 as a limit point. This is very natural in terms of the structure as a com-

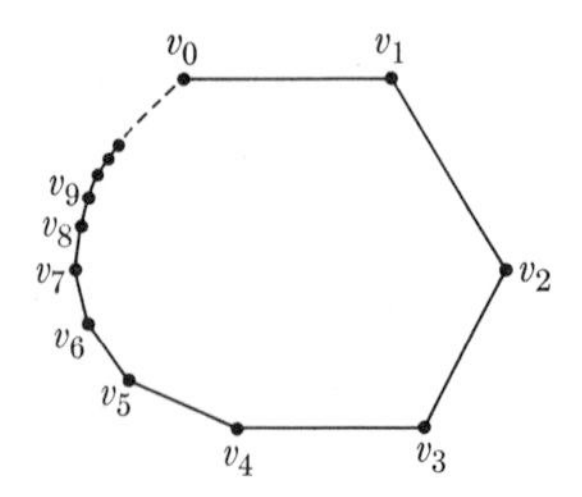

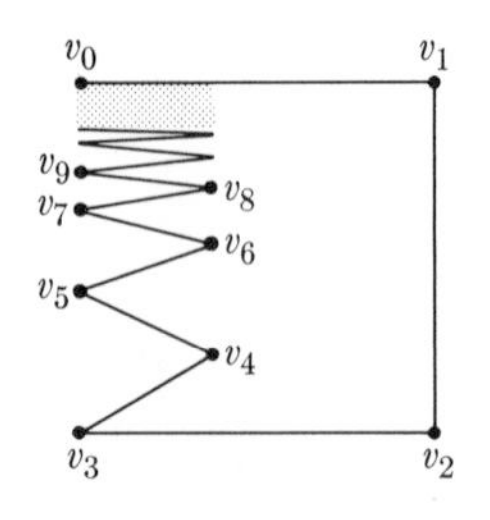

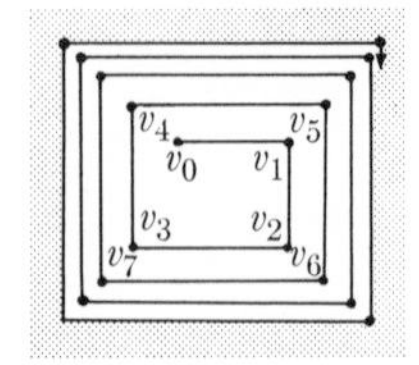

FIGURE 5–4

plex. But consider the several spaces in Fig. 5–4 with the indicated structure as complexes. As complexes, these all have the same structure as does K; as subspaces of E^2, they are quite different.

We will next define a topology for a particular class of complexes, the *star-finite* complexes. These are complexes with the property that each simplex is a face of only a finite number of simplexes. The name comes from the definition of the *star of a simplex* σ, which consists of all simplexes of which σ is a face. This is denoted by $\text{St}(\sigma)$, and we note specifically that σ is contained in $\text{St}(\sigma)$. The term *star* is often applied to the union of the carriers of $\text{St}(\sigma)$ instead of the collection of simplexes, but the meaning is usually obvious from context.

The *star topology* of a star-finite complex K is defined by taking as a basis all subsets X that intersect at most a finite number of simplexes of K and intersect these in relatively open sets, that is, in sets that are open in each simplex. The space so obtained is denoted by the symbol $|K|$ and is called the *geometric carrier* of the complex K.

Our next results indicate the nature of the spaces obtained as the geometric carriers of star-finite complexes. First, we will need the concept of the *open star* of a simplex. Given a simplex σ in K, the open star of σ, $\mathring{\text{S}}\text{t}(\sigma)$, is the open subset of the geometric carrier $|K|$, which is the interior of the carrier of the star of σ. That is, we consider the star of a simplex in the complex K, look at the carrier of these simplexes in $|K|$, and, using the star topology, take the interior of this carrier to be the open star. Observe that the open star is a subset of the *carrier* $|K|$, while the star is a subcollection of simplexes of the *complex*.

THEOREM 5–10. The geometric carrier of a star-finite topological geometric complex K is a locally compact Hausdorff space.

Proof: Each point of the carrier $|K|$ lies in the carrier of some simplex of K. Taking the star of this simplex, we have a finite number of simplexes whose carriers obviously form a compact union containing the given point of $|K|$ as an interior point. Hence $|K|$ is locally compact.

Given two points x and y of $|K|$, consider first the case in which x and y lie in the carrier of the same simplex of K. The existence of disjoint open subsets U and V, with x in U and y in V, clearly follows from the fact that a geometric simplex is Hausdorff (and more). If there is no simplex of K whose carrier contains both x and y, then the two points must lie in different open stars of vertices, say U and V. Letting x be in U and y be in V, we know that x is in $U - \overline{V}$ and y is in $V - \overline{U}$ (else x and y would necessarily be in the same simplex). These two open sets satisfy the Hausdorff condition. □

Corollary 5–11. The geometric carrier of a star-finite topological geometric complex K is metric.

Proof: First, each such carrier $|K|$ is paracompact by Theorem 2–67. Then since $|K|$ is obviously locally metrizable, Theorem 2–69 applies to complete the proof. □

The outline of an alternative proof is as follows. Let each simplex σ that is not a proper face of any other simplex be assigned a metric d_σ. If two points p and q lie in different components of $|K|$, define the distance between p and q to be unity. If p and q lie in the same component of $|K|$, then there exist many sequences $p = x_0, x_1, \ldots, x_n = q$ of points such that for each i, x_i and x_{i+1} lie in the same maximal simplex. Let

$$d(p, q) = glb \sum_{i=0}^{n} d_\sigma(x_i, x_{i+1}).$$

There are two major categories of complexes, the finite and the infinite complexes. These terms refer to the number of simplexes in the complex and not to the dimension of the complex. The *dimension of a complex* K is the largest integer n such that K contains an n-simplex. If no such integer exists, then K has infinite dimension. We leave the proofs of the following lemmas as simple exercises.

Lemma 5–12. Every finite complex has finite dimension.

Lemma 5–13. A complex of infinite dimension is infinite.

It is not hard to see that the carrier of a finite complex may be taken to be a subset of some Euclidean space and hence that *the carrier of a finite complex is a compact metric space.* (In Section 5–8, we show that the carrier of a finite complex of dimension n is homeomorphic to a subset of E^{2n+1}.) In this context, we may easily prove the following result.

Theorem 5–14. Let K be a finite complex with vertices $v_1, v_2, \ldots, v_n$. Then the collection of open stars $\{\mathring{\mathrm{St}}(v_i)\}$ is a finite open covering of the carrier $|K|$.

Proof: We need only point out that each point of $|K|$ lies in the open star of some vertex since each simplex of K lies in the star of some vertex. □

It might be hoped that a "good" topology could be found for the complexes which are not star-finite, that is, a topology in which they would be locally compact Hausdorff spaces, but this is not possible. There is no Hausdorff topology for the Cantor star, the join of a Cantor set and a point, such that (1) the interior of a 1-simplex is open, (2) the carrier is connected, and (3) the carrier is locally compact.

The geometric carrier of a star-finite complex K is called a *polytope*. A topological space X that is homeomorphic to a polytope $|K|$ is called a *triangulated space* and the complex K is a *triangulation* of the space X. Although much of the remainder of this book is devoted to a study of this important class of spaces, we cannot characterize the class. That is, necessary and sufficient topological conditions that a space have a triangulation are not known. This "Triangulation Problem" has only been partially answered to date, but many widely studied and useful spaces are known to have triangulations. (Among these are all 3-dimensional manifolds and all differentiable manifolds.) At present, we consider only a few elementary properties of the triangulated spaces.

Two simplexes s_1 *and* s_2 of a complex K *are connected in* K if there exists a chain of 1-simplexes in K joining s_1 and s_2 in the following sense. There are simplexes s_i^1, $i = 1, 2, \ldots, k$, such that (1) $s_1 \cap s_1^1$ is a vertex of s_1, (2) $s_2 \cap s_k^1$ is a vertex of s_2, and, (3) for each $i = 1, 2, \ldots, k - 1$, $s_i^1 \cap s_{i+1}^1$ is a vertex of each simplex. We leave to the reader the easy proof of the fact that we may add a fourth condition, (4) for $j \neq i - 1$, i, or $i + 1$, $s_j^1 \cap s_j^1 = \emptyset$. A chain of 1-simplexes satisfying conditions (3) and (4) is called a *simple chain*. The above connectedness relation between simplexes of a complex K can be shown to be an equivalence relation, and the resulting equivalence classes of K are called its *combinatorial components*. The *complex is connected* if it has just one combinatorial component. The proofs of the next results are left as simple exercises.

THEOREM 5–15. The geometric carrier of a connected complex is arcwise connected.

THEOREM 5–16. In a finite polytope, components and the carriers of the combinatorial components are identical.

5–5 Barycentric subdivision.

This section introduces a standard technique used for producing a triangulation of a given polytope such that the new triangulation is "finer" than the original. This *subdivision* is presented first for a complex consisting of a single simplex $s^n = \langle p_0 p_1 \cdots p_n \rangle$ together with all of its faces. Such a complex is called the *closure of a simplex* and is denoted by $\mathrm{Cl}(s^n)$.

We recall that the vertices p_i of s_n are assumed to be pointwise independent and that the points of s^n are those points of E^n which have nonnegative barycentric coordinates with respect to the vertices p_i. In particular there is a point, which we will denote by $\dot{s}^n$, whose barycentric coordinates with respect to the vertices p_i are all equal. Similarly, for each face $s^k = \langle p_{i_0} \cdots p_{i_k} \rangle$ of s^n, there is a point $\dot{s}^k$ whose barycentric coordinates with respect to the subset of vertices $p_{i_0}, \ldots, p_{i_k}$ are all equal. Note that $s_i^0 = \langle p_i \rangle$ has the corresponding point $\dot{s}_i^0 = p_i$. The collection of all points $\dot{s}_j^k$, $k = 0, 1, \ldots, n, j = 1, 2, \ldots, \alpha_k$, where α_k is the number of k-simplexes in $\mathrm{Cl}(s^n)$, will be the vertices of a new complex K', the *first barycentric subdivision of* $K = \mathrm{Cl}(s^n)$. We must say how the simplexes of K' are formed. To do so, we introduce a definition.

Let K be any geometric complex, and let s_1 and s_2 be simplexes of K. Then we will write $s_1 \prec s_2$ if and only if s_1 is a proper face of s_2. It is easily verified that under this relation "$\prec$," the complex K is a partially ordered set.

Now returning to the vertices $\dot{s}_j^k$, we will take a subset of these points to be vertices of a simplex in K', $\langle \dot{s}_1 \dot{s}_2 \cdots \dot{s}_i \rangle$, if and only if $s_1 \prec s_2 \prec s_3 \prec \cdots \prec s_i$ in K. Figure 5-5 indicates the essentially simple construction that has been described above.

The point $\dot{s}^n$ is called the *barycenter* of the simplex s^n and is the centroid of the vertices p_i with equal weights assigned to each.

This subdivision may now be done for each simplex of any geometric complex K and defines a new complex K', *the first barycentric subdivision of* K. It is evident that the geometric carriers of K and K' are identical.

Lemma 5-17. The diameter of the convex hull of any set A is equal to the diameter of A itself.

Proof: Let δ be any number such that if x and y are two points of A, then $d(x, y) < \delta$. Let a and b be any two points of the convex hull of A. We will show that $d(a, b) \leqq \delta$. To do so, consider any point z of A. Clearly, the spherical neighborhood $S(z, \delta)$ contains A. The closure $\overline{S(z, \delta)}$ is a closed convex set containing A and hence contains the convex hull of A, by definition. Thus $d(z, a) \leqq \delta$. Conversely, then, the point z lies in $\overline{S(a, \delta)}$. This is true for each point z of A; hence A lies in $\overline{S(a, \delta)}$. It follows that the convex hull of A lies in $\overline{S(a, \delta)}$, so $d(a, b) \leqq \delta$. The lemma is now immediate. □

Corollary 5-18. The diameter of a geometric simplex is the length of its longest edge (or 1-face).

For a geometric complex K, we define the *mesh of* K to be the supremum of the diameters of all simplexes of K. In view of Corollary 5-18, this supremum may be taken over all 1-simplexes of K. The principal result of this section may be expressed in these terms.

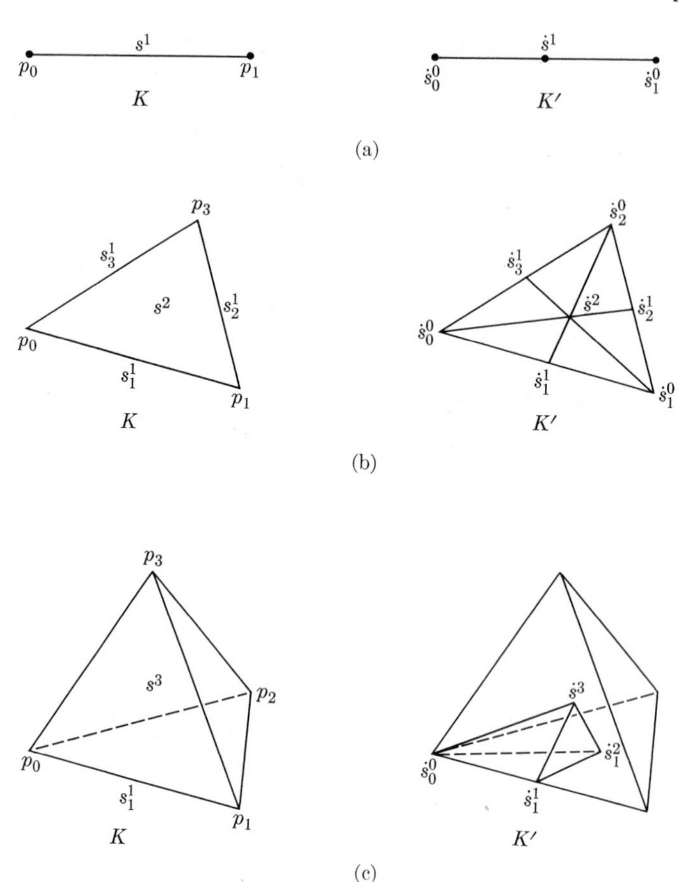

FIG. 5–5. Barycentric subdivisions. (a) One-dimensional. (b) Two-dimensional. (c) Three-dimensional. For simplicity's sake, only one of the twenty-four 3-simplexes in the subdivision of s^3 is shown.

THEOREM 5–19. If a geometric complex K has finite dimension n and has a finite mesh λ, then the mesh of its first barycentric subdivision K' does not exceed the number $n/(n+1) \cdot \lambda$.

Proof: Let s^k be any k-simplex of K. The barycenter $\dot{s}^k$ of s^k has barycentric coordinates (with respect to the vertices of s^k), each equal to $1/(k+1)$. Let $\langle \dot{s}^i \dot{s}^k \rangle$ be any 1-simplex of K' in the subdivision of s^k. Recall that, by definition, s^i is a face of s^k in K. Let the length of $\langle \dot{s}^i \dot{s}^k \rangle$ be μ. If the vertices $p_0, p_1, \ldots, p_i$ of s^k form s^i, then the remaining vertices in s^k, say $p_{i+1}, \ldots, p_k$, form a face s^{k-i-1} opposite s^i. Now the line segment from $\dot{s}^i$ to $\dot{s}^{k-i-1}$ clearly contains the simplex $\langle \dot{s}^i \dot{s}^k \rangle$. The barycenter $\dot{s}^i$ may be considered to have weight $(i+1)/(k+1)$, while

$\dot{s}^{k-i-1}$ has weight $(k - i)/(k + 1)$ (as centroids of the vertices). Then the barycenter $\dot{s}^k$ is the centroid of these two particles. If the length of the line segment from $\dot{s}^i$ to $\dot{s}^{k-i-1}$ is ρ, then we may apply the elementary law of levers to obtain

$$\left(\frac{i+1}{k+1}\right)\mu = \left(\frac{k-i}{k+1}\right)(\rho - \mu)$$

or

$$\left(\frac{i+1}{k+1} + \frac{k-i}{k+1}\right)\mu = \mu = \frac{k-i}{k+1}\cdot\rho.$$

Since ρ does not exceed the diameter of s^k, we have

$$\mu \leqq \frac{k-i}{k+1}\cdot\lambda \leqq \frac{k}{k+1}\cdot\lambda \leqq \frac{n}{n+1}\cdot\lambda.$$

Hence no 1-simplex of K' has diameter exceeding $n/(n + 1)\cdot\lambda$, and hence the mesh of K' cannot exceed $n/(n + 1)\cdot\lambda$. □

Having one barycentric subdivision K' of a complex K, we may continue the process and subdivide K', etc. Making k successive subdivisions, we arrive at the *kth barycentric subdivision of* K, which is denoted by $K^{(k)}$.

THEOREM 5-20. If the mesh λ of an n-dimensional geometric complex K is finite, then the mesh of $K^{(k)}$ approaches zero as k increases indefinitely.

Proof: From the proof of Theorem 5-19, we note that the mesh $\lambda^{(k)}$ of $K^{(k)}$ must satisfy the inequality

$$\lambda^{(k)} \leqq \left(\frac{n}{n+1}\right)^k\lambda.$$

But $[n/(n + 1)]^k$ approaches zero as k increases indefinitely. □

We observe that, since the mesh of a finite complex is obviously finite, the results of this section automatically apply to finite polytopes.

EXERCISE 5-1. Construct an infinite star-finite complex whose mesh remains unchanged by barycentric subdivision.

EXERCISE 5-2. Construct an infinite star-finite geometric complex K whose finite mesh is not the diameter of any simplex in K.

5-6 Simplicial mappings and the simplicial approximation theorem. We next look at a special class of continuous mappings of one polytope into another, namely, those mappings which carry simplexes linearly onto simplexes. Let $|K|$ and $|L|$ be two polytopes with triangulations K and L, respectively. Denote by f a (possibly many-to-one) transformation from the vertices of K into those of L, satisfying the condition that if $\langle p_0 \cdots p_n\rangle$ is a simplex of K, then the points $f(p_0), \ldots, f(p_n)$ (not all necessarily distinct) are the vertices of a simplex of L. We make use of a standard device called *barycentric extension* to extend this correspondence into a continuous mapping, still called f, of the polytope $|K|$ into the polytope $|L|$.

Let $s^n = \langle p_0 \cdots p_n \rangle$ be a simplex of K. Each point x of s^n is referred to (nonnegative) barycentric coordinates with respect to the vertices p_i. Thus we can represent x as $(x_0, x_1, \ldots, x_n)$ or, in vector notation (see Section 5–3), as

$$x = \sum_{i=0}^{n} x_i \cdot p_i, \qquad \sum x_i = 1, \quad x_i \geqq 0.$$

The continuous extension f can now be defined by setting

$$f(x) = \sum_{i=0}^{n} x_i \cdot f(p_i).$$

That is, we use the barycentric coordinates of the point x as the coordinates of its image point $f(x)$ by assigning x_i to the vertex $f(p_i)$. If it happens that $f(p_i) = f(p_j)$, $i \neq j$, then the barycentric coordinate of $f(x)$ with respect to the vertex $f(p_i)$ is $x_i + x_j$, and so on.

It is easy to verify that the extended mapping f is well-defined at every point of the polytope $|K|$. And since the barycentric coordinates of a point are continuous functions of that point, it follows that the extended mapping is continuous. The mapping f is called a *simplicial mapping* and, as we shall see shortly, such mappings constitute an important class.

In the arguments to follow, we will use the following lemmas, the proofs of which are left as exercises.

LEMMA 5–21. In a Euclidean space E^k, let $\{p_n\}$ and $\{q_n\}$ be two sequences of points converging to points p and q respectively. Denote by $\overline{p_n q_n}$ the length of the line segment between p_n and q_n. For each n, let x_n be a point on $[p_n, q_n]$. If the limit of $d(x_n, p_n)$ as $n \to \infty$ exists, then there is a point x on $\overline{pq}$ such that (1) $\lim_{n\to\infty} d(x_n, p_n) = d(x, p)$ and (2) the sequence $\{x_n\}$ converges to x.

LEMMA 5–22. Let $v_0, v_1, \ldots, v_k$ be vertices of a star-finite complex K, and let $\mathring{\mathrm{St}}(v_i)$ be the open stars of these vertices in $|K|$. Then the vertices $v_0, v_1, \ldots, v_k$ form a simplex of K if and only if the intersection $\bigcap_{i=0}^{k} \mathring{\mathrm{St}}(v_i)$ is not empty.

The chief result of this section is stated next, but its proof will be the end product of several steps.

THEOREM 5–23 (Simplicial approximation). Let $|K|$ and $|L|$ be two finite polytopes with triangulations K and L respectively, and let f be a continuous mapping of $|K|$ into $|L|$. Then, given any positive number

ϵ, there exist barycentric subdivisions K^* and L^* of K and L respectively, and a continuous mapping s of $|K|$ into $|L|$ such that

(1) s is a simplicial mapping of $|K^*|$ into $|L^*|$,
(2) for every point x of $|K|$, $d(f(x), s(x)) < \epsilon$, and
(3) s is homotopic to f.

This result will be seen to have important consequences as we proceed. For the present, we observe that this theorem implies that the *simplicial mappings are dense in every homotopy class of one finite polytope into another.* Use will be made of this property shortly.

If K and L are triangulations of the polytopes $|K|$ and $|L|$ respectively, and if f is a continuous mapping of $|K|$ into $|L|$, then we say that *K is star-related to L relative to f* provided that for every vertex p_i of K there is a vertex v_j of L such that the image $f(\mathring{\text{S}}\text{t}(p_i))$ is contained in $\mathring{\text{S}}\text{t}(v_j)$.

THEOREM 5–24. Let $|K|$ and $|L|$ be finite polytopes with triangulations K and L respectively, and let f be a continuous mapping of $|K|$ into $|L|$. If K is star-related to L relative to f, then there exists a mapping s of $|K|$ into $|L|$ such that

(1) s is a simplicial mapping of K into L,
(2) if x is any point of $|K|$, there is a vertex v_j of L such that both $f(x)$ and $s(x)$ lie in $\mathring{\text{S}}\text{t}(v_j)$, and
(3) s is homotopic to f.

Proof: It is assumed that for each vertex p_i of K there is at least one vertex $v_{j(i)}$ of L such that $f(\mathring{\text{S}}\text{t}(p_i))$ lies in $\mathring{\text{S}}\text{t}(v_{j(i)})$. We may thus define a correspondence s between the vertices of K and those of L by setting

$$s(p_i) = v_{j(i)}.$$

By assumption, each vertex of K has an image under s [we choose any one of the possible vertices $v_{j(i)}$].

Now let $\langle p_0 \cdots p_k \rangle$ be any simplex of K. By Lemma 5–22, the intersection $\bigcap_{i=0}^{k} \mathring{\text{S}}\text{t}(p_i)$ is not empty. Since $f(\mathring{\text{S}}\text{t}(p_i))$ lies in $\mathring{\text{S}}\text{t}(s(p_i))$ by our definition of $s(p_i)$, it follows that $\bigcap_{i=0}^{k} \mathring{\text{S}}\text{t}(s(p_i))$ is not empty. Again from Lemma 5–22, the vertices $s(p_0), \ldots, s(p_k)$ are those of a simplex of L. Thus s is simplicial, and by barycentric extension we obtain a continuous mapping, still called s, of $|K|$ into $|L|$. It is claimed that this mapping s also satisfies conditions (2) and (3) of the conclusion of the theorem.

First, every point x in $|K|$ lies in the interior of some simplex s^k of K, s^k taken to be of minimum dimension. This implies that every barycentric coordinate of x with respect to the vertices of s^k is positive, while the other coordinates of x are zero. If p is any vertex of s^k, then x lies in the open star of p. By definition, the image point $f(x)$ is a point of $f(\mathring{\mathrm{S}}\mathrm{t}(p))$, and this lies in $\mathring{\mathrm{S}}\mathrm{t}(s(p))$. Therefore $f(x)$ is a point of $\mathring{\mathrm{S}}\mathrm{t}(s(p))$. But also $s(x)$ lies in $\mathring{\mathrm{S}}\mathrm{t}(s(p))$, for the barycentric coordinate of $s(x)$ with respect to the vertex $s(p)$ is nonzero. Thus the mapping s satisfies condition (2).

It remains to show that s is homotopic to f. That is, we must define a mapping $h:|K| \times I^1 \to |L|$ such that $h(x, 0) = f(x)$ and $h(x, 1) = s(x)$, for each point x of $|K|$. Let x be a point of a simplex $s^k = \langle p_0 \cdots p_k \rangle$ in K. Since s is simplicial, $s(p_0), \ldots, s(p_k)$ are vertices of a simplex s^* in L. Having that $f(x)$ lies in $\mathring{\mathrm{S}}\mathrm{t}(s(p_i))$ for each vertex of s^k, it follows that $f(x)$ is a point of $\bigcap_{i=0}^k \mathring{\mathrm{S}}\mathrm{t}(s(p_i))$, which is precisely the simplex s^* of L. Having both $f(x)$ and $s(x)$ in the same simplex s^* of L, we make use of the convexity of s^* and join $f(x)$ to $s(x)$ by a (unique) line segment in s^*. Properly metrized, this line segment will be the image under a homotopy h of the line segment $x \times I^1$ in the homotopy cylinder $|K| \times I^1$. In particular, letting $d(f(x), s(x)) = 1$, we write in vector notation

$$h(x, t) = (1 - t) \cdot f(x) + t \cdot s(x).$$

The continuity of h as defined here is a consequence of Lemma 5–21. □

To prove the simplicial approximation theorem (5–23), we need only remove the hypothesis in Theorem 5–24 that K and L are star-related relative to f. To do this, we next replace the triangulations K and L by barycentric subdivisions K^* and L^*, L^* being chosen to yield the desired accuracy of approximation and K^* being chosen so as to be star-related to L^* relative to f.

Proof of Theorem 5–23: Given any positive number ϵ, Theorem 5–20 assures the existence of an integer n such that the mesh of the nth barycentric subdivision $L^{(n)}$ of L is less than $\epsilon/2$. This implies that each $\mathring{\mathrm{S}}\mathrm{t}(v)$, v a vertex of $L^{(n)}$, has diameter $<\epsilon$. The collection of open stars $\{\mathring{\mathrm{S}}\mathrm{t}(v)\}$ of vertices of $L^{(n)}$ is a finite open covering of the compact metric space $|L|$. By Theorem 1–32, there is a positive number η such that if A is any subset of $|L|$ of diameter less than η, then A lies in $\mathring{\mathrm{S}}\mathrm{t}(v)$ for some vertex v of $L^{(n)}$.

Next, $|K|$ is a compact metric space, and hence the mapping f is uniformly continuous. Thus there exists a positive number δ such that, whenever $d(x, y) < \delta$, we have $d(f(x), f(y)) < \eta$. Again using Theorem 5–20, we find that there is an integer m such that $K^{(m)}$ has mesh $< \delta/2$. Every star $\mathring{\mathrm{S}}\mathrm{t}(p)$ of a vertex of $K^{(m)}$ therefore has diameter $< \delta$ and, by construction, the diameter of $f(\mathring{\mathrm{S}}\mathrm{t}(p))$ is less than η. Hence $f(\mathring{\mathrm{S}}\mathrm{t}(p))$ is contained in $\mathring{\mathrm{S}}\mathrm{t}(v)$ for some vertex v of $L^{(n)}$. That is, $K^{(m)}$ and $L^{(n)}$ are star-related to f. Taking K^* to be $K^{(m)}$ and L^* to be $L^{(n)}$, the proof of Theorem 5–23 follows immediately from Theorem 5–24. □

5-7 Abstract simplicial complexes. One of the chief reasons for the introduction of simplicial complexes will be found in the next chapter. Briefly, a simplicial complex supports an algebraic structure (homology theory) that has proved to be very valuable. Therefore, with the goal in mind of utilizing this powerful mechanism in situations not involving a polytope, topologists were led to a definition of an abstract simplicial complex. We will not *use* this concept until Chapter 8, but because we can and must do so, much of the development of the next two chapters will be phrased in terms of abstract simplicial complexes.

An *abstract simplicial complex* K is a pair $(\mathcal{V}, \Sigma)$ where $\mathcal{V}$ is a set of (abstract) elements called *vertices*, and Σ is a collection of finite subsets of $\mathcal{V}$ with the property that each element of $\mathcal{V}$ lies in some element of Σ and, if σ is any element of Σ, then every subset of σ is again an element of Σ.

Again we distinguish between finite and infinite abstract simplicial complexes, depending upon whether the set $\mathcal{V}$ is finite or infinite. The dimension of a simplex σ is one less than the number of vertices in σ. The dimension of the abstract simplicial complex K is defined to be the maximum dimension of the elements of Σ if such exists; otherwise, K is of infinite dimension.

We may define the *star of a simplex* σ, $\mathrm{St}(\sigma)$, to be the collection of all elements in Σ of which σ is a subset. Hence we again speak of star-finite complexes. The reader may encounter definitions of a complex (not usually considered as "simplicial") in which the elements of the collection Σ are not assumed to be necessarily finite. If this is the case, then one usually defines a *closure-finite* complex as follows. The *closure of a simplex* σ, denoted by $\mathrm{Cl}(\sigma)$, is the subcomplex of K consisting of σ and all the faces of σ. Then a complex K is *closure-finite* if each closure $\mathrm{Cl}(\sigma)$ of a simplex of K is finite. A complex which is both closure-finite and star-finite is said to be *locally finite*. We will not need these latter terms with our definition (except for *star-finite*).

The reader will easily prove that every geometric simplicial complex satisfies the above definition. And a sort of converse theorem may also be established. To state it, we need a definition. Two (abstract) simplicial complexes are said to be *isomorphic complexes* if there is a one-to-one simplicial mapping φ of one onto the other such that the inverse mapping φ^{-1} is also simplicial.

THEOREM 5-25. Every finite abstract simplicial complex is isomorphic to a geometric simplicial complex (called a *geometric realization*).

Proof: Let $K = (\mathcal{V}, \Sigma)$ be a finite abstract simplicial complex with vertices $v_0, v_1, \ldots, v_n$. In Euclidean n-space E^n, let p_0 denote the origin and p_i denote the unit point on the ith axis. Clearly, the points $p_0, \ldots, p_n$

are pointwise independent and any subcollection of these points forms a geometric simplex, indeed forms a face of what we might call the *standard n-simplex* in E^n.

Consider the one-to-one correspondence $v_i \leftrightarrow p_i$, $i = 0, \ldots, n$. If a subcollection $v_{i_0}, \ldots, v_{i_k}$ of vertices of K forms a k-simplex in Σ, then there corresponds a geometric k-simplex with vertices $p_{i_0}, \ldots, p_{i_k}$. In this way we build in E^n a complex which is easily seen to be isomorphic to K by its very construction. □

EXAMPLE. Let X be a compact Hausdorff space, and let $\mathfrak{U}$ be a finite covering of X by open sets. Define $K = (\mathcal{V}, \Sigma)$ by taking $\mathcal{V}$ to be the collection $\mathfrak{U}$ and by saying that a subset $U_0, U_1, \ldots, U_p$ of elements of $\mathfrak{U}$ is a simplex in Σ if and only if the intersection $\bigcap_{i=0}^{p} U_i$ is not empty. Then K is an abstract simplicial complex. (Actually K is not "abstract," it is quite concrete.) To see this, we need only note that if $\bigcap_{i=0}^{p} U_i$ is not empty, then any subcollection $U_{i_0}, \ldots, U_{i_k}$ of the open sets $U_0, \ldots, U_p$ also have a nonempty intersection and, by definition, must constitute an element of Σ. This is not a contrived example, the idea here is at the base of the Čech homology theory (see Chapter 8).

EXERCISE 5–3. Let M be a compact metric space with metric d, and let ϵ be a positive number. Define $K_\epsilon = (\mathcal{V}, \Sigma)$, where the elements of $\mathcal{V}$ are the points of M and where a finite subset of such points constitutes a simplex in Σ if and only if the diameter of the finite subset is less than ϵ. Show that K_ϵ is a simplicial complex. This example is also useful and will be seen in Vietoris homology theory (again see Chapter 8).

5–8 An imbedding theorem for polytopes. We prove here that any n-dimensional polytope may be imbedded rectilinearly in Euclidean $(2n + 1)$-space. This is a special case of the imbedding theorem for n-dimensional separable metric spaces which we quoted in Section 3–9.

A set of points is said to be in *general position* in E^m if no $r + 2$ of the points lie on an r-dimensional hyperplane, $r = 1, 2, \ldots, m - 1$. That is, every subset with less than $m + 2$ points is geometrically independent.

THEOREM 5–26. Let $\{x_1, x_2, \ldots\}$ be any finite or countably infinite set of points in E^m, and let ϵ be any positive number. Then there is a set $\{y_1, y_2, \ldots\}$ of points in general position in E^m such that, for each $i = 1, 2, \ldots$, the distance $d(x_i, y_i) < \epsilon$.

Proof: Take $y_1 = x_1$. Suppose that $y_1, y_2, \ldots, y_{k-1}$ have been chosen to satisfy the desired conditions. Then one may choose y_k to be any point in the spherical neighborhood $S(x_k, \epsilon)$ such that y_k does not lie on any of the finitely many hyperplanes determined by all subsets of

$$\{y_1, y_2, \ldots, y_{k-1}\}. \square$$

EXERCISE 5–4. Prove that there exists a dense set of points in general position in E^m.

The following lemma is needed in our ensuing argument, but we may leave the proof as an easy exercise.

LEMMA 5–27. If two star-finite geometric complexes K_1 and K_2 are isomorphic, then their polytope carriers $|K_1|$ and $|K_2|$ are homeomorphic.

THEOREM 5–28. Let $|K|$ be an n-dimensional polytope with a triangulation K. Then $|K|$ may be imbedded rectilinearly in E^{2n+1}. Furthermore, $|K|$ may be imbedded as a closed subset of E^{2n+1}.

Proof: In view of Lemma 5–27, it suffices to construct in E^{2n+1} a geometric complex isomorphic to K. If this is done in such a way that no bounded region in E^{2n+1} contains more than a finite number of vertices of the new complex, then the resulting polytope will be closed in E^{2n+1}.

Let $v_1, v_2, \ldots$ be an arbitrary ordering of the vertices of K. To each vertex v_k we assign the point $(k, 0, \ldots, 0)$ in E^{2n+1}. Applying Theorem 5–26 with $\epsilon = \frac{1}{3}$, we obtain a set $\{y_1, y_2, \ldots\}$ in general position in E^{2n+1}. These points will be the vertices of a complex isomorphic to K, and the manner in which they are chosen clearly implies that at most a finite number of these vertices lies in any bounded region.

Now if $\langle v_{i_0} \cdots v_{i_p}\rangle$ is a p-simplex of K, $p \leqq n$, we form the p-simplex $\langle y_{i_0} \cdots y_{i_p}\rangle$ in E^{2n+1}. If we show that the resulting collection of simplexes forms a complex, then the argument will be complete, for the isomorphism will be obvious. To this end, let

$$s^q = \langle y_{i_0} \cdots y_{i_{p-1}} y_{i_p} \cdots y_{i_q}\rangle$$

and

$$s^r = \langle y_{i_0} \cdots y_{i_{p-1}} y_{j_p} \cdots y_{j_r}\rangle$$

be two simplexes with the face

$$s^{p-1} = \langle y_{i_0} \cdots y_{j_{p-1}}\rangle$$

in common. Suppose that there is a point x in $s^q \cap s^r - s^{p-1}$. Then x is not in the hyperplane containing s^{p-1} because the intersection of a simplex with the hyperplane containing one of its faces is that face alone. Hence we may construct the p-dimensional simplex $\langle x y_{i_0} \cdots y_{i_{p-1}}\rangle$ which, by convexity, must lie in both s^q and s^r. The hyperplanes containing s^q and s^r, which are q-dimensional and r-dimensional respectively, contain in their intersection the p-dimensional hyperplane determined by $\langle x y_{i_0} \cdots y_{i_{p-1}}\rangle$. Hence the $(q + 1) + (r + 1) - p$ points $y_{i_0}, \ldots, y_{i_q}, y_{j_p}, \ldots, y_{j_r}$ all lie on a hyperplane with dimension not exceeding $q + r -$ p. Then, since $q + r + 2 - p \leq q + r + 2 \leq 2n + 2$, we have a contradiction of the fact that the vertices y_i were taken to be in general position. □

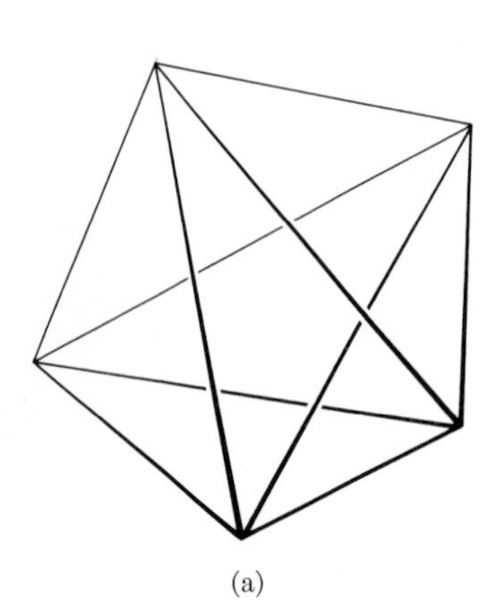
(a)

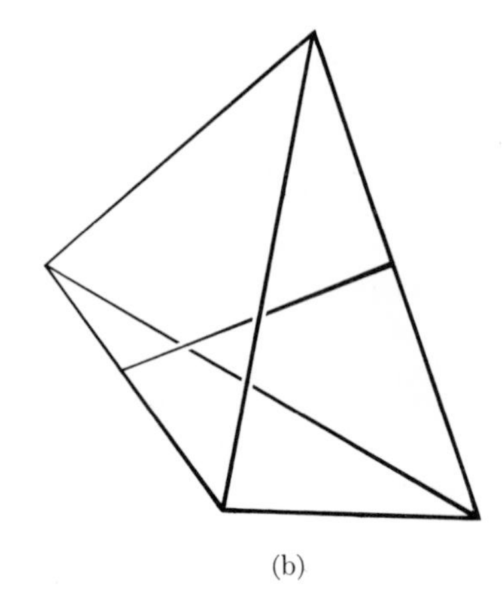
(b)

FIG. 5–6. Primitive skew curves.

Flores [77] has shown that the complex consisting of all faces of dimension $\leqq n$ of a $(2n + 2)$-simplex s^{2n+2} (the n-skeleton of s^{2n+2}) cannot be imbedded in E^{2n}. This example indicates that the dimension $2n + 1$ in Theorem 5–28 cannot, in general, be reduced.

In this connection, it is of interest to note that Kuratowski [94] has shown that a polytope may be imbedded in the plane if and only if it does not contain a subset homeomorphic to either of the primitive skew curves which we picture in Fig. 5–6. Note that Fig. 5–6(a) is precisely Flores' example for $n = 1$.

EXERCISE 5–5. If the points $p_0, \ldots, p_k$ are pointwise independent in E^n, then show that every subset of these points is also pointwise independent.

EXERCISE 5–6. Show that the points $p_i = (x_1^i, x_2^i, \ldots, x_n^i)$, $i = 0, 1, \ldots, k \leq n$, are pointwise independent if and only if the following matrix has rank $k + 1$:

$$\begin{pmatrix} x_1^0, & \ldots, & x_n^0, & 1 \\ \vdots & & & \\ x_1^k, & \ldots, & x_n^k, & 1 \end{pmatrix}.$$

EXERCISE 5–7. Let s^k be a k-simplex with vertices $p_0, \ldots, p_k$. Let $a_0, \ldots, a_k$ be distinct points of E^n, and let $f{:}s^k \to E^n$ be the barycentric extension of the correspondence $p_i \to a_i$. Show that f is an imbedding if and only if the points $a_0, \ldots, a_k$ are pointwise independent.

EXERCISE 5–8. Prove that the intersection of two finite polytopes in some Euclidean space is again a finite polytope.

EXERCISE 5–9. Prove that the convex hull of the difference of two finite polytopes in some Euclidean space is again a finite polytope.

EXERCISE 5–10. Let A and B be convex regions in E^m and E^n, respectively. Show that $A \times B$ is a convex region in E^{m+n}.

EXERCISE 5–11. Show that a minimum triangulation of I^n contains $n!$ n-simplexes.

EXERCISE 5–12. Show that the 3-sphere S^3 is a union of two solid tori.

EXERCISE 5–13. Prove that the barycentric subdivision of a pseudomanifold (see Exercise 6–15) is a pseudomanifold.

EXERCISE 5–14. Prove that every nonempty open set in E^n is an infinite complex.

EXERCISE 5–15. Apply the simplicial approximation theorem to prove that there is only a countable number of homotopy classes of mappings of one finite polytope into another.

CHAPTER 6

SIMPLICIAL HOMOLOGY THEORY

6–1 Introduction. Homology theory is essentially an algebraic study of the connectivity properties of a space. In Chapter 4, we introduced one such device, the homotopy groups, $\pi_n(Y)$. Although they are appealing intuitively, the homotopy groups are difficult to calculate even for comparatively simple spaces. The simplicial homology groups developed in this chapter permit us to answer questions about connectivity similar to those answered by means of homotopy groups. And the simplicial homology groups are computed by almost mechanical methods. On the other hand, the difficulties in homology theory are found in the underlying structures and the combinatorial approach which, for the beginning student, seems to disguise the motivation for an inordinate length of time. We try to alleviate this situation with this lengthy introduction.

Historically, the study of topology developed in two major branches, the point-set topology, which we have examined already, and the combinatorial study of connectivity, which we are about to begin and which was originated by Poincaré [113]. The unification of the two areas of interest has been under way for a generation and is still not complete. Even today, one hears of point-set topologists as distinguished from algebraic topologists.

This book follows a pattern derived from history. Having a background in point-set topology, we now introduce a radical change in our approach. Where we have studied certain point-set invariants of topological mapping, we now turn to algebraic invariants. For a brief and well-organized history of this topic, the reader is referred to "The sphere in topology" by R. L. Wilder [132].

To help the beginner keep sight of the forest, we will discuss at some length the 2-dimensional torus T pictured in Fig. 6–1. Our aim in this discussion is to explain the geometric significance of the purely algebraic concepts to be formulated shortly. First, look at T from the point-set standpoint. Clearly, this surface is a compact, connected and locally Euclidean metric space. It is also locally connected, etc. Of course, all such information above does not characterize the torus. All of these facts are also true of the 2-dimensional sphere as well. Suppose that our goal is modest, namely, that it is to distinguish topologically between T and S^2. How might it be done?

An immediate answer can be given by computing the fundamental groups of T and S^2. It turns out that the group $\pi_1(S^2)$ is a trivial group whereas $\pi_1(T)$ is not (such a curve as Z in Fig. 6–1 cannot be shrunk to a point in T). Thus we already have knowledge that suffices to distinguish

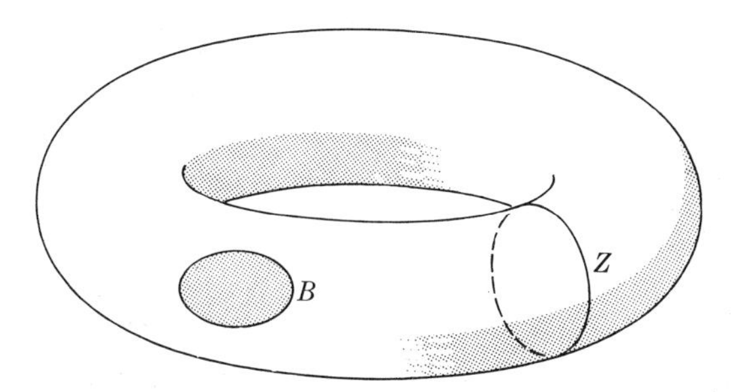

FIGURE 6–1

between a torus and a 2-sphere. Let us proceed, however, to give further study to the torus.

Envisioning a 2-sphere, it is intuitively obvious that any closed curve on the surface forms the boundary of a portion of the sphere. Or in equivalent terms, any closed curve on S^2 disconnects S^2. The same is not true of the torus. For cutting along the curve Z in Fig. 6–1 does not disconnect the torus. This implies that the curve Z is not the boundary of a portion of T. Of course, there are closed curves, such as B in Fig. 6–1, which are boundaries. The curve B may be considered as the boundary of either the shaded disc or of the complement of that disc in T.

Because the intuitive idea of a closed curve includes the notion that it "goes around something" and because it is 1-dimensional, we will temporarily and imprecisely refer to any closed curve such as B or Z in Fig. 6–1 as a *1-dimensional cycle* on T. Note that while we have pictured only *simple* closed curves on T, we do not so restrict our cycles. Those special cycles, such as B, that bound a portion of the torus T do not tell us much about the structure of the torus in the large. We will merely call them *bounding 1-cycles* and ignore them. It is the nonbounding 1-cycles, such as Z, that interest us.

There is obviously an uncountable number of such nonbounding 1-cycles on the torus. By utilizing simple notions, we will reduce this cardinality drastically. First, the two cycles Z_1 and Z_2 shown in Fig. 6–2 are not

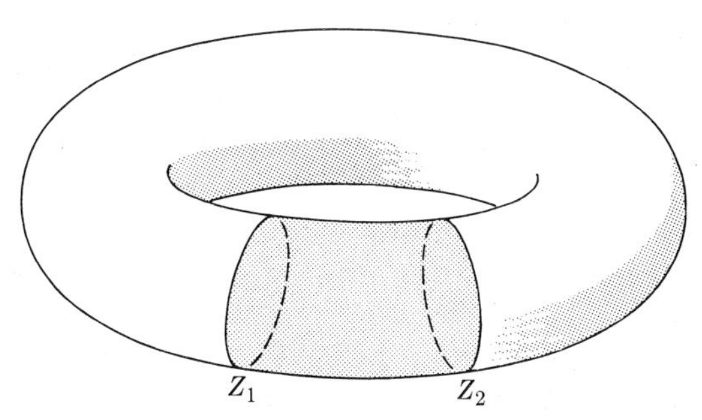

FIGURE 6–2

intrinsically different since they both go around the torus once latitudinally. More to the point, however, is the fact that taken together they form the boundary of a portion of the torus (e.g., the shaded cylinder).

The idea of taking two cycles together should suggest that we can introduce an operation of addition of cycles. For our temporary purposes, then, let us extend the definition of a 1-cycle to include the point-set unions of finitely many closed curves. In this way the operation of union gives us a well-defined addition of two cycles. Looking at Fig. 6–2 in this light, we see that $Z_1 + Z_2$ is a bounding 1-cycle. We are led to a natural method of expressing such a relation between two 1-cycles, namely, by means of an equivalence relation. We will say that a cycle Z_1 *is homologous* to a cycle Z_2 (abbreviated $Z_1 \sim Z_2$) if $Z_1 + Z_2$ is a bounding cycle. The reader will find it difficult to do precisely, but forgetting rigor he may verify that this is indeed an equivalence relation.

It should be noted that this definition of the addition of cycles implies that for any 1-cycle Z, $Z + Z$ is a bounding 1-cycle. For having two copies of such a cycle Z, as in Fig. 6–3, we may use one as the boundary of each "side" of the cut made along Z in the torus, as in Fig. 6–3. The reader is warned that this situation is *not* true in the general definition given in Section 6–4.

Furthermore, a closed curve such as Z_1 in Fig. 6–4, which passes around the torus twice, is a bounding cycle in the present situation. And if Z is any 1-cycle passing around (latitudinally) just once, then we have $Z_1 \sim Z + Z$. To see this, look at the 1-cycles Z' and Z'', both homologous to Z, in Fig. 6–4.

All this implies that any purely latitudinal cycle on the torus is either a boundary (if it passes around T an even number of times) or is homologous to Z in Fig. 6–1 (if it passes around T an odd number of times). Similar reasoning applies to the purely longitudinal cycles, so we now have two major equivalence classes of cycles (three, if we wish to include the trivial class of all boundaries). But as the cycle Z in Fig. 6–5 illustrates, a cycle can pass around the torus both latitudinally and longitudinally.

Such a cycle, however, is not new. It is homologous to the sum of two cycles, Z_1 and Z_2, one from each of the previously discussed classes. To show this, we have cut the torus along a latitudinal cycle Z_1 and a longitudinal cycle Z_2 to form a rectangle as in Fig. 6–6. The cycle Z of Fig. 6–5 now appears as the diagonal labeled Z. We readily see that $Z \sim Z_1 + Z_2$ since the sum $Z + Z_1 + Z_2$ is a boundary (of the shaded triangle, for instance).

Arguments such as these show us that we need consider but two essentially different 1-dimensional cycles on the torus, a result which implies that the 1-dimensional *Betti number* of the torus is 2. This corresponds to the intuitive notion that there are two "holes" in a torus.

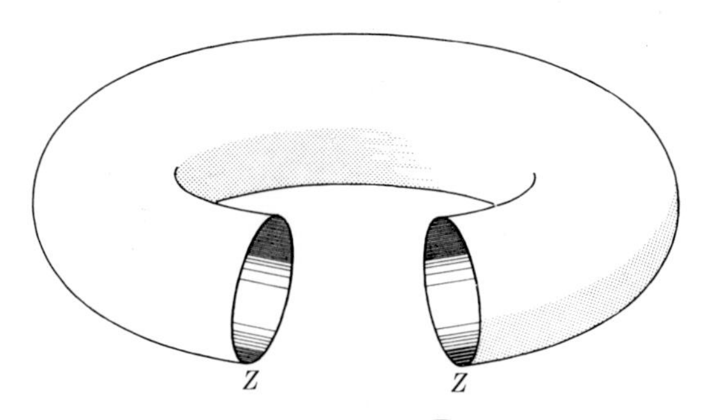

FIGURE 6–3

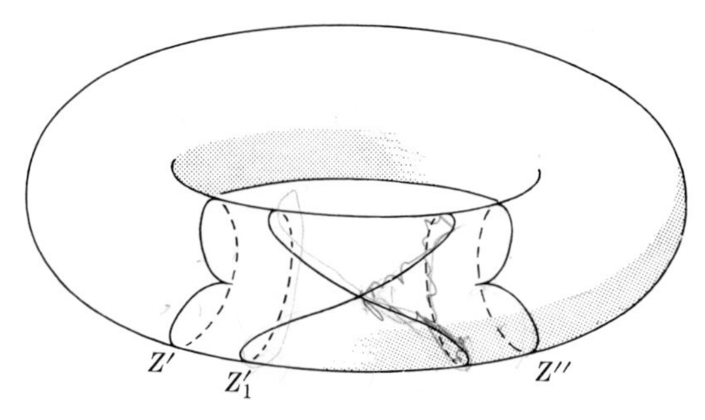

FIGURE 6–4

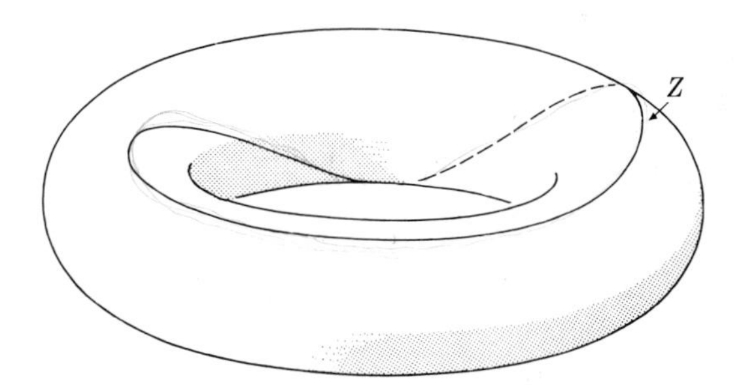

FIGURE 6–5

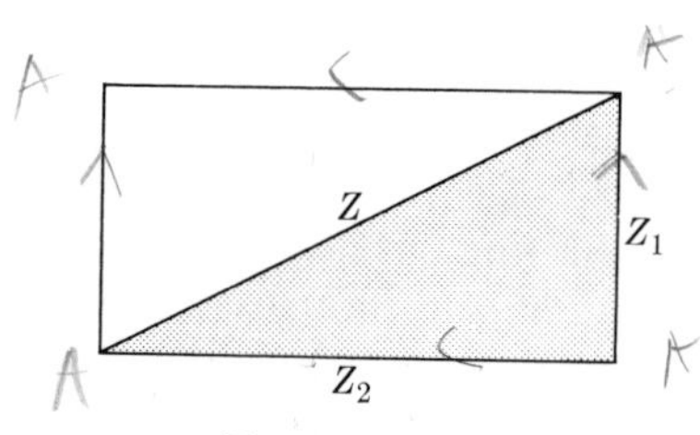

FIGURE 6–6

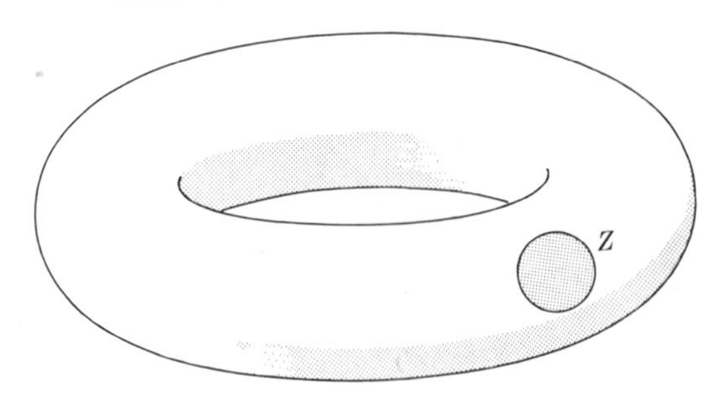

FIGURE 6–7

It might be well to note that apparently the "generators" of the homology classes of cycles discussed above are among those of the homotopy classes of the fundamental group of the torus. Indeed, this is true. However, the two types of equivalence classes, homotopy and homology, are not equivalent in general. To see this, consider the following example. In Fig. 6–7, we picture a torus from which an open disc has been removed. Now, as far as our homology classes are concerned, that "hole" is invisible. That is, there are still just two homology classes of 1-cycles.

The 1-cycle Z bounding the hole is actually the boundary of the rest of the torus as well. Hence Z is a bounding cycle. On the other hand, this hole is visible from the homotopy standpoint. For, although it is not easy to visualize, it is impossible to deform the simple closed curve Z in Fig. 6–7 to a point, while staying on the torus. It should be clear then that, in some way, the homotopy groups are more discerning than are the homology groups. More precise relations between the homotopy and the homology groups will be mentioned in Section 8–5.

In our development in this chapter, homology theory will be based upon an algebraic structure placed upon a simplicial complex. This seems to limit us to a consideration of polytopes only. For the present this will be the case, but the restriction is not so severe as it may seem. Many interesting spaces admit of a triangulation, and for many others we have suitable limiting processes (e.g., Čech homology, Chapter 8) that yield a homology theory.

6–2 Oriented complexes. As we know from analytic geometry, the concept of a directed (oriented) line segment allows the introduction of algebraic methods into geometry. In an analogous manner, the *oriented simplex* permits the use of algebraic tools in our study of complexes. We will gain generality by phrasing our definitions in terms of abstract simplicial complexes, but most of our early examples will be taken from the geometric complexes. This is done to attain our double goal of explaining the geometry underlying homology theory while being sufficiently general to permit the necessary extensions later.

An *oriented simplex* is obtained from an abstract p-simplex

$$\langle v_0 \cdots v_p \rangle = \sigma^p$$

(see Section 5-7) as follows. We choose some arbitrary fixed ordering of the vertices $v_0, v_1, \ldots, v_p$. The equivalence class of even permutations of this fixed ordering is the *positively oriented simplex*, which we denote by $+\sigma^p$, and the equivalence class of odd permutations of the chosen ordering is the *negatively oriented simplex*, $-\sigma^p$. For example, if $\langle v_0 v_1 \rangle = +\sigma^1$, then $\langle v_1 v_0 \rangle = -\sigma^1$. For a geometric simplex $s^1 = \langle p_0 p_1 \rangle$, orientation is equivalent to a choice of a positive direction on the line segment. Again, if we have chosen to let $\langle v_0 v_1 v_2 \rangle$ represent $+\sigma^2$, then $\langle v_1 v_2 v_0 \rangle$ and $\langle v_2 v_0 v_1 \rangle$ also represent $+\sigma^2$, while $\langle v_1 v_0 v_2 \rangle$, $\langle v_0 v_2 v_1 \rangle$, and $\langle v_2 v_1 v_0 \rangle$ each represents $-\sigma^2$. For a geometric simplex $s^2 = \langle p_0 p_1 p_2 \rangle$, orientation is equivalent to choosing a positive direction of traversing the three 1-faces of s^2. We note that $\langle p_0 p_1 p_2 \rangle$ and $\langle p_1 p_0 p_2 \rangle$ are opposite cyclic orderings of the vertices p_0, p_1, and p_2 and correspond to opposite directions of traversing the boundary of the 2-simplex.

An *oriented simplicial complex* is obtained from an abstract simplicial complex by choosing an arbitrary fixed orientation for each simplex in the complex. This may be done without considering how the individual simplexes are joined or whether one simplex is a face of another. One automatic method of orienting a complex (which is not necessarily the most efficient method) is to decide upon a fixed ordering of the vertices of the complex and let this ordering induce the positive orientation of the simplexes in the natural way. We will use this method in several examples.

Basic assumption. Every complex we consider henceforth will be assumed to be oriented whether or not the adjective *oriented* is used.

6-3 Incidence numbers. Given an oriented simplicial complex K, we associate with every pair of simplexes σ^m and σ^{m-1}, which differ in dimension by unity, an *incidence number* $[\sigma^m, \sigma^{m-1}]$ defined as follows:

$$[\sigma^m, \sigma^{m-1}] = 0 \qquad \text{if} \quad \sigma^{m-1} \text{ is not a face of } \sigma^m \text{ in } K;$$
$$[\sigma^m, \sigma^{m-1}] = \pm 1 \qquad \text{if} \quad \sigma^{m-1} \text{ is a face of } \sigma^m \text{ in } K.$$

To decide between $+1$ and -1 in the case where σ^{m-1} is a face of σ^m, we note that if $\sigma^m = \langle v_0 \cdots v_m \rangle$, then $+\sigma^{m-1} = \pm\langle v_0 \cdots \hat{v}_i \cdots v_m \rangle$ (recall that the circumflex accent denotes the omission of the vertex v_i), where the orientation of σ^{m-1} determines the sign. If $+\sigma^{m-1} = +\langle v_0 \cdots \hat{v}_i \cdots v_m \rangle$, consider the oriented simplex $\langle v_i v_0 \cdots \hat{v}_i \cdots v_m \rangle$. This is either $+\sigma^m$ or $-\sigma^m$; if it is $+\sigma^m$, we take the incidence number $[\sigma^m, \sigma^{m-1}]$ to be $+1$, and if $\langle v_i v_0 \cdots \hat{v}_i \cdots v_m \rangle = -\sigma^m$, we take $[\sigma^m, \sigma^{m-1}] = -1$. Again, if $\langle v_0 v_1 \cdots \hat{v}_i \cdots v_m \rangle = -\sigma^{m-1}$, then $[\sigma^m, \sigma^{m-1}] = -1$ if $\langle v_i v_0 \cdots \hat{v}_i \cdots v_m \rangle = +\sigma^m$, and $[\sigma^m, \sigma^{m-1}] = +1$ if $\langle v_i v_0 \cdots \hat{v}_i \cdots v_m \rangle = -\sigma^m$.

If $[\sigma^m, \sigma^{m-1}] = +1$, then σ^{m-1} is a *positively oriented face* of σ^m, and if the incidence number is negative, then σ^{m-1} is a *negatively oriented face* of σ^m. The choice of a positive ordering of the vertices of σ^m clearly induces a natural ordering of the vertices in each face of σ^m. Thus an orientation of σ^m induces a natural orientation of its faces. The definition above amounts to this: if σ^{m-1} is a face of σ^m, then the incidence number $[\sigma^m, \sigma^{m-1}]$ is positive or negative depending upon whether the chosen orientation of σ^{m-1} agrees or disagrees with the orientation of σ^{m-1} induced by that of σ^m.

EXAMPLE. If $+\sigma^2 = \langle v_0 v_1 v_2 \rangle$ and $+\sigma^1 = \langle v_1 v_2 \rangle$, then it is easily verified that $[\langle v_0 v_1 v_2 \rangle, \langle v_1 v_2 \rangle] = +1$. But if $+\sigma^1 = \langle v_2 v_1 \rangle$, then we have $[\langle v_0 v_1 v_2 \rangle, \langle v_2 v_1 \rangle] = -1$. For, inserting the missing vertex v_0 in front of σ^1, we have $\langle v_0 v_1 v_2 \rangle = +\sigma^2$ in the first case and $\langle v_0 v_1 v_2 \rangle = -\sigma^2$ in the second. The reader should work out a number of similar examples for higher-dimensional simplexes.

The oriented simplicial complex K, together with the system of incidence numbers $[\sigma^m, \sigma^{m-1}]$, constitutes the basic structure supporting a simplicial homology theory. We develop this next. First, however, note that for each dimension m, we may associate with K a matrix $([\sigma_i^m, \sigma_j^{m-1}])$ of incidence numbers, where the index i runs over all m-simplexes of K and the index j runs over all $(m - 1)$-simplexes. A study of this system of *incidence matrices* would yield the connectivity properties we wish to investigate. This technique was commonly used in the early days of "combinatorial" topology, but we do not develop it. The group-theoretic formulation to be introduced below evolved slowly during the decade 1925–1935 and seems to have been first suggested by E. Noether.

One basic property of the incidence numbers is needed.

THEOREM 6–1. Given any particular simplex σ_0^m of an oriented simplicial complex K, the following relationship among the incidence numbers holds:

$$\sum_{i,j} [\sigma_0^m, \sigma_i^{m-1}] \cdot [\sigma_i^{m-1}, \sigma_j^{m-2}] = 0.$$

Proof: Every $(m - 2)$-simplex $\langle v_0 \cdots \hat{v}_k \cdots \hat{v}_l \cdots v_m \rangle$ in σ^m is a face of exactly two $(m - 1)$-faces of σ^m. Hence the sum

$$\sum_{i=0}^{m} [\langle v_0 \cdots v_m \rangle, \langle v_0 \cdots \hat{v}_i \cdots v_m \rangle] \cdot [\langle v_0 \cdots \hat{v}_i \cdots v_m \rangle, \langle v_0 \cdots \hat{v}_k \cdots \hat{v}_l \cdots v_m \rangle]$$

$$= [\langle v_0 \cdots v_m \rangle, \langle v_0 \cdots \hat{v}_k \cdots v_m \rangle] \cdot [\langle v_0 \cdots \hat{v}_k \cdots v_m \rangle, \langle v_0 \cdots \hat{v}_k \cdots \hat{v}_l \cdots v_m \rangle]$$

$$+ [\langle v_0 \cdots v_m \rangle, \langle v_0 \cdots \hat{v}_l \cdots v_m \rangle] \cdot [\langle v_0 \cdots \hat{v}_l \cdots v_m \rangle, \langle v_0 \cdots \hat{v}_k \cdots \hat{v}_l \cdots v_m \rangle].$$

There are several cases to be considered. First, if

$$+\langle v_0 \cdots \hat{v}_k \cdots v_m \rangle = \langle v_l v_0 \cdots \hat{v}_k \cdots \hat{v}_l \cdots v_m \rangle$$

and

$$+\langle v_0 \cdots v_m \rangle = \langle v_k v_l v_0 \cdots \hat{v}_k \cdots \hat{v}_l \cdots v_m \rangle,$$

then the first term of the above sum is $(+1)(+1)$. Then there are two subcases:

(i) If

$$+\langle v_0 \cdots \hat{v}_l \cdots v_m \rangle = \langle v_k v_0 \cdots \hat{v}_k \cdots \hat{v}_l \cdots v_m \rangle,$$

then we have

$$\langle v_l v_k v_0 \cdots \hat{v}_k \cdots \hat{v}_l \cdots v_m \rangle = -\langle v_0 \cdots v_m \rangle$$

and the second term in the sum is $(-1)(+1)$.

(ii) If

$$\langle v_k v_0 \cdots \hat{v}_k \cdots \hat{v}_l \cdots v_m \rangle = -\langle v_0 \cdots \hat{v}_l \cdots v_m \rangle,$$

then the second term in the sum is $(+1)(-1)$.

Thus in either subcase the sum is zero. The remaining cases are handled similarly. □

6–4 Chains, cycles, and groups. Let K denote an arbitrary oriented simplicial complex, finite or not, and let G denote an arbitrary (additively written) abelian group. (There will be no essential loss of generality if the reader always thinks of the additive group Z of integers whenever we say "arbitrary abelian group.") We make the following definitions. An *m-dimensional chain on the complex K with coefficients in the group G* is a function c_m on the oriented m-simplexes of K with values in the group G such that if $c_m(+\sigma^m) = g$, g an element of G, then $c_m(-\sigma^m) = -g$. If K is infinite, then $c_m(\sigma^m) = 0$, the identity element of G, for all but a finite number of m-simplexes of K. The collection of all such m-dimensional chains on K will be denoted by the symbol $C_m(K, G)$.

We introduce an addition of m-chains by means of the usual functional addition. That is, we define

$$(c_m^1 + c_m^2)(\sigma^m) = c_m^1(\sigma^m) + c_m^2(\sigma^m),$$

where the addition on the right is the group operation in G.

THEOREM 6–2. *Under the operation just defined, $C_m(K, G)$ is an abelian group, the m-dimensional chain group of K with coefficients in G.*

The reader may prove Theorem 6–2 merely by verifying the axioms for an abelian group.

If the complex K has no m-simplexes, we take $C_m(K, G)$ to be the trivial group consisting of the identity element 0 alone and write $C_m(K, G) = 0$.

An *elementary m-chain* on K is an m-chain c_m such that $c_m(\pm\sigma_0^m) = \pm g_0$ for some particular simplex σ_0^m in K and $c_m(\sigma^m) = 0$ whenever $\sigma^m \neq \pm\sigma_0^m$. Such an elementary m-chain will be denoted by a formal product $g_0 \cdot \sigma_0^m$. Then an arbitrary m-chain c_m on K can be written as a formal linear combination $\sum g_i \cdot \sigma_i^m$, where $g_i = c_m(+\sigma_i^m)$ and all but a finite number of the coefficients g_i are zero. This notation explains the use of the word *coefficient*. Actually, this notation conveniently tabulates the function c_m in such a way that the addition of such functions is the addition of linear combinations. We use this presentation of chains throughout our subsequent development.

THEOREM 6–3. If K is a finite complex and α_m is the number of m-simplexes in K, then the chain group $C_m(K, G)$ is isomorphic to the direct sum of α_m groups, each isomorphic to the coefficient group G. If K is infinite, then $C_m(K, G)$ is isomorphic to the weak direct sum of infinitely many isomorphic copies of G.

Proof: If K is finite, then the correspondence

$$\sum_{i=1}^{\alpha_m} g_i \cdot \sigma_i^m \leftrightarrow (g_i, \ldots, g_{\alpha_m})$$

is the desired isomorphism, as is readily checked. A similar argument will handle the infinite case, simply recalling the definition of a weak direct sum. □

The result describes the chain groups completely, but so far there seems to be little if any geometric meaning in our development. This will be corrected shortly, both by the subsequent definitions and by examples. First, we introduce an algebraic mechanism that corresponds to determining the boundary of a portion of a complex. The *boundary operator* ∂ is defined first on elementary chains by the formula

$$\partial(g_0 \cdot \sigma_0^m) = \sum_{\sigma^{m-1}} [\sigma_0^m, \sigma^{m-1}] \cdot g_0 \cdot \sigma^{m-1},$$

where $[\sigma_0^m, \sigma^{m-1}]$ is the incidence number. We note that $\partial(g_0 \cdot \sigma_0^m)$ is an $(m - 1)$-chain which has nonzero coefficients only on the $(m - 1)$-faces of the simplex σ_0^m. The above definition of ∂ is extended linearly to arbitrary m-chains by setting

$$\partial\left(\sum_i g_i \cdot \sigma_i^m\right) = \sum_i \partial(g_i \cdot \sigma_i^m).$$

It is easy to see that the boundary of an m-chain is an $(m - 1)$-chain which depends only upon the m-chain itself and not upon the complex on which the m-chain is taken. (The situation here is just opposite to that found in cohomology theory, as we point out in Section 7–9.)

The fundamental property of this boundary operator is expressed in the next result.

THEOREM 6–4. *For any chain c_m in $C_m(K, G)$, $\partial(\partial c_m) = 0$. That is, $\partial(\partial c_m)$ is the $(m - 2)$-chain with value zero on each $(m - 2)$-simplex.*

Proof: It suffices to prove the theorem for an arbitrary elementary m-chain $g_0 \cdot \sigma_0^m$. For such a chain,

$$\begin{aligned}\partial(\partial(g_0 \cdot \sigma_0^m)) &= \partial\left(\sum_i [\sigma_0^m, \sigma_i^{m-1}] \cdot g_0 \cdot \sigma_i^{m-1}\right)\\ &= \sum_i \partial[\sigma_0^m, \sigma_i^{m-1}] \cdot g_0 \cdot \sigma_i^{m-1}\\ &= \sum_i \left(\sum_j [\sigma_0^m, \sigma_i^{m-1}][\sigma_i^{m-1}, \sigma_j^{m-2}] \cdot g_0 \cdot \sigma_j^{m-2}\right)\\ &= \sum_{i,j} [\sigma_0^m, \sigma_i^{m-1}][\sigma_i^{m-1}, \sigma_j^{m-2}] \cdot g_0 \cdot \sigma_j^{m-2}.\end{aligned}$$

Then Theorem 6–1 applies to complete the proof. □

The reader may prove the next result easily.

THEOREM 6–5. *The boundary operator ∂ defines a homomorphism, which we still denote by ∂, of the group $C_m(K, G)$ into the group $C_{m-1}(K, G)$.*

This result holds for each dimension $m > 0$ if we take ∂ to be the obvious trivial homomorphism in dimensions for which K has no simplexes. The case $m = 0$ will be treated later in Section 6–6.

In analogy to the intuitive discussion of cycles in Section 6–1, we now define for $m > 0$ an *m-dimensional cycle on K with coefficients in G* to be a chain z_m in $C_m(K, G)$ with the property that $\partial(z_m) = 0$, the $(m - 1)$-chain $\sum 0 \cdot \sigma_i^{m-1}$. The collection of all such m-cycles is precisely the kernel of the homomorphism ∂ in the group $C_m(K, G)$ and hence is a subgroup of $C_m(K, G)$. This subgroup is the *m-dimensional cycle group of K with coefficients in G* and is denoted by $Z_m(K, G)$. Also we define a chain b_m in $C_m(K, G)$ to be an *m-boundary* if there is an $(m + 1)$-chain c_{m+1} in $C_{m+1}(K, G)$ such that $\partial(c_{m+1}) = b_m$. The collection of all m-boundaries is the image $\partial C_{n+1}(K, G)$ of the group $C_{m+1}(K, G)$ in $C_m(K, G)$ under the homomorphism ∂. This subgroup of $C_m(K, G)$ is denoted by $B_m(K, G)$, the *group of m-boundaries of K with coefficients in G.*

Since, for any chain c_{m+1}, the $(m - 1)$-chain $\partial(\partial c_{m+1}) = 0$, it follows that any m-boundary b_m has boundary $\partial(b_m) = 0$ and hence b_m is an m-cycle. This implies that *$B_m(K, G)$ is a subgroup of $Z_m(K, G)$.* As subgroups of the abelian group $C_m(K, G)$, both $B_m(K, G)$ and $Z_m(K, G)$ are abelian groups. Therefore we may define the (additively written) factor group $Z_m(K, G) - B_m(K, G)$, which is called the *mth homology group of K over G* and is denoted by $H_m(K, G)$.

Each element of $H_m(K, G)$ is an equivalence class $[z_m]$ of m-cycles where z_m^1 and z_m^2 are in the same class if and only if the chain $z_m^1 - z_m^2$ is an m-boundary. This equivalence relation is called *homology* and is written $z_m^1 \sim z_m^2$. (We could have defined this equivalence relation first and then taken $H_m(K, G)$ to consist of the collection of equivalence classes under the natural addition.)

We have reached our first goal in this section, namely the general definition of the homology groups of a complex K. Next, we give examples to illuminate the geometric content of these algebraic formulations. We will be quite precise in these examples.

EXAMPLE 1. Let K be the complex consisting of a single 3-simplex σ^3 together with all of its faces. [This is the *closure of a simplex* σ^3 and is denoted by "$\mathrm{Cl}(\sigma^3)$."] We will orient the complex K by choosing a fixed ordering of its vertices, v_0, v_1, v_2, and v_3, and letting this induce the positive orientation of the simplexes. In this way, we have the following list of (representatives of) the oriented simplexes of K:

$$
\begin{array}{lll}
+\sigma_1^1 = \langle v_2v_3\rangle, & +\sigma_1^2 = \langle v_1v_2v_3\rangle, & +\sigma^3 = \langle v_0v_1v_2v_3\rangle, \\
+\sigma_2^1 = \langle v_1v_3\rangle, & +\sigma_2^2 = \langle v_0v_2v_3\rangle, & \\
+\sigma_3^1 = \langle v_0v_3\rangle, & +\sigma_3^2 = \langle v_0v_1v_3\rangle, & \\
+\sigma_4^1 = \langle v_1v_2\rangle, & +\sigma_4^2 = \langle v_0v_1v_2\rangle, & \\
+\sigma_5^1 = \langle v_0v_2\rangle, & & \\
+\sigma_6^1 = \langle v_0v_1\rangle. & &
\end{array}
$$

(We omit consideration of dimension zero temporarily.)

Now let G be any abelian group. The only 3-chains on K are the elementary chains $g \cdot \sigma^3$, hence the chain group $C_3(K, G)$ is isomorphic to G. Since there are no 4-simplexes in K, $C_4(K, G) = 0$, and hence $B_3(K, G) = \partial[C_4(K, G)] = 0$. It follows that $H_3(K, G) = Z_3(K, G)$. But let $g \cdot \sigma^3$ be any 3-chain. Computing its boundary we have

$$
\begin{aligned}
\partial(g \cdot \sigma^3) &= \sum_{i=1}^{4} [\sigma^3, \sigma_i^2] \cdot g \cdot \sigma_i^2 \\
&= g \cdot \sigma_1^2 - g \cdot \sigma_2^2 + g \cdot \sigma_3^2 - g \cdot \sigma_4^2.
\end{aligned}
$$

[It is easy to show that, in the present case, $[\sigma^3, \sigma_i^2] = (-1)^{i+1}$.] This chain is the zero 2-chain if and only if $g = 0$. Therefore, the only 3-cycle on K is the trivial 3-cycle $0 \cdot \sigma^3$. Hence $Z_3(K, G) = H_3(K, G)$ is trivial. This illustrates one situation in which we obtain a trivial homology group, namely, by having no cycles except the trivial cycle.

Another situation that results in a trivial homology group occurs when every cycle is a boundary. For if $Z_m(K, G) = B_m(K, G)$, then $Z_m - B_m = H_m = 0$.

This situation can be illustrated with this same example. Suppose that the 2-chain $\sum_{i=1}^{4} g_i \cdot \sigma_i^2$ is a 2-cycle. Computing its boundary, we have

$$\begin{aligned}\partial\left(\sum_{i=1}^{4} g_i \cdot \sigma_i^2\right) &= \sum_{i=1}^{4} \partial(g_i \cdot \sigma_i^2)\\ &= \sum_{i=1}^{4}\sum_{j=1}^{6} [\sigma_i^2, \sigma_j^1] g_i \cdot \sigma_j^1\\ &= \sum_{j=1}^{6}\left(\sum_{i=1}^{4} [\sigma_i^2, \sigma_j^1] g_i\right) \cdot \sigma_j^1.\end{aligned}$$

If this is to be the zero 1-chain, then for each fixed index j the sum

$$\sum_{i=1}^{4} [\sigma_i^2, \sigma_j^1] \cdot g_i$$

must be zero. For instance, for $j = 1$ we have

$$[\sigma_1^2, \sigma_1^1]g_1 + [\sigma_2^2, \sigma_1^1]g_2 + [\sigma_3^2, \sigma_1^1]g_3 + [\sigma_4^2, \sigma_1^1]g_4 = 0.$$

But σ_1^1 is not a face of σ_3^2 and σ_4^2, so the last two terms are zero. Furthermore, $[\sigma_1^2, \sigma_1^1] = +1$ and $[\sigma_2^2, \sigma_1^1] = +1$, and hence this equation reduces to nothing more than $g_1 + g_2 = 0$ or $g_2 = -g_1$. Similarly, working with σ_6^1 we obtain $g_4 = -g_3$, and working with σ_3^1 we show that $g_3 = g_2$. This means that $\sum_{i=1}^{4} g_i\sigma_i^2$ can be a 2-cycle only if $g_1 = g_3 = -g_2 = -g_4$; that is, the only 2-cycles are of the form $g \cdot \sigma_1^2 - g \cdot \sigma_2^2 + g \cdot \sigma_3^2 - g \cdot \sigma_4^2$. But we have already seen that such a 2-cycle is the boundary of the 3-chain $g \cdot \sigma^3$. Hence every 2-cycle on K is a 2-boundary, and it follows that $H_2(K, G) = 0$.

By an analogous but much longer method, the reader may prove that $Z_1(K, G) = B_1(K, G)$ and thereby show that $H_1(K, G)$ is also trivial. [We will consider $H_0(K, G)$ in Section 6–6.] Geometrically, the complex K is carried by a homeomorph of the 3-cube I^3 and is a 3-cell. Granting that the homology groups are topological invariants, we have found that the homology groups of a 3-cell are trivial for dimensions greater than zero.

EXAMPLE 2. Consider the complex consisting of all 2-simplexes, 1-simplexes and 0-simplexes that are faces of a single 3-simplex (which is not in our complex). Geometrically, this is the surface of a tetrahedron, a homeomorph of the 2-sphere, and we will denote it by S^2. The complex S^2 is precisely the 2-skeleton of the complex K in Example 1, and we orient it just as we did before, simply omitting σ^3. Since the property of being a 2-cycle does not depend upon the existence of 3-simplexes at all, the work in Example 1 shows that the only 2-cycles on S^2 are chains of the form

$$g \cdot \sigma_1^2 - g \cdot \sigma_2^2 + g \cdot \sigma_3^2 - g \cdot \sigma_4^2.$$

This implies that $Z_2(S^2, G)$ is isomorphic to G. Since there are no 3-simplexes in S^2, the chain group $C_3(S^2, G) = 0$, and hence $B_2(S^2, G) = \partial[C_3(S^2, G)] = 0$. Therefore $H_2(S^2, G) = Z_2(S^2, G)$ is isomorphic to G.

It has already been pointed out that $H_1(K, G) = 0$ in Example 1. Now any 1-chain on S^2 is certainly a 1-chain on K, and the boundary relations are the same in both cases if we deal with dimension 1. Hence we have also $H_1(S^2, G) = 0$. (Again we temporarily omit dimension zero.) We have proved, then, that for $m \neq 0$ or 2, $H_m(S^2, G) = 0$, while for $m = 2$, $H_m(S^2, G)$ is isomorphic to G. The cases for $m > 2$ follow from the fact that the chain groups are trivial for dimensions above two.

EXAMPLE 3. We return to the torus and make rigorous the intuitive discussion of Section 6–1. First, the surface must be triangulated. That is, we must construct a geometric simplicial complex T whose carrying polytope $|T|$ is homeomorphic to a torus. We make use of the process of identification here. If we identify a pair of opposite edges of a rectangle as shown in Fig. 6–8, we obtain a cylinder (Fig. 6–8b). Then we identify the opposite ends of the cylinder, maintaining orientation of these circles as we do so (Fig. 6–8c). Then, as in Fig. 6–8, we obtain a torus.

This leads us to construct the complex T as a plane rectangle and to so label the vertices that this identification process is clearly indicated. Figure 6–9 is such a complex T.

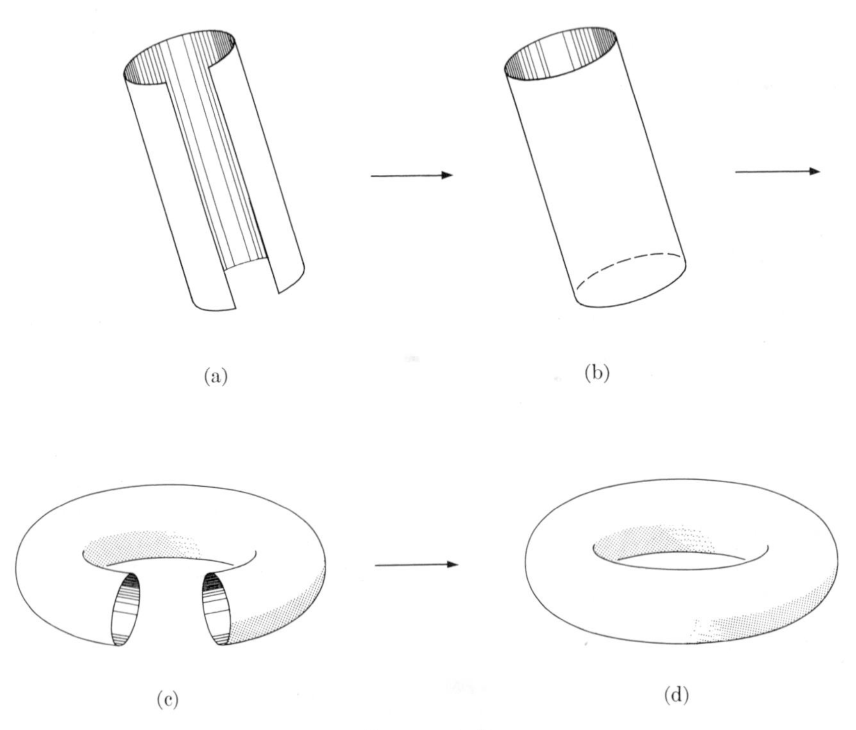

FIGURE 6–8

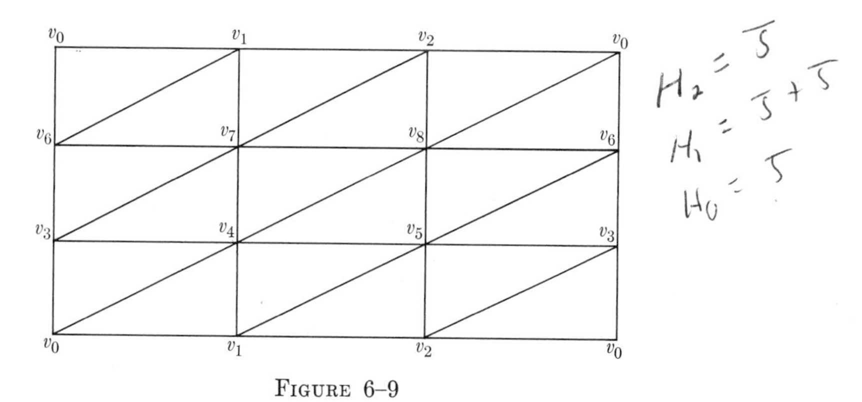

FIGURE 6-9

Remark: The above is not a *minimal* triangulation of the torus. (See Exercise 6-36.)

To proceed with our example, let T have the orientation induced by the given ordering of the vertices. There are eighteen 2-simplexes, which we number as follows:

$$
\begin{aligned}
+\sigma_1^2 &= \langle v_0v_1v_4\rangle, & +\sigma_2^2 &= \langle v_0v_3v_4\rangle, & +\sigma_3^2 &= \langle v_1v_4v_5\rangle,\\
+\sigma_4^2 &= \langle v_1v_2v_4\rangle, & +\sigma_5^2 &= \langle v_2v_3v_5\rangle, & +\sigma_6^2 &= \langle v_0v_2v_3\rangle,\\
+\sigma_7^2 &= \langle v_3v_6v_7\rangle, & +\sigma_8^2 &= \langle v_3v_4v_2\rangle, & +\sigma_9^2 &= \langle v_4v_7v_8\rangle,\\
+\sigma_{10}^2 &= \langle v_4v_5v_8\rangle, & +\sigma_{11}^2 &= \langle v_5v_6v_8\rangle, & +\sigma_{12}^2 &= \langle v_3v_5v_6\rangle,\\
+\sigma_{13}^2 &= \langle v_0v_1v_6\rangle, & +\sigma_{14}^2 &= \langle v_1v_6v_7\rangle, & +\sigma_{15}^2 &= \langle v_2v_7v_8\rangle,\\
+\sigma_{16}^2 &= \langle v_2v_7v_8\rangle, & +\sigma_{17}^2 &= \langle v_0v_2v_8\rangle, & +\sigma_{18}^2 &= \langle v_0v_6v_8\rangle.
\end{aligned}
$$

Now suppose that the 2-chain $\sum_{i=1}^{18} g_i \cdot \sigma_i^2$ is a cycle. Then it must have zero boundary, and computing the boundary we have

$$
\begin{aligned}
\partial\left(\sum_{i=1}^{18} g_i \cdot \sigma_i^2\right) &= \sum_{i=1}^{18} \partial(g_i \cdot \sigma_i^2)\\
&= g_i \cdot \langle v_iv_4\rangle - g_i \cdot \langle v_0v_4\rangle + g_1 \cdot \langle v_0v_1\rangle + \cdots\\
&\qquad + g_{13} \cdot \langle v_1v_6\rangle - g_1 \cdot \langle v_0v_6\rangle + g_1 \cdot \langle v_0v_1\rangle + \cdots
\end{aligned}
$$

Since the 1-simplex $\langle v_0v_1\rangle$ is a face of only the 2-simplexes σ_1^2 and σ_{13}^2, its coefficient in this sum is precisely $g_1 + g_{13}$. But for the 2-chain to be a cycle, its boundary must assign coefficient zero to each 1-simplex. Therefore, we must have $g_{13} = -g_1$. In a similar way, we can show that each g_i, $i \neq 1$, is either $+g_1$ or $-g_1$. It follows that $Z_2(T, G)$ is isomorphic to G. Since there are no 3-simplexes in T, we have $B_2(T, G) = 0$, and hence $H_2(T, G) = Z_2(T, G) = G$.

A computation of $Z_1(T, G)$ and $B_1(T, G)$ is tediously long and we will omit it. The reader may easily verify the following facts, however. (1) All chains of the

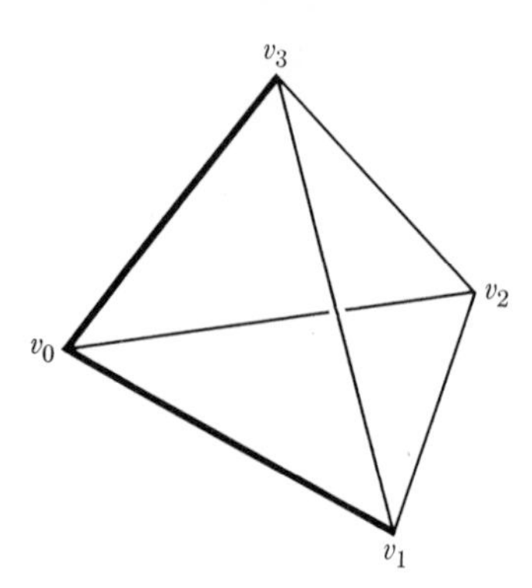

FIGURE 6–10

form $g \cdot \langle v_0v_1\rangle + g \cdot \langle v_1v_2\rangle - g \cdot \langle v_0v_2\rangle$ are nonbounding 1-cycles, and (2) all chains of the form $g \cdot \langle v_0v_3\rangle + g \cdot \langle v_3v_6\rangle - g \cdot \langle v_0v_6\rangle$ are nonbounding 1-cycles. A more difficult exercise involves proving that (3) a cycle of the form (1) is not homologous to one of the form (2) and that (4) every nonbounding 1-cycle on T is homologous to a sum of two 1-cycles, one of the form (1) and the other of the form (2). As will be seen in Section 6–5, these facts imply that $H_1(T, G)$ is isomorphic to the direct sum $G \oplus G$.

Before studying the structure of homology groups, we may profitably examine a particular coefficient group. The use of Z_2, the group of integers modulo 2, as the coefficient group permits a strongly geometric interpretation of homology theory. To retain this geometric flavor, let us temporarily limit our consideration to a geometric complex K. As it turns out, we need not orient K for *mod 2 homology theory*.

As should be expected, a p-chain mod 2 on K is a function on the p-simplexes of K with values 0 and 1, the value 1 occurring only for a finite number of p-simplexes. But now we may picture such a p-chain simply as the point-set union of those p-simplexes of K that are assigned the value 1 by the p-chain. For instance, Fig. 6–10 corresponds to the 1-chain $1\langle v_0v_1\rangle + 1\langle v_0v_3\rangle + 0\langle v_0v_2\rangle + 0\langle v_1v_2\rangle + 0\langle v_1v_3\rangle + 0\langle v_2v_3\rangle$, the heavy segments being the 1-simplexes which have the coefficient 1 in this chain.

Addition of p-chains is again done componentwise, the addition of the coefficients taken modulo 2. This operation has a geometric interpretation, too. For example, if c_1 and c_1' are 1-chains mod 2 and correspond to point sets C and C' in the 1-skeleton of K, then the sum $c_1 + c_1'$ of the two 1-chains corresponds to the closure of the symmetric difference of the sets C and C'. [The *symmetric difference* of two sets A and B is the set $(A - B) \cup (B - A) = (A \cup B) - (A \cap B)$.] In Fig. 6–11, we show a pictorial equation illustrating this addition.

Since -1 is congruent to $+1$ modulo 2, the incidence numbers $[s_i^p, s_j^{p-1}]$ may be taken to have only the values 0 and 1, the value 1 occurring if

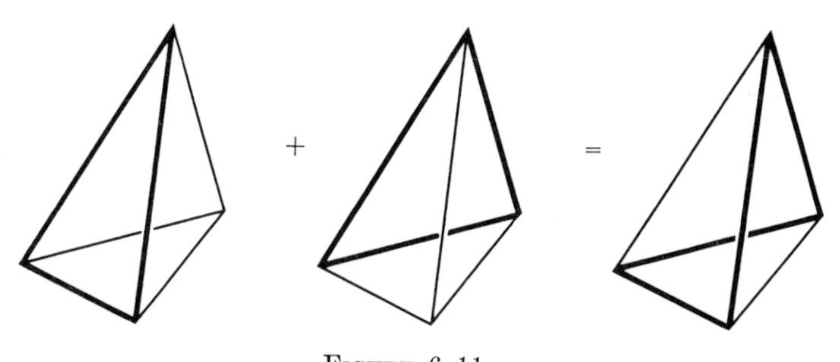

FIGURE 6-11

s_j^{p-1} is a face of s_i^p and the value 0 occurring otherwise. More simply yet, the boundary operator ∂ can be defined directly without using the incidence numbers. (This is why we need not have oriented the complex K.) For an elementary p-chain mod 2 we set

$$\partial(0 \cdot s^p) = 0 \qquad \text{and} \qquad \partial(1 \cdot s^p) = \sum_i \eta_i \sigma_i^{p-1},$$

where $\eta_i = 1$ only if s_i^{p-1} is a face of s^p. It is easily seen that the chain $\partial(1 \cdot s^p)$ corresponds geometrically to the point-set boundary of the simplex s^p. Extending this definition linearly to arbitrary p-chains mod 2, a boundary $\partial(c_p)$ corresponds to the point-set boundary of the union of p-simplexes corresponding to c_p.

The required property $\partial\partial = 0$ of this boundary operator is even easier to establish than for the general boundary operator. Each $(p - 2)$-simplex s^{p-2} in a p-simplex s^p is a face of exactly two $(p - 1)$-simplexes, say s_i^{p-1} and s_j^{p-1}. In $\partial(1 \cdot s^p)$, both s_i^{p-1} and s_j^{p-1} are assigned coefficient 1. Thus in $\partial\partial(1 \cdot s^p)$, s^{p-2} is assigned the value 1 from s_i^{p-1} and from s_j^{p-1}. The coefficient of s^{p-2} is therefore $1 + 1 \equiv 0 \bmod 2$. Geometrically, this corresponds to the fact that the (point-set) boundary of a boundary is empty.

We proceed as before to define the mod 2 cycle group $Z_p(K, Z_2)$ to be the kernel of ∂ and the mod 2 boundary group $B_p(K, Z_2)$ to be the image $\partial[C_{p+1}(K, Z_2)]$. Since $\partial\partial = 0$, $B_p(K, Z_2)$ is a subgroup of $Z_p(K, Z_2)$, and hence we may define the *mod 2 homology group* $H_p(K, Z_2) = Z_p(K, Z_2) - B_p(K, Z_2)$.

It might well be asked why, if mod 2 homology theory is so very geometric, we do not use this theory exclusively. The reasons for not limiting ourselves to coefficients mod 2 will appear in remarks and examples in the next section. Meanwhile, the reader may verify that our intuitive discussion in Section 6-1 is precisely mod 2 homology theory. Exercises on mod 2 homology theory will appear in Section 6-5.

EXERCISE 6-1. Let S^1 denote the complex consisting of all 1-simplexes and 0-simplexes in a given 2-simplex. Determine $H_p(S^1, G)$ for all $p > 0$.

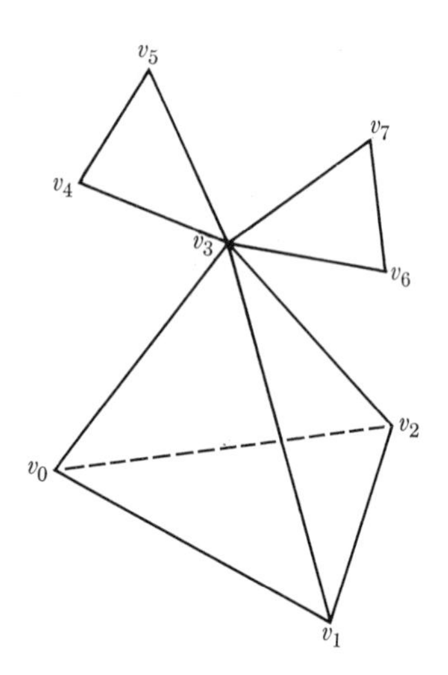

FIGURE 6–12

EXERCISE 6–2. Let C be a finite cylinder formed by identifying a pair of opposite edges of a rectangle (as was done in Fig. 6–8 for the torus). Compute $H_p(C, G)$ for all $p > 0$.

EXERCISE 6–3. Let K denote the complex consisting of a tetrahedral surface meeting two triangular simple closed curves in a common vertex, as in Fig. 6–12. Compute $H_p(K, G)$ for all $p > 0$.

6–5 The decomposition theorem for abelian groups. Betti numbers and torsion coefficients. The structure of the homology groups and the dependence of that structure upon the connectivity properties of the complex must be investigated. We will assume throughout this section that our homology groups are taken over the additive group Z of integers. We will write $C_m(K)$ for $C_m(K, Z)$, $Z_m(K)$ for $Z_m(K, Z)$, etc. We must use the decomposition theorem for finitely generated abelian groups. In fact, we paraphrase this theorem as our first result. For a proof, the reader may consult Lefschetz's *Algebraic Topology* [20], Chapter 2.

Let K be a finite complex with α_m m-simplexes. Then Theorem 6–3 says that $C_m(K)$ is (isomorphic to) the group $Z \oplus \cdots \oplus Z$ (α_m summands). Such a direct sum of infinite cyclic groups is known as a *free group.* Since any subgroup of a free group is again a free group, both $Z_m(K)$ and $B_m(K)$ are free groups with a number of generators not exceeding α_m. Finally, the factor group (or difference group) $H_m(K) = Z_m(K) - B_m(K)$ is known to be an abelian group with a finite number of generators. In such a factor group, there may be relations among the generators so that in general $H_m(K)$ is not a free group. The decomposition theorem for finitely generated abelian groups applies, however, and yields the following result.

THEOREM 6–6. For finite complex K, the integral homology group $H_m(K)$ is isomorphic to a direct sum $G_0 \oplus G_1 \oplus \cdots \oplus G_{k_m}$, where G_0 is a free group and each G_i, $i = 1, 2, \ldots, k_m$, is a finite cyclic group.

The number of generators of the free group G_0 (the *rank* of G_0) in the above decomposition of $H_m(K)$ is called the *mth Betti number of* K and is denoted by $p_m(K)$. The number of elements (the *order*) of the finite cyclic group G_i, $i > 0$, is an *mth torsion coefficient* and may be denoted by $t_m^i(K)$, $i = 1, 2, \ldots, k_m$. It is known that the groups G_i can be arranged in such an order that $t_m^i(K)$ divides $t_m^{i+1}(K)$, $0 < i < k_m$. The direct sum $G_1 \oplus G_2 \oplus \cdots \oplus G_{k_m}$ is frequently called the *torsion group* of K. This group tells us something about the manner in which the complex K is "twisted." More of this idea will appear in an example shortly.

If we were to use an arbitrary (abelian) coefficient group G, the decomposition of $H_m(K, G)$ would yield a direct sum $G \oplus \cdots \oplus G \oplus H$, where there are $p_m(K)$ summands G and where the group H depends upon G in a manner to be determined later. As we shall indicate in Section 6–9, if we know the integral groups $H_m(K)$, we can always obtain the groups $H_m(K, G)$ for any group G. For this reason, the integers Z are known as a *universal coefficient group*.

The Betti number $p_m(K)$ may be considered intuitively as the numbers of "m-dimensional holes" in the complex K. Or, in other words, $p_m(K)$ is the number of $(m + 1)$-dimensional chains which must be added to K so that every free m-cycle on K is a boundary. (A *free* cycle is one that is not due to torsion.) Thus $p_m(K)$ can often be ascertained nonrigorously simply by inspecting the complex. For instance, consider the torus again.

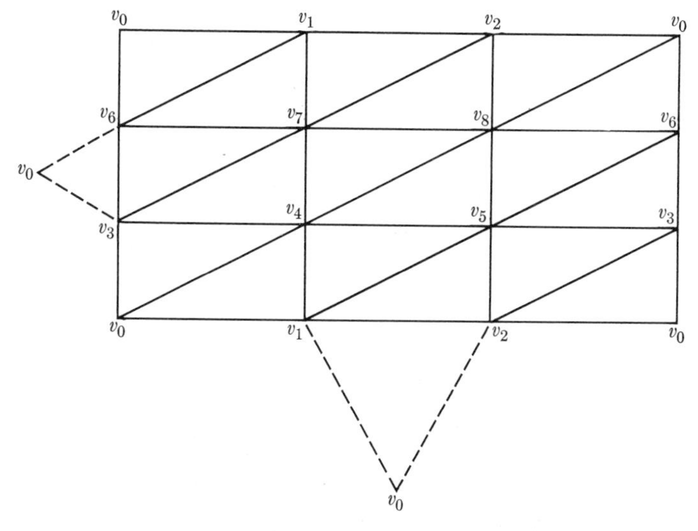

FIGURE 6–13

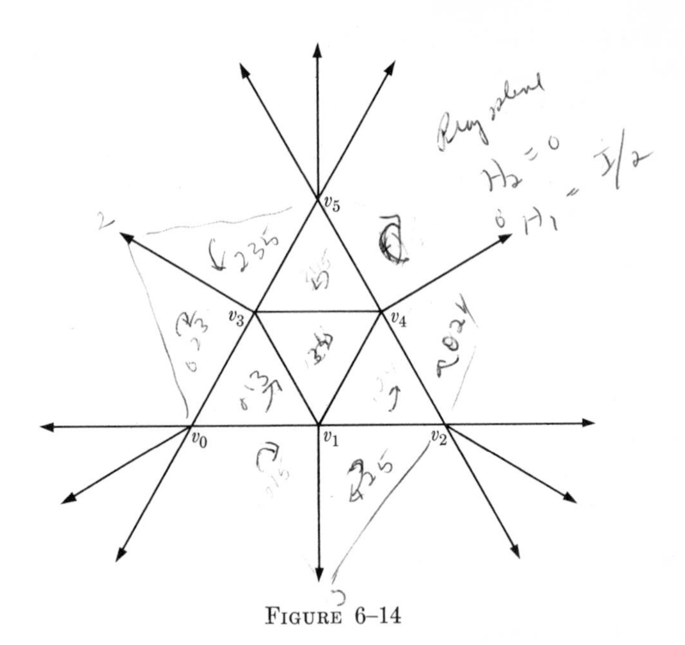

FIGURE 6–14

If we add the 2-simplex $\langle v_0v_1v_2\rangle$ (see Fig. 6–13), every longitudinal 1-cycle will bound, and if we add a 2-simplex $\langle v_0v_3v_6\rangle$, every latitudinal 1-cycle will bound. Thus we can conclude that $p_1(T) = 2$. Similarly, we may think of adding 3-simplexes to fill in the interior of T and have intuitive reason for believing that $p_2(T) = 1$.

EXAMPLE 1. To envision something of the geometry underlying the torsion group, let us examine the projective plan P^2. This may be taken to be a 2-sphere with antipodal points identified. A triangulation of P^2 may be obtained if we think of P^2 as the ordinary plane with opposite directions identified (Fig. 6–14). Let the orientation of P^2 be that induced by the given ordering of the vertices, and consider the integral 2-chain

$$\begin{aligned} c_2 = {} & 1\langle v_0v_1v_5\rangle - 1\langle v_1v_3v_4\rangle + 1\langle v_1v_2v_4\rangle + 1\langle v_3v_4v_5\rangle - 1\langle v_0v_4v_5\rangle \\ & - 1\langle v_0v_2v_4\rangle + 1\langle v_0v_1v_3\rangle - 1\langle v_0v_2v_3\rangle + 1\langle v_1v_2v_5\rangle + 1\langle v_2v_3v_5\rangle. \end{aligned}$$

The boundary of c_2 is easily computed to be the 1-chain

$$2\langle v_0v_1\rangle + 2\langle v_1v_2\rangle - 2\langle v_1v_2\rangle = 2(1\langle v_0v_1\rangle + 1\langle v_1v_2\rangle - 1\langle v_0v_2\rangle).$$

A routine calculation proves that

$$z_1 = 1\langle v_0v_1\rangle + 1\langle v_1v_2\rangle - 1\langle v_0v_2\rangle$$

is a 1-cycle. But z_1 is not a boundary! For the only 2-chain that z_1 can bound is $\frac{1}{2}c_2$, which is not an integral 2-chain at all.

A lengthy computation is needed but it can be shown that every integral 1-cycle on P^2 is either a boundary or is homologous to a multiple of z_1. But if $z_1' \sim 2kz_1$, then $z_1' = \partial(kc_2)$; that is, $z_1' \sim 0$. And if $z_1' \sim (2k+1)z_1 = 2kz_1 + z_1 = z_1 + \partial(kc_2)$, then $z_1' \sim z_1$. These facts imply that the integral homology group $H_1(P^2)$ is isomorphic to Z_2, the integers mod 2. Thus the first Betti number of P^2 is zero, and the only nonbounding 1-cycles on P^2 are *torsion cycles*.

We leave as an exercise the verification that the only 2-cycle on P^2 is the trivial one and that hence $H_2(P^2)$ is trivial.

EXAMPLE 2. The *Klein bottle* is obtained from a finite cylinder by identifying the opposite ends with the orientation of the two circles reversed. It cannot be constructed in 3-space without self-intersection, in which case it appears as in Fig. 6-15.

A triangulation B of the Klein bottle may be given as we did for the torus. In Fig. 6-16, the labeling of the vertices indicates the identification of the opposite edges of a rectangle used to obtain the Klein bottle. We will give merely the results of the calculations, namely, the integral homology groups are

$$H_2(B) = 0 \quad \text{and} \quad H_1(B) = Z \oplus Z_2.$$

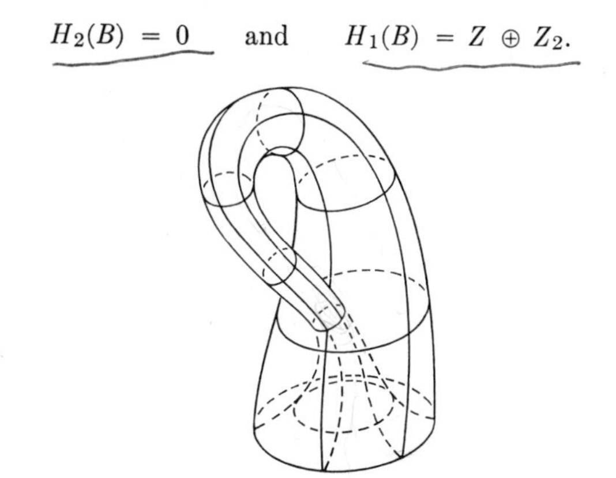

FIG. 6-15. The Klein bottle.

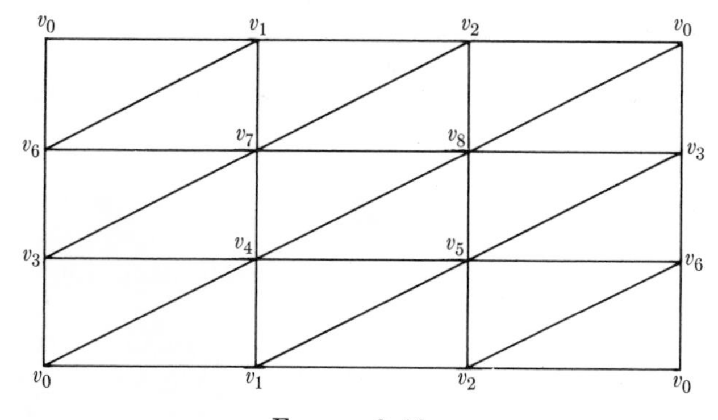

FIGURE 6-16

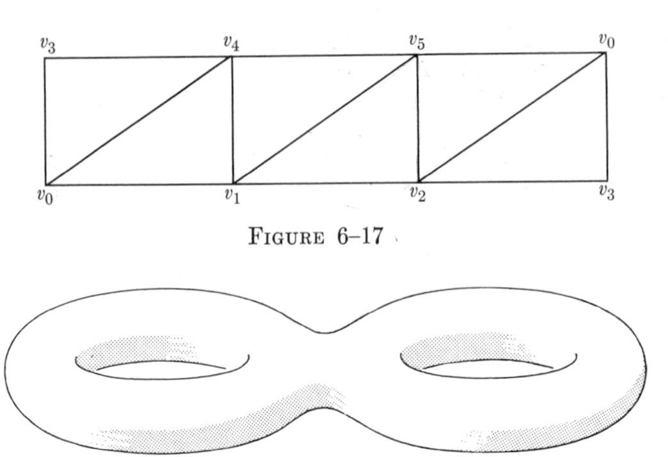

FIGURE 6–17

FIGURE 6–18

Thus, in dimension 1, we have cycles of two types, free cycles and torsion cycles. Such a cycle as $m\langle v_0v_1\rangle + m\langle v_1v_2\rangle - m\langle v_0v_2\rangle$ is a free cycle, while $m\langle v_0v_3\rangle + m\langle v_3v_6\rangle - m\langle v_0v_6\rangle$ is a torsion cycle such that if m is even, this cycle bounds and if m is odd, it does not bound.

It is easily verified that the mod 2 homology group $H_1(B, Z_2)$ of the Klein bottle is isomorphic to $Z_2 \oplus Z_2$. And herein lies the reason for not using coefficients mod 2 exclusively! There is no torsion group, or at least no way to recognize it as a part of the homology group, if we use coefficients mod 2. (See Exercise 6–7 below.)

EXERCISE 6–4. The *Möbius band* is constructed by identifying opposite ends of a rectangle after giving it a twist of 180°. This can be done with a strip of paper. A triangulation M of the Möbius band is shown in Fig. 6–17. Compute the integral homology groups $H_2(M)$ and $H_1(M)$. Compare these with the mod 2 homology groups $H_2(M, Z_2)$ and $H_1(M, Z_2)$.

EXERCISE 6–5. Use the device of inserting additional simplexes to give the Betti numbers of the surface of genus 2 pictured in Fig. 6–18.

EXERCISE 6–6. Compute the mod 2 homology groups of the projective plane P^2.

EXERCISE 6–7. For any finite complex K, prove that $H_p(K, Z_2)$ is always a direct sum of cyclic groups of order 2.

6–6 Zero-dimensional homology groups.

We have delayed consideration of the 0-dimensional homology groups until now because there are two different ways to define the boundary of a 0-chain. Introducing these earlier would only have added to the difficulties.

Since there are no simplexes of dimension -1 in a complex K, the natural definition of the boundary of a 0-chain is given by setting

$$\partial(g_i \cdot \sigma_i^0) = 0$$

for every elementary 0-chain. This clearly implies that every 0-chain is a 0-cycle and hence that $Z_0(K, G) = C_0(K, G)$. Since every 0-boundary is certainly a 0-chain, $B_0(K, G) = \partial[C_1(K, G)]$ is a subgroup of $C_0(K, G)$. Thus, in this case, we define

$$H_0(K, G) = C_0(K, G) - B_0(K, G).$$

This is called the *nonaugmented 0th homology group*. (The reason for the term *nonaugmented* will be obvious shortly.)

Another method of defining the boundary of a 0-chain is obtained as follows. In dimension theory (see Section 3–9), we consider the empty set to have dimension -1, and to be a subset of every set. Analogously, we may *augment* the complex K by adding the single (-1)-dimensional simplex ϕ and take it to be a face of every vertex. If this is done, we define a new boundary of an elementary 0-chain by setting

$$\tilde{\partial}(g_i \cdot \sigma_i^0) = g_i \cdot \phi,$$

and we use linear extension as usual to define the boundary of an arbitrary 0-chain. Thus we have

$$\tilde{\partial}(\sum g_i \cdot \sigma_i^0) = \sum \tilde{\partial}(g_i \cdot \sigma_i^0) = (\sum g_i)\phi.$$

It follows that *in the augmented case* a 0-chain $\sum g_i \cdot \sigma_i^0$ is a 0-cycle if and only if the sum $\sum g_i$ of its coefficients is zero. The sum $\sum g_i$ of the coefficients of a 0-chain is sometimes called its *Kronecker index*. The group $\tilde{Z}_0(K, G)$ is now defined as the kernel of the new homomorphism $\tilde{\partial}$, and $B_0(K, G) = \partial[C_1(K, G)]$ as usual. It must be shown that $B_0(K, G)$ is a subgroup of this new cycle group. But this is easily done, for if $\sigma^1 = \langle v_i v_j \rangle$ and $g \cdot \sigma^1$ is an elementary 1-chain, then $\partial(g\sigma^1) = g \cdot \langle v_j \rangle - g \cdot \langle v_i \rangle$. Hence the sum of the coefficients in the boundary of any elementary 1-chain is zero, and the same is true for any 1-chain, by linear extension. This means that we can define the *augmented 0th homology group*

$$\tilde{H}_0(K, G) = \tilde{Z}_0(K, G) - B_0(K, G).$$

We investigate the relation between the augmented and the nonaugmented groups shortly.

Recalling the definition of a combinatorial component of a complex K (see Section 5–4), we may state the following result.

THEOREM 6–7. *Let K be a finite complex with k combinatorial components, $K_1, K_2, \ldots, K_k$. Then, for any group G, $H_0(K, G)$ is isomorphic to the direct sum $G \oplus \cdots \oplus G$ (k summands).*

Proof: Choose a particular 0-simplex $\sigma_{i_0}^0$ in each component K_i of K. We will show that every 0-chain on K is homologous to a chain of the form

$\sum g_i \cdot \sigma^0_{i_0}$. To do this, let σ^0_{ij} be any 0-simplex in K_i. Given an element g of the group G, we first prove that the elementary 0-chain $g \cdot \sigma^0_{ij}$ is homologous either to $g \cdot \sigma^0_{i_0}$ or to $-g \cdot \sigma^0_{i_0}$. By assumption, there exists in K_i a sequence of 1-simplexes connecting σ^0_{ij} to $\sigma^0_{i_0}$. We construct a 1-chain which assigns $+g$ or $-g$ to each of these 1-simplexes (depending upon their orientations) such that this 1-chain has boundary either $g \cdot \sigma^0_{ij} - g \cdot \sigma^0_{i_0}$ or $g \cdot \sigma^0_{ij} + g \cdot \sigma^0_{i_0}$.

Now suppose that we have a 0-chain $\sum_{i,j} g_{ij} \cdot \sigma^0_{ij}$, σ^0_{ij} a vertex of K_i for each j. Let $\bar{g}_i = \sum_j g_{ij}$. Then from the argument above, $\sum_{i,j} g_{ij} \cdot \sigma^0_{ij}$ is homologous to $\sum \bar{g}_i \cdot \sigma^0_{i_0}$, which is the first and crucial fact we need.

We showed that a 0-chain is a boundary only if the sum of its coefficients is zero. It remains to be proved that a 0-chain $\sum g_i \cdot \sigma^0_{i_0}$ is a boundary only if all the coefficients g_i are zero. To do so, suppose that there exists a 1-chain c_1 such that $\partial c_1 = \sum g_i \cdot \sigma^0_{i_0}$. Clearly, we can write $c_1 = \sum_i c^i_1$, where each c^i_1 is a 1-chain on K_i so that ∂c^i_1 is also on K_i. It follows that $\partial c_1 = \sum \partial c^i_1 = \sum g_i \cdot \sigma^0_{i_0}$ or $\partial c^i_1 = g_i \cdot \sigma^0_{i_0}$. But then $g_i \cdot \sigma^0_{i_0}$ is a boundary, and the sum of its coefficients, namely, g_i itself, must be zero.

We have shown that every element of $C_0(K, G)$ is homologous to a 0-chain of the form $\sum g_i \cdot \sigma^0_{i_0}$ and that two elementary chains $g_i \cdot \sigma^0_{i_0}$ and $g_j \cdot \sigma^0_{j_0}$, $i \neq j$, are homologous only if $g_i = g_j = 0$. Thus the correspondence $\sum g_i \cdot \sigma^0_{i_0} \leftrightarrow (g_1, \ldots, g_k)$ is an isomorphism of $H_0(K, G)$ onto the direct sum $G \oplus \cdots \oplus G$, k summands. □

COROLLARY 6–8. The integral group $H_0(K)$ of a complex K with k combinatorial components is a free group on k generators. Thus $p_0(K) = k$, and there is no torsion in dimension zero.

Using the notation of the above proof, establish the following facts for the *augmented case* of the complex K.

EXERCISE 6–8. A cycle of the form $g \cdot \sigma^0_{1_0} - g \cdot \sigma^0_{i_0}$ is homologous to zero if and only if $g = 0$.

EXERCISE 6–9. Every 0-cycle is homologous to one of the form $\sum g_i \cdot \sigma^0_{i_0}$.

EXERCISE 6–10. Every 0-cycle $\sum g_i \cdot \sigma^0_{i_0}$ is the sum of $k - 1$ 0-cycles of the form $g \cdot \sigma^0_{1_0} - g \cdot \sigma^0_{i_0}$.

EXERCISE 6–11. $\widetilde{H}_0(K, G)$ is isomorphic to $G \oplus \cdots \oplus G$, $k - 1$ summands.

It is evident from the above exercises that the augmented 0th homology group of a connected complex is trivial, while from Corollary 6–8 we see that the nonaugmented integral homology group is infinite cyclic. This is the reason for the frequent use of augmented homology theory in connectivity problems. For instance, it is quicker and easier to say that "the homology groups of the n-cell are all trivial" than it is to make the exception for dimension zero that would be necessary if we did not use augmented theory.

6–7 The Euler-Poincaré formula. In 1752 the great mathematician Leonhardt Euler discovered a simple geometric fact that had escaped notice by geometers for two thousand years. Let P denote a simple polyhedron (a homeomorphic image of the 2-sphere S^2), and let V, E, and F denote the number of vertices, edges, and faces, respectively, of P. Euler's discovery was the relation between these numbers, which is expressed in the formula

$$V - E + F = 2.$$

The reader may verify this himself for the cases of a tetrahedon, a cube, etc. The formula applies, however, to irregular simple polyhedra as well. For instance, a pyramid with a trapezoidal base has $V = 5$, $E = 8$, and $F = 5$; and $V - E + F = 2$.

We prove a generalization of the Euler formula, the generalization being due to Poincaré. Let α_m, $m = 0, 1, \ldots, n$, denote the number of m-simplexes in a finite complex K of dimension n. Therefore the rank of each free integral chain group $C_m(K)$ equals α_m. Let ξ_m and β_m denote the ranks of the free groups $Z_m(K)$ and $B_m(K)$, respectively. Since the boundary operator is a homomorphism of $C_m(K)$ onto $B_{m-1}(K)$ with kernel $Z_m(K)$, it follows that (cf. Theorem 6–10)

$$\alpha_m - \xi_m = \beta_{m-1} \qquad \text{for} \quad m > 0. \tag{1}$$

Using nonaugmented homology in dimension zero, we have $Z_0(K) = C_0(K)$, so

$$\alpha_0 - \xi_m = 0. \tag{2}$$

Then, since $H_m(K) = Z_m(K) - B_m(K)$, it follows that

$$\xi_m - \beta_m = p_m(K) \tag{3}$$

(cf. Theorem 6–12; in this case $s = 0$). Combining relations (1), (2), and (3), we obtain

$$\alpha_m - p_m(K) = \beta_m + \beta_{m-1} \qquad \text{for} \quad m > 0$$

and

$$\alpha_0 - p_0(K) = \beta_0. \tag{4}$$

If we now take the alternating sum of the equations (4) over all values of m, we obtain

$$\sum_{m=0}^{n} (-1)^m(\alpha_m - p_m(K)) = \sum_{m=0}^{n} (-1)^m(\beta_m + \beta_{m-1}).$$

But it is obvious that

$$\sum_{m=0}^{n} (-1)^m(\beta_m + \beta_{m-1}) = \pm\beta_m,$$

and since $C_{n+1}(K) = 0$ (the complex K has dimension n), we have $B_m(K) = 0$ and $\beta_m = 0$. Therefore we have the famous *Euler-Poincaré formula*,

$$\sum_{m=0}^{n} (-1)^m \alpha_m = \sum_{m=0}^{n} (-1)^m p_m(K).$$

The number $\sum_{m=0}^{n} (-1)^m \alpha_m$ is called the *Euler characteristic* of the complex K and is denoted by $\chi(K)$.

Despite the noninvariant mechanism used in their definition, the simplicial homology groups of a finite (geometric) complex K are actually topological invariants of the carrying polytope $|K|$. This will be established in Section 8–2. If we assume this fact for now, then we see that the Euler characteristic $\chi(K)$ is also a topological invariant. And we can compute $\chi(K)$ simply by counting the simplexes in any triangulation whatsoever of the polytope $|K|$!

In simple cases, the number $\chi(K)$ affords a useful means for determining the Betti numbers of a complex. For instance, consider a 2-sphere triangulated as a tetrahedron. Simple enumeration yields $\alpha_0 - \alpha_1 + \alpha_2 = 4 - 6 + 4 = 2$. Since the 2-sphere is connected, $p_0(S^2) = 1$; and we have already seen that $p_2(S^2) = 1$. Therefore we have

$$p_0(S^2) - p_1(S^2) + p_2(S^2) = 1 - p_1(S^2) + 1 = 2$$

or

$$p_1(S^2) = 0.$$

As another example, consider the torus T as triangulated in Fig. 6–9. Enumerating simplexes, we obtain $\chi(T) = \alpha_0 - \alpha_1 + \alpha_2 = 9 - 27 + 18 = 0$. Since T is connected, $p_0(T) = 1$; and we showed earlier that $p_2(T) = 1$. Thus

$$\chi(T) = p_0(T) - p_1(T) + p_2(T) = 1 - p_1(T) + 1 = 0,$$

or

$$p_1(T) = 2.$$

EXERCISE 6–12. Determine the Betti numbers of the Klein bottle as triangulated in Fig. 6–16.

EXERCISE 6–13. Triangulate the surface of genus 2 in Fig. 6–18, and compute its Betti number in dimension 1.

EXERCISE 6–14. What is the second Betti number of the projective plane? of the Möbius band?

Remark: It is quite natural to ask if the homology groups solve the problem of characterizing polytopes. We mentioned above (and will prove in Section 8–2) that if $|K_1|$ and $|K_2|$ are homeomorphic polytopes, then the homology groups $H_p(K_1, G)$ and $H_p(K_2, G)$ are isomorphic for each dimension p. What we ask here is whether the converse is true. That is, if $H_p(K_1, G)$ and $H_p(K_2, G)$

FIGURE 6–19

are isomorphic for each dimension p and for all coefficient groups G, are the polytopes $|K_1|$ and $|K_2|$ necessarily homeomorphic? The answer is, "No!" In fact, we have already seen an example to refute this conjecture. In Exercise 6–3, we asked for the homology groups of a complex whose carrier is the space pictured in Fig. 6–19, a 2-sphere with two tangent circles. It is easy to compute the groups of this example and see that they are precisely the same as those of the torus. But this space has a cut point and the torus does not, so the two cannot be homeomorphic!

The reader may construct an example of a space with the same homology groups as the surface of genus 2 but that is not homeomorphic to that surface.

6–8 Some general remarks. When one starts with an oriented complex, the construction of a homology theory is a purely algebraic process. The chain groups and the boundary operators are defined, and then the homology groups follow easily. Let us review and abstract this process.

Given an oriented complex K and a coefficient group G, we consider the weak direct sums $\sum_{i=1}^{\alpha_p} G$, where α_p is the number of p-simplexes in K. These are the chain groups $C_p(K, G)$. Then the boundary operators ∂_p are defined in such a way that $\partial_{p-1}\,\partial_p = 0$ for each positive integer p. The result is that we associate a *chain complex* with the oriented complex K. Abstractly, a chain complex is a sequence $\{C_p, \partial_p\}$ of free groups C_p and homomorphisms $\partial_p : C_p \to C_{p-1}$ such that $\partial_{p-1}\,\partial_p = 0$ for each positive integer p. Given such a chain complex, we define the *cycle* group Z_p to be the kernel of ∂_p and the *boundary* group B_p to be the image $\partial_{p+1}(C_{p+1})$. Since $\partial_{p-1}\,\partial_p = 0$, B_p is a subgroup of Z_p, and because both are free groups, we may define the *homology* group H_p to be the (additively written) factor group $Z_p - B_p$.

Thus, starting with an oriented complex as we did in the development of simplicial homology theory, the above process yields unique homology groups. Also, whenever and however we can associate a chain complex or an oriented complex with a topological space X, we can develop a homology theory for the space X. This assertion has been shown to be true for triangulated spaces, of course, and the automatic process whereby it is done constitutes the chief reason for considering simplicial homology theory first. There are techniques for associating an oriented complex with more general spaces, moreover, and hence there are homology theories for more general spaces. We describe some of these in Chapter 8.

There is another concept that is often used in the literature and that we can mention here. The weak direct sum of a sequence of free abelian groups is again a free abelian group. Using this fact, we may form *the* chain group

$$C(K, G) = C_0(K, G) \oplus C_1(K, G) \oplus \cdots \oplus C(K, G) \oplus \cdots$$

by taking the weak direct sum of the individual chain groups of an oriented complex K over a coefficient group G. By definition, each element of $C(K, G)$ is a sequence $(c_0, c_1, \ldots, c_n, \ldots)$, where c_p is a p-chain of K with values in G and where all but a finite number of the components c_p are zero. Such a weak direct sum as $C(K, G)$ is often called a *graded group.*

On the graded group $C(K, G)$ we have the boundary operator $\partial = \{\partial_p\}$, which is easily seen to be an endomorphism of $C(K, G)$ into itself with the property $\partial\partial = 0$. Abstracting this situation, we arrive at the following definition. If F is an abelian group and d is an endomorphism of F into itself such that $d^2 = 0$, then F is called a *differential group* with *differential operator* d. In these terms, the chain group $C(K, G)$ is a graded differential group with differential operator ∂.

Now given a differential group F, we may define the *cycle* group $Z(F)$ to be the kernel of the differential operator d, and the *boundary* group $B(F)$ to be the image $d(F)$, and finally the *derived group* of F to be $H(F) = Z(F) - B(F)$. Thus *the* homology group $H(K, G)$ is the derived group of the graded differential chain group $C(K, G)$.

Our discussion here serves to exhibit the essential algebraic constructs insofar as we can at this point. We will return to these and other abstract formulations as we become prepared to carry them further. This approach has been introduced here because it is widely used and because the reader should be aware of it early in his study of algebraic topology.

6–9 Universal coefficients. We indicate in this section how the homology groups $H_p(K, G)$ of a finite complex K with coefficient group G can be determined if we know the group G and the integral homology groups $H_p(K)$. This is done by decomposing the integral chain groups $C_p(K)$ in a particular way and showing how this leads to the structure of $H_p(K)$ as was given in Section 6–5. Then the same technique is applied to the chain groups $C_p(K, G)$ to obtain the desired result. To fill in the details of the admittedly sketchy arguments below, the reader may consult Chapter II of Lefschetz's *Algebraic Topology* [20].

The following algebraic results are needed, but will be stated without proof.

THEOREM 6–9. Every subgroup of a free group is a free group.

THEOREM 6–10. A subgroup B of a free group A is a direct summand of A if the factor group (difference group) A/B is a free group, and then A/B is isomorphic to the complementary group B' in $A = B \oplus B'$.

Let B be a subgroup of a free group A. Then an element x of A belongs to the *rational closure* $[B]$ of B if some multiple of x is in B.

THEOREM 6–11. The rational closure of any subgroup B of a free group A is a free group. Moreover, $A/[B]$ is a free group.

THEOREM 6–12. If B is a subgroup of a free group A of finite rank, then there exists a basis (generating elements) $x_1, \ldots, x_n$ of A, and there are integers r and s, both nonnegative, with $r + s \leqq n$ and integers $t_1, \ldots, t_s$ greater than unity such that t_i divides t_{i+1}, $i < s$, with the property that $(x_1, \ldots, x_{r+s})$ is a basis for $[B]$, while $(x_1, \ldots, x_r, t_1x_{r+1}, \ldots, t_sx_{r+s})$ is a basis for B. The factor group A/B is then isomorphic to the direct sum of the free group generated by the elements $x_{r+s+1}, \ldots, x_n$ and cyclic groups of orders t_i.

The elements $x_1, \ldots, x_r$ in the basis of B in the above theorem are called the *free elements* of B; the elements t_ix_{r+i} are the *torsion elements* of B; and the integers t_i are the *torsion coefficients*.

The following notation is useful. If G is an abelian group and t is an integer, then tG is that subgroup of G consisting of all elements tg where g is in G; G^t is that subgroup of G consisting of all elements g such that $tg = 0$; and G_t denotes the factor group G/tG. For instance, a cyclic group of order t is $Z_t = Z/tZ$, where Z, as usual, denotes the group of integers.

The integral chain group $C_p(K)$ is a free group on α_p generators, where α_p is the number of p-simplexes in K. Since the homomorphism ∂ on $C_p(K)$ has image $B_{p-1}(K)$ and kernel $Z_p(K)$, the fundamental theorem on homomorphisms applies to show that $C_p(K)/Z_p(K)$ is isomorphic to $B_{p-1}(K)$. By Theorem 6–9, $B_{p-1}(K)$ is a free group, so $C_p(K)/Z_p(K)$ is also free. Since $Z_p(K)$ is free, Theorem 6–10 permits us to write $C_p(K)$ as a direct sum $C_p(K) = Z_p(K) \oplus X_p(K)$, where the boundary operator ∂ throws $X_p(K)$ isomorphically onto $B_{p-1}(K)$.

We next examine $Z_p(K)$. It is easily seen that each element of the rational closure $[B_p(K)]$ is a p-cycle, so $[B_p(K)]$ is a subgroup of $Z_p(K)$. By Theorem 6–11, $Z_p(K)/[B_p(K)]$ and $[B_p(K)]$ are free groups. Hence we may use Theorem 6–10 again to decompose $Z_p(K)$ into $Z_p(K) = [B_p(K)] \oplus W_p(K)$.

If we apply Theorem 6–12, we may write the group $B_p(K)$ as a direct sum $B_p(K) = \Delta_p(K) \oplus \theta_p(K)$, where $\Delta_p(K)$ is the subgroup generated by the free elements of $B_p(K)$ and $\theta_p(K)$ is the subgroup generated by the torsion elements of $B_p(K)$. This then yields a decomposition of the rational closure of $B_p(K)$ as $[B_p(K)] = \Delta_p(K) \oplus [\theta_p(K)]$. Finally, since $X_p(K)$ is isomorphic to $B_{p-1}(K) = \Delta_{p-1}(K) \oplus \theta_{p-1}(K)$, we can write $X_p(K)$ as a direct sum $\Gamma_p(K) \oplus \Phi_p(K)$, where ∂ throws $\Gamma_p(K)$ onto $\Delta_{p-1}(K)$ and $\Phi_p(K)$ onto $\theta_{p-1}(K)$. Gathering this up, we may write the chain group $C_p(K)$ as

$$C_p(K) = \Delta_p(K) \oplus [\theta_p(K)] \oplus W_p(K) \oplus \Gamma_p(K) \oplus \Phi_p(K).$$

Let a_i^p, b_i^p, c_i^p, d_i^p, and e_i^p be bases for these groups, so chosen that $\theta_p(K)$ has the basis $t_i^p b_i^p$ and ∂ throws d_i^p on a_i^{p-1} and throws e_i^p on $t_i^{p-1} b_i^{p-1}$.

Since

$$Z_p(K) = \Delta_p(K) \oplus [\theta_p(K)] \oplus W_p(K)$$

and

$$B_p(K) = \Delta_p(K) \oplus \theta_p(K),$$

it follows that

$$H_p(K) = \frac{Z_p(K)}{B_p(K)} \cong W_p(K) \oplus \frac{[\theta_p(K)]}{\theta_p(K)},$$

where $W_p(K)$ is a free group whose rank r_p is the pth Betti number of K and where the torsion group $T_p(K) = [\theta_p(K)]/\theta_p(K)$ is isomorphic to the direct sum of cyclic groups $Z_{t_1} \oplus \cdots \oplus Z_{t_s}$.

Next let us introduce an arbitrary coefficient group G. It is clear that $C_p(K, G)$ may be considered as the group of all linear combinations of the basis elements of $C_p(K)$ with coefficients in G. The same remark also applies to the groups in the decomposition of $C_p(K)$ given above. (This is an example of a tensor product, which we define below.) If we use this idea, it is obvious that we may write

$$C_p(K, G) = \Delta_p(K, G) \oplus [\theta_p(K, G)] \oplus W_p(K, G) \oplus \Gamma_p(K, G) \oplus \Phi_p(K, G).$$

Now the boundary operator ∂ carries

$$\Delta_p(K, G) \oplus [\theta_p(K, G)] \oplus W_p(K, G)$$

onto zero because each of their basis elements is so mapped. Also ∂ carries $\Gamma_p(K. G)$ isomorphically onto $\Delta_{p-1}(K, G)$, which means there are no cycles in $\Gamma_p(K, G)$. Then ∂ carries $\Phi_p(K, G)$ into $[\theta_{p-1}(K, G)]$ by the formula $\partial g \cdot e_i^p = t_i^{p-1} g \cdot b_i^{p-1}$. The kernel of ∂ in $\Phi_p(K, G)$ is therefore the direct sum $\sum G^{t_i^{p-1}} \cdot e_i$. The kernel of ∂ is the direct sum of the individual kernels so that

$$Z_p(K, G) = \Delta_p(K, G) \oplus [\theta_p(K, G)] \oplus W_p(K, G) \oplus \sum G^{t_i^{p-1}} \cdot e_i.$$

Similarly,

$$\begin{aligned} \partial C_{p+1}(K, G) &= \partial\Gamma_{p+1}(K, G) \oplus \partial\Phi_{p+1}(K, G) \\ &= \Delta_p(K, G) \oplus \theta_p(K, G) = B_p(K, G). \end{aligned}$$

Therefore, the homology group $H_p(K, G)$ is given as

$$H_p(K, G) = \frac{Z_p(K, G)}{B_p(K, G)} = \frac{[\theta_p(K, G)]}{\theta_p(K, G)} \oplus W_p(K, G) \oplus \sum G^{t_i^{p-1}} \cdot e_i.$$

The first term on the right is exactly $\sum G_{t_i^p} \cdot b_i$, so we have

$$H_p(K, G) = W_p(K, G) \oplus \sum G_{t_i^p} \cdot b_i \oplus \sum G^{t_i^{p-1}} \cdot e_i.$$

Abstractly, then,

$$H_p(K, G) = \sum_1^{r_p} G \oplus \sum G_{t_i^p} \oplus \sum G^{t_i^{p-1}}.$$

Thus it appears that if we know the Betti number r_p and the torsion coefficients t_i^p and t_i^{p-1}, as determined from the integral homology groups $H_p(K)$ and $H_{p-1}(K)$, then we can determine $H_p(K, G)$ precisely. For this reason, the integers are often called a *universal coefficient group*.

The development above has been rephrased recently by introducing a new concept. Let A and B be two modules over a ring R. The *tensor product* $A \otimes B$ of A and B is the module generated by all pairs (a, b), a in A and b in B, with the relations

$$\begin{aligned} &(a_1 + a_2, b) - (a_1, b) - (a_2, b) = 0, && a_1, a_2 \text{ in } A, && b \text{ in } B, \\ &(a, b_1 + b_2) - (a, b_1) - (a, b_2) = 0, && a \text{ in } A, && b_1, b_2 \text{ in } B, \\ &(ra, b) - r(a, b) = 0, \quad a \text{ in } A, && b \text{ in } B, && r \text{ in } R, \end{aligned}$$

and

$$(a, rb) - r(a, b) = 0, \qquad a \text{ in } A, \qquad b \text{ in } B, \qquad r \text{ in } R.$$

An equivalent definition of the tensor product may be given as follows. Let $X(A, B)$ be the free module generated by the set of all pairs (a, b), and let $Y(A, B)$ be the least subgroup of $X(A, B)$ containing all elements of the four forms

$$(a_1 + a_2, b) - (a_1, b) - (a_2, b), \qquad (a, b_1 + b_2) - (a, b_1) - (a, b_2),$$

$$(ra, b) - r(a, b), \qquad \text{and} \qquad (a, rb) - r(a, b).$$

Then

$$A \otimes B = \frac{X(A, B)}{Y(A, B)}.$$

Next, given two modules A and B over a ring R, we define Hom(A, B), the *module of all homomorphisms* φ of A into B. The addition of two homomorphisms is the usual functional addition. That is,

$$(\varphi_1 + \varphi_2)(a) = \varphi_1(a) + \varphi_2(a),$$

and the multiplication of a homomorphism φ by an element of the ring R is given by

$$(r\varphi)(a) = r \cdot \varphi(a) = \varphi(ra).$$

Now if A is a free group, if B is a subgroup of A, and G is any abelian group, then the inclusion isomorphism $i:B \to A$ induces a homomorphism

$$i_*:B \otimes G \to A \otimes G.$$

The kernel of i_* is essentially a function of the groups $H = A/B$ and G. We denote this kernel by Tor$(A/B, G)$.

In these terms, we can write the "universal coefficient theorem" given above as

$$H_p(K, G) = H_p(K) \otimes G \oplus \text{Tor}(H_{p-1}(K), G).$$

Again we must say that the preceding discussion is merely an indication of the more modern and abstract approach to homology theory. If he wishes to pursue this development later, the reader is referred to Chapter V of Eilenberg and Steenrod, *Introduction to Axiomatic Homology Theory* [7].

6–10 Simplicial mappings again. In Section 5–6, we proved that any continuous mapping of one polytope into another can be approximated arbitrarily closely by a homotopic simplicial mapping defined on suitably chosen triangulations of the two polytopes. We could not investigate some useful properties of simplicial mappings at that time, however, since these properties involve the homology groups. In particular, a simplicial mapping of one complex into another induces homomorphisms of the homology groups of the first complex into those of the second. This is in direct analogy to the induced homomorphisms on the homotopy groups given in Theorem 4–28.

Simplicial mappings are redefined here in terms of abstract complexes, and hence there is no mention of continuity in connection with such "mappings." This generality permits a wider application of the results (for instance, see Section 8-1 on Čech homology theory) and also leads to a consideration of certain transformations, the chain-mappings, defined directly upon the chain groups of a complex.

Let K_1 and K_2 be abstract simplicial complexes, and let φ be a single-valued transformation of the vertices of K_1 into the vertices of K_2. The transformation φ is a *simplicial mapping* of K_1 into K_2 provided that if $\sigma^p = \langle v_0 v_1 \cdots v_p \rangle$ is any simplex of K_1, the collection of vertices $\varphi(v_0)$, $\varphi(v_1), \ldots, \varphi(v_p)$ in K_2 forms a simplex σ^q of K_2. Since φ is not assumed to be one-to-one, it may happen that, for some $i \neq j$, we have $\varphi(v_i) = \varphi(v_j)$. The simplex σ^q in K will then be of lower dimension than σ^p, and in such a case we say that *φ collapses σ^p*.

We now proceed to show how such a simplicial mapping φ of K_1 into K_2 induces a homomorphism of the group $H_p(K_1, G)$ into $H_p(K_2, G)$ for each dimension p. To begin with, we define a transformation φ_p of the chain group $C_p(K_1, G)$ into the chain group $C_p(K_2, G)$ as follows. Let $g\sigma^p$ be an elementary p-chain on K_1. We set

$$\varphi_p(g\sigma^p) = 0 \qquad \text{if } \varphi \text{ collapses } \sigma^p$$

and

$$\varphi_p(g\sigma^p) = g \cdot \varphi(\sigma^p) \qquad \text{if } \varphi \text{ does not collapse } \sigma^p.$$

That is, if $\sigma^p = \langle v_0 \cdots v_p \rangle$, then $\varphi_p(g\sigma^p) = g\langle \varphi(v_0) \cdots \varphi(v_p) \rangle$ only if the image vertices $\varphi(v_i)$ are all distinct. This definition is extended linearly to arbitrary p-chains by means of the formula

$$\varphi_p\left(\sum g_i\sigma_i^p\right) = \sum \varphi_p(g_i\sigma_i^p).$$

The proof of Lemma 6-13 is left as an easy exercise.

LEMMA 6-13. The transformations $\varphi_p : C_p(K_1, G) \to C_p(K_2, G)$ are homomorphisms.

The key property of the collection $\{\varphi_p\}$ of homomorphisms is the content of the next theorem. It is convenient to let ∂ denote the boundary operator in both K_1 and K_2.

THEOREM 6-14. For any p-chain c_p in $C_p(K_1, G)$, $p > 0$,

$$\partial(\varphi_p(c_p)) = \varphi_{p-1}(\partial c_p).$$

Proof: It suffices to prove the desired relation for an elementary chain $g\sigma^p$. If φ_p does not collapse σ^p, the proof is easy. For by definition,

$$\begin{aligned}\partial(\varphi_p(g\sigma^p)) &= \partial(g\langle\varphi(v_0)\cdots\varphi(v_p)\rangle)\\ &= \sum_{i=0}^{p}(-1)^i g\langle\varphi(v_0)\cdots\widehat{\varphi(v_i)}\cdots\varphi(v_p)\rangle,\end{aligned}$$

where the symbol $\widehat{\varphi(v_i)}$ means that this vertex is deleted. On the other hand, we have

$$\begin{aligned}\varphi_{p-1}(\partial(g\sigma^p)) &= \varphi_{p-1}\left(\sum_{i=0}^{p}(-1)^i g\langle v_0\cdots\hat{v}_i\cdots v_p\rangle\right)\\ &= \sum_{i=0}^{p}(-1)^i\varphi_{p-1}(g\langle v_0\cdots\hat{v}_i\cdots v_p\rangle)\\ &= \sum_{i=0}^{p}(-1)^i g\langle\varphi(v_0)\cdots\widehat{\varphi(v_i)}\cdots\varphi(v_p)\rangle,\end{aligned}$$

which establishes the desired relation in this case.

If φ collapses σ^p, then $\varphi(v_i) = \varphi(v_j)$ for some $i \neq j$, and it follows from the definitions that

$$\partial(\varphi_p(g\sigma^p)) = \partial(0) = 0.$$

There is no loss of generality in assuming that the two vertices v_i and v_j are v_0 and v_1 (though not necessarily in this order), for the class of orderings of the vertices that gives the orientation of $\langle v_0\cdots v_p\rangle$ contains either $\langle v_iv_jv_1v_0\cdots\hat{v}_i\cdots\hat{v}_j\cdots v_p\rangle$ or $\langle v_jv_iv_0v_1\cdots\hat{v}_i\cdots\hat{v}_j\cdots v_p\rangle$. With this agreement,

$$\begin{aligned}\varphi_{p-1}(\partial(g\sigma^p)) &= \varphi_{p-1}\left(\sum_{k=0}^{p}(-1)^k g\langle v_0\cdots\hat{v}_k\cdots v_p\rangle\right)\\ &= \varphi_{p-1}(g\langle v_1v_2\cdots v_p\rangle) - \varphi_{p-1}(g\langle v_0v_2\cdots v_p\rangle)\\ &\qquad + \sum_{k=2}^{p}(-1)^k\varphi_{p-1}(g\langle v_0v_1\cdots\hat{v}_k\cdots v_p\rangle)\\ &= g\langle\varphi(v_1)\varphi(v_2)\cdots\varphi(v_p)\rangle - g\langle\varphi(v_0)\varphi(v_2)\cdots\varphi(v_p)\rangle\\ &\qquad + \sum_{k=2}^{p}(-1)^k g\langle\varphi(v_0)\varphi(v_1)\cdots\widehat{\varphi(v_k)}\cdots\varphi(v_p)\rangle.\end{aligned}$$

Since $\varphi(v_0) = \varphi(v_1)$, each term in the summation is zero by definition and also the first two terms cancel. If more than the face $\langle v_0v_1\rangle$ of σ^p is collapsed, of course, the first two terms may already be zero. □

The collection $\{\varphi_p\}$ of homomorphisms induced by a simplicial mapping φ is often denoted by the same symbol φ, and the basic property established in Theorem 6–14 is then given by the symbolic formula

$$\partial\varphi = \varphi\partial.$$

One says that the induced mapping φ on chain groups *commutes* with the boundary operator ∂. A schematic representation of this relationship is helpful in remembering it.

$$\begin{array}{ccc} C_p(K_1, G) & \xrightarrow{\partial} & C_{p-1}(K_1, G) \\ \varphi_p \big\downarrow & & \big\downarrow \varphi_{p-1} \\ C_p(K_2, G) & \xrightarrow[\partial]{} & C_{p-1}(K_2, G) \end{array}$$

The relationship is now given by requiring *commutativity in the diagram.* Such diagrams of groups and homomorphisms are very useful, and the reader will see them often, both in this book and in the current literature.

Let us examine the consequences of the relation $\partial\varphi = \varphi\partial$.

LEMMA 6–15. *If z_p is a p-cycle of K_1, then $\varphi_p(z_p)$ is a p-cycle of K_2.*

Proof: We need only show that $\partial\varphi_p(z_p) = 0$. But by Theorem 6–14, $\partial\varphi_p(z_p) = \varphi_{p-1}\,\partial(z_p) = \varphi_{p-1}(0) = 0$ since, by definition, $\partial z_p = 0$. □

LEMMA 6–16. *If b_p is a p-boundary of K_1, then $\varphi_p(b_p)$ is a p-boundary of K_2.*

Proof: If $b_p = \partial c_{p+1}$, then $\varphi_p(b_p) = \varphi_p(\partial c_{p+1}) = \partial\varphi_{p+1}(c_{p+1})$ by Theorem 6–14. Thus $\varphi_{p+1}(c_{p+1})$ is a $(p + 1)$-chain of K_2 with boundary $\varphi_p(b_p)$. □

LEMMA 6–17. *If z_p^1 and z_p^2 are homologous p-cycles of K_1, then $\varphi_p(z_p^1)$ and $\varphi_p(z_p^2)$ are homologous p-cycles of K_2.*

Proof: If z_p^1 and z_p^2 are homologous, then $z_p^1 - z_p^2 = \partial c_{p+1}$. Then we have $\varphi_p(z_p^1) - \varphi_p(z_p^2) = \varphi_p(z_p^1 - z_p^2) = \varphi_p(\partial c_{p+1}) = \partial\varphi_{p+1}(c_{p+1})$. □

THEOREM 6–18. *The homomorphism $\varphi_p{:}C_p(K_1, G) \to C_p(K_2, G)$ induces a homomorphism φ_{p*} of $H_p(K_1, G)$ into $H_p(K_2, G)$.*

Proof: For an element of a homology group, that is, a coset in the cycle group, we use our customary notation $[z_p]$, where z_p is any representative of the homology element. We define the desired homomorphism φ_{p*} by setting

$$\varphi_{p*}([z_p]) = [\varphi_p(z_p)].$$

It must be shown that φ_{p*} is well-defined. This entails proving that if z_p' is any other representative of $[z_p]$, then $\varphi_p(z_p')$ is a representative of $[\varphi_p(z_p)]$. But this is precisely the content of Lemma 6–17. The fact that φ_{p*} is a homomorphism follows immediately since φ_p is a homomorphism. □

It should be obvious that *the identity simplicial mapping of a complex K onto itself induces the identity isomorphisms on the homology group of K.* Two complexes K_1 and K_2 are said to be *isomorphic complexes* if there exists a *one-to-one* simplicial mapping φ of K_1 onto K_2 such that φ^{-1} is also simplicial. (We made use of this situation in the realization theorem in Section 5–7.) Again it should be evident that such a simplicial isomorphism induces isomorphisms of the groups $H_p(K_1, G)$ onto $H_p(K_2, G)$.

A special case of a simplicial mapping occurs when the complex K_1 is a subcomplex of K_2. The simplicial mapping i defined by

$$i(v) = v$$

for each vertex v of K_1 is called the *injection mapping* of K_1 into K_2. The induced homomorphisms $i_p:C_p(K_1, G) \to C_p(K_2, G)$ can easily be shown to be isomorphisms into. But it does *not* follow that the induced homomorphisms $i_{p*}:H_p(K_1, G) \to H_p(K_2, G)$ are isomorphisms into! To see why this may be so, consider a p-cycle on K_1 which does not bound on K_1. There may be $(p + 1)$-simplexes in K_2 that are not in K_1, and these may give a $(p + 1)$-chain on which the p-cycle bounds in K_2. A simple instance of this situation is obtained by injecting the 1-skeleton of a 2-simplex into the closure of the 2-simplex. Since this 1-skeleton is a 1-sphere, its first homology group is isomorphic to the coefficient group G, whereas the first homology group of the closure of a 2-simplex is zero. And there is no *isomorphism* of a nontrivial group G onto zero.

Several instances of the use of simplicial mappings will occur in our subsequent developments, so no examples are given here. The exercises below should prove rewarding, however, and the reader is urged to complete them before proceeding.

A *closed n-pseudomanifold* is a finite complex K with the following properties:

(a) K is homogeneously n-dimensional in the sense that every simplex of K is a face of some n-simplex of K.

(b) Every $(n - 1)$-simplex is a face of exactly two n-simplexes.

(c) Given two n-simplexes σ_1^n and σ_2^n of K, there is a finite chain of n-simplexes and $(n - 1)$-simplexes, beginning with σ_1^n and ending with σ_2^n, such that any two successive elements of the chain are incident.

We define K to be *orientable* if the integral homology group $H_n(K)$ is not trivial; otherwise K is *nonorientable.*

EXERCISE 6–15. Prove that if K is an orientable n-pseudomanifold, then the n-simplexes σ_i^n can be so oriented that the integral n-chain $x_n = \sum 1\sigma_i^n$ is an n-cycle. Also prove that every integral n-cycle on K is then a multiple of x_n and hence $H_n(K)$ is infinite cyclic.

EXERCISE 6–16. Let $K = S^n$ be the boundary complex of an $(n + 1)$-simplex σ^{n+1}. Show that S^n is an n-pseudomanifold.

EXERCISE 6–17. Let T^2, P^2, and B be triangulations of the torus, the projective plane, and the Klein bottle, respectively. Show that each is a 2-pseudomanifold.

EXERCISE 6–18. Prove that if K_1 and K_2 are two n-pseudomanifolds with fundamental n-cycles x_n^1 and x_n^2, and if $\varphi:K_1 \to K_2$ is a simplicial mapping of K_1 into K_2, then $\varphi_*(x_n^1) = k \cdot x_n^2$. (The number k is called the *degree* of φ.)

EXERCISE 6–19. Let $\varphi:S^2 \to T^2$ be a simplicial mapping of some triangulation S^2 of the 2-sphere onto some triangulation T^2 of the torus. Prove that the degree of φ must be zero.

EXERCISE 6–20. Construct a simplicial mapping $\varphi:T^2 \to S^2$ whose degree is n, for each $n = \ldots, -3, -2, -1, 0, 1, 2, 3, \ldots$

EXERCISE 6–21. Prove that if K is a nonorientable n-pseudomanifold, then $H_{n-1}(K)$ is cyclic of order 2.

EXERCISE 6–22. Prove that if K is any complex and K^p is its p-skeleton, then for each i, $0 \leq i < p$, $H_i(K)$ and $H_i(K^p)$ are isomorphic. What can be said about $H_p(K^p)$ in relation to $H_p(K)$?

6–11 Chain-mappings. The construction of the homomorphisms $\varphi_p:C_p(K_1, G) \to C_p(K_2, G)$ from a simplicial mapping $\varphi:K_1 \to K_2$ was the key feature of Section 6–10. To permit even greater generality, we now consider merely the algebraic structure and assume that we are given a collection $\{\varphi_p\}$ of homomorphisms $\varphi_p:C_p(K_1, G) \to C_p(K_2, G)$. That is, the collection $\{\varphi_p\}$ is not necessarily induced by some simplicial mapping. Such a collection $\{\varphi_p\}$ is called a *chain-mapping* of K_1 into K_2 if the commutative relation,

$$\partial(\varphi_p(c_p)) = \varphi_{p-1}(\partial c_p),$$

holds for each chain c_p in $C_p(K_1, G)$.

In these terms, every simplicial mapping induces a chain-mapping. But there are chain-mappings that are not induced by any simplicial mapping. As a simple example, consider the complex K consisting of three vertices v_0, v_1, and v_2 and three 1-simplexes $\langle v_0v_1\rangle$, $\langle v_0v_2\rangle$, and $\langle v_1v_2\rangle$. We will give homomorphisms φ_0 and φ_1 of the mod 2 chain groups of K into themselves which satisfy the commutative relation and yet are not induced by a simplicial mapping. As usual, it suffices to define φ_0 and φ_1 on elementary chains. Letting a denote either 0 or 1, we define

$$\varphi_0(a\langle v_i\rangle) = a\langle v_i\rangle \qquad (i = 0, 1, 2),$$

$$\varphi_1(a\langle v_0v_1\rangle) = a\langle v_0v_2\rangle + a\langle v_1v_2\rangle,$$

$$\varphi_1(a\langle v_0v_2\rangle) = a\langle v_0v_1\rangle + a\langle v_1v_2\rangle,$$

and

$$\varphi_1(a\langle v_1v_2\rangle) = a\langle v_0v_1\rangle + a\langle v_0v_2\rangle.$$

The only simplicial mapping that could induce φ_0 is the identity simplicial mapping i, but i cannot induce φ_1, so no simplicial mapping induces (φ_0, φ_1). It is easily shown that φ_0 and φ_1 are homomorphisms and that $\partial\varphi_1 = \varphi_0\partial$. Thus the pair (φ_0, φ_1) is a chain-mapping induced by no simplicial mapping.

We can apply the same definition of chain-mapping to the abstract chain complexes of Section 6–8. Given two chain complexes (C_p, ∂_p) and (C'_p, ∂'_p), a mapping $f:(C_p, \partial_p) \to (C'_p, \partial'_p)$ is a sequence of homomorphisms $f_p:C_p \to C'_p$ such that $\partial'_p f_p = f_{p-1}\partial_p$, $p > 0$. This permits all our algebraic constructions to be applied to chain complexes.

Since Lemmas 6–15, 6–16, and 6–17 and Theorem 6–18 depend entirely upon the commutative relation $\partial\varphi = \varphi\partial$, these results apply to arbitrary chain-mappings as well as to those induced by simplicial mappings. Therefore we know that *a chain-mapping* $\varphi = \{\varphi_p\}$ *induces homomorphisms* φ_{p*} *of the homology groups of* K_1 *into those of* K_2. We also let φ_* denote the entire collection $\{\varphi_{p*}\}$.

Our chief problem is one of comparing two chain-mappings as to their induced homomorphisms on homology groups. In particular, we introduce a relationship between two chain-mappings φ^1 and φ^2 which assures us that the induced homomorphisms φ^1_* and φ^2_* are the same. This relationship is a combinatorial analogue of the homotopic relation between continuous mappings.

Let φ^1 and φ^2 be chain-mappings of the integral chain groups $C_p(K_1)$ of a complex K_1 into the integral chain groups $C_p(K_2)$ of a complex K_2. Then φ^1 and φ^2 are *chain-homotopic* provided that there is a collection $\mathfrak{D} = \{\mathfrak{D}_p\}$ of homomorphisms $\mathfrak{D}_p:C_p(K_1) \to C_{p+1}(K_2)$ such that for every chain c_p in $C_p(K_1)$ the following relation holds:

$$\partial\mathfrak{D}_p(c_p) = \varphi^2_p(c_p) - \varphi^1_p(c_p) - \mathfrak{D}_{p-1}(\partial c_p), \quad \mathfrak{D}_{-1}(0) = 0.$$

That is, $\mathfrak{D}_p(c_p)$ is a $(p + 1)$-chain on K_2 whose boundary is given by the above equation. The collection $\mathfrak{D}$ is called a *deformation operator* (see Lefschetz [96]), and the fundamental relation is often given by the symbolic formula

$$\partial\mathfrak{D} = \varphi^2 - \varphi^1 - \mathfrak{D}\partial.$$

At first sight this definition looks more complicated than it actually is. Several examples will help to clarify the basic idea. Let $1\sigma^0$ be an elementary 0-chain. Then $\mathfrak{D}_0(1 \cdot \sigma^0)$ is a 1-chain on K_2, and we have

$$\partial\mathfrak{D}_0(1 \cdot \sigma^0) = \varphi^2(1 \cdot \sigma^0) - \varphi^1(1 \cdot \sigma^0) - \mathfrak{D}_{-1}\partial(1 \cdot \sigma^0).$$

In the nonaugmented case, $\partial(1 \cdot \sigma^0) = 0$, so

$$\partial \mathfrak{D}_0(1 \cdot \sigma^0) = \varphi^2(1 \cdot \sigma^0) - \varphi^1(1 \cdot \sigma^0).$$

Thus $\mathfrak{D}_0(1 \cdot \sigma^0)$ is a 1-chain on K_2 whose boundary is $\varphi^2(1 \cdot \sigma^0) - \varphi^1(1 \cdot \sigma^0)$. That is, if φ^1 and φ^2 are chain-homotopic, then every 0-cycle on K_2 of the form $\varphi^2(1 \cdot \sigma^0) - \varphi^1(1 \cdot \sigma^0)$ is a 0-boundary.

Now let $1 \cdot \sigma^1$ be an elementary 1-chain on K_1. Then $\mathfrak{D}_1(1 \cdot \sigma^1)$ is a 2-chain on K_2, and

$$\partial \mathfrak{D}_1(1 \cdot \sigma^1) = \varphi^2(1 \cdot \sigma^1) - \varphi^1(1 \cdot \sigma^1) - \mathfrak{D}_0 \partial(1 \cdot \sigma^1).$$

We know that $\partial\partial = 0$, and we had better check this for the chain $\mathfrak{D}_1(1 \cdot \sigma^1)$. Computing, we have

$$\begin{aligned} \partial[\partial \mathfrak{D}_1(1 \cdot \sigma^1)] &= \partial[\varphi^2(1 \cdot \sigma^1) - \varphi^1(1 \cdot \sigma^1) - \mathfrak{D}_0 \partial(1 \cdot \sigma^1)] \\ &= \partial\varphi^2(1 \cdot \sigma^1) - \partial\varphi^1(1 \cdot \sigma^1) - \partial \mathfrak{D}_0 \partial(1 \cdot \sigma^1). \end{aligned}$$

Applying the fundamental relation to the last term on the right, we have

$$\partial \mathfrak{D}_0 \partial(1 \cdot \sigma^1) = \varphi^2 \partial(1 \cdot \sigma^1) - \varphi^1 \partial(1 \cdot \sigma^1) - \mathfrak{D}_{-1} \partial\partial(1 \cdot \sigma^1).$$

Since $\partial\partial(1 \cdot \sigma^1) = 0$ and both φ^1 and φ^2 commute with ∂, we have

$$\begin{aligned} \partial\partial \mathfrak{D}_1(1 \cdot \sigma^1) &= \partial\varphi^2(1 \cdot \sigma^1) - \partial\varphi^1(1 \cdot \sigma^1) - \varphi^2\partial(1 \cdot \sigma^1) + \varphi^1\partial(1 \cdot \sigma^1) \\ &= \partial\varphi^2(1 \cdot \sigma^1) - \varphi^2\partial(1 \cdot \sigma^1) - \partial\varphi^1(1 \cdot \sigma^1) \\ &\qquad + \varphi^1\partial(1 \cdot \sigma^1) = 0. \end{aligned}$$

Thus the 2-chain $\mathfrak{D}_1(1 \cdot \sigma^1)$ has $\varphi^2(1 \cdot \sigma^1)$ and $-\varphi^1(1 \cdot \sigma^1)$ in its boundary as well as the two 1-chains $\mathfrak{D}_0 \partial(1 \cdot \sigma^1)$. In a sense, this 2-chain plays the role of the homotopy cylinder in this combinatorial situation. Figure 6–20 is a picture of a typical chain $\mathfrak{D}_1(1 \cdot \sigma^1)$.

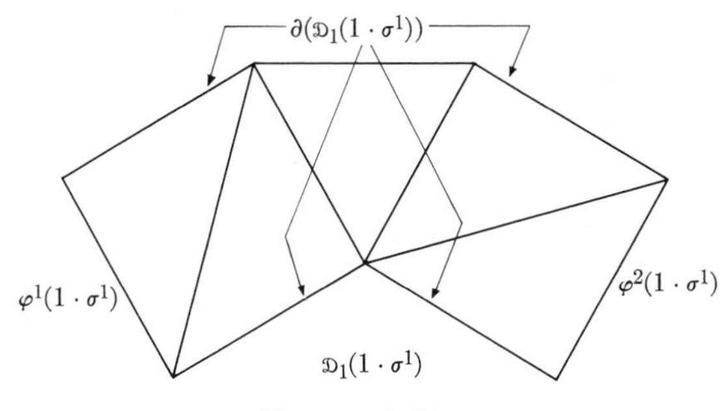

FIGURE 6–20

Our primary interest is in cycles, of course, so let the chain c_p be a p-cycle on K_1. Then the fundamental formula tells us that the $(p+1)$-chain $\mathfrak{D}_p(c_p)$ has boundary

$$\partial \mathfrak{D}_p(c_p) = \varphi^2(c_p) - \varphi^1(c_p)$$

because $\mathfrak{D}_{p-1}\,\partial c_p = \mathfrak{D}_{p-1}(0) = 0$. Thus for any p-cycle z_p on K_1, the chain $\mathfrak{D}_p(z_p)$ has boundary $\varphi^2(z_p) - \varphi^1(z_p)$, and this is precisely the condition which says that $\varphi^2(z_p)$ is homologous to $\varphi^1(z_p)$. In other words, if the chain-mappings φ^1 and φ^2 are chain-homotopic, then every cycle z_p on K_1 is mapped onto homologous cycles $\varphi^1(z_p)$ and $\varphi^2(z_p)$ on K_2. This statement concludes the proof of the following result.

THEOREM 6–19. If φ^1 and φ^2 are chain-homotopic chain-mappings of a complex K_1 into a complex K_2, then the induced homomorphisms φ^1_* and φ^2_* on the integral homology groups coincide.

Let K_1 and K_2 be two complexes, and let i_1 and i_2 be the identity chain-mappings of K_1 onto itself and K_2 onto itself, respectively. Then K_1 and K_2 are *chain-equivalent complexes* if there are chain-mappings $\varphi{:}K_1 \to K_2$ and $\psi{:}K_2 \to K_1$ such that the composite mapping $\varphi\psi{:}K_2 \to K_2$ is chain-homotopic to i_2 and $\psi\varphi{:}K_1 \to K_1$ is chain-homotopic to i_1. It should be noted that chain-equivalent complexes need not be isomorphic, although the converse is true. This definition is the combinatorial analogue of the concept of homotopically equivalent spaces (see Section 4–4), and hence the following analogue of Corollary 4–29 is not surprising.

THEOREM 6–20. Chain-equivalent complexes have isomorphic integral homology groups.

Proof: As was remarked following Theorem 6–18, the induced homomorphisms i_{1*} and i_{2*} are the identity isomorphisms and hence, by Theorem 6–19, the induced mappings $(\psi\varphi)_*$ and $(\varphi\psi)_*$ are isomorphisms onto. It is easy to show that $(\psi\varphi)_* = \psi_*\varphi_*$ and that $(\varphi\psi)_* = \varphi_*\psi_*$ (see Exercise 6–23 below). Since $\psi_*\varphi_*$ is an isomorphism onto, it follows that φ_* must be an isomorphism and that ψ_* must be onto. Similarly, since $\varphi_*\psi_*$ is an isomorphism onto, ψ_* is an isomorphism and φ_* is onto. Therefore both φ_* and ψ_* are isomorphisms onto. □

EXERCISE 6–23. Let $\varphi{:}K_1 \to K_2$ and $\psi{:}K_2 \to K_3$ be chain-mappings. Show that the composite mapping $\psi\varphi$ is a chain-mapping of K_1 into K_3, and show that the induced homomorphism $(\psi\varphi)_*$ is the composite $\psi_*\varphi_*$ of the induced homomorphisms.

6–12 Cone-complexes. Let K be a finite complex, and let v be a vertex not in K. The *cone at v over K* is the complex vK consisting of (1) all simplexes of K, (2) the vertex v, and (3) all simplexes of the form $\langle vv_0 \cdots v_p \rangle$, where $\langle v_0 \cdots v_p \rangle$ is a simplex of K. In the case of a geometric complex K, the carrier $|vK|$ is precisely the join of v and K as given in Section 4–3. We of course orient vK.

In Section 4–3, we showed that every join of a point and a space is contractible and hence is homotopically trivial, that is, all homotopy groups of such a space vanish. We now apply Theorem 6–20 to prove that every cone-complex vK is homologically trivial, meaning that the integral homology groups $H_p(vK)$, $p > 0$, are all trivial, whereas $H_0(vK)$ is infinite cyclic. This will be done by showing that the cone-complex vK is chain-equivalent to the complex consisting of the single vertex v. In view of Theorem 6–20, this will complete the proof since a single vertex is certainly homologically trivial.

An auxiliary result will be needed. Consider an oriented simplex $\sigma^p = \langle v_0 \cdots v_p \rangle$ in K. Let $v\sigma^p$ denote the oriented simplex $\langle vv_0 \cdots v_p \rangle$ in vK. Similarly, if c_p is a chain in $C_p(K)$, then vc_p denotes the obvious $(p+1)$-chain in $C_{p+1}(vK)$.

LEMMA 6–21. Let $1 \cdot v\sigma^p$ be an elementary $(p+1)$-chain on vK. Then

$$\partial 1 \cdot v\sigma^p = 1 \cdot \sigma^p - v\partial(1 \cdot \sigma^p).$$

Proof: We need only compute.

$$\begin{aligned}\partial 1 \cdot \langle vv_0 \cdots v_p \rangle &= 1 \cdot \langle v_0 \cdots v_p \rangle - 1 \cdot \langle vv_1 \cdots v_p \rangle + \cdots \\ &\qquad + (-1)^p 1 \cdot \langle vv_0 \cdots v_{p-1} \rangle \\ &= 1 \cdot \langle v_0 \cdots v_p \rangle - 1 \cdot v\langle v_1 \cdots v_p \rangle + \cdots \\ &\qquad + (-1)^p 1 \cdot v\langle v_0 \cdots v_{p-1} \rangle \\ &= 1 \cdot \sigma^p - v\partial(1 \cdot \sigma^p). \ \square\end{aligned}$$

To show that vK and v are chain-equivalent, let φ be the simplicial mapping which carries each vertex v_i of vK onto v. Also let φ denote the associated chain-mapping. Let the map ψ of the definition of chain-homotopy be the injection chain-mapping i of v into vK. It is evident that the composite mapping $\varphi\psi$ is the identity mapping of v onto itself, and hence the requirement that $\varphi\psi$ be chain-homotopic to the identity is automatically satisfied. To complete the argument, we must construct a deformation operator connecting the composite mapping $\psi\varphi = i\varphi$ and the identity mapping j on vK. For brevity's sake, let us denote $i\varphi$ by τ. We wish the operator $\mathfrak{D} = \{\mathfrak{D}_p\}$ to be such that

$$\partial\mathfrak{D}_p(c_p) = j(c_p) - \tau(c_p) - \mathfrak{D}_{p-1}(\partial c_p).$$

We define $\mathfrak{D}$ on elementary chains as follows:

$$\begin{aligned} \mathfrak{D}_p(1 \cdot \sigma^p) &= 0 && \text{if } v \text{ is a vertex of } \sigma^p && (p \geqq 0) \\ &= 1 \cdot v\sigma^p && \text{if } v \text{ is not a vertex of } \sigma^p && (p \geqq 0). \end{aligned}$$

This definition is extended linearly to arbitrary integral p-chains, as usual.

Now as a simplicial mapping, τ throws every vertex of vK onto v, and hence τ collapses every simplex σ^p, $p > 0$, in vK. Thus, as a chain-mapping, $\tau(1 \cdot \sigma^p) = 0, p > 0$. The operator $\mathfrak{D}$ should satisfy

$$\begin{aligned} \partial\mathfrak{D}_p(1 \cdot \sigma^p) &= j(1 \cdot \sigma^p) - \mathfrak{D}_{p-1}\partial(1 \cdot \sigma^p) && (p > 0) \\ &= 1 \cdot \sigma^p - \mathfrak{D}_{p-1}\partial(1 \cdot \sigma^p) && (p > 0). \end{aligned}$$

Or, by definition of $\mathfrak{D}$, we must show that

$$\partial(1 \cdot v\sigma^p) = 1 \cdot \sigma^p - v(\partial\sigma^p).$$

But this is precisely the conclusion of Lemma 6–21 and hence is a deformation operator. In view of Theorem 6–20, this completes a proof of the desired result.

THEOREM 6–22. Any cone-complex is homologically trivial.

As an example of a cone-complex, consider the following situation. In Euclidean $(n + 1)$-space E^{n+1}, let v_0 denote the origin and v_i, $i = 1, 2, \ldots, n + 1$, be the unit points on the axes. The $n + 1$ points $v_0, v_1, \ldots, v_n$ determine an n-dimensional geometric simplex s^n in the hyperplane $x_{n+1} = 0$. The boundary of s^n is (topologically) an $(n - 1)$-sphere. Let K denote the $(n - 1)$-skeleton of $\mathrm{Cl}(s^n)$, and let $v_{n+1} = v$. Construct the cone-complex vK. By a projection parallel to the line through the points $(0, 0, \ldots, 0, 1)$ and $(1/(n + 1), \ldots, 1/(n + 1), 0)$, we obtain a homeomorphism of the carrier $|vK|$ onto s^n. Thus the carrier $|vK|$ is also a topological n-cell. If we accept the topological invariance of simplicial homology groups (see Section 8–3), then this construction proves that any n-cell is homologically trivial.

6–13 Barycentric subdivision again. In Section 5–5, we developed the method of refining a geometric complex K known as *barycentric subdivision*. It was noted that the barycentric subdivision K' of K has the same geometric carrier as does K. If the simplicial homology groups are to have geometric significance, they certainly should be invariant under barycentric subdivision. That is, we should be able to show that $H_p(K', G)$ and $H_p(K, G)$ are isomorphic for each p. It is this fact which we prove in this section. We will eventually do more than this. Our actual goal is to prove that the homology groups depend only upon the coefficient group G and the *space* $|K|$. In short, the homology groups are topological invariants.

For purposes of generalization, we redefine barycentric subdivision in terms of abstract simplicial complexes. To do this, recall that an abstract simplicial complex K is a pair $(\mathcal{V}, \Sigma)$, where $\mathcal{V}$ is a collection of abstract elements called *vertices*, and $\Sigma = \{\sigma\}$ is a collection of finite subsets σ of $\mathcal{V}$ with the property that if σ is an element of Σ, then every subset of σ is again an element of Σ. Now the *barycentric subdivision* of $K = (\mathcal{V}, \Sigma)$ is a complex $K' = (\mathcal{V}', \Sigma')$, where (1) $\mathcal{V}' = \Sigma$; that is, the vertices of K' are the simplexes of K (if σ is a simplex of K, we will denote it by $\dot{\sigma}$ when thinking of it as a vertex of K'), and (2) the simplexes $'\sigma$ in Σ' are defined by saying that $(\dot{\sigma}_0, \ldots, \dot{\sigma}_p)$ constitutes a simplex $'\sigma^p = \langle \dot{\sigma}_0 \cdots \dot{\sigma}_p \rangle$ in Σ' if for some permutation $(i_0, \ldots, i_p)$ of $(0, 1, \ldots, p)$, it is true that σ_{i_j} is a face of $\sigma_{i_{j+1}}$, for each $j < p$, as simplexes of K. As in the geometric case, the simplex σ_{i_p} of highest dimension is called the *carrier of* $'\sigma^p$. It will profit the reader to return to Section 5–5 and compare the above definition with the geometric case. By identifying the vertex $\dot{\sigma}$ with the centroid of σ in the geometric simplicial complex, he will see that the two definitions agree in that they yield isomorphic abstract complexes.

To proceed with the chief business of this section, we will prove that the barycentric subdivision K' of a complex K is chain-equivalent to K itself. In view of Theorem 6–20, this will prove that the integral homology groups $H_p(K')$ and $H_p(K)$ are isomorphic. The proof of the following theorem is quite long and involved, so we will not hesitate to digress in order to illustrate the situations we meet.

THEOREM 6–23. A finite complex K and its barycentric subdivision K' are chain-equivalent.

Proof: We must define chain-mappings u of K into K' and u' of K' into K such that both the composite mappings uu' and $u'u$ are chain-homotopic to the identity chain-mappings. We define u' first and as a simplicial mapping. If $\dot{\sigma}$ is a vertex of K', we let $u'(\dot{\sigma})$ be any vertex of the carrier σ of $\dot{\sigma}$. The choice of the vertex of σ to be used as $u'(\dot{\sigma})$ is arbitrary but, once made, it is fixed. It is easily seen that u' is indeed a simplicial mapping and hence induces a chain-mapping, also called u', of K' into K. Furthermore, as a chain-mapping, u' has the following effect on an elementary chain $1 \cdot {}'\sigma$ of K':

$$u'(1 \cdot {}'\sigma) = \eta \cdot \sigma, \tag{1}$$

where $\eta = 0, \pm 1$, and where σ is some face of the carrier of $'\sigma$. If u' collapses $'\sigma$, then $\eta = 0$; otherwise $\eta = \pm 1$, depending upon the relative orientations of $'\sigma$ and σ.

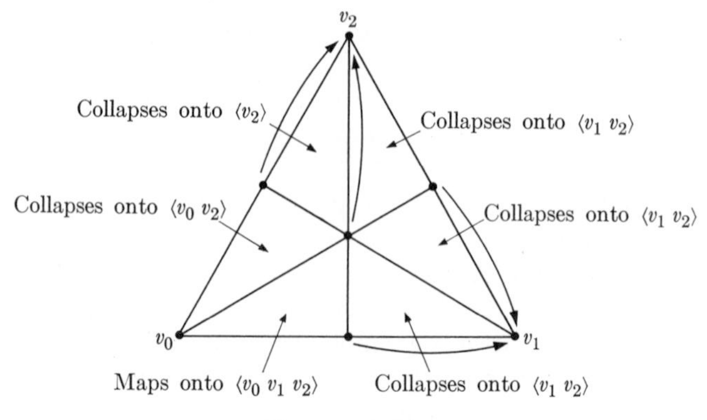

FIGURE 6–21

Figure 6–21 shows an example of a u'. The arrows indicate where each vertex of K' is sent by u'. Note that $u'(\dot{\sigma}^0) = \sigma^0$ as required. It should be noted that u' sends one 2-simplex of K' onto the 2-simplex of K. Such must always be the case (see Sperner's lemma as in Lefschetz [22]).

The chain-mapping u of K into K' does *not* come from a simplicial mapping. For an elementary chain $1 \cdot \sigma^p$ of K, u will yield a chain of K' consisting of terms $\pm 1 \cdot {}'\sigma^p$, where ${}'\sigma^p$ ranges over the p-simplexes in the subdivision of σ^p, the sign being determined by orientation, so that the boundary of $u(1 \cdot \sigma^p)$ is in the subdivision of the boundary of σ^p. We give an inductive definition of u as follows. For a vertex σ^0 of K, we let

$$u(1 \cdot \sigma^0) = 1 \cdot \dot{\sigma}^0.$$

The vertices σ^0 and $\dot{\sigma}^0$ are the same point, but we are regarding $\dot{\sigma}^0$ as a vertex of K'. It is obvious that, so far, we have the necessary commutativity with the boundary operator, since $\partial u(1 \cdot \sigma^0) = u\partial(1 \cdot \sigma^0) = 0$. For a 1-simplex $\sigma^1 = \langle \sigma_0^0 \sigma_1^0 \rangle$ of K, oriented as indicated, we set

$$u(1 \cdot \sigma^1) = 1 \cdot \langle \dot{\sigma}_0^0 \dot{\sigma}^1 \rangle + 1 \cdot \langle \dot{\sigma}^1 \dot{\sigma}_1^0 \rangle.$$

Verifying the commutative relation, we find that

$$\partial u(1 \cdot \sigma^1) = 1 \cdot \dot{\sigma}^1 - 1 \cdot \dot{\sigma}_0^0 + 1 \cdot \dot{\sigma}_1^0 - 1 \cdot \dot{\sigma}^1 = 1 \cdot \dot{\sigma}_1^0 - 1 \cdot \dot{\sigma}_0^0$$

and

$$u\partial(1 \cdot \sigma^1) = u(1 \cdot \sigma_1^0 - 1 \cdot \sigma_0^0) = 1 \cdot \dot{\sigma}_1^0 - 1 \cdot \dot{\sigma}_0^0.$$

Hence the relation $\partial u = u\partial$ holds so far.

Now assume that u has been defined for all elementary chains of dimension $q < p$ such that $\partial u = u\partial$. We then define u on an elementary p-chain $1 \cdot \sigma^p$ by setting

$$u(1 \cdot \sigma^p) = \dot{\sigma}^p u\partial(1 \cdot \sigma^p);$$

that is, $u(1 \cdot \sigma^p)$ is the p-chain on the cone at $\dot{\sigma}^p$ over the chain $u\partial(1 \cdot \sigma^p)$. (We are using the notation of the previous section here.) Checking the commutative relation, we find that

$$\partial u(1 \cdot \sigma^p) = u\partial(1 \cdot \sigma^p) \pm \dot{\sigma}^p \partial u(\partial(1 \cdot \sigma^p)),$$

as was shown in Section 6–12. But we have $\partial u = u\partial$ for dimension $p - 1$, so that $\partial u(\partial(1 \cdot \sigma^p)) = u\partial\partial(1 \cdot \sigma^p) = u(0) = 0$. This completes the inductive definition of the chain-mapping u.

The composite mapping $u'u$ *is* the identity chain-mapping on K. For $u'u$ carries each elementary 0-chain $1 \cdot \sigma^0$ onto itself. Thus we need only prove that uu' is chain-homotopic to the identity chain-mapping on K' to complete the proof. This means that we must construct a deformation operator $\mathfrak{D} = \{\mathfrak{D}_p\}$ such that for any elementary p-chain $1 \cdot {}'\sigma^p$ of K' we have

$$\partial \mathfrak{D}_p(1 \cdot {}'\sigma^p) = 1 \cdot {}'\sigma^p - uu'(1 \cdot {}'\sigma^p) - \mathfrak{D}_{p-1}\partial(1 \cdot {}'\sigma^p). \tag{2}$$

The definition is again inductive.

Given any vertex $\dot{\sigma}$ in K', we define $\mathfrak{D}_0$ so that

$$\partial \mathfrak{D}_0(1 \cdot \dot{\sigma}) = 1 \cdot \dot{\sigma} - uu'(1 \cdot \dot{\sigma})$$

[because $\mathfrak{D}_{-1}(\partial(1 \cdot \dot{\sigma})) = 0$]. That is, we must define $\partial\mathfrak{D}_0(1 \cdot \dot{\sigma})$ to be a 0-chain on a pair of vertices of K'. But $u'(1 \cdot \dot{\sigma})$ is a 0-chain on σ^0, where σ^0 is some vertex of the carrier σ of $\dot{\sigma}$. Then $uu'(1 \cdot \dot{\sigma}) = 1 \cdot \dot{\sigma}^0$, since u takes each elementary 0-chain onto itself (essentially). Thus there is a 1-simplex $\langle \dot{\sigma}^0 \dot{\sigma} \rangle$ in the subdivision of σ, the carrier of $\dot{\sigma}$. We take $\mathfrak{D}_0(1 \cdot \dot{\sigma})$ to be the chain $1 \cdot \langle \dot{\sigma}^0 \dot{\sigma} \rangle$, and the desired relation will hold.

Assuming that the homomorphisms $\mathfrak{D}_0, \ldots, \mathfrak{D}_{p-1}$ have been defined so as to satisfy relation (2) above, consider an elementary p-chain $1 \cdot {}'\sigma^p$ of K'. We wish to define $\mathfrak{D}_p(1 \cdot {}'\sigma^p)$ so that relation (2) holds. The chain $1 \cdot {}'\sigma^p - uu'(1 \cdot {}'\sigma^p) - \mathfrak{D}_{p-1}\partial(1 \cdot {}'\sigma^p)$ is on a cone $u(\sigma^q) = \dot{\sigma}^q u(\partial\sigma^q)$ for some simplex σ^q in K. Such a cone is homologically trivial by Theorem 6–22, and hence every p-cycle on $\dot{\sigma}^q u(\partial\sigma^q)$ bounds a $(p + 1)$-chain on this cone. If we show that the chain $1 \cdot {}'\sigma^p - uu'(1 \cdot {}'\sigma^p) - \mathfrak{D}_{p-1}(\partial(1 \cdot {}'\sigma^p))$ is actually a p-cycle, then it will bound some $(p + 1)$-chain of K', which can then be taken as $\mathfrak{D}_p(1 \cdot {}'\sigma^p)$. Thus we compute

$$\begin{aligned}
&\partial[1 \cdot {}'\sigma^p - uu'(1 \cdot {}'\sigma^p) - \mathfrak{D}_{p-1}\partial(1 \cdot {}'\sigma^p)] \\
&\qquad = \partial(1 \cdot {}'\sigma^p) - \partial uu'(1 \cdot {}'\sigma^p) - \partial\mathfrak{D}_{p-1}\partial(1 \cdot {}'\sigma^p) \\
&\qquad = \partial(1 \cdot {}'\sigma^p) - \partial uu'(1 \cdot {}'\sigma^p) \\
&\qquad\qquad - [\partial(1 \cdot {}'\sigma^p) - uu'\partial(1 \cdot {}'\sigma^p) - \mathfrak{D}_{p-2}\partial\partial(1 \cdot {}'\sigma^p)]
\end{aligned}$$

since $\mathfrak{D}_{p-1}$ satisfies relation (2). Clearly, $\mathfrak{D}_{p-1}\partial\partial(1 \cdot {}'\sigma^p) = 0$, so the right-hand side of this equation reduces to $uu'\partial(1 \cdot {}'\sigma^p) - \partial uu'(1 \cdot {}'\sigma^p)$.

But both u and u' are chain-mappings and commute with ∂, so

$$uu'\partial(1 \cdot '\sigma^p) = \partial uu'(1 \cdot '\sigma^p).$$

This shows that $1 \cdot '\sigma^p - uu'(1 \cdot '\sigma^p) - \mathfrak{D}_{p-1}\partial(1 \cdot '\sigma^p)$ is a p-cycle. Hence by our remark above, this is also a boundary. We take $\mathfrak{D}_p(1 \cdot '\sigma^p)$ to be a $(p + 1)$-chain on the cone $u(\sigma^q)$ which this cycle bounds. This completes the inductive definition of the deformation operator $\mathfrak{D}$ and proves that uu' is chain-homotopic to the identity on K'. □

In view of Theorem 6–20, we can immediately state the following corollary to Theorem 6–23.

THEOREM 6–24. Let K' be the barycentric subdivision of a finite complex K. Then for each dimension p, the integral homology groups $H_p(K)$ and $H_p(K')$ are isomorphic.

This result is a formal statement of the *invariance of simplicial homology groups under barycentric subdivisions.* Repeated applications of Theorem 6–24 afford an obvious proof of the following corollary.

COROLLARY 6–25. Let $K^{(n)}$ be the nth barycentric subdivision of a finite complex K. Then for each dimension p, the integral homology groups $H_p(K)$ and $H_p(K^{(n)})$ are isomorphic.

Finally we may apply the "universal coefficient theorem," Section 6–9, to prove the next result.

COROLLARY 6–26. Let $K^{(n)}$ be the nth barycentric subdivision of a finite complex K, and let G be an arbitrary abelian group. Then for each dimension p, the homology groups $H_p(K, G)$ and $H_p(K^{(n)}, G)$ are isomorphic.

The continuous barycentric simplicial mappings, defined in Section 5–6, on geometric complexes certainly induce homomorphisms of the homology groups just as do the abstract simplicial mappings. Let us gather some information which should be quite suggestive. The key facts are the simplicial approximation theorem (Theorem 5–23) and Corollary 6–25. From the first of these results, we know that any continuous mapping of one finite polytope into another can be approximated arbitrarily closely by a simplicial mapping on a suitably chosen triangulation of the two polytopes. This simplicial mapping induces homomorphisms of the homology groups of these subdivisions. But in view of Corollary 6–25, we may consider that the induced homomorphisms are on the homology groups of the original polytopes. This strongly suggests that a *continuous mapping of a polytope induces homomorphisms of the homology groups.* This conjecture is true and could be proved by carrying out a program based upon this line of thought [27]. We will not carry out such a program, but we will use another approach to attain the same end (see Section 8–4).

6–14 The Brouwer degree. Consider two n-spheres S^n and Σ^n and a continuous mapping $f{:}S^n \to \Sigma^n$. With every such mapping f we associate an integer $\rho(f)$, called the *degree of* f. Intuitively, the degree $\rho(f)$ is the algebraic number of times that the image $f(S^n)$ wraps around Σ^n.

Each n-sphere, S^n and Σ^n, has a (curvilinear) triangulation isomorphic to the boundary complex of a geometric $(n + 1)$-simplex. Let K and L denote these triangulations of S^n and Σ^n, respectively. In proving the simplicial approximation theorem (5–23) we showed that for each mapping f there is a barycentric subdivision $K^{(k)}$ of K that is star-related to L relative to f.

We know that the integral homology group $H_n(K)$ is infinite cyclic, which means that there is a *fundamental n-cycle* z_n on K such that every integral n-cycle on K is a multiple $m \cdot z_n$, m an integer. By Corollary 6–25, $H_n(K^{(k)})$ is also infinite cyclic. Furthermore, it is easy to prove that if ${z_n}^{(i)}$ denotes the fundamental n-cycle on $K^{(i)}$, then $u({z_n}^{(i)})$ is the fundamental n-cycle on $K^{(i+1)}$. (Here u is the chain-mapping associated with barycentric subdivision as in Theorem 6–23.) This last statement follows from the fact that the induced homomorphism u_* of $H_n(K^{(i)})$ into $H_n(K^{(i+1)})$ is actually an isomorphism onto (Corollary 6–25).

Just as in the proof of the simplicial approximation theorem, we may now construct a simplicial mapping φ of $K^{(k)}$ into L such that φ is homotopic to f (we need not be concerned about the accuracy of the approximation). This simplicial mapping φ induces a homomorphism φ_{n*} of $H_n(K^{(k)})$ into $H_n(L)$. The image $\varphi(z_n^{(k)})$ of the fundamental n-cycle on $K^{(k)}$ is certainly an n-cycle on L. If we denote the fundamental n-cycle on L by γ_n, it follows that $\varphi(z_n^{(k)}) = \rho \cdot \gamma_n$ for some integer ρ. We define ρ to be the *degree of the mapping* f and will abbreviate it by deg (f).

We must show that deg (f) does not depend upon the simplicial mapping φ, as it seems to do. To accomplish this, we will consider just how the mapping φ is defined. Recall that $K^{(k)}$ is star-related to L relative to f provided that for every vertex v_i or $K^{(k)}$, there is a vertex w_j of L such that $f(\mathring{\mathrm{S}}\mathrm{t}(v_i))$ is contained in $\mathring{\mathrm{S}}\mathrm{t}(w_j)$. It is possible, however, that more than one vertex of L contains $f(\mathring{\mathrm{S}}\mathrm{t}(v_i))$ in its star and hence there may be several choices for $\varphi(v_i)$ in defining the simplicial mapping φ. If such is the case, it is clear that any admissible choice of φ can be changed into any other by means of a sequence of admissible choices each differing from its predecessor at only one vertex. Thus we may consider only the effect of changing φ at a single vertex v_i to form a new mapping φ'.

There are exactly $n + 2$ vertices in L. The image $f(\mathring{\mathrm{S}}\mathrm{t}(v_i))$ lies in at most $n + 1$ stars $\mathring{\mathrm{S}}\mathrm{t}(w_j)$, since the intersection of all $n + 2$ stars $\mathring{\mathrm{S}}\mathrm{t}(w_j)$ is empty, which $f(\mathring{\mathrm{S}}\mathrm{t}(v_i))$ certainly is not. Thus there is at least one vertex, say w, of L that is not an admissible image $\varphi'(v_i)$. Therefore no simplex of $K^{(k)}$ having v_i as a vertex can be mapped onto a simplex of L having

w as a vertex, no matter what choice we may take for $\varphi'(v_i)$. Conversely, if σ^n is an n-simplex of $K^{(k)}$ and is mapped by φ' onto a simplex of L having w as a vertex, then $f(\sigma^n)$ lies in $\mathring{\mathrm{S}}\mathrm{t}(w)$. Therefore a change in the mapping φ at the vertex v_i cannot alter the coefficient in the chain $\varphi(z_n^{(k)})$ on any n-simplex having w as a vertex. Since $\deg(f) = \rho$ is taken to be the coefficient in $\varphi(z_n^{(k)})$ assigned to *each* n-simplex of L, this argument proves that a change in φ at one vertex, and hence at any number of vertices, does not alter $\deg(f) = \rho$.

Next, suppose that we had used the barycentric subdivision $K^{(k+1)}$ instead of $K^{(k)}$ to define φ. Again we let u be the chain-mapping of $K^{(k)}$ into $K^{(k+1)}$ associated with subdivision. The reverse chain-mapping u' is induced by the simplicial mapping assigning to each simplex of $K^{(k+1)}$ a face of the simplex of $K^{(k)}$ that contains σ in the point-set sense. Therefore the composite mapping $\varphi u'$ assigns to each vertex of $K^{(k+1)}$ a vertex of L that is admissible from the standpoint of approximating f. It follows that $\varphi u'$ assigns to the fundamental n-cycle $u(z_n^{(k)}) = z_n^{(k+1)}$ an n-cycle $\rho' \cdot \gamma_n$. But now u_* and u'_* are inverse isomorphisms, so $\varphi u' u(z_n^{(k)}) = \varphi(z_n^{(k)}) = \rho \cdot \gamma_n$, which proves that $\rho' = \rho$. This implies that we may use any suitably fine subdivision of K in defining φ and hence $\deg(f)$.

The next step is to consider a subdivision of the complex L and see if the computation of $\deg(f)$ using this complex gives the same integer ρ. It suffices, of course, to consider only the first barycentric subdivision L'. We may choose a subdivision $K^{(m)}$ of K such that $K^{(m)}$ is star-related to L' relative to f. This obviously implies that $K^{(m)}$ is also star-related to L relative to f, since the stars of vertices of L contain the stars of vertices of L'. Let ρ be the degree of f computed using $K^{(m)}$ and L, and let ρ' be that computed using $K^{(m)}$ and L'. If γ_n is the fundamental n-cycle on L, then $u(\gamma_n)$ is the fundamental n-cycle on L'. Letting z_n be the fundamental n-cycle on $K^{(m)}$, we have

$$\varphi(z_n) = \rho \cdot \gamma_n$$

and

$$\varphi'(z_n) = \rho' \cdot u(\gamma_n).$$

From the remarks made above, given any simplex σ^n in $K^{(m)}$, the mapping $u'\varphi'(\sigma^n)$ is a star-mapping of $K^{(m)}$ into L which is homotopic to f. We must have $u'\varphi'(z_n) = \rho \cdot \gamma_n$. But $u'\varphi'(z_n) = u'(\rho' \cdot u(\gamma_n))$, and since u_* and u'_* are inverse isomorphisms, it follows that $\rho' = \rho$. This shows that $\deg(f)$ does not depend upon the subdivision of the complex L.

We need an important result before showing that the number $\deg(f)$ does not depend at all upon the triangulations K and L of the n-spheres S^n and Σ^n. We will prove that $\deg(f)$, as defined by means of K and L, depends only upon the homotopy class of f. To this end, let f and g be homotopic mappings of S^n into Σ^n. Thus there is a mapping $h{:}S^n \times$

$I^1 \to \Sigma^n$ such that $h(s, 0) = f(s)$ and $h(s, 1) = g(s)$ for each point s in S^n. We let ϵ denote the Lebesgue number (see Theorem 1–32) of the covering of Σ^n by the open stars $\mathring{\mathrm{S}}\mathrm{t}(w_j)$. Since $S^n \times I^1$ is compact, the mapping h is uniformly continuous. Hence there is a positive number δ such that if A is any subset of S^n and B is any subset of I^1, each of diameter less than δ, then the diameter of $h(A \times B)$ is less than ϵ.

Let K' be a subdivision of K with mesh less than $\delta/2$ and choose numbers $0 = t_0 < t_1 < t_2 < \cdots < t_k = 1$ such that $t_i - t_{i-1} < \delta$ for each i. Each open star $\mathring{\mathrm{S}}\mathrm{t}(v_i)$ in K' has diameter less than δ as has each open interval (t_{i-1}, t_i). Thus each set $h[\mathring{\mathrm{S}}\mathrm{t}(v_i) \times (t_{i-1}, t_i)]$ has diameter less than ϵ and hence lies in the star of some vertex w_j of L. Now if t is any number satisfying $t_{i-1} \leqq t \leqq t_i$, the star-mapping approximating the restricted mapping $h|S^n \times t{:}S^n \to \Sigma^n$ may be defined by setting $\varphi(v_i \times t) = w_j$, using the same simplicial mapping for any such t. It now follows that we obtain the same number $\rho = \deg\,(h|S^n \times t)$ for all values of t, $t_{i-1} \leqq t \leqq t_i$. Passing from one such subinterval to the next must also give the same number ρ because the approximating star-mappings agree at the end point t_i. Therefore $\deg\,(h|S^n \times t)$ must be the same for all values of t in I^1, and we have proved that $\deg\,(f) = \deg\,(g)$; that is, $\deg\,(f)$ is invariant under homotopy.

Last, we show that the degree of f does not depend upon the triangulations K and L of the n-spheres S^n and Σ^n. We will use two lemmas in this argument.

LEMMA 6–27. *If H, K, and L are three finite geometric complexes, and $f{:}|H| \to |K|$ and $g{:}|K| \to |L|$ are continuous mappings on the indicated carriers, and if H is star-related to K relative to f, and K is star-related to L relative to g, then H is star-related to L relative to the composite mapping gf.*

Proof: If $f(\mathring{\mathrm{S}}\mathrm{t}(v_i))$ lies in some $\mathring{\mathrm{S}}\mathrm{t}(w_j)$ and $g(\mathring{\mathrm{S}}\mathrm{t}(w_j))$ lies in some $\mathring{\mathrm{S}}\mathrm{t}(u_k)$, then $gf(\mathring{\mathrm{S}}\mathrm{t}(v_i))$ lies in $\mathring{\mathrm{S}}\mathrm{t}(u_k)$. □

LEMMA 6–28. *If S^n, Σ^n, and X^n are n-spheres, if $f{:}S^n \to \Sigma^n$ and $g{:}\Sigma^n \to X^n$ are continuous, and if H, K, and L are triangulations of S^n, Σ^n, and X^n, respectively, which are admissible for defining $\deg\,(f)$ and $\deg\,(g)$, then H and L are admissible triangulations for defining $\deg\,(gf)$, and $\deg\,(gf) = (\deg\,(f)) \cdot (\deg\,(g))$.*

Proof: If ρ is the algebraic sum of the number of n-simplexes of H that are mapped onto a simplex of K by the star-mapping φ approximating f, and if ρ' is the algebraic sum of the number of n-simplexes of K mapped onto a simplex of L by the star-mapping φ' approximating g (and this is the geometric significance of degree), then $\rho \cdot \rho'$ is certainly the algebraic sum of the n-simplexes of H mapped by the admissible star-mapping $\varphi'\varphi$ (previous lemma) onto an n-simplex of L. □

Returning to the main problem, let L and L_0 be any two triangulations of Σ^n. Choose a subdivision L_0' of L_0 such that L_0' is star-related to L relative to the identity mapping $i:\Sigma^n \to \Sigma^n$. The associated star-mapping φ of L_0' into L is continuous and is homotopic to i. It would seem obvious that this implies that deg $(\varphi) = \pm 1$, but this has not been shown. The degree of φ and hence of i depends, as far as we know, upon the choices of L and L_0. In fact, the statement that deg $(i) = \pm 1$ is a special case of the theorem we are trying to prove. It is convenient to prove this special case first.

Let L' be a refinement of L such that L' is star-related to L_0' relative to i, and let φ' be the approximating star-mapping. Applying Lemma 6–28, take each n-sphere to be Σ^n, the mappings to be the identity mapping, and take H to be L', K to be L_0', and L to be L. We then have deg $(\varphi'\varphi) =$ (deg (φ')) $\cdot$ (deg (φ)). The number deg $(\varphi'\varphi)$ is defined with respect to L' and L, which is the vital point here. Since $\varphi'\varphi$ is homotopic to the identity, we know that deg $(\varphi'\varphi) =$ deg (i), where deg (i) is defined relative to L' and L. For the complexes L' and L, the iterated simplicial mapping u' of the subdivision process is a star-mapping approximating the identity i. Since u' induces an isomorphism of $H_n(L')$ onto $H_n(L)$, we know that deg $(u') =$ deg $(i) = 1$. Thus we have deg $(\varphi') \cdot$ deg $(\varphi) = 1$, so each of these numbers is either $+1$ or -1. Since L and L_0 may have had opposite orientations, we could have deg $(\varphi) = -1$, but in this case we may merely reorient L_0 so that deg $(\varphi) = +1$. This implies that the identity mapping $i:\Sigma^n \to \Sigma^n$ can always be taken to have degree $+1$ regardless of the triangulations L and L_0 used in defining deg (i).

Continuing with the general case, choose a subdivision K' of K so fine that K' is star-related to both L and L_0 relative to the mapping f of S^n into Σ^n. Clearly $f = if$, where i is the identity mapping on Σ^n. The value of deg (f) as defined relative to K' and L is equal to the product deg $(i) \cdot$ deg (f), where now deg (f) is defined relative to K' and L_0. Since we may take deg $(i) = +1$, the two definitions of deg (f) agree. Hence we may conclude that the degree of f is independent of the particular triangulation of Σ^n. Finally, letting i' be the identity mapping of S^n onto itself, we have $f = fi'$, and the same argument shows that deg (f) is independent of the triangulation of S^n. This lengthy argument has proved the following result.

THEOREM 6–29. *The degree of a continuous mapping f of an n-sphere S^n into an n-sphere Σ^n depends only upon the homotopy class of f.*

This means that any two homotopic mappings f and g of S^n into Σ^n have the same degree. The converse theorem was proved by H. Hopf, namely, *if* f *and* g *are two mappings of* S^n *into* Σ^n *and if* deg (f) = deg (g), *then* f *and* g *are homotopic*. These two results yield a succinct classification of the continuous mappings of one n-sphere into another. In particular, we may conclude that the homotopy classes of such mappings are in one-to-one correspondence with the integers. This serves to show that the nth homotopy group of an n-sphere $\pi_n(S^n)$ *is infinite cyclic*. We do not include Hopf's proof here, but will give the following indicative result.

THEOREM 6–30. The n-sphere is not contractible, i.e., the identity mapping of S^n onto itself is essential.

Proof: A constant mapping certainly has degree zero and hence cannot be homotopic to the identity whose degree we proved to be unity. □

Oddly enough, one of the most important facts about degree is also one of the most obvious.

THEOREM 6–31. If $f:S^n \to E^n$ is continuous and deg $(f) \neq 0$, then each point of Σ^n lies in the image $f(S^n)$.

Proof: Suppose that p is a point of Σ^n and that p is not in the compact set $f(S^n)$. Let 2ϵ be the distance $d(p, f(S^n))$ from p to $f(S^n)$, and choose a simplicial subdivision L of Σ^n of mesh less than ϵ. Choose a subdivision K of S^n which is star-related to L relative to f, and let $\varphi:K \to L$ be the associated star-mapping. Then no simplex σ^n in K is mapped onto an n-simplex of L containing the point p. This implies that deg $(f) = 0$, contrary to our assumption. □

We may point out that this theorem may be proved in another way by reference to Theorem 4–13. Our proof above is included as an example in using the tools of this section. We will exhibit an application of the above result in the next section.

Next we extend the concept of degree to include mappings of the closure of an open set in S^n into another n-sphere Σ^n. The theory differs from that above in that the degree is defined locally and may vary from point to point. Indeed, this new concept of degree fails to be defined at some points.

Let D be a connected open set in S^n, and let $f:\overline{D} \to \Sigma^n$ be continuous. Let p be a point of Σ^n not in $f(\overline{D} - D)$. Let U be a spherical neighborhood of p so small that $U \cap f(\overline{D} - D)$ is empty. By the Tietze extension theorem (2–31) there is an extension $f':S^n - D \to \Sigma^n$ of $f|(\overline{D} - D)$ into the n-cell $\Sigma^n - U$. Let $\tilde{f}:S^n \to \Sigma^n$ be defined by $\tilde{f}(x) = f(x)$ if x is in $\overline{D}$, and $\tilde{f}(x) = f'(x)$ if x is in $S^n - D$. We define the *degree of* f *on* $\overline{D}$ *at* p, deg $(f, \overline{D}, p)$, to be the degree of $\tilde{f}$.

As an example, let $\overline{D}$ be the closed unit disc in E^2 given by the complex coordinates $|z| \leqq 1$. Let $f:\overline{D} \to E^2$ be defined by $f(z) = (z - \frac{1}{2})^2$. The

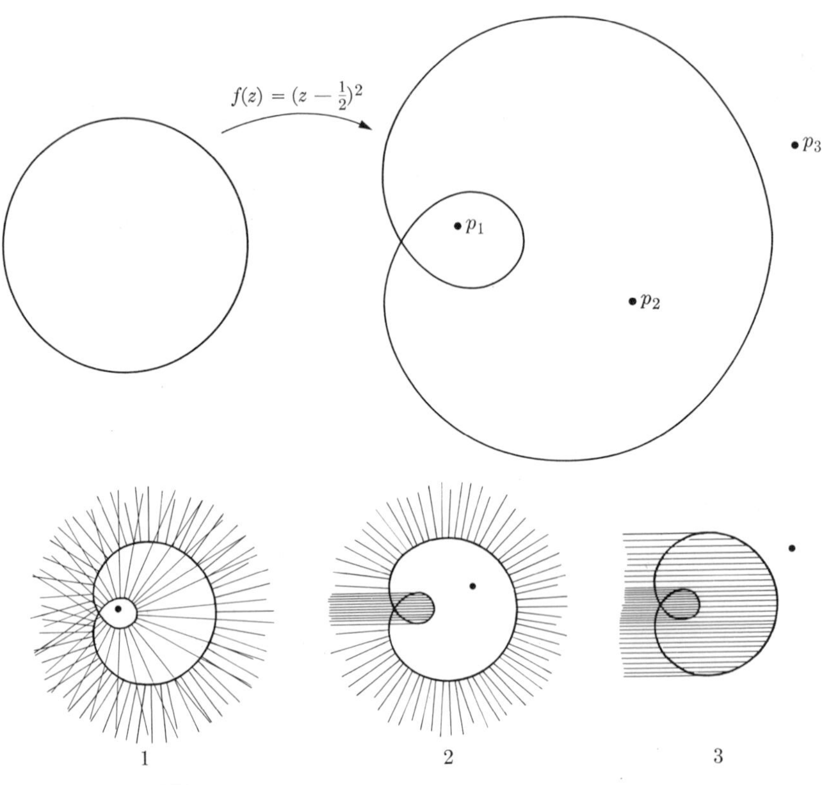

FIGURE 6–22

unit circle is mapped as indicated in Fig. 6–22, and we show as shaded areas in the remaining figures the three extensions f' required to compute the degree of f at p_2, p_1, and p_0.

Once we have shown that deg $(f, \overline{D}, p)$ is independent of the particular extension f', we can use the previous theorems on degree to obtain corresponding results for this new theory. But this independence is almost obvious. For in computing deg (f), we can choose a triangulation L of Σ^n so fine that some n-simplex σ^n of L lies entirely in the spherical neighborhood U. Then we choose a triangulation K of S^n star-related to L relative to $\tilde{f}$ and such that the approximating star-mapping φ carries a simplex of K onto σ^n only if that simplex lies in D. This is possible simply by making the approximation sufficiently accurate. The degree of $\tilde{f}$ may be computed merely as the coefficient of σ^n on the image $\varphi(z_n)$ of the fundamental n-cycle z_n of K. This coefficient is determined only by those simplexes of K that are mapped by φ onto σ^n and, in fact, is the number mapped with positive orientation minus the number mapped with negative orientation. It is now evident that the choice of the extension f' is immaterial.

The homotopy invariance of degree has the following formulation in this setting.

THEOREM 6–32. Let $\overline{D}$ be the closure of an open set D in S^n, let Σ^n be an n-sphere, let $f_0:\overline{D} \to \Sigma^n$ and $f_1:\overline{D} \to \Sigma^n$ be continuous, and let p be a point of $\Sigma^n - f_0(\overline{D} - D) - f_1(\overline{D} - D)$. If there is a homotopy $h:\overline{D} \times I^1 \to \Sigma^n$ between f_0 and f_1 such that $h[(\overline{D} - D) \times I^1]$ does not contain p, then $\deg(f_0, \overline{D}, p) = \deg(f_1, \overline{D}, p)$.

Proof: Let U be a spherical neighborhood of p that does not meet the compact set $h[(\overline{D} - D) \times I^1]$. By the Tietze theorem again, there is an extension of $h|(\overline{D} - D) \times I^1$, say $h':(S^n - D) \times I^1 \to \Sigma^n - U$, and we can combine this with h to obtain a mapping $\tilde{h}:S^n \times I^1 \to \Sigma^n$. This clearly gives a homotopy between an extension $\tilde{f}_0$ of f_0 and an extension $\tilde{f}_1$ of f_1. By Theorem 6–29, $\deg(\tilde{f}_0) = \deg(\tilde{f}_1)$. But these are, respectively, $\deg(f_0, \overline{D}, p)$ and $\deg(f_1, \overline{D}, p)$. □

Similarly, Theorem 6–31 has the following formulation.

THEOREM 6–33. If $\overline{D}$ is the closure of an open set D in S^n, if $f:\overline{D} \to \Sigma^n$ is a continuous mapping of $\overline{D}$ into an n-sphere, and if p is a point of Σ^n such that $\deg(f, \overline{D}, p) \neq 0$, then p is in $f(\overline{D})$.

Proof: Suppose that p is not in $f(\overline{D})$. Let U be a spherical neighborhood of p that does not meet the compact set $f(\overline{D})$. Then p is not in $f'(S^n - D)$, either, and so is not in $f(S^n)$. This contradicts Theorem 6–31. □

The theory of degree has also been approached by using differentiable mappings instead of simplicial mappings as the basic approximations. The reader who is interested in such a development is referred to Nagumo [110, 111].

6–15 The fundamental theorem of algebra, an existence proof.

The theorem to which this section's heading refers is the following result.

THEOREM 6–34. Every polynomial $P(z) = a_0 + a_1 z + \cdots + z^n$, the coefficients a_i being complex numbers, and $n > 0$, has at least one zero.

There are many proofs of this result, one of which we give here as an application of Theorem 6–29. First, we remark that we may consider P as a mapping $P:E^2 \to E^2$ and if we set $P(\infty) = \infty$, we have a continuous mapping $P:S^2 \to S^2$.

LEMMA 6–35. The polynomial $P(z)$ is homotopic to the mapping $f(z) = z^n$.

Proof: We define the homotopy explicitly by setting

$$h(z, t) = z^n + (1 - t)(a_0 + a_1 z + \cdots + a_{n-1} z^{n-1}), \qquad \text{for } z \text{ finite},$$

and
$$h(\infty, t) = \infty.$$

for all t and all finite z, h is continuous by elementary theorems in function theory. It is easy to show that $\lim_{z \to \infty} h(z, t) = \infty$ for all t and hence that h is continuous on $S^2 \times I^1$. □

LEMMA 6–36. The degree of $f(z) = z^n$ is n.

Proof: We indicate in Fig. 6–23 two triangulations of the 2-sphere, considered as the plane plus a point at infinity, on which $f(z)$ is actually simplicial. In this mapping, for example, the n shaded pieces of Fig. 6–23(a) are mapped onto the shaded piece of Fig. 6–23(b) in a sense-preserving fashion. It is now clear that the degree is n. □

Proof of Theorem 6–34: From Lemmas 6–35 and 6–36, it follows that the degree of $P(z) = n$. Then by Theorem 6–31, each point of S^2 is the image of some point of S^2. In particular, there is at least one point z_0 such that $P(z_0) = 0$. □

It is tempting, but incorrect, to say that because deg $(f) = n$, each point is the image of at least n points. The function $f(z) = z^n$ is a counterexample since only zero is mapped onto zero. At the time of this writing, the following question cannot be answered. If $f{:}S^n \to \Sigma^n$ is continuous and deg $(f) = k$, is the set of points x in Σ^n, such that $f^{-1}(x)$ has at least k points, nonempty? No example is known for which this set fails to be dense in Σ^n.

The proof given in this section is typical of the use of degree theory in constructing existence proofs. The method can be summarized as follows. (a) So phrase the problem that for some mapping f, deg $(f, \overline{D}, p) \neq 0$ implies the existence of the desired quantity (in our case, a root of a

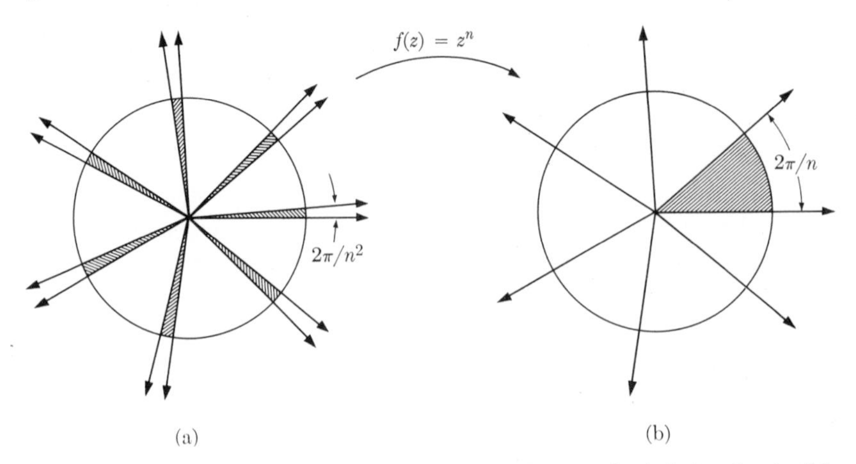

FIG. 6–23. n shaded simplexes in (a) map onto the one shaded simplex in (b).

polynomial), (b) find a simple function f^* homotopic to f such that $\deg(f^*, \overline{D}, p)$ can be computed, and (c) do the computation.

It is possible to obtain a degree theory in more general spaces than we have considered here. A first generalization is to n-dimensional orientable manifolds. With restrictions placed upon the mappings, a degree theory can be set up even in function spaces. The usual approach is to consider a function f on a set $\overline{D}$ in a Banach space B, f being completely continuous (that is, f carries bounded sets into compact sets) such that $f:\overline{D} \to B$. This condition allows one to make approximations by means of mappings of Euclidean spaces for which degree theory can be defined. If the degree thus obtained for f is not zero, we have the existence of a solution of certain functional equations. This is the Leray-Schauder method [97, 98]. For a self-contained account, see Rado and Reichelderfer [29].

6-16 The no-retraction theorem and the Brouwer fixed-point theorem. We recall that a *retraction* of a space X onto a subset A of X is a continuous mapping $r:X \to A$ such that $r(a) = a$ for each point a in A. In other words, the restriction $r|A$ of r to A is the identity mapping. The following "no-retraction theorem" seems to be intuitively obvious.

THEOREM 6-37. There is no retraction of an n-cell onto its boundary, $n > 0$.

Proof: There is no loss of generality in taking the n-cell to be the set of points in E^n satisfying the inequality $\sum_{i=0}^n x_i^2 \leqq 1$ whose boundary is the sphere S^{n-1}. Suppose that there is a retraction r of this n-cell onto S^{n-1}. Define the mapping

$$h(x, t) = r[(1 - t) \cdot x], \qquad x \text{ in } S^{n-1},$$

where x is taken to be a unit vector in E^n. Clearly, we have $h(x, 1) = r(0 \cdot x) = r(0)$ for each point x in S^{n-1}, so $h(x, 1)$ is a constant mapping of S^{n-1} onto the point $r(0)$. But $h(x, 0) = r(x) = x$ is the identity mapping of S^{n-1} onto itself. Thus $h(x, t)$ is a homotopy between a constant mapping which has degree zero and the identity mapping which has degree 1. By Theorem 6-29, this is impossible, so the retraction r cannot exist. □

The following result, equivalent to Theorem 6-37, may be proved as an exercise.

THEOREM 6-38. There exists no retraction $r:S^n \times I^1 \to S^n$ such that $r(x, 0) = x$ and $r(x, 1) = p_0$, a point in S^n, for all points x in S^n.

Intuitively, this says that it is impossible to peel an orange without breaking the skin.

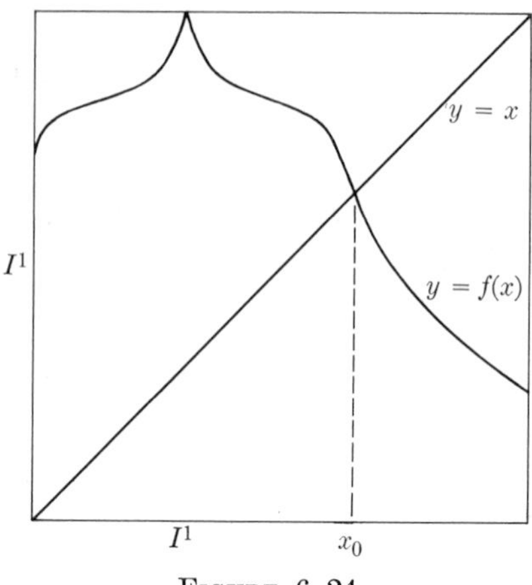

FIGURE 6–24

Looking in another direction, let f be a continuous mapping of the unit interval I^1 into itself. It is quite obvious that there is at least one point x_0 in I^1 for which $f(x_0) = x_0$. To see this, we merely note that the graph of $y = f(x)$, must cross the graph of $y = x$ at least once (see Fig. 6–24).

It might not be quite so obvious that the same result is true for a continuous mapping of an n-cell into itself. The reader may ponder over the problem of using the truth of the theorem for the case $n = 1$ together with the fact that an n-cell is (topologically) the product of n unit intervals to prove the general case. The proof given below is short and easy only because we bring to bear some formidable machinery.

THEOREM 6–39. Given any continuous mapping f of an n-cell into itself, there is at least one point x_0 for which $f(x_0) = x_0$.

Proof: Again we may take the n-cell as in the proof of Theorem 6–37. Now suppose there is a mapping f of this n-cell into itself which has no fixed point. For each point x in this n-cell, let $L(x)$ be the directed ray from $f(x)$ to x. Since there is no fixed point, there is a unique ray $L(x)$ for each point x in the n-cell, and hence a unique point $L(x) \cap S^{n-1}$. Let r be the mapping defined by $r(x) = L(x) \cap S^{n-1}$. That is, we map x onto $f(x)$ and then back along $L(x)$ until we meet S^{n-1}. It is easy to see that r is continuous, and clearly $r(x) = x$ for each point x in S^{n-1}. This means that r is a retraction of the n-cell onto its boundary, which contradicts Theorem 6–37. Thus the unique rays $L(x)$ cannot exist as claimed and there must be a fixed point. □

The *Brouwer fixed-point theorem* above also implies the no-retraction theorem. For if we assume the fixed-point theorem and suppose that r is a retraction of the n-cell, we obtain a contradiction as follows. There are mappings of S^{n-1} onto itself with no fixed points. One of these is the antipodal mapping g, which interchanges antipodal points. Then the composite mapping gr is a continuous mapping of an n-cell into itself having no fixed point. This contradicts Theorem 6-39 and proves the two theorems to be equivalent.

We often express the Brouwer fixed-point theorem by saying that *the n-cell has the fixed-point property.* Many other spaces also have the same property. Results in this direction may be found in Lefschetz [22], Young [135], and others.

Fixed-point theorems also have been used for existence proofs, particularly in function spaces. The use of the fixed-point property in function spaces is due to Birkhoff and Kellogg [64], who approximated the spaces by suitable mappings of an n-cell into itself. (Also see Birkhoff [4].) Schauder later refined the method and proved that a continuous mapping of a convex subset C of a Banach space into a compact subset of C has a fixed point. This result is the basis of many existence proofs in analysis.

6-17 Mappings into spheres. Some very important topological properties of Euclidean spaces can be established by a study of mappings into spheres. We collect some of these results in this section, the methods being largely those of Borsuk [68].

LEMMA 6-40. Let K be a finite geometric complex of dimension $<n$. Then every mapping $f:|K| \to S^n$ is inessential.

Proof: Let L be a triangulation of S^n with dimension n. By the simplicial approximation theorem (6-23), there is a mapping $g:|K| \to S^n$ such that g is homotopic to f and is simplicial on suitably chosen subdivisions of K and L. Since a simplicial mapping does not raise the dimension of simplexes, g cannot map $|K|$ *onto* S^n. Hence Theorem 4-13 applies to show that g, and therefore f, is inessential. □

COROLLARY 6-41. If $m < n$, then every mapping $f:S^m \to S^n$ is inessential and admits an extension $\tilde{f}$ to the $(m+1)$-cell bounded by S^m.

Proof: That f is inessential follows from Lemma 6-40, and the extension $\tilde{f}$ is given by Theorem 4-5. □

Note that this result can be used to prove that *for $m < n$, the homotopy group $\pi_m(S^n)$ is trivial.*

LEMMA 6-42. Let K be a finite geometric complex with dimension $\leqq n$, let A be a closed subset of $|K|$, and let $f:A \to S^n$ be continuous. Then f has an extension $\tilde{f}:|K| \to S^n$.

Proof: By Theorem 2-35, there is an open set U in $|K|$ such that A lies in U and such that there is an extension $\bar{f}:U \to S^n$ of f. If we take a suitably fine barycentric subdivision $K^{(k)}$ of K, we find a subcomplex L of $K^{(k)}$ such that $|L|$ contains A and is contained in U. Let $K_p^{(k)}$ denote the p-skeleton of $K^{(k)}$. It is clear that $\bar{f}$ has an extension $\bar{f}^0$ mapping $|K_0^{(k)}| \cup |L|$ into S^n. [We need only assign images to the vertices of $K^{(k)} - L$.] Then $\bar{f}^0$ is also an extension of f.

Suppose that for $p < \dim K \leqq n$, there is an extension $\bar{f}^p:|K_p^{(k)}| \cup |L| \to S^n$ of f. Then $\bar{f}^p$ is defined on the boundary of each $(p+1)$-simplex $s_1, \ldots, s_r$ in $K^{(k)} - L$. But then by Corollary 6–41, $\bar{f}^p$ can be extended over each s_i, thus yielding $\bar{f}^{p+1}:|K_{p+1}^{(k)}| \cup |L| \to S^n$. Since $\dim K = \dim K^{(k)} \leqq n$, this proves the lemma. □

LEMMA 6–43. Let K be a finite geometric complex of dimension $\leqq n+1$, let A be a closed subset of $|K|$, and let $f:A \to S^n$ be continuous. Then there is a finite set F in $|K| - A$ such that f has an extension $\bar{f}:|K| - F \to S^n$.

Proof: Carrying on with the proof of Lemma 6–42, let $s_1, \ldots, s_q$ be the $(n+1)$-simplexes of $K^{(k)} - L$, and let F be the set of barycenters $\hat{s}_i$ of these simplexes. A radial projection of each $s_i - \hat{s}_i$ onto the boundary of s_i yields a retraction $r:|K| - F \to |K_n^{(k)}| \cup |L|$. The composite mapping $\bar{f}^n r:|K| - F \to S^n$ is the desired extension of f. □

LEMMA 6–44. Let A be a closed subset of S^n, and let B be a set consisting of exactly one point from each component of $S^n - A$. For every mapping $f:A \to S^{n-1}$ there is a finite subset F of B and an extension $\bar{f}:S^n - F \to S^{n-1}$ of f.

Proof: From Lemma 6–43, there is a finite subset $(x_1, \ldots, x_q)$ of $S^n - A$ and an extension $\bar{f}:S^n - (x_1, \ldots, x_q) \to S^{n-1}$ of f. For each x_i, let b_i be the point in B lying in the same component of $S^n - A$ as does x_i, and take $F = (b_1, \ldots, b_q)$. To prove that f has the desired extension, we use induction, showing that if f has an extension

$$\bar{f}^{i-1}:S^n - (b_1, \ldots, b_{i-1}, x_1, \ldots, x_q) \to S^{n-1},$$

then f also has an extension

$$\bar{f}^i:S^n - (b_1, \ldots, b_i, x_{i+1}, \ldots, x_q) \to S^{n-1}.$$

Since we may take $\bar{f}^0 = \bar{f}$, this will complete the proof.

Since x_i and b_i lie in the same component of $S^n - A$, there is a finite sequence of points $x_i = y_0, y_1, \ldots, y_m = b_i$ and also a sequence of convex n-cells $I_1, \ldots, I_m$ in $S^n - A$ such that y_{j-1} and y_j lie in I_j for each $j = 1, 2, \ldots, m$ and such that the boundary S_j of I_j contains none of the points x_i, b_i, or y_j. It now suffices to show that if f has an extension

$$\bar{f}^i_{j-1}:S^n - (b_1, \ldots, b_{i-1}, y_{j-1}, x_{i+1}, \ldots, x_q) \to S^{n-1},$$

then f also has an extension

$$\bar{f}_j^i: S^n - (b_1, \ldots, b_{i-1}, y_j, x_{i+1}, \ldots, x_q) \to S^{n-1}.$$

Let r be a retraction of $S^n - y_j$ onto $\overline{S^n - I_j}$. Then by setting $\bar{f}_j^i(x) = \bar{f}_{j-1}^i(r(x))$ for each point x in $S^n - (b_1, \ldots, b_{i-1}, y_j, x_{i+1}, \ldots, x_q)$, we have the desired extension. □

THEOREM 6–45. Let K be a finite geometric complex. For K to have dimension $\leqq n$, it is necessary and sufficient that for every closed subset A of $|K|$ and every mapping $f: A \to S^n$, there exists an extension $\bar{f}: |K| \to S^n$.

Proof. The necessity of the condition is precisely Lemma 6–42. Suppose then that $\dim K > n$. Let s be an $(n+1)$-simplex in K. Take A to be the boundary of s and $f: A \to S^n$ to be a homeomorphism. If f has an extension $\bar{f}: |K| \to S^n$, then $f^{-1}\bar{f}: |K| \to A$ is a retraction. In particular, this mapping retracts the $(n+1)$-cell s onto its boundary, contradicting Theorem 6–37. □

COROLLARY 6–46. If P is a finite polytope and K_1 and K_2 are two triangulations of P, then $\dim K_1 = \dim K_2$.

A proof of Corollary 6–46 can be supplied by the reader.

THEOREM 6–47 (Borsuk separation). Let X be a compact subset of E^n, and let x_0 be a point in $E^n - X$. For x_0 to lie in the unbounded component of $E^n - X$, it is necessary and sufficient that the mapping $f: X \to S^{n-1}$ defined by

$$f(x) = \frac{x - x_0}{||x - x_0||}$$

be inessential. (We are using vector notation for points of E^n.)

Proof: By means of a translation, we may consider x_0 to be the origin in E^n. Since X is compact, it lies inside some sufficiently large spherical neighborhood $S(0, r)$ of the origin. The similarity mapping sending each point x onto x/r maps E^n homeomorphically onto itself, with X being carried into the n-cell bounded by S^{n-1}. Therefore we could have assumed this condition on X originally; also, if $x_0 = 0$, the mapping f is given by

$$f(x) = \frac{x}{||x||} \cdot$$

Suppose that 0 lies in the unbounded component C of $E^n - X$. Since C is arcwise-connected (Theorem 3–5), there is a mapping $p: I^1 \to C$ with $p(0) = 0$ and $p(1) = x_1$, where x_1 is a point having norm $||x_1||$ greater than unity. Consider the mapping $H:X \times I^1 \to S^{n-1}$ defined by

$$H(x, t) = \frac{x - p(t)}{||x - p(t)||}.$$

Clearly,

$$H(x, 0) = f(x),$$

while

$$H(x, 1) = \frac{x - x_1}{||x - x_1||}.$$

Since each point x in X is inside S^{n-1} and x_1 is not, it is easily seen that for no x is

$$H(x, 1) = \frac{x - x_1}{||x - x_1||} = \frac{x_1}{||x_1||},$$

which is a point in S^{n-1}. For this equation would imply that

$$x = \frac{||x - x_1|| + ||x_1||}{||x_1||} \cdot x_1,$$

and x would have norm exceeding that of x_1. Therefore $H(x, t)$ is a homotopy between f and a mapping $H(x, 1)$ that does not cover S^{n-1}. By Theorem 4–13, $H(x, 1)$, and hence f, is inessential.

Conversely, assume that the component C of $E^n - X$ that contains the origin is bounded. Then $C \cup X$ is a closed subset of the n-cell bounded by S^{n-1}. If the mapping f is inessential, then by Theorem 4–5 there is an extension $\tilde{f}:C \cup X \to S^{n-1}$. Define the mapping r of the n-cell bounded by S^{n-1} by setting

$$r(x) = \tilde{f}(x), \qquad x \text{ in } C \cup X,$$

and

$$r(x) = \frac{x}{||x||}, \qquad x \text{ not in } C \cup X.$$

The two definitions agree on X, so r is continuous; and for points in S^{n-1}, $x = x/||x||$, so $r(x) = x$. Thus r is a retraction of the n-cell onto its boundary, contradicting Theorem 6–37. □

The next result is also due to Borsuk.

THEOREM 6–48. Let X be a closed subset of S^n. Then $S^n - X$ is connected if and only if every mapping $f:X \to S^{n-1}$ is inessential.

Proof: Suppose that $S^n - X$ is connected, and let $f:X \to S^{n-1}$ be any mapping. Let x_0 be any point in $S^n - X$. By Lemma 6–44, there is an extension $\tilde{f}:S^n - x_0 \to S^{n-1}$. But $S^n - x_0$ is contractible, so $\tilde{f}$ is inessential and therefore f is also inessential.

On the other hand, if $S^n - X$ is not connected, let x_1 and x_2 be points in different components of $S^n - X$. If we regard $S^n - x_1$ as E^n, it follows that x_2 lies in a bounded component of $E^n - X$. Therefore Theorem 6–47 applies to give an essential mapping of X into S^{n-1}. □

Since the property expressed in Theorem 6–48 is topological, i.e., is preserved by homeomorphisms, we have the next result as a corollary.

THEOREM 6–49. If X is a closed set in S^n that separates S^n, and Y is any homeomorphic image of X, then Y separates S^n.

Since S^{n-1} separates S^n, we immediately have the following portion of the *generalized Jordan curve theorem.*

THEOREM 6–50. If Σ is a set in S^n that is homeomorphic to S^{n-1}, then Σ separates S^n.

We do not know as yet that the set Σ in Theorem 6–50 separates S^n into exactly two connected open sets, but it is easy to show that each component of $S^n - \Sigma$ has all of Σ as its boundary. For if not, then some component C has $\overline{C} - C$ lying in a topological $(n - 1)$-cell Γ^{n-1} in Σ. But every mapping of Γ^{n-1} into S^{n-1} is inessential, which proves the following result.

THEOREM 6–51. No homeomorph of an $(n - 1)$-cell separates S^n.

However, it is conceivable that a set Σ be the common boundary of three connected open sets in S^n. Earlier, we saw an example of a continuum with this property (Section 3–8), and we know that homeomorphs of spheres can be wildly imbedded. It turns out that, although a 2-sphere can be so wildly imbedded in S^3 that neither of its complementary domains is a 3-cell, no sphere S^{n-1} can be so badly imbedded in S^n that its complement has more than two components. We shall give a proof of this later. The following result is as far as we can go at present.

THEOREM 6–52. Let A and B be subsets of S^n such that (1) B is the boundary of A, (2) A is homeomorphic to the n-cell I^n, and (3) B is homeomorphic to S^{n-1}. Then $S^n - B$ has two components $S^n - A$ and $A - B$, and in particular, $A - B$ is open in S^n.

Proof: Since I^n does not admit essential mappings into S^{n-1}, it follows from Theorem 6–48 that $S^n - A$ is connected. The set $A - B$ is homeomorphic to an open n-cell and so is connected. On the other hand, S^{n-1} does have an essential mapping into itself (e.g., the identity), hence by Theorem 6–48, $S^n - B$ is not connected. We need only note that $S^n - B = (S^n - A) \cup (A - B)$ to complete the proof. □

THEOREM 6–53 (Invariance of domain). If U_1 and U_2 are homeomorphic subsets of S^n, and if U_1 is open, then U_2 is open.

Proof: Let $h:U_1 \to U_2$ be a homeomorphism. Let x_2 be a point of U_2, and take $x_1 = h^{-1}(x_2)$. Let V_1 be a spherical neighborhood of x_1 such that V_1 lies in U_1. Then $\overline{V}_1$ and $\overline{V}_1 - V_1$ satisfy the conditions of Theorem 6–52, and $h(\overline{V}_1) - h(\overline{V}_1 - V_1)$ is open. Since x_2 lies in this open set, it follows that U_2 is a union of open sets and is open. □

Query: Why is Theorem 6–53 not obviously true?

A space M is *locally Euclidean of dimension* n if each point of M lies in a subset of M which is homeomorphic to E^n.

THEOREM 6–54. Let M_1 and M_2 be two locally Euclidean spaces of dimension n. If U is an open subset of M_1, and if $h:U \to M_2$ is a homeomorphism of U into M_2, then $h(U)$ is open in M_2.

Proof: Let x_2 be a point of $h(U)$, and take $x_1 = h^{-1}(x_2)$. Select open sets V_1 and V_2 containing x_1 and x_2, respectively, such that both V_1 and V_2 are homeomorphic to E^n and such that V_1 lies in U and $h(V_1)$ lies in V_2. Since E^n is homeomorphic to an open subset of S^n, we may choose homeomorphisms $g_1:V_1 \to S^n$ and $g_2:V_2 \to S^n$, where $g_1(V_1)$ and $g_2(V_2)$ are open subsets of S^n. Then $g_2hg_1^{-1}$ maps $g_1(V_1)$ homeomorphically onto a subset of $g_2(V_2)$. By Theorem 6-53, this subset of $g_2(V_2)$ is open in S^n and hence is open in $g_2(V_2)$. Thus the set $hg_1^{-1}g_1(V_1) = h(V_1)$ is open in V_2 and hence is open in M_2. Since x_2 lies in $h(V_1)$ and $h(V_1)$ lies in U_2, it follows that U_2 is a union of open sets. □

One of the many "intuitively obvious" results in topology is the following consequence of the above theorem. (See Brouwer [70].)

COROLLARY 6–55. Two locally Euclidean spaces M_1 and M_2 of different dimensions cannot be homeomorphic.

Proof: If $\dim M_1 = m_1$ and $\dim M_2 = m_2$, and if $m_1 = m_2 + k$, $k > 0$, then $M_2 \times E^k$ is also locally Euclidean of dimension m_1. The nonopen set $M_2 \times 0$ in $M_2 \times E^k$ is homeomorphic to M_2. If M_1 were homeomorphic to M_2, then it would be homeomorphic to a nonopen subset of a locally Euclidean space of dimension m_1, contradicting Theorem 6–54. □

The point of Corollary 6–55, the reason that it is not so easy as it is "obvious," is that just because there is a way of describing a space with, say, 83 parameters does not mean there is no way of describing the same space with 79 parameters. The unit square, for example, can be described with two parameters and also with one, using a Peano mapping. The second parametrization maps several points onto one, but it is not obvious that this must happen for all mappings of I^{79} onto I^{83}.

EXERCISE 6–24. Let S be the surface of genus 2 (Fig. 6–18). Assume S to be imbedded in E^3, and let U be the bounded component of $E^3 - S$. Triangulate the solid $\overline{U} = U \cup S$, and compute its integral homology groups.

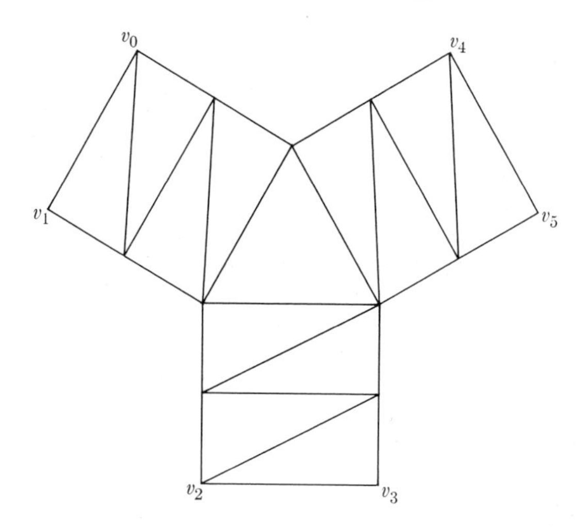

FIGURE 6–25

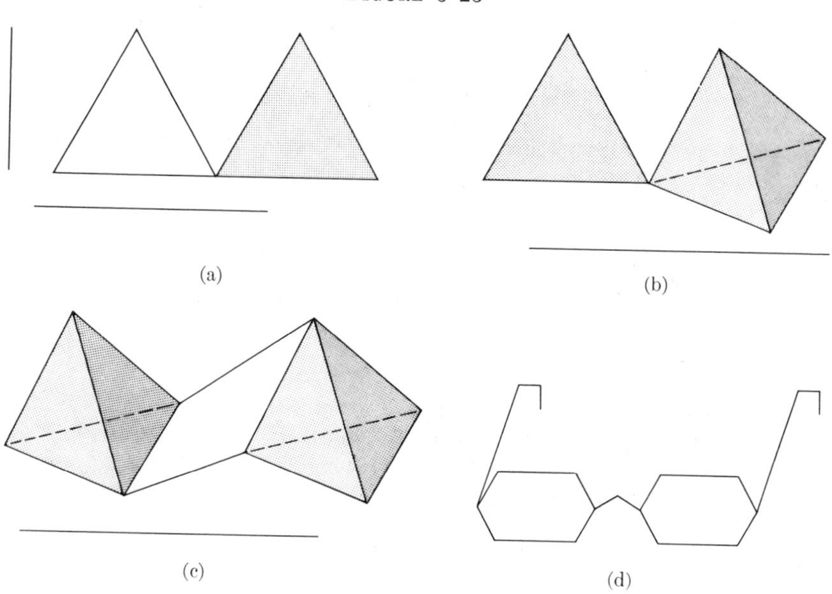

FIGURE 6–26

EXERCISE 6–25. Identify the three 1-simplexes $\langle v_0v_1\rangle$, $\langle v_2v_3\rangle$, and $\langle v_4v_5\rangle$ in the complex pictured in Fig. 6–25 by identifying vertices v_0, v_2, and v_4 and by identifying vertices v_1, v_3, and v_5. Determine the integral homology groups and the mod 2 homology groups of the resulting surface.

EXERCISE 6–26. Without calculation, determine the integral homology groups of the complexes pictured in Fig. 6–26. (Assume that there are no 3-simplexes in any of these complexes.)

EXERCISE 6–27. In the triangulated torus of Fig. 6–9, replace one 2-simplex with the complement of the simplex $\langle v_1 v_3 v_4 \rangle$ in the projective plane pictured in Fig. 5–14. Determine the integral homology groups of the resulting surface.

EXERCISE 6–28. Identify n distinct points of S^2. Find a triangulation of the resulting surface, and compute the integral homology groups.

EXERCISE 6–29. Let S^4 denote the 4-skeleton of the closure of a 5-simplex. Show that the 2-skeleton of S^4 consists of two projective planes triangulated as in Fig. 6–14, each containing every 1-simplex of S^4.

EXERCISE 6–30. Identify the opposite faces of a cube in two different ways (there are more than two ways, of course). Triangulate the resulting solids and find their integral homology groups.

EXERCISE 6–31. Projective n-space P^n may be obtained by identifying antipodal points of S^n. Prove that P^n may be triangulated as an n-pseudomanifold which is orientable if n is odd and nonorientable if n is even.

EXERCISE 6–32. Let $S_1, \ldots, S_k$ be any finite set of spheres. Prove that $\mathbb{P}_{i=1}^{k} S_i$ can be triangulated as an orientable pseudomanifold.

EXERCISE 6–33. Let K be a 2-pseudomanifold, let α_i, $i = 0, 1$, and 2, denote the number of i-simplexes in K, and let $\chi(K)$ be the Euler characteristic of K. Prove that

$$3\alpha_2 = 2\alpha_1, \qquad \alpha_1 = 3(\alpha_0 - \chi(K)), \qquad \alpha_0 \geqq \tfrac{1}{2}(7 + \sqrt{49 - 24\chi(K)}).$$

EXERCISE 6–34. Using the results of Exercise 6–35 and assuming that any triangulation of the 2-sphere S^2, the projective plane P^2, and the torus T must be 2-pseudomanifolds, show that the following inequalities are satisfied:

$$\begin{array}{lllll} \text{for } S^2, & \alpha_0 \geqq 4, & \alpha_1 \geqq 6, & \text{and} & \alpha_2 \geqq 4, \\ \text{for } P^2, & \alpha_0 \geqq 6, & \alpha_1 \geqq 15, & \text{and} & \alpha_2 \geqq 10, \\ \text{and for } T, & \alpha_0 \geqq 7, & \alpha_1 \geqq 21, & \text{and} & \alpha_2 \geqq 14. \end{array}$$

In particular, find a minimal triangulation of the torus.

EXERCISE 6–35. Construct a triangulation of S^n which is symmetric with respect to the origin in E^{n+1}. Define the antipodal mapping f carrying each vertex into its antipodal vertex. Show that f is simplicial and that for any element h_n of the integral homology group $H_n(S^n)$, we have $f_*(h_n) = (-1)^n h_n$.

EXERCISE 6–36. Show that any mapping of the $(n + 1)$-disc into S^n maps at least one pair of antipodal boundary points onto a single point.

EXERCISE 6–37. Let $f_i(x) = f_i(x_1, x_2, \ldots, x_n)$, $i = 1, 2, \ldots, n$, be real-valued continuous functions on the n-disc. If

$$\frac{f_1(x)}{x_1} = \cdots = \frac{f_n(x)}{x_n} > 0$$

for no boundary point x, prove that the system of equations

$$f_i(x_1, \ldots, x_n) = 0 \qquad (i = 1, \ldots, n)$$

has at least one solution in the n-disc.

EXERCISE 6–38. A *linear graph* is a finite connected 1-dimensional complex. A vertex of a linear graph is *odd* or *even* provided that it is a face of an odd or even number of 1-simplexes. Prove that there is an even number of odd vertices in any linear graph.

EXERCISE 6–39. An *Euler line* in a linear graph is a line drawn without lifting the pencil and without retracing any 1-simplex (crossing at a vertex is permitted). Show that a linear graph may be traced with an Euler line if and only if there are no more than two odd vertices in the graph. Furthermore, prove that if there are two odd vertices, the tracing Euler line must begin at one of the odd vertices and will terminate at the other.

CHAPTER 7

FURTHER DEVELOPMENTS IN ALGEBRAIC TOPOLOGY

This chapter consists of two major parts, the first devoted to relative homology theory and the second to cohomology theory. We introduce the two subjects separately.

7-1 Relative homology groups. It is often found that we know the homology groups of a complex and wish to deduce from this knowledge information about the groups of some subcomplex. And conversely, knowing the groups of a subcomplex, we may want to obtain some knowledge of the groups of the entire complex. It is the relations between such groups that form a goal in studying relative homology theory, an invention of S. Lefschetz [95]. We will use several sections in reaching for this goal.

Throughout the first part of this chapter, we will be considering an abstract simplicial complex K and a closed subcomplex L contained in K. A p-chain c_p on K is called a *p-cycle of K modulo L* provided that ∂c_p is a chain on L, that is, ∂c_p has nonzero coefficients only on simplexes of L. We will set up the relative homology groups of K modulo L and discuss the geometric interpretations as we proceed.

Let i denote the identity simplicial mapping, the *injection*, of L into K defined by $i(v) = v$ for each vertex of L. As we have remarked before, i induces (or is) an isomorphism of the chain groups $C_p(L, G)$ into the chain groups $C_p(K, G)$, but this does *not* mean that the induced homomorphism i_* on homology groups is an isomorphism. For the remainder of this discussion, we will use only the group Z of integers as coefficients and will write $C_p(L)$ for $C_p(L, Z)$, etc.

In view of the isomorphism i, the chain group $C_p(L)$ may be considered as a subgroup of $C_p(K)$ and, since both are free groups, we may define the *relative chain group of K modulo L* (with integral coefficients) as the difference groups

$$C_p\left(\frac{K}{L}\right) = C_p(K) - C_p(L).$$

An element $\bar{c}_p$ of $C_p(K/L)$ is a *relative p-chain of K modulo L* and, of course, is a coset in $C_p(K)$. For such a coset, it will be convenient to write

$$\bar{c}_p = c_p \oplus C_p(L),$$

where c_p is any chain of K in the coset $\bar{c}_p$. Clearly, $\bar{c}_p$ is an equivalence class $[c_p]$ and consists of all chains of K of the form $c_p + k_p$, where k_p is a p-chain of L. As usual, we say that c_p is a *representative* of $\bar{c}_p$.

We now have a graded group $C(K/L) = C_0(K/L) \oplus C_1(K/L) \oplus \cdots \oplus C_p(K/L) \oplus \cdots$, and all we need for a homology theory is the boundary operator (see Section 6–8). The new boundary operator is defined by

$$\overline{\partial c_p} = \bar{\partial}(c_p \oplus C_p(L)) = \partial c_{p-1} \oplus C_{p-1}(L),$$

where ∂ is the usual boundary operator on $C_p(K)$. To say this in a slightly different way, to form the boundary of a relative chain, one takes the usual boundary of any one of its representatives and then considers the coset of this boundary. Of course, it must be shown that $\bar{\partial}$ is well-defined. To do so, let c_p and c'_p be two representatives of the same relative chain $\bar{c}_p$. By definition, $c'_p = c_p + k_p$ for some chain k_p on L. Therefore, $\partial c'_p = \partial c_p + \partial k_p$. Since L is a closed subcomplex, the boundary ∂k_p is in $C_{p-1}(L)$ [of course, ∂k_p is actually in $B_{p-1}(L)$]. Thus $\partial c'_p$ and ∂c_p lie in the same coset in $C_{p-1}(K)$, that is, $\partial c'_p \oplus C_{p-1}(L) = \partial c_p \oplus C_{p-1}(L)$.

The fundamental requirement for a boundary operator is that it be of order 2. We show that $\bar{\partial}(\overline{\partial c_p}) = 0$ for any relative p-chain $\bar{c}_p$. If $\bar{c}_p = c_p \oplus C_p(L)$, then $\bar{\partial}(\overline{\partial c_p}) = \bar{\partial}(\partial c_p \oplus C_{p-1}(L)) = \partial\partial c_p \oplus C_{p-2}(L)$, by definition. But $\partial\partial c_p = 0$. Thus $\bar{\partial}(\overline{\partial c_p}) = C_{p-2}(L)$, and $C_{p-2}(L)$ *is* the zero element of $C_{p-2}(K/L)$.

Once the property $\overline{\partial\partial} = 0$ is established, we may apply the usual method of obtaining a homology theory. Thus we define the *relative cycle groups of K mod L* as

$$Z_p\left(\frac{K}{L}\right) = \text{the kernel of } \bar{\partial} \text{ in } C_p\left(\frac{K}{L}\right) = \bar{\partial}^{-1}(0),$$

and we define the *relative boundary groups of K mod L* as

$$B_p\left(\frac{K}{L}\right) = \bar{\partial}C_{p+1}\left(\frac{K}{L}\right).$$

The fact that $\overline{\partial\partial} = 0$ implies that $B_p(K/L)$ is a subgroup of $Z_p(K/L)$. Hence, since both of these groups are abelian, we may define the *relative homology groups of K mod L* as

$$H_p\left(\frac{K}{L}\right) = Z_p\left(\frac{K}{L}\right) - B_p\left(\frac{K}{L}\right).$$

By these definitions, a relative chain $\bar{z}_p$ is a relative cycle if and only if $\overline{\partial z_p} = 0$. This means that $\bar{z}_p = z_p \oplus C_p(L)$ is in $Z_p(K/L)$ if and only if ∂z_p lies in $C_{p-1}(L)$. That is, a chain on K represents a relative cycle if its boundary lies in L. Of course, a true cycle on K, a chain with zero boundary, is also a representative of some relative cycle. Similarly,

$\bar{b}_p = b_p \oplus C_p(L)$ is a relative boundary if and only if there is a chain d_{p+1} in $C_{p+1}(K)$ such that $b_p - \partial d_{p+1}$ lies in $C_p(L)$, that is, b_p together with some chain on L constitutes the boundary of a chain on K. It will benefit the reader to draw some sketches illustrating this concept geometrically.

EXERCISE 7–1. Let K be a finite complex, and let v be a vertex of K. Determine the relative integral homology groups $H_p(K/v)$, $p = 0, 1, 2, \ldots$

EXERCISE 7–2. Let S^2 denote the boundary complex of a 3-simplex σ^3, and let S^1 denote the boundary complex of one 2-simplex σ^2 of S^2. Determine the relative integral homology groups $H_p(S^2/S^1)$, $p = 0, 1, 2$.

7–2 The exact homology sequence. An economical and very suggestive way to gather the interrelations between the homology groups $H_p(K)$, $H_p(L)$, and $H_p(K/L)$ is in the form of the exact homology sequence. This algebraic construct was first formally recognized by Hurewicz [86] in 1941, although the various parts were known earlier. Let us look at the individual parts first.

We have mentioned the injection mapping i of the subcomplex L into the complex K and the resulting induced homomorphism i_* of the groups $H_p(L)$ into $H_p(K)$. There is also the canonical homomorphism j of $C_p(K)$ onto $C_p(K/L)$ given by

$$j(c_p) = c_p \oplus C_p(L).$$

By definition, $j(\partial c_p) = \partial c_p \oplus C_{p-1}(L)$ and $\bar{\partial}(j(c_p)) = \bar{\partial}(c_p \oplus C_p(L)) = \partial c_p \oplus C_{p-1}(L)$. That is, we have $j\partial = \bar{\partial}j$, so j is a chain-mapping. In view of Section 6–10, there is an induced homomorphism j_* of the groups $H_p(K)$ into $H_p(K/L)$.

A more complicated homomorphism is defined next. Let $\bar{z}_p$ be a relative cycle with representative z_p in $C_p(K)$. Then $\overline{\partial z_p} = C_{p-1}(L)$, by definition. But if $\bar{\partial}(z_p \oplus C_p(L)) = C_{p-1}(L)$, then ∂z_p must lie in $C_{p-1}(L)$. Furthermore, $\partial(\partial z_p) = 0$, so the chain ∂z_p is actually a cycle, that is, ∂z_p lies in $Z_{p-1}(L)$. As an element of $Z_{p-1}(L)$, the chain ∂z_p determines a unique element of the homology group $H_{p-1}(L)$. We define a transformation *on homology classes* by setting

$$\partial_*([\bar{z}_p]) = [\partial z_p],$$

where we are using our usual notation for equivalence classes.

We first show that ∂_* is well-defined, that is, if $\bar{z}'_p$ is some other representative of $[\bar{z}_p]$, then ∂z_p and $\partial z'_p$ are homologous. To do this, let $\bar{z}'_p$ be homologous to $\bar{z}_p$, that is, $\bar{z}_p - \bar{z}_p$ is a relative boundary. Thus if z'_p and z_p represent $\bar{z}'_p$ and $\bar{z}_p$, respectively, then there is a chain d_{p+1} in $C_{p+1}(K)$ such that $z'_p - z_p - \partial d_{p+1} = x_p$, where x_p is a chain in $C_p(L)$. Then we have that $\partial z'_p - \partial z_p - \partial\partial d_{p+1} = \partial z'_p - \partial z_p = \partial x_p$, which implies that

$\partial z'_p$ and ∂z_p are homologous. Therefore ∂_* is well-defined. It is merely a routine verification of the definition to show that ∂_* is a homomorphism of $H_p(K/L)$ into $H_{p-1}(L)$, and this may be left as an exercise.

The necessary mechanism to set up the *homology sequence of the pair* (K, L) is now at hand. This is the sequence of groups and homomorphisms symbolized in the following diagram:

$$\cdots \xrightarrow{i_*} H_p(K) \xrightarrow{j_*} H_p\left(\frac{K}{L}\right) \xrightarrow{\partial_*} H_{p-1}(L) \xrightarrow{i_*} H_{p-1}(K) \xrightarrow{j_*} \cdots \xrightarrow{i_*} H_0(K).$$

The important relations between the homology groups $H_p(K)$, $H_p(L)$, and $H_p(K/L)$ are collected in the theorem following this definition: a sequence of groups and homomorphisms $\psi_i : G_i \to G_{i-1}$ (usually $i = 1, 2, 3, \ldots$, or $i = \cdots, -2, -1, 0, 1, 2, \ldots$) is said to be an *exact sequence* if, for each i, the image under ψ_i of G_i is the same subgroup of G_{i-1} as is the kernel of ψ_{i-1}. That is, we have $\psi_i(G_i) = \psi_{i-1}^{-1}(0)$, where 0 is the identity element of G_{i-2}. If the sequence terminates in a first group the map into that group is required to be onto.

THEOREM 7–1. The homology sequence of a pair (K, L) is exact.

Proof: There are three parts to this proof: (1) kernel of $j_* =$ image under i_*, (2) kernel of $\partial_* =$ image under j_*, and (3) kernel of $i_* =$ image under ∂_*. Since these arguments are quite typical of those found in relative homology theory, we give them below in some detail. However, they are not difficult, and the reader may prefer to prove them himself.

Part (1), kernel of $j_* =$ image under i_*. Let z_p be a cycle on K such that $j(z_p)$ is homologous to zero. Since $j(z_p) = z_p \oplus C_p(L)$, this means that the relative cycle $z_p \oplus C_p(L)$ is assumed to be a relative boundary. Therefore there is a chain d_{p+1} on K such that $z_p - \partial d_{p+1} = x_p$, where x_p is a chain in $C_p(L)$. This in turn implies that the coset $j(z_p) = z_p \oplus C_p(L)$ contains the element x_p of $C_p(L)$. Therefore $z_p - x_p = \partial d_{p+1}$, or z_p is homologous to x_p, and this is the same as saying that $[z_p] = [x_p]$. Since i is the injection mapping, we have $i_*([x_p]) = [x_p] = [z_p]$. Thus if $j_*([z_p]) = 0$, then $[z_p]$ is the image of an element of $H_p(L)$, so we have the kernel of j_* contained in the image under i_*.

Now assume that the element $[z_p]$ of $H_p(K)$ is the image under i_* of an element $[x_p]$ of $H_p(L)$. This says that z_p is homologous to $i(x_p) = x_p$, or that $z_p - x_p = \partial d_{p+1}$ for some chain d_{p+1} on K. Therefore $z_p - \partial d_{p+1} = x_p$, where x_p is in $C_p(L)$. This implies that z_p is homologous to ∂d_{p+1} mod L. But ∂d_{p+1} is on L, so z_p is homologous to zero mod L. Therefore $j(z_p) = z_p \oplus C_p(L)$ is homologous to zero mod L, and this implies that the kernel of j_* contains the image under i_*, completing the proof of part (1).

Part (2), kernel of ∂_* = image under j_*. Let $\bar{z}_p = z_p \oplus C_p(L)$ be a relative p-cycle representing a homology class $[\bar{z}_p]$ such that $\partial_*([\bar{z}_p]) = 0$, that is, $[\bar{z}_p]$ is in the kernel of ∂_*. By the definition of ∂_*, this implies that $\overline{\partial z_p}$ is homologous to zero mod L or that ∂z_p is in $C_{p-1}(L)$. If ∂z_p is homologous to zero in L, then there is a chain d_p on L such that $\partial d_p = \partial z_p$. Consider then the chain $z_p - d_p$. Since z_p and d_p have the same boundary, this chain is a cycle. Thus $z_p - d_p$ is a representative of the coset $\bar{z}_p$, and $z_p - d_p$ is an absolute cycle on K. It follows that $j(z_p - d_p) = \bar{z}_p$, and we have that the kernel of ∂_* is contained in the image under j_*.

On the other hand, assume that $\bar{z}_p = j(z_p)$ for some absolute cycle z_p on K. Then $\bar{z}_p = z_p \oplus C_p(L)$ and $\overline{\partial z_p} = \partial z_p \oplus C_{p-1}(L) = C_{p-1}(L)$ since $\partial z_p = 0$. Thus $\partial_*([\bar{z}_p]) = [\partial z_p]$, which is the zero coset in $H_{p-1}(L)$. This implies that the kernel of ∂_* contains the image under j_* and completes the proof of part (2).

Part (3), kernel of i_* = image under ∂_*. Assume that z_p is a representative of a homology element in $H_p(L)$ which lies in the kernel of i_*, that is, $i_*([z_p]) = 0$ in $H_p(K)$. This means that $i(z_p) = z_p$ is homologous to zero on K, or there is a chain d_{p+1} on K such that $z_p = \partial d_{p+1}$. Now $j(d_{p+1}) = d_{p+1} \oplus C_{p+1}(L) = \bar{d}_{p+1}$ is a relative chain in $C_{p+1}(K/L)$. We show that $\bar{d}_{p+1}$ is actually a relative cycle. For $\overline{\partial d_{p+1}} = \partial d_{p+1} \oplus C_p(L) = z_p \oplus C_p(L)$, and z_p is itself a chain on L. Thus $\overline{\partial d_{p+1}} = C_p(L)$, the zero element of $C_p(K/L)$. Then we have that $\partial_*([\bar{d}_{p+1}]) = [\partial d_{p+1}] = [z_p]$, or $[z_p]$ is the image under ∂_* of an element of $H_{p+1}(K/L)$. Therefore the kernel of i_* is contained in the image under ∂_*.

Last, if $[z_p]$ is any element of $H_p(L)$ for which there is a relative cycle $\bar{d}_{p+1}$ such that $\partial_*([\bar{d}_{p+1}]) = [z_p]$, then we show that $i_*([z_p]) = 0$. By definition, $\partial_*([\bar{d}_{p+1}]) = [\partial d_{p+1}]$ for some chain d_{p+1} on K. If $[z_p] = [\partial d_{p+1}]$, as assumed, then z_p is homologous to ∂d_{p+1} on K, which says that $i_*([z_p]) = [z_p]$ is the zero element of $H_p(K)$. Therefore the kernel of i_* contains the image under ∂_*, completing the proof of part (3). □

LEMMA 7–2. In the exact sequence $\psi_i : G_i \to G_{i-1}$, suppose that the subsequence of four groups

$$G_{i+2} \xrightarrow{\psi_{i+2}} G_{i+1} \xrightarrow{\psi_{i+1}} G_i \xrightarrow{\psi_i} G_{i-1}$$

is such that both G_{i+2} and G_{i-1} are trivial. Then ψ_{i+1} is an isomorphism of G_{i+1} onto G_i.

Proof: If $G_{i+2} = 0$, then the image $\psi_{i+2}(G_{i+2}) = 0$ in G_{i+1}. By the exactness of the sequence, the kernel of ψ_{i+1} is zero, and thus ψ_{i+1} is an isomorphism. Then since G_{i-1} is trivial, the entire group G_i is in the kernel of ψ_i. By exactness then, the entire group G_i is in the image under ψ_{i+1}, and ψ_{i+1} is onto. □

The above lemma is a very useful tool in working with exact sequences.

As a simple instance, consider the case in which K is the closure of an n-simplex, $n > 1$, and let L be the $(n-1)$-skeleton of K, an $(n-1)$-sphere. Setting up the homology sequence of this pair, we have

$$\cdots \to H_n(K) \to H_n\left(\frac{K}{L}\right) \to H_{n-1}(L) \to H_{n-1}(K) \to \cdots.$$

Since $H_n(K)$ and $H_{n-1}(K)$ are trivial for $n > 1$, Lemma 7–2 applies to show that $H_n(K/L)$ is isomorphic to $H_{n-1}(L)$, which is infinite cyclic.

EXERCISE 7–3. Let σ^2 be any 2-simplex in the triangulation of the torus T. Show that $H_2(T)$ and $H_2(T/\sigma^2)$ are isomorphic. What can be said about $H_1(T)$ and $H_1(T/\sigma^2)$?

EXERCISE 7–4. Let s be a meridian circle on the torus T. Determine the relations between $H_p(T)$ and $H_p(T/s)$, $p = 1, 2$.

7–3 Homomorphisms of exact sequences. If (G_i, ψ_i) and (H_i, φ_i) are two exact sequences, then a collection of homomorphisms $g = (g_i)$, $g_i{:}G_i \to H_i$, is a homomorphism $g{:}(G_i, \psi_i) \to (H_i, \varphi_i)$ if the property $g_{i-1}\psi_i = \varphi_i g_i$ holds for all i. This means that both "paths" from G_i to H_{i-1} are the same homomorphism or that we have *commutativity in the diagram:*

$$\begin{array}{ccccccc} \cdots \to & G_i & \xrightarrow{\psi_i} & G_{i-1} & \to \cdots \\ & \downarrow{\scriptstyle g_i} & & \downarrow{\scriptstyle g_{i-1}} & \\ \cdots \to & H_i & \xrightarrow{\varphi_i} & H_{i-1} & \to \cdots \end{array}$$

Let (K_1, L_1) and (K_2, L_2) be two pairs of complexes and closed subcomplexes. A mapping $\varphi{:}(K_1, L_1) \to (K_2, L_2)$ is a *simplicial mapping of the pair* (K_1, L_1) into the pair (K_2, L_2) provided that $\varphi{:}K_1 \to K_2$ is a simplicial mapping of K_1 into K_2 and that $\varphi(L_1)$ is contained in L_2. It follows from Section 6–10 that φ induces a homomorphism $\varphi_* = \{\varphi_{p*}\}$ of the groups $H_p(K_1)$ into $H_p(K_2)$ and that $\varphi|L_1$ (φ restricted to L_1) induces a homomorphism $(\varphi|L_1)_* = \{(\varphi|L_1)_{p*}\}$ of the groups $H_p(L_1)$ into $H_p(L_2)$. Furthermore, since $\varphi(L_1)$ is contained in L_2, it is easy to see that φ also induces a homomorphism $\bar{\varphi}_* = \{\bar{\varphi}_{p*}\}$ of the relative groups $H_p(K_1/L_1)$ into $H_p(K_2/L_2)$. We wish to show that these homomorphisms constitute a homomorphism of the homology sequence of (K_1, L_1) into that of (K_2, L_2) in the sense of the above definition. That is, we wish to prove commutativity in each square in the following diagram:

$$\begin{array}{ccccccccc} \cdots \to & H_p(L_1) & \xrightarrow{i^1_*} & H_p(K_1) & \xrightarrow{j^1_*} & H_p(K_1/L_1) & \xrightarrow{\partial^1_*} & H_{p-1}(L_1) & \to \cdots \\ & \downarrow{\scriptstyle (\varphi|L_1)_{p*}} & & \downarrow{\scriptstyle \varphi_{p*}} & & \downarrow{\scriptstyle \bar{\varphi}_{p*}} & & \downarrow{\scriptstyle (\varphi|L_1)_{p-1*}} & \\ \cdots \to & H_p(L_2) & \xrightarrow{i^2_*} & H_p(K_2) & \xrightarrow{j^2_*} & H_p(K_2/L_2) & \xrightarrow{\partial^2_*} & H_{p-1}(L_2) & \to \cdots \end{array}$$

The reader may easily verify that $\varphi_{p^*} i_*^1 = i_*^2 (\varphi | L_1)_{p^*}$ and that $\bar{\varphi}_{p^*} j_*^1 = j_*^2 \varphi_{p^*}$. We prove only that $(\varphi|L_1)_{p-1^*}\, \partial_*^1 = \partial_*^2 \bar{\varphi}_{p^*}$. To do this, let $\bar{z}_p$ be a relative cycle with representative z_p. Then $\partial^1 z_p$ is a representative of $\partial_*^1([z_p])$, and $\varphi\partial^1 z_p$ is a representative of $(\varphi|L_1)_{p^*}\, \partial_*^1([\bar{z}_p])$. Similarly, $\partial^2\varphi(z_p)$ is a representative of $\partial_*^2 \varphi_{p^*}(\bar{z}_p) = \partial_*^2[\varphi(z_p) \oplus C_p(L_2)]$. Since φ is simplicial, we have $\partial^2 \varphi = \varphi \partial^1$, which proves the desired commutativity.

EXERCISE 7–5. If $\varphi:(K_1, L_1) \to (K_2, L_2)$ is such that $\varphi(K_1)$ is contained in L_2, show that φ_* is the trivial homomorphism.

EXERCISE 7–6. Let v be a vertex of a 2-simplex σ^2 in the sphere S^2. Define the mapping $\varphi:(S^2, \sigma^2) \to (S^2, v)$ by setting $\varphi(v_j) = v_j$ for v_j not in σ^2 and $\varphi(v_i) = v$ for v_i in σ^2. Discuss the induced homomorphisms φ_*, $(\varphi|\sigma^2)_*$, and $\overline{\varphi}_*$.

7–4 The excision theorem. If L is a closed subcomplex of K, then we say that $K - L$ is an *open subcomplex* of K.

Consider three complexes, M, L, and K, where M is an open subcomplex of L, and L is a closed subcomplex of K. Both $K - M$ and $L - M$ are closed subcomplexes, and clearly $K - M$ contains $L - M$. We may construct the relative homology groups $H_p[(K - M)/(L - M)]$. The excision theorem states that these groups are isomorphic to the groups $H_p(K/L)$ for each dimension p. Intuitively, this means that the interior of L is unimportant as far as homology modulo L is concerned.

We will approach a proof of the excision theorem indirectly. Let L and K_1 be two closed subcomplexes of a complex K, and define L_1 to be $K_1 \cap L$. Let i be the injection mapping of K_1 into K; then $i|L_1$ (i restricted to L_1) is the injection of L_1 into L. These give us isomorphisms, still called i and $i|L_1$, of $C_p(K_1)$ into $C_p(K)$ and of $C_p(L_1)$ into $C_p(L)$. Furthermore, we also have an induced isomorphism $\bar{i}$ of $C_p(K_1/L_1)$ into $C_p(K/L)$ since i maps L_1 into L. As usual, we use the star subscript to denote the corresponding induced homomorphisms on homology groups.

A remark which has not been made before but which should be evident is that such an induced homomorphism will be an isomorphism onto if it comes from a chain-mapping that is an isomorphism onto. The geometric property that will ensure that $\bar{i}$ is an isomorphism onto is that K_1 contains $K - L$. For if this is true, let $c_p \oplus C_p(L_1)$ be a relative chain in $C_p(K_1/L_1)$. This chain is mapped by $\bar{i}$ onto $c_p \oplus C_p(L)$. Now given $c_p' \oplus C_p(L)$ in $C_p(K/L)$, there is a representative c_p'' of $c_p' \oplus C_p(L)$, where c_p'' is a chain on $K - L$. Since K_1 contains $K - L$, c_p'' is also on K_1. Therefore $c_p'' \oplus C_p(L_1)$ is a relative chain in $C_p(K_1/L_1)$ which is mapped by $\bar{i}$ onto the given chain $c_p' \oplus C_p(L)$. Thus $\bar{i}$ is onto. Since $\bar{i}$ is already an isomorphism, we have that $\bar{i}$ is an isomorphism of $C_p(K_1/L_1)$ onto $C_p(K/L)$ if K_1 contains $K - L$, and in this case $\bar{i}_*$ is an isomorphism of $H_p(K_1/L_1)$ onto $H_p(K/L)$.

The above situation may be rephrased as follows. If K_1 contains $K - L$, then $K - K_1$ is contained in L. Letting M be the open subcomplex $K - K_1$, we have $K_1 = K - M$ and $L_1 = K_1 \cap L = (K - M) \cap L = (K \cap L) - (M \cap L) = L - M$. Therefore the argument above has established the next result.

THEOREM 7–3 (Excision). If L is a closed subcomplex of a complex K, and if M is an open subcomplex of L, then the injection mapping of $K - M$ into K induces an isomorphism of $H_p[(K - M)/(L - M)]$ onto $H_p(K/L)$ for each dimension p.

As an example of the use of exact sequences and the excision theorem, we give an inductive proof of the following result.

THEOREM 7–4. The integral homology group $H_n(S^n)$ of the n-sphere is infinite cyclic.

Proof: Let s^{n+1} be a geometric $(n + 1)$-simplex. The boundary complex of s^{n+1} [the n-skeleton of $\mathrm{Cl}(s^{n+1})$] is a triangulation of the n-sphere S^n. (We prove only that this triangulation has infinite cyclic nth homology group, of course.) Let s^n be one of the n-simplexes in S^n, and denote by S^{n-1} the boundary complex of s^n. Using the cone construction (see Section 6–12), we have shown that the n-cell $\mathrm{Cl}(s^n)$ is homologically trivial. We also know that $H_n(s^n/S^{n-1})$ is isomorphic to $H_{n-1}(S^{n-1})$, $n > 1$.

Now let T^n be the closed subcomplex of S^n consisting of all the simplexes of S^n except the *open* simplex $\mathring{s}^n$. It should be clear that T^n may be considered as a cone over S^{n-1} at the vertex opposite s^n and also that $T^n \cap s^n$ is precisely S^{n-1}. Letting $K = S^n$, $L = T^n$, and $M = S^n - s^n$, we may apply the excision theorem to prove that $H_n(s^n/S^{n-1})$ is isomorphic to $H_n(S^n/T^n)$. Setting up the exact homology sequence of the pair (S^n, T^n), we have

$$\cdots \to H_n(T^n) \to H_n(S^n) \to H_n\left(\frac{S^n}{T^n}\right) \to H_{n-1}(T^n) \to \cdots.$$

Since T^n is a cone and is homologically trivial, we know that both $H_n(T^n)$ and $H_{n-1}(T^n)$ are trivial for $n > 1$. By Lemma 7–2, this implies that $H_n(S^n)$ is isomorphic to $H_n(S^n/T^n)$. Combining this with the fact that $H_n(s^n/S^{n-1})$ is isomorphic to $H_{n-1}(S^{n-1})$, we have that $H_n(S^n)$ is isomorphic to $H_{n-1}(S^{n-1})$ for $n > 1$. Having previously shown that $H_2(S^2)$ is infinite cyclic (see Section 6–4), we have completed an inductive proof. □

EXERCISE 7–7. Let K be a finite complex, and let v and w be two vertices not in K. Let vwK denote the double cone over K at the vertices v and w. Show that

$$H_p(vwK) \text{ is isomorphic to } H_{p-1}(K) \qquad (p > 0)$$

and that

$$H_0(vwK) \text{ is infinite cyclic.}$$

7–5 The Mayer-Vietoris sequence. In a somewhat different direction but with similar arguments we now set up the Mayer-Vietoris exact sequence, which exhibits the relationships between the groups of the union and intersection of two complexes. Let K be a complex which is the union of two closed subcomplexes K_1 and K_2 where we assume, in general, that $K_1 \cap K_2$ is a nonempty subcomplex of K also. The Mayer-Vietoris sequence is as follows:

$$\cdots \xrightarrow{s_*} H_{p+1}(K_1 \cup K_2) \xrightarrow{v_*} H_p(K_1 \cap K_2) \xrightarrow{j_*} H_p(K_1) \oplus H_p(K_2) \xrightarrow{s_*} H_p(K_1 \cup K_2) \xrightarrow{v_*} \cdots.$$

After defining the homomorphisms s_*, v_*, and j_*, we will show that this sequence is also exact.

Let j be the mapping of the chain group $C_p(K_1 \cap K_2)$ into the direct sum $C_p(K_1) \oplus C_p(K_2)$ defined by

$$j(c_p) = (c_p, -c_p).$$

This is possible since $K_1 \cap K_2$ is contained in both K_1 and K_2. Thus j is the injection of $C_p(K_1 \cap K_2)$ into $C_p(K_1)$ and the negative injection of $C_p(K_1 \cap K_2)$ into $C_p(K_2)$. The mapping j induces a homomorphism j_* of $H_p(K_1 \cap K_2)$ into $H_p(K_1) \oplus H_p(K_2)$ in the usual way. Note again that while j is an isomorphism into, j_* need not be an isomorphism.

We easily define a mapping s of $C_p(K_1) \oplus C_p(K_2)$ into $C_p(K_1 \cup K_2)$ by setting

$$s(c_p^1, c_p^2) = c_p^1 + c_p^2.$$

This is possible since a chain on either K_1 or K_2 is on $K_1 \cup K_2$. The chain-mapping s induces a homomorphism s_* of $H_p(K_1) \oplus H_p(K_2)$ into $H_p(K_1 \cup K_2)$. Once again a nontrivial element of $H_p(K_1) \oplus H_p(K_2)$ may be mapped by s_* onto the zero element of $H_p(K_1 \cup K_2)$, for a cycle may fail to bound on K_1, for instance, and yet bound on $K_1 \cup K_2$.

The construction of v_* is more laborious. Let c_p be a chain on $K = K_1 \cup K_2$. This implies that c_p may be written as $c_p = c_p^1 + c_p^2$, where c_p^1 is on K_1 and c_p^2 is on K_2. These chains c_p^1 and c_p^2 are determined only modulo $K_1 \cap K_2$, that is, $c_p^1 + c_p^2 = k_p^1 + k_p^2$ if and only if $c_p^1 - k_p^1 = c_p^2 - k_p^2 = d_p$, where d_p is a chain on $K_1 \cap K_2$. This in turn implies that $C_p[(K_1 \cup K_2)/(K_1 \cap K_2)]$ is isomorphic to

$$C_p\left(\frac{K_1}{K_1 \cap K_2}\right) \oplus C_p\left(\frac{K_2}{K_1 \cap K_2}\right).$$

Now if z_p is a cycle on $K_1 \cup K_2$, we may write $z_p = z_p^1 + z_p^2$, and hence $\partial z_p = \partial z_p^1 + \partial z_p^2 = 0$ or $\partial z_p^1 = -\partial z_p^2$. Both of these chains are in

$Z_{p-1}(K_1 \cap K_2)$. Clearly, both are cycles and, since ∂z_p^1 is on K_1 and $-\partial z_p^2$ is on K_2 and they are equal, it follows that they both must be on $K_1 \cap K_2$. Since z_p^1 is determined modulo $C_p(K_1 \cap K_2)$, the cycle ∂z_p^1 is determined modulo $B_{p-1}(K_1 \cap K_2)$. Thus we may define v_* directly by setting

$$v_*([z_p]) = [\partial z_p^1].$$

We must show that v_* is well-defined, which entails showing that if z_p is homologous to $'z_p$ on K, then ∂z_p^1 is homologous to $\partial' z_p$ on $K_1 \cap K_2$. If $z_p \sim {}'z_p$, then $z_p - {}'z_p = \partial t_{p+1}$, and writing $t_{p+1} = t_{p+1}^1 + t_{p+1}^2$, $z_p = z_p^1 + z_p^2$, and $'z_p = {}'z_p^1 + {}'z_p^2$, we have

$$z_p^1 + z_p^2 = {}'z_p^1 + {}'z_p^2 + \partial t_{p+1}^1 + \partial t_{p+1}^2.$$

Then we have

$$z_p^1 = {}'z_p^1 + \partial t_{p+1}^1 + d_p^1 \qquad \text{and} \qquad z_p^2 = {}'z_p^2 + \partial t_{p+1}^2 + d_p^2,$$

where d_p^1 and d_p^2 are chains on $K_1 \cap K_2$. Now the relation $z_p \sim {}'z_p$ clearly implies that $\partial z_p^1 \sim \partial' z_p^1$ on $K_1 \cap K_2$, so v_* is well-defined. Again it is an easy exercise to show that v_* is a homomorphism.

THEOREM 7–5. The Mayer-Vietoris sequence is exact.

Proof: As in Theorem 7–1, we must prove the three equalities (1) kernel of s_* = image under j_*, (2) kernel of v_* = image under s_*, and (3) kernel of j_* = image under v_*.

Part (1), kernel of s_* = image under j_*. Let d_p be a chain on $K_1 \cap K_2$, and consider the image $j(d_p) = (d_p, -d_p)$. Then $s(d_p, -d_p) = d_p - d_p = 0$. Thus $j(d_p)$ is in the kernel of s, and this suffices to show that the image under j_* is contained in the kernel of s_*.

On the other hand, let (c_p^1, c_p^2) be an element of $C_p(K_1) \oplus C_p(K_2)$ such that $s(c_p^1, c_p^2) = c_p^1 + c_p^2 = 0$. Then $c_p^1 = -c_p^2$, and since c_p^1 is on K_1 and c_p^2 is on K_2, it follows that c_p^1 can only be on $K_1 \cap K_2$. Thus $j(c_p^1) = (c_p^1, -c_p^1) = (c_p^1, c_p^2)$. Therefore the kernel of s_* is contained in the image under j_*.

Part (2), kernel of v_* = image under s_*. Let (z_p^1, z_p^2) represent an element of $H_p(K_1) \oplus H_p(K_2)$. Then $s(z_p^1, z_p^2) = z_p^1 + z_p^2$ is a cycle on $K_1 \cup K_2$. By definition, $v_*([z_p^1 + z_p^2]) = [\partial z_p^1]$. But z^1 is a cycle on K_1, so $\partial z_p^1 = 0$ on K_1. Also, of course, $\partial z_p^1 = -\partial z_p^2 = 0$ on K_2, so $\partial z_p^1 = 0$ on $K_1 \cap K_2$ as well. This shows that an image under s_* of an element of $H_p(K_1) \oplus H_p(K_2)$ lies in the kernel of v_*.

To prove the converse, we let z_p denote a cycle on $K_1 \cup K_2$ such that $v_*([z_p]) = 0$. By definition, $v_*([z_p]) = v_*([z_p^1 + z_p^2]) = [\partial z_p^1]$ for some decomposition $z_p = z_p^1 + z_p^2$, where z_p^1 is on K_1 and z^2 is on K_2. Since z_p is a cycle, $\partial z_p^1 + \partial z_p^2 = 0$ or $\partial z_p^1 = -\partial z_p^2$. By assumption then, there is a chain on $K_1 \cap K_2$ which is bounded by both z_p^1 and z_p^2. This proves that z_p^1 and z_p^2 are absolute cycles on K_1 and K_2, respectively, and hence the pair (z_p^1, z_p^2) represents some element of $H_p(K_1) \oplus H_p(K_2)$. Then $s(z_p^1, z_p^2) = z_p^1 + z_p^2 = z_p$. Thus the kernel of v_* is contained in the image under s_*.

Part (3), kernel of j_* = image under v_*. Let x_p be a cycle on $K_1 \cap K_2$ such that $j(x_p)$ is homologous to zero, that is, $x_p \sim 0$ on K_1 and $x_p \sim 0$ on K_2. Then there exist chains c_{p+1}^1 on K_1 and c_{p+1}^2 on K_2 such that $x_p = \partial c_{p+1}^1 = \partial c_{p+1}^2$. Consider the chain $z_{p+1} = c_{p+1}^1 - c_{p+1}^2$ on $K_1 \cup K_2$. Clearly, z_{p+1} is an absolute cycle on $K_1 \cup K_2$ since $\partial z_{p+1} = \partial c_{p+1}^1 - \partial c_{p+1}^2 = x_p - x_p = 0$. Then $v_*([z_{p+1}]) = v_*([c_{p+1}^1 + c_{p+1}^2]) = [\partial c_{p+1}^1] = [x_p]$. This shows that the kernel of j_* is contained in the image under v_*.

Conversely, let x_p be a representative of a homology element $v_*([z_{p+1}])$. Then if $z_{p+1} = z_{p+1}^1 + z_{p+1}^2$, as before, we mean that $x_p = \partial z_{p+1}^1 + \partial d_{p+1}$, where d_{p+1} is some chain on $K_1 \cap K_2$. Since $\partial z_{p+1} = \partial z_p^1 + \partial z_p^2 = 0$, we have that $\partial z_{p+1}^1 = -\partial z_{p+1}^2$. Hence also we have that $x_p = -\partial z_p^2 + \partial d'_{p+1}$, where d'_{p+1} is also a chain on $K_1 \cap K_2$. Thus $x_p = \partial(z_{p+1}^1 + d_{p+1})$ is homologous to zero, on K_1, and $x_p = \partial(z_{p+1}^2 + d'_{p+1})$ is also homologous to zero, on K_2. This implies that $j(x_p) = (x_p, -x_p)$ is homologous to zero and therefore the image under v_* is contained in the kernel of j_*. □

We give another inductive proof of the fact that $H_n(S^n)$ is infinite cyclic, as an example of the use of the Mayer-Vietoris sequence. In Euclidean $(n+1)$-space E^{n+1}, the $n+1$ points $(0, 0, \ldots, 0)$, $(1, 0, \ldots, 0)$, $\ldots$, $(0, \ldots 0, 1, 0)$ determine an n-simplex s^n in the hyperplane $x_{n+1} = 0$. The boundary complex of s^n is a triangulation of S^{n-1}. Let $v_+ = (0, \ldots, 0, 1)$ and $v_- = (0, \ldots, 0, -1)$, and construct the two cone complexes $K_1 = v_+S^{n-1}$ and $K_2 = v_-S^{n-1}$. It is obvious that $K_1 \cup K_2$ is a triangulation of S^n, while $K_1 \cap K_2 = S^{n-1}$. Setting up the Mayer-Vietoris sequence, we have

$$\cdots \to H_n(K_1) \oplus H_n(K_2) \xrightarrow{s_*} H_n(S^n) \xrightarrow{v_*} H_{n-1}(S^{n-1}) \xrightarrow{j_*} H_{n-1}(K_1) \oplus H_{n-1}(K_2) \to \cdots.$$

Both K_1 and K_2 are homologically trivial, so for $n > 1$ we have that $H_n(K_1) \oplus H_n(K_2)$ and $H_{n-1}(K_1) \oplus H_{n-1}(K_2)$ are trivial. It follows from Lemma 7-2 that v_* is an isomorphism onto.

EXERCISE 7-8. Let L be a closed subcomplex of a complex K, and suppose that the simplicial mapping $\varphi: K \to L$ has the property that its restriction to L is the identity mapping i (that is, φ is a simplicial retraction). Show that

$$H_p(K) = \text{image of } i_{p*} \oplus \text{kernel of } \varphi_{p*}$$

or

$$H_p(K) = H_p(L) \oplus H_p\left(\frac{K}{L}\right).$$

7–6 Some general remarks. As was done in Section 6–8, we may take the weak direct sum of the homology groups $H_p(K)$, $H_p(L)$, and $H_p(K/L)$. In this way, we obtain the graded groups $H(K)$, $H(L)$, and $H(K/L)$. The homology sequence may then be diagrammed briefly as

$$\begin{array}{ccc} H(K) & \xrightarrow{j_*} & H(K/L) \\ {\scriptstyle i_*}\nwarrow & & \swarrow{\scriptstyle \partial_*} \\ & H(L) & \end{array}$$

Similarly, the Mayer-Vietoris sequence is often diagrammed as

$$\begin{array}{ccc} H(K_1) \oplus H(K_2) & \xrightarrow{s_*} & H(K_1 \cup K_2) \\ {\scriptstyle j_*}\nwarrow & & \swarrow{\scriptstyle v_*} \\ & H(K_1 \cap K_2) & \end{array}$$

These simplified diagrams make it easy to remember the relationships involved in these exact sequences.

Another remark may be made. It would not be difficult to retrace our steps in Sections 7–1 through 7–5 and use an arbitrary abelian group G as coefficients in place of the integers Z. In this way, we would obtain the more general homology sequence of the pair (K, L),

$$\begin{array}{ccc} H(K, G) & \xrightarrow{j_*} & H(K/L, G) \\ {\scriptstyle i_*}\nwarrow & & \swarrow{\scriptstyle \partial_*} \\ & H(L, G) & \end{array}$$

and the general Mayer-Vietoris sequence,

$$\begin{array}{ccc} H(K_1, G) \oplus H(K_2, G) & \xrightarrow{s_*} & H(K_1 \cup K_2, G) \\ {\scriptstyle j_*}\nwarrow & & \swarrow{\scriptstyle v_*} \\ & H(K_1 \cap K_2, G) & \end{array}$$

7–7 The Eilenberg-Steenrod axioms for homology theory. As we have seen, a meaningful homology theory is a complicated mechanism. To construct such a theory, one must start with a topological space, and from the space obtain a complex. Then from the complex, we obtain an oriented complex, from the oriented complex obtain the groups of chains, and finally from the groups of chains construct the homology groups. In our development of simplicial homology theory, we were quite vague about the crucial step from a space to a complex. Indeed, we essentially started with the complex. Furthermore, the simplicial complex is a specialized type of complex in that the problem of orientation is easily solved. Thus, in a sense, simplicial homology theory carefully avoids two difficult stages in the development of a homology theory.

Many attempts to construct homology groups for general spaces have been successful, and we will study some of these in Chapter 8. In an effort to unify these many theories, Eilenberg and Steenrod [7] were led to an axiomatic treatment of homology theory. We state below the axiom system which they have shown to characterize a homology theory. These axioms apply to much more general categories of spaces and mappings, but we will find a valid interpretation of the axioms if we think of *simplicial complex* whenever the word *space* is used and of *simplicial mapping* whenever *mapping* is used.

According to Eilenberg and Steenrod, a homology theory on an admissible category of spaces and mappings is a collection of three functions:

1. A function $H_p(X, A)$, defined for each pair of spaces (X, A) where A is a closed subspace of X and for each integer p, whose value is an abelian group, the p-dimensional relative homology group of X modulo A.

2. A function f_{*p}, defined for each mapping $f:(X, A) \to (Y, B)$ such that $f(A)$ is contained in B and for each integer p, whose value is a homomorphism of $H_p(X, A)$ into $H_p(Y, B)$. This is the homomorphism induced by f.

3. A function $\partial(p, X, A)$, defined for each pair (X, A) and each integer p, whose value is a homomorphism of $H_p(X, A)$ into $H_p(A, \emptyset)$. This is the boundary operator.

In practice, we reduce the symbol $\partial(p, X, A)$ to ∂ and drop the index p on f_{*p} since these will be understood from the context. Now the three functions above are required to satisfy the following axioms:

Axiom 1. If f is the identity mapping of (X, A) onto itself, then f_* is the identity isomorphism of $H_p(X, A)$ onto itself for each p.

Axiom 2. If $f:(X, A) \to (Y, B)$ and $g:(Y, B) \to (Z, C)$, then the composition of f_* and g_* is $(gf)_*$. Briefly, $(gf)_* = g_*f_*$.

Axiom 3. If $f:(X, A) \to (Y, B)$, with $f|A:A \to B$, then the compositions ∂f_* and $(f|A)_*\partial$ coincide. Briefly, $\partial f_* = (f|A)_*\partial$.

Axiom 4. If $i:A \to X$ and $j:(X, \emptyset) \to (X, A)$ are injection mappings, then the sequence

$$\cdots \to H_p(A) \xrightarrow{i_*} H_p(X) \xrightarrow{j_*} H_p(X, A) \xrightarrow{\partial} H_{p-1}(A) \xrightarrow{i_*} \cdots$$

is exact. (Note that we write $H_p(A)$ for $H_p(A, \emptyset)$, etc.)

Axiom 5 (homotopy). If f and g are homotopic mappings of (X, A) into (Y, B), then for each p, f_{*p} and g_{*p} coincide.

Axiom 6 (excision). If U is an open subset of X whose closure $\overline{U}$ is contained in the interior of A, then the injection mapping of $(X - U, A - U)$

into (X, A) induces isomorphisms of $H_p(X - U, A - U)$ onto $H_p(X, A)$ for each p.

Axiom 7. If P is a space consisting of a single point, then $H_p(P) = 0$ for $p \neq 0$.

A proof that this system of axioms characterizes a homology theory is very lengthy. The Eilenberg-Steenrod book [7] contains not only this proof but a wealth of detail on the general problems of homology theory. We will merely note that each of these axioms has appeared either as a theorem or as a remark in our treatment of simplicial homology theory. It follows that at least these properties must be assumed for a homology theory. The remarkable fact is that these few axioms are enough. We leave this topic with the urgent advice, read Eilenberg and Steenrod!

7-8 Relative homotopy theory. In this section, we intend merely to call attention to the subject of relative homotopy theory. A few basic results are quoted without proof in hopes of arousing interest in the references that are given.

Consider a triple (X, A, x), where X is a space and A is a closed subspace of X containing the point x. Again we look at certain mappings of the unit cube I^n, $n \geqq 2$, into X, but now we do not insist, as we did in Section 4-7, that all of the boundary $\beta(I^n)$ of I^n map onto the point x. Let B^{n-1} be the set $[I^1 \times \beta(I^{n-1})] \cup (0 \times I^{n-1})$. That is, B^{n-1} is the boundary of I^n minus the *open* top face. We could take B^{n-1} to be the closure of $\beta(I^n) - (1 \times I^{n-1})$. We consider the function space $F_n(X, A, x)$ consisting of all mappings $f: I^n \to X$ such that $f(\beta(I^n))$ lies in A and $f(B^{n-1}) = x$. Note that all of $\beta(I^n)$ except the open top face maps onto the base point x. We use the compact-open topology in $F_n(X, A, x)$.

As in Section 4-7, we define $\pi_n(X, A, x)$ to be the collection of arcwise-connected components of $F_n(X, A, x)$, and we call it the nth homotopy group of X modulo A.

THEOREM 7-6. $\pi_n(X, A, x)$ is a group for $n \geqq 2$.

The juxtaposition of two elements of $F_n(X, A, x)$ is defined precisely as was done in Section 4-7, and the arguments establishing the group structure of $\pi_n(X, A, x)$ are almost identical to those for $\pi_n(X, x)$. Rather than give this proof, it might be more valuable to see why the theorem fails for $n = 1$. When $n > 1$, the set B^{n-1} and the top face $1 \times I^{n-1}$ intersect in I^n. But in the case $n = 1$, B^0 is just the point 0, whereas $0 \times I^0$ is the point 1, and these do not meet in I^2. Juxtaposing two mappings of I^n essentially means fastening the two together along the hyperplane $x_1 = 1$ of the first mapping and $x_1 = 0$ of the second. For $n > 1$, this can be done because B^{n-1} meets both these hyperplanes. But when

$n = 1$, the point 1 does not have to be mapped onto the point x, and we cannot fasten $g(0) = x$ to $f(1) \neq x$.

As was done for absolute homotopy, one may define $\pi_n(X, A, x)$ to be the fundamental group $\pi_1(F_{n-1}(X, A, x), e_x)$. Since $(F_{n-1}(X, A, x), e_x)$ can be shown to be a Hopf space whenever $n - 1 \geqq 2$ or $n \geqq 3$, the following result is implied by Theorem 4–18.

THEOREM 7–7. $\pi_n(X, A, x)$ is abelian for $n \geqq 3$.

That the relative homology group is a true generalization of the absolute homotopy group follows from the next lemma, whose proof is very easy.

LEMMA 7–8. If A is the single point x, then $\pi_n(X, A, x) = \pi_n(X, x)$.

The succeeding development is sketched to show the similarity of the two relative theories, homotopy and homology.

THEOREM 7–9. If $f:(X, A, x) \to (Y, B, y)$ is continuous, then there is an induced homomorphism $f_* : \pi_n(X, A, x) \to \pi_n(Y, B, y)$, $n \geqq 2$. If i is the identity mapping of (X, A, x) onto itself, then i_* is the identity isomorphism. If $f:(X, A, x) \to (Y, B, y)$ and $g:(Y, B, y) \to (Z, C, z)$, then $(gf)_* = g_* f_*$.

This result simply says that these groups and homomorphisms satisfy Axioms 1 and 2 in the Eilenberg-Steenrod axiom system.

For $n \geqq 2$, the *boundary function* $\partial : F_n(X, A, x) \to F_{n-1}(X, A, x)$, defined by $(\partial f)(t_1, t_2, \ldots, t_n) = f(1, t_2, \ldots, t_n)$, can be shown to induce a homomorphism ∂_* of $\pi_n(X, A, x)$ into $\pi_n(A, x)$, the absolute homology group of A modulo x. Then the following can be established.

THEOREM 7–10. If $f:(X, A, x) \to (Y, B, y)$ is continuous, then $\partial_* f_* = (f|A)_* \partial_*$.

Thus the Eilenberg-Steenrod Axiom 3 is satisfied. Furthermore, although it is understandably more difficult to do, the fourth axiom can also be proved as a theorem in relative homotopy theory.

THEOREM 7–11. Let $i:(A, x) \to (X, x)$ and $j:(X, x, x) \to (X, A, x)$ be the identity injection mappings. Then the sequence

$$\cdots \to \pi_{n+1}(X, A, x) \xrightarrow{\partial_*} \pi_n(A, x) \xrightarrow{i_*} \pi_n(X, x) \xrightarrow{j_*} \pi_n(X, A, x) \to \cdots$$

is exact.

It will be noted that we stop short of the excision axiom here. In fact, the excision theorem is not true for relative homotopy theory. The lack of this property seems to be the chief difficulty in computing homology groups. Such a difficulty is always stimulating, of course, and much effort is now being put into a study of homotopy theory. For a clear exposition of this important topic, the reader is referred to the excellent books by Hilton [13] and Hu [14(a)] where many further references will be found.

7-9 Cohomology groups. Certain duality theorems in the homology theory of manifolds were discovered early and seemed to reflect the existence of a theory dual to homology theory. Although the genesis of this dual theory, now called *cohomology*, is cloudy, it occurred during the decade 1925–1935 concurrently with the change in emphasis away from the numerical invariants (Betti numbers and torsion coefficients) toward the group structures. Lefschetz [19] was the first to use cocycles under the name *pseudocycle* and the *co-* terminology was introduced by Whitney [131] in 1938. Pontrjagin [115] laid the algebraic foundations for the duality theorems. Other founders of the theory include Alexander, Alexandroff, Čech, and Vietoris.

We will not give a preliminary intuitive explanation of cohomology theory, but we will try to clarify the development as it proceeds. In this section, we construct the cohomology groups of a simplicial complex. While doing so, we review homology theory as well so that we may exhibit the many parallels between the two theories.

Let K be an oriented abstract simplicial complex (see Section 6-2). The orientation of K permits the definition of the incidence numbers $[\sigma^p, \sigma^{p-1}]$, and we recall Theorem 6-1, which states that for a fixed n-simplex σ_0^n, $n > 1$,

$$\sum_{i,j} [\sigma_0^n, \sigma_i^{n-1}][\sigma_i^{n-1}, \sigma_j^{n-2}] = 0.$$

We define an *integral p-chain* c_p to be a function from the oriented p-simplexes of K to the integers, which is nonzero for at most a finite number of p-simplexes and which satisfies the condition

$$c_p(-\sigma^p) = -c_p(\sigma^p).$$

The p-chain which has the value $+1$ on a particular simplex σ_0^p (and value -1 on $-\sigma_0^p$, of course) and zero elsewhere is called an *elementary p-chain* and is denoted by $1 \cdot \sigma_0^p$. With this definition, one may write an arbitrary p-chain c_p as a formal polynomial

$$c_p = \sum^f \eta_i \sigma_i^p,$$

where the η_i are integers and the superscript f denotes a finite sum. Chains are added in the natural way (componentwise if one thinks of them as sums, by functional addition if one thinks of them as functions) and hence form a free group $C_p(K)$, the p-dimensional integral chain group of K. Allowing the chains to have values in an abelian group G, we would obtain the groups $C_p(K, G)$ in the same manner.

There is a very slight generalization when we define cochains. An *integral p-cochain* c^p is an arbitrary (not necessarily zero almost everywhere) function from the oriented p-simplexes of K to the integers, satisfying the condition

$$c^p(-\sigma^p) = -c^p(\sigma^p).$$

Considering an elementary cochain to be the same as an elementary chain, a p-cochain may be written as a possibly infinite linear combination of p-simplexes with integral coefficients. Thus we may consider that every chain is a cochain but not conversely. In a finite complex, the two concepts are identical.

The addition of cochains is done in the natural manner either by components or by functional addition. We thus obtain the group $C^p(K)$, the *p-dimensional integral cochain group* of K. Here too it can be assumed that the values of the cochains are in some arbitrary abelian group G, and we thereby obtain the groups $C^p(K, G)$. As is true for any abelian group, the groups $C^p(K, G)$ admit of the integers as a ring of operators, that is, $C^p(K, G)$ is a module over the integers. Note that if there are no p-simplexes in K, then we set $C^p(K) = C^p(K, G) = 0$.

As mentioned above, the two groups $C^p(K)$ and $C_p(K)$ are identical in the case of a finite complex K. If K is an infinite complex, however, the cochain group $C^p(K)$ is the direct sum of infinitely many infinite cyclic groups, whereas $C_p(K)$ is the *weak* direct sum. Thus, in general, the chain group is a subgroup of the cochain group.

We recall that for an elementary chain $1 \cdot \sigma_0^p$, the boundary operator ∂ is defined by

$$\partial(1 \cdot \sigma_0^p) = \sum_i [\sigma_0^p, \sigma_i^{p-1}]\sigma_i^{p-1},$$

where $[\sigma_0^p, \sigma_i^{p-1}]$ is an incidence number. Since every simplex has only a finite number of faces, $\partial(1 \cdot \sigma_0^p)$ is a finite sum and is a $(p - 1)$-chain. Also the chain $\partial(1 \cdot \sigma_0^p)$ depends only upon the simplex σ_0^p and not upon the complex K in which σ_0^p is located.

The above definition is extended linearly to arbitrary p-chains by means of the formula

$$\partial c_p = \partial\left(\sum{}^f \eta_i \sigma_i^p\right) = \sum{}^f \eta_i \cdot \partial(1 \cdot \sigma_i^p).$$

Since the sums are finite, the result is a $(p - 1)$-chain. Again we remark that ∂c_p depends only upon the chain c_p and not upon the complex K. Finally, using the property expressed in Theorem 6–1, we have already shown that for any chain c_p,

$$\partial(\partial c_p) = 0.$$

In an analogous manner, we now define the *coboundary operator* δ. First, for an elementary cochain $1 \cdot \sigma_0^p$, we define

$$\delta(1 \cdot \sigma_0^p) = \sum_i [\sigma_i^{p+1}, \sigma_0^p] \cdot \sigma_i^{p+1},$$

where $[\sigma_i^{p+1}, \sigma_0^p]$ is an incidence number. This says that the coboundary of $1 \cdot \sigma_0^p$ is a function assigning nonzero coefficients only to those $(p+1)$-simplexes that have σ_0^p as a face. But this implies that $\delta(1 \cdot \sigma_0^p)$ depends not only upon σ_0^p but on how σ_0^p lies in the complex K. This is a fundamental difference between the two operators ∂ and δ. Furthermore, it is possible that σ_0^p is a face of infinitely many $(p+1)$-simplexes. Thus $\delta(1 \cdot \sigma_0^p)$ is not necessarily finite even though it is a cochain.

Again the coboundary operator is extended linearly to arbitrary cochains by setting

$$\delta(c^p) = \delta\left(\sum_i \eta_i \cdot \sigma_i^p\right) = \sum \eta_i \, \delta(1 \cdot \sigma_i^p).$$

This is a $(p+1)$-cochain, of course, and depends upon the complex K as well as the cochain c^p.

We remark that we are following current practice in using subscripts to indicate the dimension of chains and superscripts to give the dimension of cochains. This may be construed as a mnemonic device, the subscript on a chain reminding us that ∂ lowers dimension while the superscript recalls that δ raises dimension.

Theorem 6–1 may also be used to show that the boundary operator is of order 2.

THEOREM 7–12. For every integral cochain c^p, $\delta(\delta c^p) = 0$.

Proof: It suffices to prove this for an elementary cochain $1 \cdot \sigma_0^p$. To do so, consider

$$\delta(1 \cdot \sigma_0^p) = \sum_i [\sigma_i^{p+1}, \sigma_0^p]\sigma_i^{p+1}$$

and its coboundary

$$\begin{aligned}\delta\left(\sum_i [\sigma_i^{p+1}, \sigma_0^p]\sigma_i^{p+1}\right) &= \sum_i [\sigma_i^{p+1}, \sigma_0^p] \cdot \delta\,(1 \cdot \sigma_i^{p+1}) \\ &= \sum_i [\sigma_i^{p+1}, \sigma_0^p] \left(\sum_j [\sigma_j^{p+2}, \sigma_i^{p+1}] \cdot \sigma_j^{p+2}\right) \\ &= \sum_{i,j} [\sigma_i^{p+1}, \sigma_0^p][\sigma_j^{p+2}, \sigma_i^{p+1}] \cdot \sigma_j^{p+2}.\end{aligned}$$

By an argument similar to that for Theorem 6–1, each coefficient in $\delta\delta(1 \cdot \sigma_0^p)$ is zero. □

From this point on, we will concentrate on cohomology theory. In the cochain group $C^p(K)$, we have the usual pair of subgroups $Z^p(K)$, the kernel of δ (or *the group of integral p-cocycles on* K), and $B^p(K) = \delta C^{p-1}(K)$, *the group of integral p-coboundaries on* K. The relation $\delta\delta = 0$ implies that $B^p(K)$ is a subgroup of $Z^p(K)$. Since both $Z^p(K)$ and $B^p(K)$ are abelian groups, we may form the difference group $H^p(K) = Z^p(K) - B^p(K)$, *the p-dimensional integral cohomology group of* K. Obviously, following the same route and using a coefficient group G, we may construct the corresponding groups $H^p(K, G)$.

If $z_1^p - z_2^p = \delta c^{p-1}$, that is, if z_1^p and z_2^p are in the same element of $H^p(K)$, then we say that z_1^p is cohomologous to z_2^p and write $z_1^p \backsim z_2^p$. (Recall that the homology relation is symbolized by $\sim$.)

EXERCISE 7–9. Show that a 0-cochain is a cocycle if and only if it assigns the same value to each vertex in a combinatorial component of K. Hence prove that $H^0(K) = Z^0(K)$ is a free group on $p_0(K)$ generators.

EXERCISE 7–10. If K is any complex and $K^{(k)}$ is its k-skeleton, prove that

$$H^p(K) = H^p(K^{(k)}) \qquad (0 \leqq p < k).$$

What can be said about $H^k(K^{(k)})$ in relation to $H^k(K)$?

7–10 Relations between chain and cochain groups. Perhaps the best approach to an understanding of cocycles and cohomology theory is to study the relationships between chains and cochains. We defined a cochain to be a function on simplexes, but by using linear extension again we can and will consider a cochain to be a function on integral chains. Actually, we may take a cochain to be a homomorphism of the integral chain group $C_p(K)$ into the group of coefficients G of the cochain group $C^p(K, G)$. For let c^p be a cochain with coefficients in an abelian group G. That is, we may write $c^p = \sum g_i \sigma_i^p$ or $c^p(\sigma_i^p) = g_i$, where each g_i is in G. Let $d_p = \sum^f \eta_j \sigma_j^p$ be an integral chain. We may then define the *value of* c^p *on* d_p by

$$c^p(d_p) = c^p\left(\sum_j^f \eta_j \sigma_j^p\right) = \sum_j \eta_j c^p(1 \cdot \sigma_j^p) = \sum_j \eta_j \cdot g_j.$$

Clearly, $\sum_j \eta_j \cdot g_j$ is an element of G since the multiplication of a group element g_j by an integer η_j has meaning.

For a fixed cochain c^p, this operation yields a homomorphism of $C_p(K)$ into G. Furthermore, every homomorphism of $C_p(K)$ into G can be obtained in this manner. (The proofs of both these statements are left as exercises.) The natural addition of homomorphisms is precisely the addition of cochains, and we therefore know that the group $C^p(K, G)$ is the group $\mathrm{Hom}(C_p(K), G)$ of homomorphisms of $C_p(K)$ into G. See Section 7–16.

In place of the functional notation $c^p(d_p)$, it is often convenient to use a product notation. That is, we will write

$$c^p(d_p) = c^p \cdot d_p.$$

The result of this "product" is called the *Kronecker index*, $\mathrm{KI}(c^p, d_p)$. In fact, this product is actually a pairing of the groups $C^p(K, G)$ and $C_p(K)$ to the group G.

In the product notation above, the basic relation between the boundary operator ∂ and the coboundary operator δ is expressed in the following result.

THEOREM 7-13. Let c^{p-1} be any element of $C^{p-1}(K, G)$ and d_p any element of $C_p(K)$. Then $\delta c^{p-1} \cdot d_p = c^{p-1} \cdot \partial d_p$.

Proof: We need only consider an elementary chain $1 \cdot \sigma_0^p$. If $c^{p-1} = \sum_i g_i \sigma_i^{p-1}$, we have, by definition,

$$\begin{aligned}\delta c^{p-1} \cdot (1 \cdot \sigma_0^p) &= \delta\left(\sum_i g_i \sigma_i^{p-1}\right) \cdot (1 \cdot \sigma_0^p) \\ &= \left(\sum_i g_i\, \delta\sigma_i^{p-1}\right) \cdot (1 \cdot \sigma_0^p) \\ &= \left[\sum_i g_i \left(\sum_j [\sigma_j^p, \sigma_i^{p-1}]\sigma_j^p\right)\right] \cdot (1 \cdot \sigma_0^p) \\ &= \sum_i g_i[\sigma_0^p, \sigma_i^{p-1}],\end{aligned}$$

and, on the other hand,

$$\begin{aligned}c^{p-1} \cdot \partial(1 \cdot \sigma_0^p) &= c^{p-1} \cdot \left(\sum_j [\sigma_0^p, \sigma_j^{p-1}] \cdot \sigma_j^{p-1}\right) \\ &= \left(\sum_i g_i \sigma_i^{p-1}\right) \cdot \left(\sum_j [\sigma_0^{p-1}, \sigma_j^{p-1}] \cdot \sigma_j^{p-1}\right) \\ &= \sum_j g_j[\sigma_0^p, \sigma_j^{p-1}]. \square\end{aligned}$$

The relation given in the above theorem can be interpreted as saying that δ and ∂ are *adjoint operators*, and it could be taken as a definition of δ. In the study of differentiable manifolds [14] one comes to recognize the relation $\delta c^{p-1} \cdot d_p = c^{p-1} \cdot \partial d_p$ as a combinatorial form of Stokes' theorem. We note that for elementary chains $1 \cdot \sigma^{p-1}$ and $1 \cdot \sigma^p$, this formula reduces to

$$\delta(1 \cdot \sigma^{p-1}) \cdot (1 \cdot \sigma^p) = (1 \cdot \sigma^{p-1}) \cdot \partial(1 \cdot \sigma^p) = [\sigma^p, \sigma^{p-1}],$$

the incidence number.

As an immediate consequence of Theorem 7–13, we have the next result.

COROLLARY 7–14.

$$(\text{coboundary}) \cdot (\text{cycle}) = 0;$$

$$(\text{cocycle}) \cdot (\text{boundary}) = 0.$$

Proof: If b^p is a coboundary and z_p is a cycle, then $b^p = \delta c^{p-1}$ and $\partial z_p = 0$. This implies that

$$b^p \cdot z_p = \delta c^{p-1} \cdot z_p = c^{p-1} \cdot \partial z_p = c^{p-1} \cdot 0 = 0.$$

And if z^p is a cocycle and b_p is a boundary, then $\delta z^p = 0$ and $b_p = \partial c_{p+1}$, whence

$$z^p \cdot b_p = z^p \cdot \partial c_{p+1} = \partial z^p \cdot c_{p+1} = 0 \cdot c_{p+1} = 0. \square$$

COROLLARY 7–15. The Kronecker index induces a pairing of $H^p(K, G)$ and $H_p(K)$ to G.

Proof: The product of a cocycle and a cycle depends only upon the cohomology and homology classes, respectively. For suppose that $z_1^p \backsim z_2^p$ and $z_p^1 \sim z_p^2$. Then we have

$$z_1^p = z_2^p + \delta c_1^{p-1} \quad \text{and} \quad z_p^1 = z_p^2 + \partial c_{p+1}^2.$$

Then

$$\begin{aligned} z_1^p \cdot z_p^1 &= (z_2^p + \delta c_1^{p-1}) \cdot (z_p^2 + \delta c_{p+1}^2) \\ &= z_2^p \cdot z_p^2 + \delta c_1^{p-1} \cdot z_p^2 + z_2^p \cdot \delta c_{p+1}^2 + \delta c_1^{p+1} \cdot \delta c_{p+1}^2. \end{aligned}$$

Then Corollary 7–14 applies to show that the last three terms on the right-hand side are zero. Hence $z_1^p \cdot z_p^1 = z_2^p \cdot z_p^2$. $\square$

As an example of the use of cohomology theory, consider the following situation. We use the coefficient group Z_2 of integers mod 2. Let K be a triangulated surface, and choose an "orientation at each vertex"; that is, choose a sense on a small circle around each vertex. For each edge $\sigma^1 = \langle v_0 v_1 \rangle$, define

$c(\sigma^1) = 0$ if the orientations at v_0 and v_1 agree in the obvious meaning of the phrase

$ = 1$ if the orientations at v_0 and v_1 disagree.

This defines a 1-cochain c with coefficients in Z_2. We show that c is a cocycle. Consider δc on any particular 2-simplex $\sigma^2 = \langle v_0 v_1 v_2 \rangle$:

$$\delta c(\langle v_0 v_1 v_2 \rangle) = c(\partial \langle v_0 v_1 v_2 \rangle) = c(\langle v_0 v_1 \rangle) + c(\langle v_0 v_2 \rangle) + c(\langle v_1 v_2 \rangle)$$

(the signs are all positive in mod 2 theory). Of the three orientations at v_0, v_1, and v_2, either all three agree, or two agree and disagree with the third. In either case, we have that $c(\langle v_0 v_1 \rangle) + c(\langle v_0 v_2 \rangle) + x(\langle v_0 v_2 \rangle) = 0 \bmod 2$.

If we were to change the orientation at some of the vertices, the change could be described by means of a function α on the vertices (a 0-chain) by setting

$$\begin{aligned} \alpha(v) &= 0 \quad \text{if the orientation is not changed} \\ &= 1 \quad \text{if the orientation is reversed.} \end{aligned}$$

With this new orientation, we form a 1-cocycle c' as we formed c above. It is easily seen that for any edge $\sigma^1 = \langle v_0 v_1 \rangle$,

$$c'(\sigma^1) = c(\sigma^1) + \alpha(v_0) + \alpha(v_1).$$

Furthermore, we have $\delta\alpha(\langle v_0 v_1 \rangle) = \alpha(v_0) + \alpha(v_1)$. Hence

$$c' = c + \delta\alpha,$$

or c and c' are in the same cohomology class. We may call this the *orientation class* of K. If c'' is another element of this class, then $c'' = c + \delta\beta$, and we may obtain c'' by a reorientation β at the vertices. The use of this concept is embodied in the following exercises.

EXERCISE 7-11. Prove that the surface K is orientable or not, depending upon whether the orientation cohomology class is zero or not.

EXERCISE 7-12. If $c(z_1) = 0$ for every integral cycle z_1 on K, then prove that K is orientable. (This means intuitively that if the orientation does not change around any closed path, then the surface is orientable.)

7-11 Simplicial and chain-mappings.

Given a simplicial mapping φ of a complex K_1 into a complex K_2, we saw in Section 6-10 that φ induces chain-mappings φ_p of $C_p(K_1, G)$ into $C_p(K_2, G)$, and we proved the commutative property

$$\varphi_{p-1} \partial c_p = \partial \varphi_p c_p.$$

This led to the induced homomorphisms φ_{p*} of $H_p(K_1, G)$ into $H_p(K_2, G)$. We now give the analogous situation for cohomology theory. To be precise, we show that the chain-mapping φ_p induces a mapping φ^p of the group $C^p(K_2, G)$ into $C^p(K_1, G)$. (Note that φ^p is opposite in direction to φ_p.) We may properly call φ^p the *adjoint* of φ_p. It is defined as follows. If c^p is a cochain in $C^p(K_2, G)$, then $\varphi^p c^p$ is that cochain on K_1 whose value on an elementary chain $g_0 \cdot \sigma_0^p$ is given by $c^p \cdot \varphi_p(g_0 \cdot \sigma_0^p) = c^p(\varphi_p(g_0\sigma_0^p))$. By linear extension, we have the formula

$$\varphi^p c^p(d^p) = c^p(\varphi_p(d_p)).$$

In the product notation we write

$$\varphi^p c^p \cdot d_p = c^p \cdot \varphi_p(d_p).$$

The next result states the necessary commutative property.

LEMMA 7–16. For any cochain c^p in $C^p(K_2, G)$,

$$\varphi^{p+1}\delta c^p = \delta\varphi^p c^p.$$

Proof: Let d_{p+1} be any chain on K_1. Then we have

$$\begin{aligned}\varphi^{p+1}\delta c^p(d_{p+1}) &= \delta c^p(\varphi_{p+1} d_{p+1})\\ &= c^p(\partial\varphi_{p+1} d_{p+1})\\ &= c^p(\varphi_p \partial d_{p+1})\\ &= \varphi^p c^p(\partial d_{p+1})\\ &= \delta\varphi^p c^p(d_{p+1}). \square\end{aligned}$$

The commutative property given in the above lemma is expressed symbolically as $\varphi\delta = \delta\varphi$ and is applied in cohomology theory just as the relation $\varphi\partial = \partial\varphi$ is applied in homology theory. First, the image under φ^p of a cocycle on K_2 is a cocycle on K_1, and the image under φ^p of a coboundary on K_2 is a coboundary on K_1. These facts are easily checked, for if c^p is a cocycle on K_2, then $\delta c^p = 0$. But $\delta\varphi^p c^p = \varphi^{p+1}\delta c^p = \varphi^{p+1}(0) = 0$, so $\varphi^p c^p$ is a cocycle on K_1. And if $b^p = \delta c^{p-1}$, then $\varphi^p b^p = \varphi^p \delta c^{p-1} = \delta\varphi^{p-1} c^{p-1}$; that is, $\varphi^p b^p$ is the coboundary of $\varphi^{p-1}c^{p-1}$.

It follows that φ^p induces a homomorphism φ^{p^*} of $H^p(K_2, G)$ into $H^p(K_1, G)$ defined by

$$\varphi^{p^*}([c^p]) = [\varphi^p c^p].$$

The homomorphism φ^{p^*} on the cohomology groups is adjoint to the homomorphism φ_{p_*} on homology groups in the sense that, if $[c^p]$ is an element of $H^p(K_2,G)$ and $[z_p]$ is an element of $H_p(K_1, G)$, then we have

$$[c^p] \cdot \varphi_{p*}[z_p] = \varphi^{p^*}[c^p] \cdot [z_p].$$

LEMMA 7–17. If $\varphi:K_1 \to K_2$ and $\psi:K_2 \to K_3$ are simplicial mappings (or chain-mappings), then the composite mapping $\psi\varphi$ is a simplicial mapping (chain-mapping) of K_1 into K_3, and the induced homomorphisms satisfy

$$(\psi\varphi)^* = \varphi^*\psi^*.$$

This may be proved by direct computation, and the proof is left as an exercise.

In Section 6–11, we introduced the concept of a chain-homotopy, but we repeat it here for convenience. Let φ and ψ be two chain-mappings of K_1 into K_2. Then φ and ψ are chain-homotopic if there exists a deformation operator $\mathfrak{D} = \{\mathfrak{D}_p\}$, a collection of homomorphisms of the integral chain groups $\mathfrak{D}_p:C_p(K_1) \to C_{p+1}(K_2)$, such that for each p,

$$\partial\mathfrak{D}_pc_p = \psi_pc_p - \varphi_pc_p - \mathfrak{D}_{p-1}\partial c_p.$$

Now in view of the foregoing use of adjoint mappings, it is natural to define an adjoint operator $\overline{\mathfrak{D}} = \{\mathfrak{D}^p\}$, where $\mathfrak{D}^p:C^p(K_2) \to C^{p-1}(K_1)$, by means of the formula

$$\mathfrak{D}^pc^p(1 \cdot \sigma^{p-1}) = c^p(\mathfrak{D}_{p-1}(1 \cdot \sigma^{p-1})),$$

for an elementary chain $1 \cdot \sigma^{p-1}$ and to extend this linearly to arbitrary chains as usual.

LEMMA 7–18. For any cochain c^p on K_2,

$$\delta\mathfrak{D}^pc^p = \psi^pc^p - \varphi^pc^p - \mathfrak{D}^{p+1}\delta c^p.$$

Proof: Let $1 \cdot \sigma^p$ be an elementary chain on K_1. Then

$$\begin{aligned}
\delta\mathfrak{D}^pc^p(1 \cdot \sigma^p) &= \mathfrak{D}^pc^p(\partial(1 \cdot \sigma^p)) \\
&= c^p(\mathfrak{D}_{p-1}\partial(1 \cdot \sigma^p)) \\
&= c^p(\psi_p(1 \cdot \sigma^p) - \varphi_p(1 \cdot \sigma^p) - \partial\mathfrak{D}_p(1 \cdot \sigma^p)) \\
&= \psi^pc^p(1 \cdot \sigma^p) - \varphi^pc^p(1 \cdot \sigma^p) - c^p(\partial\mathfrak{D}_p(1 \cdot \sigma^p)) \\
&= \psi^pc^p(1 \cdot \sigma^p) - \varphi^pc^p(1 \cdot \sigma^p) - \delta c^p(\mathfrak{D}_p(1 \cdot \sigma^p)) \\
&= \psi^pc^p(1 \cdot \sigma^p) - \varphi^pc^p(1 \cdot \sigma^p) - \mathfrak{D}^{p+1}\delta c^p(1 \cdot \sigma^p).
\end{aligned}$$

Linear extension then completes the proof. □

Finally, using arguments analogous to those in Section 6–11, we may establish the next result.

THEOREM 7–19. If φ and ψ are chain-homotopic chain-mappings of K_1 into K_2, then the induced homomorphisms φ^* and ψ^* coincide.

EXERCISE 7–13. Prove Theorem 7–19.

EXERCISE 7–14. Define the degree of a simplicial mapping of one orientable n-pseudomanifold into another in terms of cohomology classes, and prove that the two definitions give equal degrees.

7–12 The cohomology product. There is an algebraic structure, the *cohomology ring,* which appears in cohomology theory but has no analogue in homology theory. This concept is based upon a "multiplication" of cochains and has interesting consequences in the theory of manifolds. We do not discuss these applications here but simply present the structure and refer the reader to Whitney in [44].

Let the coefficients of the cochains on a complex K be taken to be elements of a ring R with unit element 1 (the integers, for example). We define the cocycle e^0 given by $e^0(v) = 1$ for each vertex v of K. It turns out that e^0 is the identity element of the cohomology ring of K.

The complex K will be oriented, as we have done before, by adopting a simple-ordering of its vertices, and all simplexes will be written in this ordering. That is, if we write $\sigma^p = \langle v_0 \cdots v_p \rangle$, it is implied that $v_0 < v_1 < \cdots < v_p$ in the given ordering.

Now let c^p and c^q be cochains on K. We define the *cohomology product,* or *cup-product,* $c^p \cup c^q$ to be the $(p+q)$-cochain whose values are determined by the formula

$$c^p \cup c^q(\langle v_0 \cdots v_p v_{p+1} \cdots v_{p+q} \rangle) = c^p(\langle v_0 \cdots v_p \rangle) \cdot c^q(\langle v_p \cdots v_{p+q} \rangle),$$

where the product on the right is multiplication in the ring R.

The cup-product satisfies the following five properties:

$$c^p \cup c^q \text{ is a bilinear function.} \tag{1}$$

$$(c^p \cup c^q) \cup c^r = c^p \cup (c^q \cup c^r). \tag{2}$$

$$c^p \cup e^0 = c^p. \tag{3}$$

$$e^0 \cup c^p = c^p. \tag{4}$$

$$\delta(c^p \cup c^q) = \delta c^p \cup c^q + (-1)^p c^p \cup \delta c^q. \tag{5}$$

We will see that the first four properties permit us to construct, from the collection of cochain groups, a ring with unit element e^0. Each of the first four properties is easily verified, and these verifications are left as exercises. The fifth property may be expressed by saying that the coboundary operator δ is an "antidifferentiation." To verify this property, consider the value of the $(p + q + 1)$-cochain $\delta(c^p \cup c^q)$ on a particular simplex $\langle v_0 \cdots v_{p+q+1}\rangle$. Straightforward computation results in

$$\begin{aligned}
\delta(c^p \cup c^q)\langle v_0 \cdots v_{p+q+1}\rangle &= c^p \cup c^q(\partial\langle v_0 \cdots v_{p+q+1}\rangle) \\
&= c^p \cup c^q\left(\sum_{j=0}^{p+q+1} (-1)^j\langle v_0 \cdots \hat{v}_j \cdots v_{p+q+1}\rangle\right) \\
&= \sum_{j=0}^{p+1} (-1)^j c^p(\langle v_0 \cdots \hat{v}_j \cdots v_{p+1}\rangle) \cdot c^q(\langle v_{p+1} \cdots v_{p+q+1}\rangle) \\
&\quad + \sum_{j=p}^{p+q+1} (-1)^j c^p(\langle v_0 \cdots v_p\rangle) \cdot c^q(\langle v_p \cdots \hat{v}_j \cdots v_{p+q+1}\rangle) \\
&= c^p(\partial\langle v_0 \cdots v_{p+1}\rangle) \cdot c^p(\langle v_{p+1} \cdots v_{p+q+1}\rangle) \\
&\quad + (-1)^p c^p(\langle v_0 \cdots v_p\rangle) \cdot c^q(\partial\langle v_p \cdots v_{p+q+1}\rangle) \\
&= \delta c^p(\langle v_0 \cdots v_{p+1}\rangle) \cdot c^q(\langle v_{p+1} \cdots v_{p+q+1}\rangle) \\
&\quad + (-1)^p c^p(\langle v_0 \cdots v_p\rangle) \cdot \delta c^q(\langle v_p + \cdots + v_{p+q+1}\rangle) \\
&= (\delta c^p \cup c^q + (-1)^p c^p \cup \delta c^q)(\langle v_0 \cdots v_{p+q+1}\rangle).
\end{aligned}$$

The fifth property, just proved, is important because of the following consequences.

LEMMA 7-20.

$$\begin{aligned}
\text{(cocycle)} \cup \text{(cocycle)} &= \text{cocycle} \\
\text{(cocycle)} \cup \text{(coboundary)} &= \text{coboundary} \\
\text{(coboundary)} \cup \text{(cocycle)} &= \text{coboundary}
\end{aligned}$$

The proof of this lemma is very similar to that of Corollary 7-14 and is left as an exercise.

THEOREM 7-21. The cohomology class of the cup-product of two cocycles depends only upon the cohomology classes of the two factors.

Proof: Let $z_1^p = z_2^p + \delta c_1^{p-1}$ and $z_1^q = z_2^q + \delta c_2^{q-1}$.

Then
$$\begin{aligned}
z_1^p \cup z_1^q &= (z_2^p + \delta c_1^{p-1}) \cup (z_2^q + \delta c_2^{q-1}) \\
&= z_2^p \cup z_2^q + \delta c_1^{p-1} \cup z_2^q + z_2^p \cup \delta c_2^{q-1} + \delta c_1^{p-1} \cup \delta c_2^{q-1}.
\end{aligned}$$

Each of the last three terms on the right is a coboundary by Lemma 7-20. □

We may now construct the *cohomology ring* $\mathcal{R}(K, R)$ of the complex K with coefficients in the ring R. A ring, we recall, is an additively written abelian group which is closed under an associative binary operation of multiplication, and the multiplication is distributive with respect to addition. To form the cohomology ring, we simply take the direct sum of all the cohomology groups $H^p(K, R)$ and use the cup-product as the operation of multiplication. It is merely a routine verification of the definition to show that $\mathcal{R}(K, R)$ is a ring with unit element e^0.

There is one point that remains to be investigated here. The cup-product on cochains was defined in terms of a particular ordering of the vertices of the complex K. This, of course, does not yield a topologically invariant definition, and we cannot force it to do so. However, we can and will show that the induced cup-product on *cohomology classes* is independent of the ordering of the vertices of K and hence that the cohomology ring is well-defined. We first note that, in defining the cup-product on cochains, we really used only an ordering in the small, that is, an ordering of the vertices of each simplex of K which is consistent in the sense that the ordering induced on any face of a simplex by the ordering of the simplex agrees with the ordering given on the face. We then point out that there is a natural ordering in the small of the barycentric subdivision K' of K, the vertices of a simplex of K' being ordered according to the dimensions of their carriers. It is this ordering of K' in the small which we adopt.

Digressing a moment, we may observe that if K_1 and K_2 are two complexes, each with vertices ordered in the small, and if φ is a simplicial mapping of K_1 into K_2 which preserves the ordering (does not invert it), then it is easily proved that the homomorphism $\varphi^p{:}C^p(K_2, R) \to C^p(K_1, R)$ preserves the cup-product of cochains and therefore induces a homomorphism of the cohomology ring $\mathcal{R}(K_2, R)$ into $\mathcal{R}(K_1, R)$.

Returning to the complex K and its barycentric subdivision K', we find that there is a natural simplicial mapping $u'{:}K' \to K$ (see Section 6–13). As defined, u' has certain arbitrary choices involved; if $\dot{\sigma}$ is a vertex of K', we set $u'(\dot{\sigma})$ equal to any vertex of the carrier σ. We can avoid this difficulty by always choosing $u'(\dot{\sigma})$ to be the highest vertex of σ in the ordering assigned to σ. It may be readily shown that if we so define u', then u' preserves ordering in the small. As remarked above, the induced homomorphism $(u')^*$ preserves the cup-product of cochains and hence induces a homomorphism of the cohomology ring $\mathcal{R}(K, R)$ into $\mathcal{R}(K', R)$.

In Section 6–13, we showed that the induced mapping u'_* of $H_p(K', R)$ into $H_p(K, R)$ is an isomorphism onto. It is evident that the adjoint mapping $(u')^*$ is also an isomorphism of $H^p(K, R)$ onto $H^p(K', R)$. Furthermore, the chain-mapping $u{:}C_p(K, R) \to C_p(K', R)$ (again see Section 6–13) induces an isomorphism u_* of $H_p(K, R)$ onto $H_p(K', R)$, so the adjoint homomorphism $u^*{:}H^p(K', R) \to H^p(K, R)$ is an isomorphism onto. Clearly, u and hence u^* do not depend upon an ordering of the vertices of K. Thus the ring $\mathcal{R}(K, R)$, based upon any ordering in the small of the vertices of K, is isomorphic to the ring $\mathcal{R}(K', R)$, which does not depend upon the ordering of K. Therefore $\mathcal{R}(K, R)$ is actually independent of the ordering of the vertices of K.

THEOREM 7–22. If φ is a simplicial mapping of a complex K_1 into a complex K_2, then φ^* induces (or is) a homomorphism of the cohomology ring $\mathcal{R}(K_2, R)$ into $\mathcal{R}(K_1, R)$.

Proof: We know that if φ preserves the ordering of vertices in the small, then φ^* is such a homomorphism. We use the fact that the cohomology rings do not depend upon the ordering of vertices and simply construct orderings in K_1 and K_2 that are preserved by φ. To do so, choose any simple-ordering of the vertices of K_2. We then order the inverses $\varphi^{-1}(w_i)$, where w_i is a vertex of K_2, just as the vertices w_i are ordered in K_2, and if $\varphi^{-1}(w_i)$ contains more than one vertex, then we choose any simple-ordering of $\varphi^{-1}(w_i)$. It is easily seen that such an ordering of K_1 is preserved by φ. □

The cohomology ring $\mathcal{R}(K, R)$ is almost but not quite a commutative ring. If the ring R of coefficients is commutative, then, *for cohomology classes only*, we have

$$a^p \cup b^q = (-1)^{pq} b^q \cup a^p.$$

We repeat that this relation holds only for cohomology classes a^p and b^q and does not hold for cochains! This relation is known as the Grassman property, and it means that if R is commutative, the $\mathcal{R}(K, R)$ is a *Grassman ring*. To establish this property, let the vertices of K be ordered in the small in any fixed way. There is a cup-product $\cup$ based upon this ordering, and there is a cup-product $\cup'$ based upon the opposite or negative ordering of the vertices of each simplex. In view of Theorem 7–22, the product of two cohomology classes is independent of ordering. Hence if z^p and z^q are cocycles, then $z^p \cup z^q$ and $z^p \cup' z^q$ are cohomologous.

Let $\sigma^{p+q} = \langle v_0 \cdots v_{p+q}\rangle$ be a $(p+q)$-simplex written in the first ordering. For the second ordering we write $\langle v_{p+q} \cdots v_0\rangle'$, the prime being used merely to denote that the simplex is ordered in the opposite of the given ordering. By definition, we have

$$z^q \cup' z^p(\langle v_{p+q} \cdots v_0\rangle') = z^q(\langle v_{p+q} \cdots v_p\rangle') \cdot z^p(\langle v_p \cdots v_0\rangle').$$

We also know that

$$\langle v_{p+q} \cdots v_0\rangle' = (-1)^{\frac{1}{2}(p+q)(p+q+1)}\langle v_0 \cdots v_{p+q}\rangle$$

since it will take exactly $\frac{1}{2}(p+q)(p+q+1)$ interchanges of the $p+q$ vertices to reverse their order. Similarly,

$$\langle v_{p+q} \cdots v_p\rangle' = (-1)^{\frac{1}{2}q(q+1)}\langle v_p \cdots v_{p+q}\rangle$$

and
$$\langle v_p \cdots v_0 \rangle' = (-1)^{\frac{1}{2}p(p+1)} \langle v_0 \cdots v_p \rangle.$$

Hence we may simply compute the result. First,

$$z^q \cup' z^p(\langle v_{p+q} \cdots v_0 \rangle') = (-1)^{\frac{1}{2}(p+q)(p+q+1)} z^q \cup' z^p(\langle v_0 \cdots v_{p+q} \rangle),$$

$$z^q(\langle v_{p+q} \cdots v_p \rangle') = (-1)^{\frac{1}{2}(q)(q+1)} z^q(\langle v_p \cdots v_{p+q} \rangle),$$

and

$$z^p(\langle v_p \cdots v_0 \rangle') = (-1)^{\frac{1}{2}(p)(p+1)} z^p(\langle v_0 \cdots v_p \rangle).$$

This implies then that

$$(-1)^{\frac{1}{2}(p+q)(p+q+1)} z^q \cup' z^p(\langle v_0 \cdots v_{p+q} \rangle)$$
$$= (-1)^{\frac{1}{2}q(q+1)} z^q(\langle v_p \cdots v_{p+q} \rangle) \cdot (-1)^{\frac{1}{2}p(p+1)} z^p(\langle v_0 \cdots v_p \rangle),$$

or

$$z^q \cup' z^p(\langle v_0 \cdots v_{p+q} \rangle) = (-1)^{pq} z^p(\langle v_p \cdots v_{p+q} \rangle) \cdot z^p(\langle v_0 \cdots v_p \rangle),$$

and, using the commutativity of the coefficient ring R, we next find that

$$z^p \cup' z^p(\langle v_0 \cdots v_{p+q} \rangle) = (-1)^{pq} z^p(\langle v_0 \cdots v_p \rangle) \cdot z^q(\langle v_p \cdots v_{p+q} \rangle).$$

Finally, since $z^q \cup z^p$ is cohomologous to $z^q \cup' z^p$, we have

$$z^q \cup z^p \sim z^q \cup' z^p = (-1)^{pq} z^p \cup z^q.$$

This then implies that, for cohomology classes,

$$[z^q \cup z^p] = (-1)^{pq} [z^p \cup z^q],$$

and the Grassman property of the cohomology ring is established.

7–13 The cap-product. Under the same conditions which permitted the definition of the cup-product of cochains, we can also define a "product" between chains and cochains. This will lead in a natural way to a product between homology and cohomology classes.

Let K be a complex, and use coefficients in a ring R with unit element 1. For any cochain c^p and chain d_q, we define the *cap-product* $c^p \cap d_q$ as follows. First,

$$c^p \cap d_q = 0 \qquad \text{whenever } p > q.$$

If $p \leqq q$, then the product $c^p \cap d_q$ is a $(q - p)$-chain. To define this chain, consider an elementary chain $g \cdot \sigma^q$, where g is in R, and $\sigma^q = \langle v_0 \cdots v_q \rangle$ is written in the given ordering in the small of K. We then set

$$c^p \cap g \cdot \langle v_0 \cdots v_q \rangle = c^p(\langle v_{q-p} \cdots v_p \rangle) \cdot g \langle v_0 \cdots v_{q-p} \rangle.$$

That is, the given product is an elementary chain with coefficient $c^p(\langle v_{q-p} \cdots v_q \rangle) \cdot g$ assigned to the simplex $\langle v_0 \cdots v_{q-p} \rangle$. We extend this definition linearly to arbitrary chains.

Again there are five important properties of the cap-product:

$$c^p \cap d_q \text{ is a bilinear function.} \tag{1}$$

$$c^p \cap (c^q \cap d_r) = (c^p \cup c^q) \cap d_r. \tag{2}$$

$$e^0 \cap d_q = d_q. \tag{3}$$

$$\mathrm{KI}(c^p \cap d_q) = c^p \cdot d_q \qquad \text{when } q = p. \tag{4}$$

$$\partial(c^p \cap d_q) = (-1)^{q-p} \delta c^p \cap d_q + c^p \cap \partial d_q. \tag{5}$$

Property 1 is easily checked, of course. Property 2 relates the cap-product to the cup-product, and we note that each side here is an $(r - p - q)$-chain as required. To establish Property 2, the reader has only to compute the value of each side on an arbitrary simplex σ^{r-p-q}. Property 3 is very easily proved. Property 4 says that the definition of the Kronecker index of a 0-chain (see Sections 6–6 and 7–10) agrees with that given above wherever it can. Property 5 for the cap-product readily implies that this product induces a cap-product of cohomology and homology classes. That is, we can prove the obvious analogues to Lemma 7–20 and Theorem 7–21. Verification of these remarks is left as an exercise.

We point out that the cap-product does *not* give rise to a ring structure. It may be considered as a pairing of the groups $H^p(K, R)$ and $H_q(K, R)$ to the group $H_{q-p}(K, R)$, and this pairing is important in certain duality theorems in the theory of manifolds (see Chapter VIII of Wilder [42]).

If the simplicial mapping φ of a complex K_1 into a complex K_2 preserves the ordering in the small of vertices, then one may prove by direct computation that

$$\varphi_*(\varphi^* c^p \cap d_q) = c^p \cap \varphi_* d_q$$

holds true for any cochain c^p on K_2 and any chain d_q on K_1. This is called the *permanence relation.*

By methods similar to those of the previous section, one may prove that the induced cap-product on cohomology and homology classes is independent of the ordering of the vertices of K.

THEOREM 7–23. If $\varphi: K_1 \to K_2$ is a simplicial mapping of the complex K_1 into the complex K_2, then the induced homomorphisms φ_* and φ^* satisfy the permanence relation

$$\varphi_*(\varphi^* a^p \cap b_q) = a^p \cap \varphi_* b_q$$

for all elements a^p of $H^p(K_2, R)$ and b_q of $H_q(K_1, R)$.

As an instance of the use of the cap-product, consider an orientable n-pseudomanifold M^n, and let z_n be the fundamental n-cycle on M^n. If c^p is any cochain on M^n, we assign an $(n - p)$-chain to c^p by applying the cap-product

$$c^p \cap z_n = c_{n-p}.$$

In view of Property 5 and its consequences, the correspondence $c^p \to c_{n-p}$ induces a homomorphism of $H^p(M^n)$ into $H_{n-p}(M^n)$. It is shown in the theory of manifolds that this homomorphism is actually an isomorphism onto. This then establishes the Poincaré duality theorem. (See Chapter VIII of Wilder [42].)

Next, consider two n-pseudomanifolds M^n and N^n with fundamental n-cycles z_n and γ_n. Let φ be a simplicial mapping of M^n into N^n, and assume that the degree ρ of φ is not zero. In view of the permanence relation,

$$\varphi_*(\varphi^*c^p \cap z_n) = c^p \cap \varphi_*z_n = \rho(c^p \cap \gamma_n) \qquad (\rho \neq 0)$$

for any cochain c^p on N^n. Now if φ^*c^p is cohomologous to zero, it follows that $c^p \cap \gamma_n$ is also homologous to zero, which is true only if c^p is cohomologous to zero. This means that φ^* is an isomorphism of $H^p(N^n)$ into $H^p(M^n)$ and hence is a ring-isomorphism of $\mathcal{R}(N^n)$ into $\mathcal{R}(M^n)$. We have proved the following result.

THEOREM 7–24. If M^n and N^n are orientable n-pseudomanifolds, and if $\varphi: M^n \to N^n$ is a simplicial mapping with nonzero degree, then the cohomology ring $\mathcal{R}(M^n)$ contains a subring isomorphic to $\mathcal{R}(N^n)$.

Making use of the duality mentioned above, Theorem 7–24 implies that the Betti numbers of M^n and N^n must satisfy the inequality

$$p_q(M^n) \geqq p_q(N^n).$$

This provides a necessary condition for the existence of a mapping of nonzero degree from one pseudomanifold into another. For instance, a mapping of nonzero degree from the 2-sphere S^2 into the torus T is impossible because $p_1(S^2) = 0$ and $p_1(T) = 2$.

7–14 Relative cohomology theory. The next two sections constitute the cohomology counterpart of Sections 7–1 through 7–7. Since most of the terminology and many of the methods are now familiar, we can be brief without loss of completeness. Actually, relative cohomology theory is conceptually the simpler of the two relative theories.

Just as was done for the chain group $C_p(K/L)$ of a complex K modulo a closed subcomplex L, we may define the *relative integral cochain group*

$C^p(K/L)$ of K modulo L to be the difference group

$$C^p\left(\frac{K}{L}\right) = C^p(K) - C^p(L).$$

This is misleading, however, because the factor group as written is really a subgroup of $C^p(K)$; speaking precisely, it is isomorphic to a subgroup of $C^p(K)$. To see this, one need only notice that every cochain modulo L is a cochain on $K - L$, the open subcomplex.

It is easier to regard the relative cochain groups as follows. Let i be the injection isomorphism of the chain group $C_p(L)$ into the chain group $C_p(K)$, and let 0_p be the zero p-cochain on L. Then the adjoint homomorphism i^* maps $C^p(K)$ into $C^p(L)$, and $(i^*)^{-1}(0_p)$ is the kernel of i^*. We then write

$$C^p\left(\frac{K}{L}\right) = (i^*)^{-1}(0_p),$$

and $C^p(K/L)$ is obviously a subgroup of $C^p(K)$. It is an easy exercise to prove that the two definitions of $C^p(K/L)$ are equivalent.

We recall that the closure $\mathrm{Cl}(\sigma)$ of a simplex σ is the complex consisting of all faces of σ (including σ itself), and that the star $\mathrm{St}(\sigma)$ of σ consists of all simplexes which have σ as a face. Since L is taken to be a closed subcomplex, it follows that for every simplex σ in L, $\mathrm{Cl}(\sigma)$ is contained in L. Similarly, $K - L$ is an open subcomplex, and if σ is in $K - L$, then $\mathrm{St}(\sigma)$ is contained in $K - L$. [Of course, $\mathrm{Cl}(\sigma)$ is not necessarily in $K - L$.] If we next define the coboundary operator $\bar{\delta}$ on $C^p(K/L)$ as we did $\bar{\partial}$ on $C_p(K/L)$, it is apparent from these remarks that for an elementary cochain $g \cdot \sigma$ on $K - L$, $\bar{\delta}(g \cdot \sigma)$ is also on $K - L$. Thus the same coboundary operator can be used for the relative theory as is used for the absolute theory; that is, δ maps $C^p(K/L)$ into $C^{p+1}(K/L)$. It follows that $\bar{\delta}$ may be taken to be δ restricted to the subgroup $C^p(K/L)$ and hence that $\overline{\delta\delta} = 0$. We may now drop the upper bar.

Following the now-familiar pattern, we define the groups

$$Z^p\left(\frac{K}{L}\right) = \text{kernel of } \delta,$$

$$B^p\left(\frac{K}{L}\right) = \delta\left(C^{p-1}\left(\frac{K}{L}\right)\right),$$

and

$$H^p\left(\frac{K}{L}\right) = Z^p\left(\frac{K}{L}\right) - B^p\left(\frac{K}{L}\right).$$

We emphasize that a cochain mod L is a cocycle mod L if and only if it is an absolute cocycle of K lying in $K - L$ and that a cochain mod L is a coboundary mod L if and only if it is the absolute coboundary of a cochain in $K - L$.

If c^p and $\bar{d}_p$ are a relative cochain and an integral chain mod L, respectively, then the Kronecker index is defined exactly as in Section 7–10,

$$c^p \cdot \bar{d}_p = \sum \eta_i g_i.$$

Also, with c^p and $\bar{d}_p$ as above, we have

$$c^p \cdot \bar{\partial} \bar{d}_{p+1} = \delta c^p \cdot \bar{d}_{p+1},$$

where $\bar{\partial}$ is the relative boundary operator. This holds because c^p has value zero on simplexes of L. Hence $c^p \cdot \bar{\partial}\bar{d}_{p+1}$ is precisely the same as $c^p \cdot \partial d_{p+1}$, where d_{p+1} is any representative of the relative chain $\bar{d}_{p+1}$. Similarly, we have $\delta c^p \cdot \bar{d}_{p+1} = \delta c^p \cdot d_{p+1}$, and the relation follows from the corresponding relation for absolute theory.

Just as in absolute cohomology, the above relation implies that the product $z^p \cdot z_p$ of a relative cocycle and a relative cycle depends only upon the cohomology class of z^p and the homology class of z_p. Thus this Kronecker index produces a pairing of the groups $H^p(K/L, R)$ and $H_p(K/L)$ to the ring R.

We can carry over the theory of the cup-product to the case of relative cohomology groups, too. To do so, we merely note that if c^p and c^q are two cochains which vanish on L, then $c^p \cup c^q$ also must vanish on L. Then by retracing the steps of Section 7–12, we obtain the relative cohomology ring $\mathcal{R}(K/L, R)$.

EXERCISE 7–15. Why does the cap-product fail to be well-defined in relative cohomology?

Next we may consider a simplicial mapping φ of the pair (K_1, L_1) into the pair (K_2, L_2). We know that there are induced mappings of the relative chain groups. Let these be $\bar{\varphi}: C_p(K_1/L_1) \to C_p(K_2/L_2)$. For any relative chain $\bar{d}_p$, we have $\bar{\partial}\bar{\varphi}(\bar{d}_p) = \bar{\varphi}\bar{\partial}(\bar{d}_p)$. For the case of cochains modulo L, we may consider the induced mapping φ^* of $C^p(K_2)$ into $C^p(K_1)$. If c^p is a cochain of K_2 mod L_2, then c^p vanishes on L_2. It follows that $\varphi^* c^p$ must vanish on L_1, for if σ^p is a simplex of L_1, then $\varphi^* c^p(\sigma^p) = c^p(\varphi(\sigma^p)) = 0$ because $\varphi(\sigma^p)$ is a simplex of L_2. Therefore φ^* may be considered as an induced mapping of $C^p(K_2/L_2)$ into $C^p(K_1/L_1)$. This then leads to induced mappings, still called φ^*, of the relative cohomology groups and the relative cohomology ring in the natural manner. Finally, we have the permanence relation

$$\varphi^* c^p \cdot \bar{d}_p = c^p \cdot \bar{\varphi}\bar{d}_p$$

for a cochain c^p of K_2 mod L_2 and a chain $\bar{d}_p$ of K_1 mod L_1. This holds for $\bar{d}_p$ because it holds for every representative d_p of $\bar{d}_p$.

7-15 Exact sequences in cohomology theory. Consider again a pair (K, L) consisting of a complex K and a closed subcomplex L. There is an exact cohomology sequence of (K, L) which is constructed in a manner similar to the construction of the homology sequence. The homomorphisms involved are as follows.

1. Any cocycle of K mod L is a cocycle of K, and therefore there is an injection mapping π^* of $H^p(K/L)$ into $H^p(K)$.

2. The adjoint mapping i^* of the injection i of L into K induces or is a homomorphism of $H^p(K)$ into $H^p(L)$.

3. If c^p is a cocycle on L, then c^p can be considered as a cochain on K by putting $c^p = 0$ on all simplexes of $K - L$. Then δc^p lies on $K - L$ since $\delta c^p = 0$ on all $(p+1)$-simplexes of L. Clearly, δc^p is a $(p+1)$-cocycle on $K - L$, and hence δ induces a homomorphism δ^* of $H^p(L)$ into $H^{p+1}(K/L)$. The *cohomology sequence* of K mod L may now be set up as

$$\cdots \overset{\delta^*}{\leftarrow} H^p(L) \overset{i^*}{\leftarrow} H^p(K) \overset{\pi^*}{\leftarrow} H^p\left(\frac{K}{L}\right) \overset{\delta^*}{\leftarrow} H^{p-1}(L) \overset{i^*}{\leftarrow} \cdots$$

To prove that this sequence is exact, one may make use of the Kronecker index $c^p \cdot d_p$ as a pairing of these groups and the corresponding homology groups, and then applying the exactness of the homology sequence. The actual proof of the following theorem is left as an exercise.

THEOREM 7-25. The cohomology sequence of a pair (K, L) is exact.

We may leave it to the reader to prove the next result, also.

THEOREM 7-26. Let $\varphi:(K_1, L_1) \to (K_2, L_2)$ be simplicial. Then φ induces a homomorphism of the cohomology sequence of (K_2, L_2) into that of (K_1, L_1).

The excision theorem also has a cohomology analogue. If M is an open subcomplex of L, and L is a closed subcomplex of K, then the excision theorem states that the identity injection i of $(K - M, L - M)$ into (K, L) induces an isomorphism i_* of $H_p[(K - M)/(L - M)]$ onto $H_p(K/L)$. In the usual dual technique, the injection i induces an isomorphism i^* of $H^p(K/L)$ onto $H^p[(K - M)/(L - M)]$, and i^* is also an isomorphism of the corresponding cohomology rings.

Next, we may mention the cohomology companion of the Mayer-Vietoris sequence. Let K be the union of two closed subcomplexes K_1 and K_2. The desired homomorphisms are as follows:

1. $j^*:C^p(K_1) \oplus C^p(K_2) \to C^p(K_1 \cap K_2)$ is given by setting $j^*(c_1^p, c_2^p) = j_1^*(c_1^p) - j_2^*(c_2^p)$, where j_1 and j_2 are the identity injections of $K_1 \cap K_2$

into K_1 and K_2, respectively. We then have the induced homomorphisms, also called j^*, on the cohomology groups.

2. $s^*:C^p(K) \to C^p(K_1) \oplus C^p(K_2)$ is defined by setting $s^*(c^p) = (i_1^*(c^p), i_2^*(c^p))$, where i_1 and i_2 are the injections of K_1 and K_2, respectively, into K. Again we use s^* to denote the induced homomorphisms on cohomology groups.

3. The homomorphism $v^*:H^p(K_1 \cap K_2) \to H^{p+1}(K)$ is more complicated. Any cochain c^p of K can be written as $c_1^p + c_2^p$ with c_i^p on K_i, and in particular any cochain c^p of K mod $K_1 \cap K_2$ has a unique decomposition $c_1^p + c_2^p$, where c_i^p lies on $K_i - K_1 \cap K_2$. In this case, $\delta c^p = \delta c_1^p + \delta c_2^p$ is also a unique decomposition. Now let d^p be a cocycle on $K_1 \cap K_2$, and let c^p be a cochain on K mod $K_1 \cap K_2$ such that $i^*(c^p) = d^p$. Then δc^p is a cocycle on K mod $K_1 \cap K_2$, which means that $\delta c^p = c_1^{p+1} + c_2^{p+1}$, with c_i^{p+1} actually being a cocycle on $K_i - K_1 \cap K_2$. By the excision isomorphism of $C^{p+1}(K_i/K_1 \cap K_2)$ onto $C^{p+1}(K/K_i)$, c_1^{p+1} may be considered to be a cochain of K mod K_2, and since $C^{p+1}(K/K_2)$ is a subgroup of $C^{p+1}(K)$, c_1^{p+1} is a cochain, actually a cocycle, of K.

We now put $v^*[d^p]$ equal to the cohomology class $[c_1^{p+1}]$. (We could have used c_2^{p+1} instead.) Note that the cochain c^p such that $i^*(c^p) = d^p$ is determined modulo the kernel of i^*, which is precisely the group $C^p(K/K_1 \cap K_2)$. Therefore, c_1^{p+1} is determined up to $B^p(K/K_1 \cap K_2)$, which is contained in $B^p(K)$, and $v^*[d^p]$ is well-defined on cohomology classes.

The Mayer-Vietoris sequence for cohomology is

$$\cdots \overset{s^*}{\leftarrow} H^{p+1}(K_1 \cup K_2) \overset{v^*}{\leftarrow} H^p(K_1 \cap K_2) \overset{j^*}{\leftarrow} H^p(K_1) \oplus H^p(K_2)$$
$$\overset{s^*}{\leftarrow} H^p(K_1 \cup K_2) \overset{v^*}{\leftarrow} \cdots.$$

The exactness of this sequence may be proved in the same manner as was suggested for the cohomology sequence, that is, by using the pairing given by the Kronecker index. Of course, it is possible to give a direct proof in both cases. The interested reader will be tempted to give two proofs in each case.

This completes our presentation of cohomology theory. A great deal has been deliberately left to the reader. The reason for the omission of proofs is twofold. First, many proofs in cohomology are dual to those in homology, and the necessary manipulatory skills should have been developed by this point. Second, we feel that this chapter will be read primarily by those who wish to go on to more advanced topics in algebraic topology, and such a reader should be required to fill in the details of the proofs for himself.

EXERCISE 7-16. Let K be a finite orientable n-pseudomanifold, and let M be an open subcomplex of K. Prove that M is connected if and only if $H^n(M)$ is infinite cyclic. Then suppose that $H^{n-1}(K) = 0$ and that L_1 and L_2 are two closed disjoint subcomplexes of K, neither of which separates K. Prove that $L_1 \cap L_2$ also does not separate K.

7-16 Relations between homology and cohomology groups. We give here a brief resumé of the Pontrjagin theory of character groups and indicate how this theory leads to a duality between homology and cohomology groups.

Let $\mathcal{R}$ denote the additive group of real numbers modulo 1, and let G be any abelian group. A homomorphism $\varphi: G \to \mathcal{R}$ of G into $\mathcal{R}$ is called a *character* of G. Given two characters φ_1 and φ_2 of G, their sum is given by the usual functional addition, i.e., for each element g of G,

$$(\varphi_1 + \varphi_2)(g) = \varphi_1(g) + \varphi_2(g),$$

the addition on the right being performed in $\mathcal{R}$, of course. Under this operation, the characters of G constitute a new abelian group, the *character group* of G, which we will denote by $\widetilde{G}$. Briefly then, $\widetilde{G} = \text{Hom}(G, \mathcal{R})$ (Sec. 7-16). The following are examples. If Z is the group of integers, then $\widetilde{Z} = \mathcal{R}$. If G is a finite group, then G and $\widetilde{G}$ are isomorphic. The reader may verify these statements as exercises.

If G is a countable group with the discrete topology, then $\widetilde{G}$ may be topologized with a convergence topology as follows. We say that the sequence of characters $\{\varphi_n\}$ converges to the character φ if, for each element g in G, the sequence $\{\varphi_n(g)\}$ converges to $\varphi(g)$ in $\mathcal{R}$. The topological group so obtained is compact and separable (see Theorem 31 of Pontrjagin [116]). Also the collection of continuous characters in $\widetilde{G}$ constitutes a subgroup isomorphic to G itself (Theorem 32 of [116]). We shall restrict attention to a countable discrete group G throughout this section, and we will assume that $\widetilde{G}$ is topologized as above.

Let H be a subgroup of G. The collection of all characters of G which map H onto zero in $\mathcal{R}$ is easily seen to be a closed subgroup of G. This subgroup is called the *annihilator* of H. Similarly, if $\widetilde{H}$ is a subgroup of $\widetilde{G}$, then all those elements of G which are mapped onto zero by each element of $\widetilde{H}$ form a subgroup of G, the *annihilator* of $\widetilde{H}$. Note that the smaller a subgroup is, the more characters there are which map it onto zero, and hence the larger is its annihilator and conversely. Precisely, if the subgroup H of G contains a subgroup H', then $\widetilde{H}'$ contains $\widetilde{H}$.

We quote several results of Pontrjagin here without proof (for the proofs see [116] again).

LEMMA 7–27. Let H be a subgroup of G and J be a closed subgroup of $\widetilde{G}$. Then J is the annihilator of H if and only if H is the annihilator of J.

LEMMA 7–28. If H is a subgroup of G with annihilator J in $\widetilde{G}$, then the difference group $\widetilde{G} - J$ is the character group of H, and J is the character group of the difference group $G - H$.

COROLLARY 7–29. If g is a fixed element of G, and if $\varphi(g) = 0$ for each character φ of G, then $g = 0$. Thus the annihilator of $\widetilde{G}$ is the identity element of G.

COROLLARY 7–30. If H is a subgroup of G with annihilator J in $\widetilde{G}$, and if L is a subgroup of H with annihilator K, then $K - J$ is the character group of $H - L$.

Let G_1 and G_2 be two countable discrete groups, and let $h:G_1 \to G_2$ be a homomorphism of G_1 into G_2. Then h induces a homomorphism h^* of $\widetilde{G}_2$ into $\widetilde{G}_1$ defined by

$$(h^*\phi)(g_1) = \phi(h(g_1)),$$

where φ is a character of G_2, and g_1 is an element of G_1. This homomorphism h^* is said to be the *dual homomorphism* of h.

LEMMA 7–31. Let $h:G_1 \to G_2$ be a homomorphism. Then the annihilator of $h(G_1)$ is the kernel of h^*, and the annihilator of $h^*(G_2)$ is the kernel of h.

We may now turn to the homology theory of a finite complex K. Letting G be any countable discrete group, the chain groups $C_p(K, G)$ are also countable, and we assign to them the discrete topology. Let φ be a character of $C_p(K, G)$. Given any simplex σ^p of K, the elementary chains $g \cdot \sigma^p$ form a group $G(\sigma^p)$ isomorphic to G. Hence φ restricted to $G(\sigma^p)$ defines a character of G through this isomorphism. We may denote this character of G by $\varphi(\sigma^p)$. Clearly we have $\varphi(-\sigma^p) = -\varphi(\sigma^p)$. But this is exactly the condition needed to make φ a p-chain on K with coefficients in $\widetilde{G}$. It is easy to show that, by linear extension, this process defines an algebraic isomorphism of $C_p(K, G)$ onto $C_p(K, \widetilde{G})$.

Furthermore, if c_p is an element of $C_p(K, G)$, and if φ_p is an element of $C_p(K, \widetilde{G})$, then there is a unique element of $\mathcal{R}$ given by

$$\varphi_p(c_p) = \sum_{i=1}^{\alpha_p} (\varphi_p(\sigma_i^p)) \cdot (c_p(\sigma_i^p)).$$

From this pairing of the two groups $C_p(K, G)$ and $C_p(K, \widetilde{G})$ to the group $\mathcal{R}$, it follows that

$$\partial\varphi_{p+1}(c_p) = \varphi_{p+1}(\delta c_p).$$

This relation shows that the homomorphism $\partial: C_{p+1}(K, \widetilde{G}) \to C_p(K, \widetilde{G})$ is dual to $\delta: C_p(K, G) \to C_{p+1}(K, G)$. (We recall that for a finite complex, the chain and cochain groups coincide.)

Applying the above lemmas, we proceed as follows. The kernel of ∂ is the group $Z_{p+1}(K, \widetilde{G})$, the kernel of δ is $Z^p(K, G)$, the image $\partial[C_{p+1}(K, \widetilde{G})]$ is $B_p(K, \widetilde{G})$, and the image $\delta(C_p(K, G)]$ is $B^{p+1}(K, G)$. Hence by Lemma 7-27, we may conclude that

$$Z_p(K, \widetilde{G}) \text{ is the annihilator of } B^p(K, G) \tag{1}$$

and

$$Z^p(K, G) \text{ is the annihilator of } B_p(K, \widetilde{G}). \tag{2}$$

Using Lemma 7-31, we have that

$$B_p(K, \widetilde{G}) \text{ is the annihilator of } Z^p(K, G). \tag{3}$$

Finally, applying Corollary 7-30 to these statements, we obtain the following duality between homology and cohomology groups.

THEOREM 7-32. *Let K be a finite complex, and let G be a countable discrete group. Then $H_p(K, \widetilde{G})$ is the character group of $H^p(K, G)$.*

By going over the same steps again, we can also obtain the dual to 7-32.

THEOREM 7-33. *With K and G as in Theorem 7-32, $H^p(K, \widetilde{G})$ is the character group of $H_p(K, G)$.*

In other terms, we have the two statements

$$H_p(K, \operatorname{Hom}(G, \mathfrak{R})) = \operatorname{Hom}(H^p(K, G), \mathfrak{R})$$

and

$$H^p(K, \operatorname{Hom}(G, \mathfrak{R})) = \operatorname{Hom}(H_p(K, G), \mathfrak{R}).$$

In particular, then, for the additive group Z of integers, we have

$$H_p(K, \mathfrak{R}) = \operatorname{Hom}(H^p(K, Z), \mathfrak{R})$$

and

$$H^p(K, \mathfrak{R}) = \operatorname{Hom}(H_p(K, Z), \mathfrak{R}).$$

These results and their converses prove that $\mathfrak{R}$, the additive group of reals modulo 1, is also a universal coefficient group (see Section 6-9).

CHAPTER 8

GENERAL HOMOLOGY THEORIES

8–1 Cech homology theory (introduction). In this first description of Čech homology theory, we follow closely the technique of Čech's original paper [72]. A more recently developed approach, together with greater generality, will be found in Section 8–3. Our purpose in this section is to construct the machinery to be used in Section 8–2 to prove the topological invariance of the simplicial homology groups of a finite polytope.

Given a compact Hausdorff space X, let $\Sigma(X)$ denote the family of all finite coverings of X by open sets. The coverings in $\Sigma(X)$ will be denoted by script letters $\mathfrak{U}$, $\mathfrak{V}$, ... and the open sets in a covering by italic capitals U, V, ... An element $\mathfrak{U}$ of $\Sigma(X)$ may be considered as a simplicial complex if we define *vertex* to mean *open set U in* $\mathfrak{U}$ and agree that a subcollection $U_0, \ldots, U_p$ of such vertices constitutes a p-simplex if and only if the intersection $\bigcap_{i=0}^{p} U_i$ is not empty (see Section 5–7). The resulting complex is known as the *nerve of the covering* $\mathfrak{U}$. No symbolic distinction will be made between a covering and its nerve; the proper interpretation should always be obvious from the context.

Alexandroff [48] introduced the concept of the nerve of a covering in 1928, and the idea has become very important. If we take a geometric realization of the nerve of a covering of the space X, then in some sense we have a triangulated approximation to X. And this "approximation" gets better as finer and finer coverings are used. This technique forms an important connection between point-set topology and the combinatorial methods of simplicial complexes. Indeed, it seems probable that Čech was motivated by a combination of Alexandroff's ideas and the earlier homology theory of Vietoris (see Section 8–6).

Since the nerves of coverings are to play an important role in our development, a few remarks about such complexes are in order. First, we may point out that even though the space X be low-dimensional it may have nerves of high dimension and these nerves may in no way resemble the space. For instance, consider the covering $\mathfrak{U}$ of the unit interval I^1 by the open sets $U_1 = [0, \frac{2}{3})$, $U_2 = (\frac{1}{3}, 1]$, $U_3 = (\frac{1}{4}, \frac{1}{2}) \cup (\frac{2}{3}, 1]$, and $U_4 = [0, \frac{1}{3}) \cup (\frac{1}{2}, \frac{3}{4})$. It is easily seen that there are six 1-simplexes $\langle U_1 U_2 \rangle$, $\langle U_1 U_3 \rangle$, $\langle U_1 U_4 \rangle$, $\langle U_2 U_3 \rangle$, $\langle U_2 U_4 \rangle$, and $\langle U_3 U_4 \rangle$ and four 2-simplexes $\langle U_1 U_2 U_3 \rangle$, $\langle U_1 U_2 U_4 \rangle$, $\langle U_1 U_3 U_4 \rangle$, and $\langle U_2 U_3 U_4 \rangle$. Since there are no points in common with all four open sets, there are no 3-simplexes in $\mathfrak{U}$. Hence a geometric realization of $\mathfrak{U}$ is a tetrahedral surface. This example points out that coverings are not necessarily so well behaved as they are often envisioned. But, as the following theorem shows, we can do much worse than the above example.

THEOREM 8–1. Let C be a compact Hausdorff space which is dense in itself, and let K be any finite simplicial complex. Then there is an open covering $\mathcal{U}$ of C such that the nerve of $\mathcal{U}$ has a subcomplex isomorphic to K.

Proof: We perform an induction on the number of vertices of K. The theorem is obvious for all complexes with one vertex. Suppose that the theorem is true for all complexes with $n - 1$ vertices, and let K have n vertices $v_1, v_2, \ldots, v_n$. Consider the subcomplex K' of K consisting of all simplexes of K not having v_n as a vertex. Let $\mathcal{U}' = \{U_i'\}$ be a finite open covering of C whose nerve contains a subcomplex isomorphic to K'. If v_n is an isolated vertex of K, we need only add an arbitrary open set U (not in $\mathcal{U}'$) to $\mathcal{U}'$ to obtain a new covering $\mathcal{U}$ with the desired property. If v_n is not an isolated vertex of K, then for each simplex $\sigma = \langle v_{i_0} \cdots v_{i_j} v_n \rangle$ in K, let $U_{i_0}', \ldots, U_{i_j}'$ denote the open sets in $\mathcal{U}'$ corresponding to $v_{i_0}, \ldots, v_{i_j}$. By definition, $\bigcap_{k=0}^{j} U_{i_k}'$ is not empty. Choose a point p_k in each U_{i_k}'. Since C is dense in itself, we can find an open set V_σ not in $\mathcal{U}'$ such that V_σ lies in $\bigcap_{k=0}^{j} U_{i_k}'$ and contains no point p_k. Let $U = \bigcup V_\sigma$, where the union is taken over all simplexes σ of K having v_n as a vertex. It is possible that U is an element of $\mathcal{U}$, but U is not one of the open sets U_i' corresponding to a vertex of K'. The new covering $\mathcal{U}$ consisting of the elements of $\mathcal{U}'$ and the open set U contains all the sets needed to construct an isomorphic image of the complex K. □

The difficulties that seem to stem from the above result are largely apparent, rather than actual. We include the theorem merely to show that such questions do exist. We now return to Čech theory.

The collection $\Sigma(X)$ of finite open coverings of a space X may be partially ordered by refinement (see Section 2–11). A covering $\mathcal{V}$ refines the covering $\mathcal{U}$, $\mathcal{U} < \mathcal{V}$, if every element of $\mathcal{V}$ is contained in some element of $\mathcal{U}$. Also given two coverings $\mathcal{U}$ and $\mathcal{V}$ in $\Sigma(X)$, we define the covering $\mathcal{U} \cap \mathcal{V}$ consisting of all nonempty intersections $U \cap V$ for U in $\mathcal{U}$ and V in $\mathcal{V}$. Clearly, $\mathcal{U} \cap \mathcal{V} > \mathcal{U}$, and $\mathcal{U} \cap \mathcal{V} > \mathcal{V}$. This establishes $\Sigma(X)$ as a *directed set* under refinement (see Section 2–14).

If $\mathcal{V} > \mathcal{U}$ in $\Sigma(X)$, then there is a simplicial mapping $\pi_{\mathcal{U}\mathcal{V}}$ of $\mathcal{V}$ into $\mathcal{U}$ called a *projection*. This is defined by taking $\pi_{\mathcal{U}\mathcal{V}}(V)$, V in $\mathcal{V}$, to be any (fixed) element U of $\mathcal{U}$ such that V lies in U. Of course, there may be several elements of $\mathcal{U}$ containing the set V and hence several choices for $\pi_{\mathcal{U}\mathcal{V}}(V)$. This means that there may be many projections of $\mathcal{V}$ into $\mathcal{U}$. To see that any such projection $\pi_{\mathcal{U}\mathcal{V}}$ is indeed a simplicial mapping, it suffices to say that if $\bigcap_{i=0}^{p} V_i$ is nonempty, then $\bigcap_{i=0}^{p} \pi_{\mathcal{U}\mathcal{V}}(V_i)$ is also nonempty because each V_i lies in $\pi_{\mathcal{U}\mathcal{V}}(V_i)$.

For reasons to be discussed later, the coefficient group G used in Čech homology theory is usually taken to be either a compact abelian topological group or a vector space over a field. Given a covering $\mathfrak{U}$ in $\Sigma(X)$, we may apply the methods of Chapter 6 to define the chain groups $C_p(\mathfrak{U}, G)$, the cycle groups $Z_p(\mathfrak{U}, G)$, etc. In view of Section 6–10, each projection $\pi_{\mathfrak{U}\mathfrak{V}}$ induces a chain-mapping, also denoted by $\pi_{\mathfrak{U}\mathfrak{V}}$, of the complex $\mathfrak{V}$ into the complex $\mathfrak{U}$. Then the chain-mapping induces homomorphisms $_*\pi_{\mathfrak{U}\mathfrak{V}}$ of the homology groups $H_p(\mathfrak{V}, G)$ of $\mathfrak{V}$ into the groups $H_p(\mathfrak{U}, G)$ of $\mathfrak{U}$. Since there may be many projections of $\mathfrak{V}$ into $\mathfrak{U}$, one difficulty must be overcome before putting this machinery to work.

THEOREM 8–2. If $\mathfrak{V} > \mathfrak{U}$ in $\Sigma(X)$, then any two projections π_1 and π_2 of $\mathfrak{V}$ into $\mathfrak{U}$ are chain-homotopic.

Proof: We must construct a deformation operator $\mathfrak{D} = \{\mathfrak{D}_p\}$ such that each $\mathfrak{D}_p$ is a homomorphism of $C_p(\mathfrak{V}, G)$ into $C_{p+1}(\mathfrak{U}, G)$ and, for any chain c_p on $\mathfrak{V}$, we have

$$\partial \mathfrak{D}_p c_p = \pi_2 c_p - \pi_1 c_p - \mathfrak{D}_{p-1} \partial c_p \tag{a}$$

(see Section 6–11).

To do this, we proceed as follows. If V is any element of $\mathfrak{V}$, and if we denote $\pi_1(V)$ by U, then $\pi_2(V)$ will be U'; that is, a prime on an element of $\mathfrak{U}$ indicates that it is an image under π_2. Let us orient $\mathfrak{V}$ by choosing a fixed ordering of its vertices, and if $i_0 < i_1 < \cdots < i_p$, let $\langle V_{i_0} \cdots V_{i_p} \rangle$ determine the positive orientation of the simplex in $\mathfrak{V}$ with these vertices. We define $\mathfrak{D}_p$ on elementary chains $g\sigma^p$ by

$$\mathfrak{D}_p g \langle V_{i_0} \cdots V_{i_p} \rangle = \sum_{j=0}^{p} (-1)^j g'_j \langle U_{i_0} \cdots U_{i_j} U'_{i_j} \cdots U'_{i_p} \rangle, \tag{b}$$

where $g'_{i_j} = 0$ if not all the sets $U_{i_0}, \ldots, U_{i_j}, U'_{i_j}, \ldots, U'_{i_p}$ are distinct, and where $g'_{i_j} = g$ if all these sets are distinct. As usual, $\mathfrak{D}_p$ is extended linearly to arbitrary chains.

To prove that relation (a) holds, it suffices to consider an elementary chain $g\langle V_{i_0} \cdots V_{i_p} \rangle$ on $\mathfrak{V}$. By sheer computation, we may show that

$$\begin{aligned} \partial \mathfrak{D}_p g \langle V_{i_0} \cdots V_{i_p} \rangle &= \partial \sum_{j=0}^{p} (-1)^j g'_j \langle U_{i_0} \cdots U_{i_j} U'_{i_j} \cdots U'_{i_p} \rangle \\ &= \sum_{j=0}^{p} (-1)^j \Bigg[\sum_{n=0}^{j} (-1)^n g'_j \langle U_{i_0} \cdots \hat{U}_{i_n} \cdots U_{i_j} U'_{i_j} \cdots U'_{i_p} \rangle \\ &\quad + \sum_{n=j}^{p} (-1)^{n+1} g'_j \langle U_{i_0} \cdots U_{i_j} U'_{i_j} \cdots \hat{U}'_{i_n} \cdots U'_{i_p} \rangle \Bigg] \end{aligned}$$

(cont.)

$$= g'_0\langle U'_{i_0} \cdots U'_{i_p}\rangle + \sum_{n=0}^{p} (-1)^{n+1} g'_0\langle U_{i_0}U'_{i_0} \cdots \hat{U}'_{i_n} \cdots U'_{i_p}\rangle$$

$$+ \sum_{j=1}^{p-1} (-1)^j \left[\sum_{n=0}^{j} (-1)^n g'_j\langle U_{i_0} \cdots \hat{U}_{i_n} \cdots U_{i_j}U'_{i_j} \cdots U'_{i_j}\rangle\right.$$

$$\left. + \sum_{n=j}^{p} (-1)^{n+1} g'_j\langle U_{i_0} \cdots U_{i_j}U'_{i_j} \cdots \hat{U}'_{i_n} \cdots U'_{i_p}\rangle\right]$$

$$+ (-1)^p \sum_{n=0}^{p} (-1)^n g'_p\langle U_{i_0} \cdots \hat{U}_{i_n} \cdots U_{i_p}U'_{i_p}\rangle$$

$$+ (-1)^p(-1)^{p+1} g'_p\langle U_{i_0} \cdots U_{i_p}\rangle. \qquad \text{(c)}$$

We note that the first and last terms in the above sum are $\pi_2 g\langle V_{i_0} \cdots V_{i_p}\rangle$ and $-\pi_1 g\langle V_{i_0} \cdots V_{i_p}\rangle$ respectively. Furthermore, the pairs of terms of the forms

$$(-1)^j(-1)^j g'_j\langle U_{i_0} \cdots \hat{U}_{i_j}U'_{i_j} \cdots U'_{i_p}\rangle,$$

$$(-1)^{j-1}(-1)^j g'_{j-1}\langle U_{i_0} \cdots U_{i_{j-1}}\hat{U}'_{i_{j-1}}U'_{i_j} \cdots U'_{i_p}\rangle$$

are opposite in sign and will cancel if $g'_{j-1} = g'_j$. If $g'_{j-1} \neq g'_j$, then one and only one of the two must be zero. This implies that either $U'_{i_{j-1}}$ is the same as one of the sets $U_{i_0}, \ldots, U_{i_{j-1}}$ or U'_{i_j} is one of the sets $U_{i_0}, \ldots, U_{i_j}$. There are many possible cases here, and we will not give a complete argument.

In case $g'_j = g$ for all j, the cancellation of terms mentioned above allows us to arrange the sum in the form

$$g\langle U'_{i_0} \cdots U'_{i_p}\rangle - g\langle U_{i_0} \cdots U_{i_p}\rangle$$

$$- \sum_{j=0}^{p} (-1)^j \left[\sum_{n=0}^{j-1} (-1)^n g\langle U_{i_0} \cdots \hat{U}_{i_n} \cdots U_{i_{j-1}}U'_{i_j} \cdots U'_{i_p}\rangle\right.$$

$$\left. + \sum_{n=j}^{p} (-1)^n g\langle U_{i_0} \cdots U_{i_{j-1}}U'_{i_j} \cdots \hat{U}'_{i_n} \cdots U'_{i_p}\rangle\right], \qquad \text{(d)}$$

which is precisely $(\pi_2 - \pi_1 - \mathfrak{D}_{p-1}\partial)g\langle V_{i_0} \cdots V_{i_p}\rangle$. We complete our argument with a simple illustration of a case in which one of the g'_j is zero.

Let $g\langle V_0V_1V_2\rangle$ be an elementary 2-chain on $\mathcal{V}$, and suppose that $\pi_2(V_0) = U'_0 = \pi_1(V_0) = U_0$, while the sets U_1, U_2, U'_1, and U'_2 are distinct. Computing $\partial\mathfrak{D}_2 g\langle V_0V_1V_2\rangle$, we obtain

$$\partial[g'_0\langle U_0U'_0U'_1U'_2\rangle - g'_1\langle U_0U_1U'_1U'_2\rangle + g'_2\langle U_0U_1U_2U'_2\rangle].$$

Since $U_0 = U_0'$, $g_0' = 0$. But $g_1' = g_2' = g$. Hence we have the chain

$$- g\langle U_1U_1'U_2'\rangle + g\langle U_0U_1'U_2'\rangle - g\langle U_0U_1U_2'\rangle + g\langle U_0U_1U_1'\rangle$$
$$+ g\langle U_1U_2U_2'\rangle - g\langle U_0U_2U_2'\rangle + g\langle U_0U_1U_2'\rangle - g\langle U_0U_1U_2\rangle.$$

The third and the seventh terms cancel, so we have

$$\begin{aligned}\partial\mathfrak{D}_2g\langle V_0V_1V_2\rangle = g\langle U_0U_1'U_2'\rangle - g\langle U_0U_1U_2\rangle - g\langle U_1U_1'U_2'\rangle \\ + g\langle U_1U_2U_2'\rangle - g\langle U_0U_1U_1'\rangle + g\langle U_0U_2U_2'\rangle. \quad \text{(e)}\end{aligned}$$

On the other hand, computing $(\pi_2 - \pi_1 - \mathfrak{D}_1\partial)g\langle V_0V_1V_2\rangle$, we obtain

$$\begin{aligned}g\langle U_0'U_1'U_2'\rangle - g\langle U_0U_1U_2\rangle - \mathfrak{D}_1[g\langle V_1V_2\rangle - g\langle V_0V_2\rangle + g\langle V_0V_1\rangle] \\ = g\langle U_0'U_1'U_2'\rangle - g\langle U_0U_1U_2\rangle \\ - [g\langle U_1U_1'U_2'\rangle - g\langle U_1U_2U_2'\rangle \\ - 0\langle U_0U_0'U_2'\rangle + g\langle U_0U_2U_2'\rangle \\ + 0\langle U_0U_0'U_1'\rangle - g\langle U_0U_1U_1'\rangle]. \quad \text{(f)}\end{aligned}$$

It is obvious that, except for the first terms, the above two chains coincide. But since $U_0 = U_0'$, $g\langle U_0U_1'U_2'\rangle = g\langle U_0'U_1'U_2'\rangle$; hence $\partial\mathfrak{D}_2g\langle V_0V_1V_2\rangle = [\pi_2 - \pi_1 - \mathfrak{D}_1\partial]g\langle V_0V_1V_2\rangle$. □

We can reword Theorem 6–19 in the present context.

THEOREM 8–3. If $\mathfrak{U} < \mathfrak{V}$ in $\Sigma(X)$, then any two projections π_1 and π_2 of $\mathfrak{V}$ into $\mathfrak{U}$ induce the same homomorphisms of $H_p(\mathfrak{V}, G)$ into $H_p(\mathfrak{U}, G)$. That is, $_*\pi_1$ and $_*\pi_2$ coincide.

The machinery needed to define a Čech cycle is now at hand. *A p-dimensional Čech cycle* of the space X is a collection $z_p = \{z_p(\mathfrak{U})\}$ of p-cycles $z_p(\mathfrak{U})$, one from each and every cycle group $Z_p(\mathfrak{U}, G)$, $\mathfrak{U}$ in $\Sigma(X)$, with the property that if $\mathfrak{U} < \mathfrak{V}$, then $\pi_{\mathfrak{U}\mathfrak{V}}z_p(\mathfrak{V})$ is homologous to $z_p(\mathfrak{U})$. (Of course, this homology takes place on the complex $\mathfrak{U}$.) Note that in view of Theorem 8–3, the particular choice of the projection $\pi_{\mathfrak{U}\mathfrak{V}}$ is immaterial. Each cycle $z_p(\mathfrak{U})$ in the collection z_p is called a *coordinate of the Čech cycle*. Hence a Čech cycle has a coordinate on every covering of the space X.

The addition of Čech cycles is defined in a natural way by setting

$$\{z_p(\mathfrak{U})\} + \{z_p'(\mathfrak{U})\} = \{z_p(\mathfrak{U}) + z_p'(\mathfrak{U})\},$$

where the addition on the right is that of chains on the complex $\mathfrak{U}$. The homology relation between Čech cycles is defined as follows. First, a Čech cycle $z_p = \{z_p(\mathfrak{U})\}$ is homologous to zero on X (or is a *bounding Čech cycle*) if each coordinate $z_p(\mathfrak{U})$ is homologous to zero on the covering $\mathfrak{U}$, for all $\mathfrak{U}$ in $\Sigma(X)$. In other words, $\{z_p(\mathfrak{U})\}$ bounds if and only if there is a $(p+1)$-chain $c_{p+1}(\mathfrak{U})$ on each covering $\mathfrak{U}$ in $\Sigma(X)$ such that the coordinate $z_p(\mathfrak{U}) = \partial c_{p+1}(\mathfrak{U})$. Then two Čech cycles z_p and z'_p are *homologous Čech cycles* if their difference $z_p - z'_p$ is homologous to zero. Note that if $z_p = \{z_p(\mathfrak{U})\}$ is a bounding Čech cycle, nothing is required of the chains $c_{p+1}(\mathfrak{U})$ beyond the fact that $\partial c_{p+1}(\mathfrak{U}) = z_p(\mathfrak{U})$. In particular, there are no "connecting homologies" between homologous Čech cycles. To put it another way, we do not attempt to define a "Čech chain" at all! An example will clarify this shortly.

The reader may prove that the homology relation defined above is an equivalence relation on the set of all Čech p-cycles. The corresponding equivalence classes $[z_p]$ of homologous Čech p-cycles are the elements of the *pth Čech homology group* $H_p(X, G)$, the group operation being defined by the expected formula

$$[z_p] + [z'_p] = [z_p + z'_p],$$

where we are using our customary symbol for an equivalence class.

Čech homology groups are topological invariants of the space X by their very definition. For these groups depend only upon the collection $\Sigma(X)$ and its structure as a directed set. If h is a homeomorphism of X onto X', then for each covering $\mathfrak{U}$ in $\Sigma(X)$, the collection $h(\mathfrak{U})$ of all images of elements of $\mathfrak{U}$ is an open covering of X' and conversely. Certainly $\mathfrak{U}$ and $h(\mathfrak{U})$ are isomorphic complexes. Also the partial ordering of $\Sigma(X)$ by refinement is preserved by h. By filling in the details, the reader may easily prove that $H_p(X, G)$ and $H_p(X', G)$ are isomorphic.

We have glossed over a difficult question here, namely the existence of Čech cycles. Given the pair (X, G), how do we know that there are cycles $z_p(\mathfrak{U}, G)$ on each open covering of X such that if $\mathfrak{U} < \mathfrak{V}$, then $\pi_{\mathfrak{U}\mathfrak{V}} z_p(\mathfrak{V})$ is homologous to $z_p(\mathfrak{U})$? We will discuss this question in Section 8–3.

Obviously, it would be a formidable task to compute the Čech homology groups of a space directly from the definition. Such work is rarely necessary, however. Our next theorem tends to simplify Čech theory, and its proof exhibits some of the standard techniques used in applying this theory.

A subcollection $\Sigma'(X)$ of $\Sigma(X)$ is called a *cofinal family of coverings* of X provided that for every covering $\mathfrak{U}$ in $\Sigma(X)$ there is some covering $\mathfrak{U}'$ in $\Sigma'(X)$ such that $\mathfrak{U}' > \mathfrak{U}$. Given such a cofinal family $\Sigma'(X)$, we may go through the development of Čech theory again, restricting the cycles, homologies, etc., to be on elements of $\Sigma'(X)$. In so doing we construct *pseudo Čech cycles* and *pseudo Čech homology groups* $H_p(X, \Sigma'(X), G)$. It is our aim to show these pseudo Čech groups to be isomorphic to the full Čech groups $H_p(X, G)$. This result is to be applied in Section 8–2 to prove the topological invariance of the simplicial homology groups of a finite polytope.

THEOREM 8–4. Let $\Sigma'(X)$ be a cofinal family of coverings of X. Then the pseudo Čech homology group $H_p(X, \Sigma'(X), G)$ is isomorphic to the Čech homology group $H_p(X, G)$ for each dimension p.

Proof: Let $z_p = \{z_p(\mathfrak{U})\}$ be a Čech p-cycle. Since $\Sigma'(X)$ is a subcollection of $\Sigma(X)$, there is a coordinate $z_p(\mathfrak{U}')$ of z_p on each element $\mathfrak{U}'$ of $\Sigma'(X)$. Hence we may define a transformation $f(z_p) = z'_p$, where $z'_p = \{z_p(\mathfrak{U}')\}$ and $z_p(\mathfrak{U}') = z_p(\mathfrak{U})$ for $\mathfrak{U}' = \mathfrak{U}$. It should be obvious that f is a homomorphism of the cycle group $Z_p(X, G)$ into the cycle group $Z_p(X, \Sigma'(X), G)$. One readily shows that f also carries $B_p(X, G)$ into $B_p(X, \Sigma'(X), G)$. Hence f induces a homomorphism f_* of $H_p(X, G)$ into $H_p(X, \Sigma'(X), G)$. We will show that f_* is actually an isomorphism onto.

We first prove that f_* is onto. To do so, let $\{z'_p(\mathfrak{U}')\}$ be a pseudo Čech cycle. Given any covering $\mathfrak{U}$ in $\Sigma(X)$, choose a covering $\mathfrak{U}'$ in $\Sigma'(X)$ such that $\mathfrak{U}' > \mathfrak{U}$. This is possible because $\Sigma'(X)$ is a cofinal family. Using any projection $\pi_{\mathfrak{U}\mathfrak{U}'}$, we define $z_p(\mathfrak{U}) = \pi_{\mathfrak{U}\mathfrak{U}'}z_p(\mathfrak{U}')$. In this way, we obtain a collection $\{z_p(\mathfrak{U})\}$ of coordinates, one on each element $\mathfrak{U}$ in $\Sigma(X)$. By taking the projection $\pi_{\mathfrak{U}'\mathfrak{U}'}$ to be the identity simplicial mapping, the resulting collection of coordinates obviously maps *onto* $\{z_p(\mathfrak{U}')\}$ under the homomorphism f. Hence if the collection so constructed is a Čech cycle, then f_* is onto.

Now each coordinate $z_p(\mathfrak{U})$ is independent, in the sense of homology, of the choice of the covering $\mathfrak{U}'$ in $\Sigma'(X)$ used to define $z_p(\mathfrak{U})$. For if $\mathfrak{V}'$ is another element of $\Sigma'(X)$ such that $\mathfrak{V}' > \mathfrak{U}$, we may show that

$$\pi_{\mathfrak{U}\mathfrak{V}'}z_p(\mathfrak{V}') \sim \pi_{\mathfrak{U}\mathfrak{U}'}z_p(\mathfrak{U}') = z_p(\mathfrak{U}).$$

To do so, choose a covering $\mathfrak{W}'$ in $\Sigma'(X)$ such that $\mathfrak{W}' > \mathfrak{U}' \cap \mathfrak{V}'$. Since we are dealing with a pseudo Čech cycle, we have both

$$\pi_{\mathfrak{U}'\mathfrak{W}'}z_p(\mathfrak{W}') \sim z_p(\mathfrak{U}')$$

and

$$\pi_{\mathfrak{V}'\mathfrak{W}'}z_p(\mathfrak{W}') \sim z_p(\mathfrak{V}').$$

Now $\pi_{\mathfrak{U}\mathfrak{U}'}\pi_{\mathfrak{U}'\mathfrak{W}'}z_p(\mathfrak{W}')$ and $\pi_{\mathfrak{U}\mathfrak{V}'}\pi_{\mathfrak{V}'\mathfrak{W}'}z_p(\mathfrak{W}')$ are homologous because each is an image in $\mathfrak{U}$ of $z_p(\mathfrak{W}')$ under two projections of $\mathfrak{W}'$ into $\mathfrak{U}$, and Theorem 8–2 applies. Thus we have

$$\pi_{\mathfrak{U}\mathfrak{V}'}z_p(\mathfrak{V}') \sim \pi_{\mathfrak{U}\mathfrak{V}'}\pi_{\mathfrak{V}'\mathfrak{W}'}z_p(\mathfrak{W}') \sim \pi_{\mathfrak{U}\mathfrak{U}'}\pi_{\mathfrak{U}'\mathfrak{W}'}z_p(\mathfrak{W}') \sim \pi_{\mathfrak{U}\mathfrak{U}'}z_p(\mathfrak{U}').$$

This proves that $z_p(\mathfrak{U})$ is well-defined up to a homology.

To show that $\{z_p(\mathfrak{U})\}$ is a Čech cycle, let $\mathfrak{V} > \mathfrak{U}$ in $\Sigma(X)$. We need only show that $\pi_{\mathfrak{U}\mathfrak{V}}z_p(\mathfrak{V})$ is homologous to $z_p(\mathfrak{V})$. From $\Sigma'(X)$, choose a covering $\mathfrak{V}' > \mathfrak{V}$. Then

$$z_p(\mathfrak{V}) \sim \pi_{\mathfrak{V}\mathfrak{V}'}z_p(\mathfrak{V}')$$

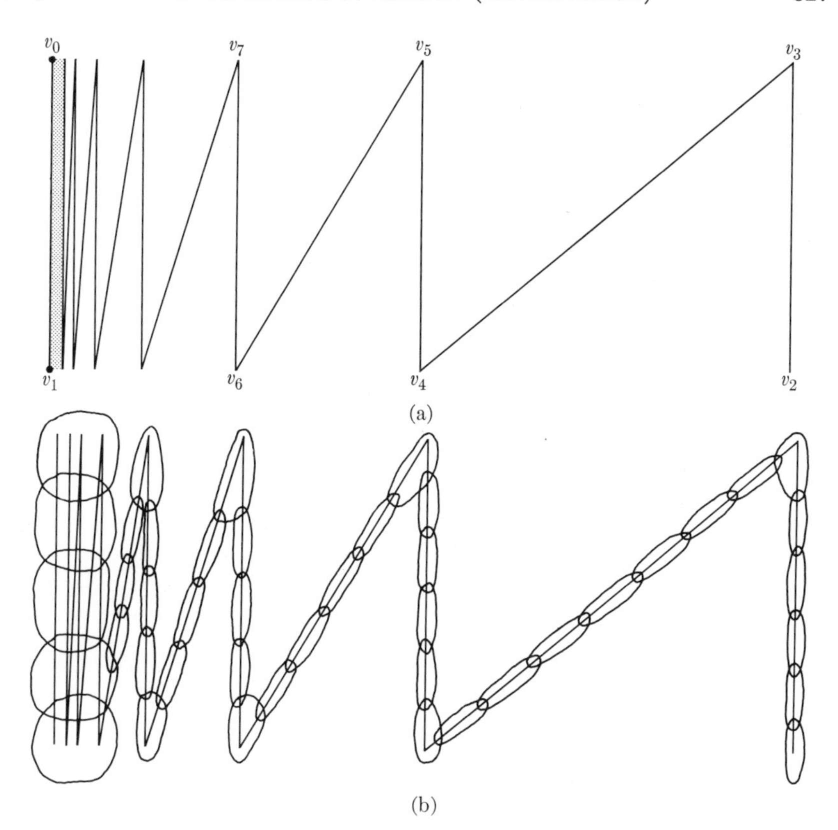

(a)

(b)

FIGURE 8–1

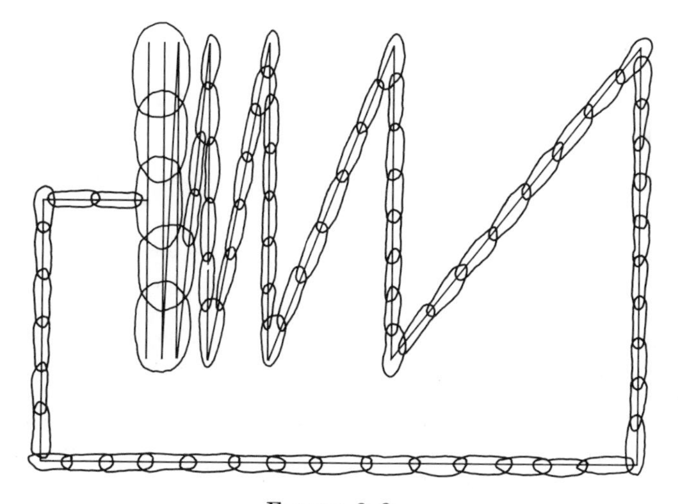

FIGURE 8–2

and $$z_p(\mathcal{U}) \sim \pi_{\mathcal{U}\mathcal{V}'} z_p(\mathcal{V}')$$

by the argument used above. Therefore we have

$$\pi_{\mathcal{U}\mathcal{V}} z_p(\mathcal{V}) \sim \pi_{\mathcal{U}\mathcal{V}} \pi_{\mathcal{V}\mathcal{V}'} z_p(\mathcal{V}') \sim \pi_{\mathcal{U}\mathcal{V}'} z_p(\mathcal{V}') \sim z_p(\mathcal{U}).$$

This proves that $\{z_p(\mathcal{U})\}$ is a Čech cycle. Since our construction is valid up to a homology, this proves that f_* is onto.

Next assume that the pseudo Čech cycle $\{z_p(\mathcal{U}')\}$ is homologous to zero. We show that the corresponding Čech cycle $\{z_p(\mathcal{U})\}$ constructed above is also homologous to zero. For any covering $\mathcal{U}$ in $\Sigma(X)$, let $\mathcal{U}'$ be an element of $\Sigma'(X)$ such that $\mathcal{U}' > \mathcal{U}$. By assumption, $z_p(\mathcal{U}') \sim 0$. Thus

$$z_p(\mathcal{U}) \sim \pi_{\mathcal{U}\mathcal{U}'} z_p(\mathcal{U}') \sim 0,$$

and each coordinate of the Čech cycle bounds. This completes the proof that f_* is an isomorphism of $H_p(X, G)$ onto $H_p(X, \Sigma'(X), G)$. □

A few examples will help to clarify the sometimes subtle differences between Čech homology theory and simplicial homology theory. First, however, we point out that the two theories agree on finite polytopes (see Section 8–2). Hence our examples must begin with infinite polytopes.

EXAMPLE 1. Consider the infinite geometric complex K indicated in Fig. 8–1(a), a triangulation of the topologist's sine curve. There is no finite sequence of 1-simplexes joining the vertices v_0 and v_2. Hence there are two combinatorial components of K even though the carrier $|K|$ is connected. It follows that the augmented simplicial homology group $\widetilde{H}_0(K, G)$ is isomorphic to G (see Section 6–6). On the other hand, the augmented Čech homology group $H_0(|K|, G)$ is trivial! To prove this directly, consider Fig. 8–1(b), in which we picture one member of a particular cofinal family of coverings. We cover the limit segment v_0v_1 with a simple chain of sets of diameter $1/n$ and then cover the rest of the set with another such simple chain.

In point of fact, the conclusion that $\widetilde{H}_0(|K|, G)$ is trivial is a consequence of the following result.

THEOREM 8-5. Let C be a Hausdorff continuum. Then the augmented Čech homology group $\widetilde{H}_0(C, G)$ is trivial.

For a proof of this theorem, the reader is referred to Section 11, Chapter V, of Wilder [42], or he may prove it himself as an exercise.

EXAMPLE 2. The infinite geometric complex K pictured in Fig. 8–2 has just one combinatorial component, but clearly it carries no simplicial 1-cycle except the trivial one. On the other hand, the carrier $|K|$, as imbedded in the plane, does have a nonbounding Čech 1-cycle. Here again a cofinal family of coverings may be constructed, each covering being the union of two simple chains as in Example 1, in such a way that the existence of the nonbounding 1-cycle is obvious.

From the above examples as well as by analogy to the simplicial theory, the reader may have inferred that a Čech 1-cycle is in some way associated with a connected subset of the space. The following example should correct such a mistaken impression.

EXAMPLE 3. Consider the annulus in Fig. 8–3 and the cofinal family of coverings $\{\mathfrak{U}_n\}$, where each $\mathfrak{U}_n$ consists of finitely many spherical neighborhoods of radius $1/n$. A Čech 1-cycle may be defined in such a way that $z_1(\mathfrak{U}_{2k})$ is determined by the covering of the outer circle in the boundary and $z_1(\mathfrak{U}_{2k+1})$ is determined by the covering of the inner circle. The necessary homologies connecting $z_1(\mathfrak{U}_n)$ and $\pi z_1(\mathfrak{U}_{n+1})$ are constructed on the entire covering $\mathfrak{U}_n$, of course. We should add that the resulting Čech cycle is in the same homology class as is one constructed on the covering of, say, the outer boundary alone.

EXAMPLE 4. Even the connecting homologies between $z_1(\mathfrak{U}_n)$ and $\pi z_1(\mathfrak{U}_{n+1})$ need not be over the same portion of the space at each covering. To illustrate this point, consider the torus and a cofinal family of coverings such as might be constructed using spherical neighborhoods of radius $1/n$. Define a Čech 1-cycle (actually a pseudo Čech cycle) on these coverings $\{\mathfrak{U}_n\}$ as follows. Let $z_1(\mathfrak{U}_{2k})$ be determined by the covering of the circle J_1, and let $z_1(\mathfrak{U}_{2k+1})$ be determined by the covering of J_2 (see Fig. 8–4). For $\mathfrak{U}_{2k}$, we construct the homology connecting $z_1(\mathfrak{U}_{2k})$ and $\pi z_1(\mathfrak{U}_{2k+1})$ on the covering of the upper half of the torus, and for $\mathfrak{U}_{2k+1}$, we construct the homology between $z_1(\mathfrak{U}_{2k+1})$ and $\pi z_1(\mathfrak{U}_{2k+2})$ on the lower half. Again the resulting cycle is in the same homology class as is the cycle obtained by considering only the coverings of, say, J_1.

To begin to clarify the restrictions that were placed upon the coefficient group G used in Čech homology theory, we consider next a more complicated example.

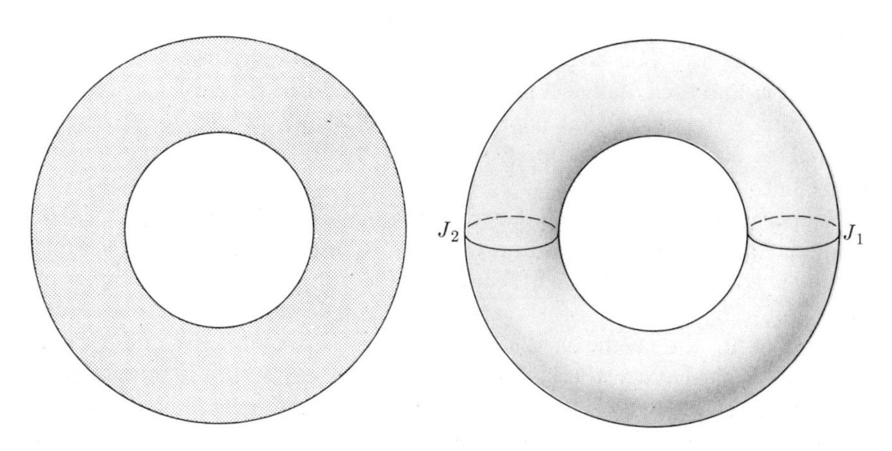

FIGURE 8–3 FIGURE 8–4

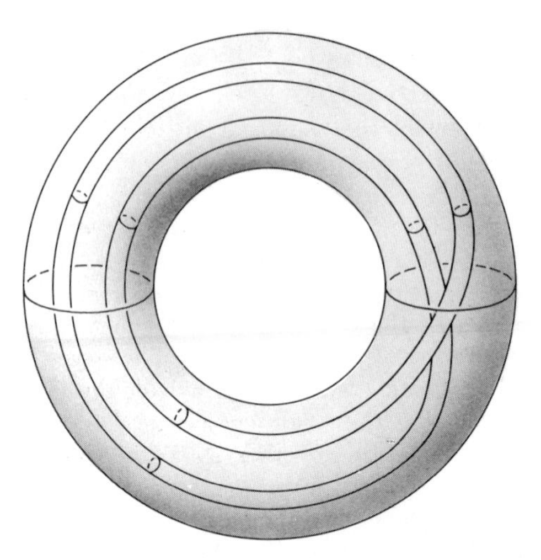

FIG. 8–5. The imbedding of tori in constructing a solenoid.

EXAMPLE 5. Let S denote the solenoid (see Section 3–8). We may consider S to be constructed as follows. Given a solid torus (or anchor ring) T_1 in E^3, let T_2 be another solid torus imbedded in the interior of T_1 as shown in Fig. 8–5. Then let T_3 be a solid torus imbedded in T_2 as T_2 is imbedded in T_1, etc. The intersection $\bigcap_{n=1}^{\infty} T_n$ is the solenoid S as we will use it here.

From an intuitive standpoint, one expects that the solenoid S must carry a nontrivial 1-cycle if for no other reason than that S certainly links such a circle as J in Fig. 8–5. But requiring that S carry a nontrivial Čech 1-cycle imposes certain restrictions upon the coefficient group used in the homology theory. To see how such restrictions come about, we construct a cofinal family (sequence) $\{\mathfrak{U}_n\}$ of coverings of S by covering each T_n with a finite number of open connected sets of diameter less than $1/n$. Furthermore, we choose these coverings in such a way that $\mathfrak{U}_{n+1} > \mathfrak{U}_n$ for each n, and each $\mathfrak{U}_n$ has a polygonal simple closed curve as a geometric realization. It follows that a projection π of $\mathfrak{U}_{n+1}$ into $\mathfrak{U}_n$ may be considered as a simplicial mapping of degree 2 carrying one simple closed curve onto another.

Suppose that z_1 is a Čech 1-cycle of S. Each coordinate $z_1(\mathfrak{U}_n)$ is a cycle on $\mathfrak{U}_n$, and we may assume $\mathfrak{U}_n$ to be so oriented that $z_1(\mathfrak{U}_n)$ assigns the same coefficient g_n to each 1-simplex of $\mathfrak{U}_n$. Consider a projection π of $\mathfrak{U}_{n+1}$ into $\mathfrak{U}_n$. By our construction, $\pi z_1(\mathfrak{U}_{n+1})$ must assign to each 1-simplex of $\mathfrak{U}_n$ the coefficient $2g_{n+1}$. Since z_1 was assumed to be a Čech cycle, we must have $\pi z_1(\mathfrak{U}_{n+1})$ homologous to $z_1(\mathfrak{U}_n)$, and it follows that we must have $g_n = 2g_{n+1}$. This being true for all n, one readily sees that if g is any element of the coefficient group G to be used as the coefficient of $z_1(\mathfrak{U}_1)$, then G must also contain all elements of the form $g/2^k$. Since this condition is not satisfied by an arbitrary group, we may conclude that for the solenoid to have the desired 1-cycle, we must exercise judgment in the choice of the coefficient group. For example, we may use a field of coefficients, or a vector space over a field, or a compact abelian topological group. Certainly we could not use the integers, nor any cyclic group of order 2^m, for then the only Čech 1-cycle on S would be the trivial one having all coefficients equal to zero.

By simple alterations in the construction of the solenoid, it should be obvious that we could obtain examples which would force us to avoid cyclic coefficient groups of orders $3^k, 5^m, \ldots$. It is in part the existence of such spaces that imposes the restrictions placed upon the coefficient groups in Čech homology theory. There are other reasons as well, and we will mention one more in Section 8–3.

EXAMPLE 6. A metric continuum M is *snakelike* if, given any positive number ϵ, there is a simple chain $U_1, U_2, \ldots, U_n$ of open sets of diameter less than ϵ covering M. (The "links" of the chain are not assumed to be connected.) An arc is obviously snakelike, the set in Example 1 above is snakelike, and so is the pseudo-arc (see Section 3–8). Three arcs with an end point in common (a triod) is not snakelike.

If M is a snakelike metric continuum, and if $\mathfrak{U}$ is any covering in $\Sigma(M)$, then by the Lebesque covering theorem (1–32), there is a positive number ϵ such that every subset of M having diameter $<\epsilon$ lies in some element of $\mathfrak{U}$. Hence the ϵ-chain assumed in the definition of M is a refinement of $\mathfrak{U}$. We may conclude that the $(1/n)$-chains, call them $\mathfrak{U}_n$, constitute a cofinal family of coverings of M. Since no such simple chain carries a nontrivial simplicial p-cycle, $p > 0$, it follows that the Čech homology groups $H_p(M, G)$, $p > 0$, are all trivial. Suppose that $z_0 = \{z_0(\mathfrak{U}_n)\}$ is an augmented pseudo Čech 0-cycle of M. Each coordinate $z_0(\mathfrak{U}_n)$ bounds on the simple chain $\mathfrak{U}_n$ whose nerve is isomorphic to a polygonal arc. Thus we may prove that the augmented Čech group $\widetilde{H}_0(M, G)$ is also trivial. This proves that each snakelike continuum has the same Čech homology structure as does the arc. But a snakelike continuum need not resemble an arc at all! Hence we see that even in such an apparently simple case, the Čech homology groups fail to characterize the space on which they are defined.

In his definitive paper [58] on snakelike continua, Bing gives the following example of a snakelike continuum with just one "end point" (see the reference for the definition of end point). Let C denote the Cantor set on the interval $0 \leq x \leq 1$ on the x-axis in E^2. Let M_0 denote the set of all closed semicircles in the upper half-plane with center at $(\frac{1}{2}, 0)$ and end points in C. For $i = 1, 2, 3, \ldots$, let M_i denote the set of all semicircles in the lower half-plane with center at $(\frac{5}{2}) \cdot 3^i$, 0 and end points

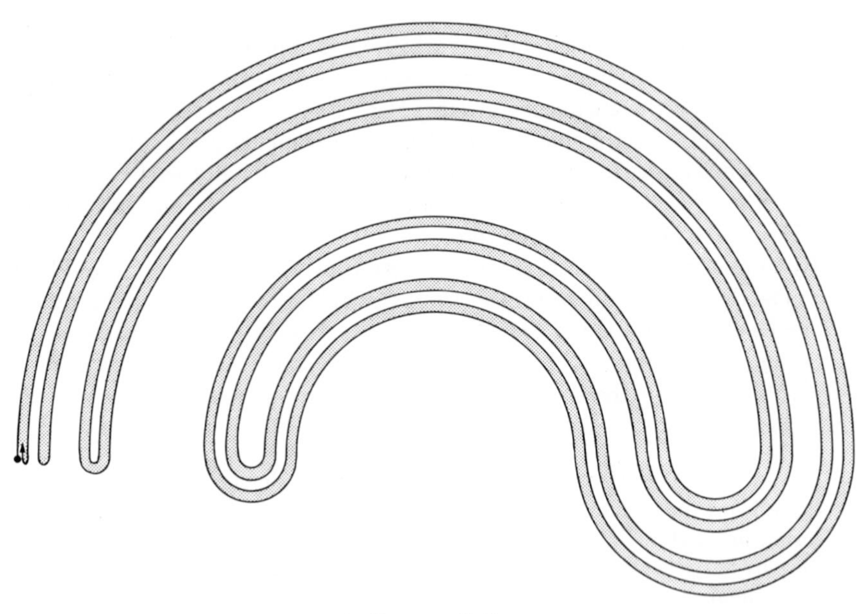

FIGURE 8–6

in C. The continuum $M = \cup_{i=0}^{\infty} M_i$ is snakelike and has just one end point, the origin. We have pictured this set in Fig. 8–6. If we add to this set its reflection in the y-axis, we obtain a snakelike continuum with no end points. On the other hand, Bing proves that *a nondegenerate snakelike continuum is a pseudo-arc if and only if each of its points is an end point!*

8–2 The topological invariance of simplicial homology groups. Since the Čech homology groups of any space are topologically invariant, we may prove the invariance of the simplicial homology groups of a finite polytope merely by exhibiting an isomorphism of the Čech groups onto the simplicial groups. Such is the goal of this section.

Let K be a finite geometric complex, and let $|K|$ be the polytope carrier of K. Let $K^{(n)}$ denote the nth barycentric subdivision of K. A vertex $v^{(n+1)}$ of $K^{(n+1)}$ is said to be *barycentrically related* to a vertex $v^{(n)}$ of $K^{(n)}$ if $v^{(n)}$ is any vertex of that simplex of $K^{(n)}$ whose barycenter is $v^{(n+1)}$. For instance, in Fig. 8–7, v' is barycentrically related to v_1 and to v_2 but not to v_0.

Then a vertex $v^{(n+k)}$ of $K^{(n+k)}$ is barycentrically related to $v^{(n)}$ if there is a sequence $v^{(n+k-1)}, \ldots, v^{(n+1)}$ such that $v^{(n+i+1)}$ is barycentrically related to $v^{(n+i)}$ for each $i = k-1, k-2, \ldots, 1, 0$.

For any vertex $v^{(n)}$ of $K^{(n)}$, let $\mathring{\mathrm{St}}(v^{(n)})$ denote the *open star* of $v^{(n)}$, that is, the collection of all open simplexes of $K^{(n)}$ having $v^{(n)}$ as a vertex. In

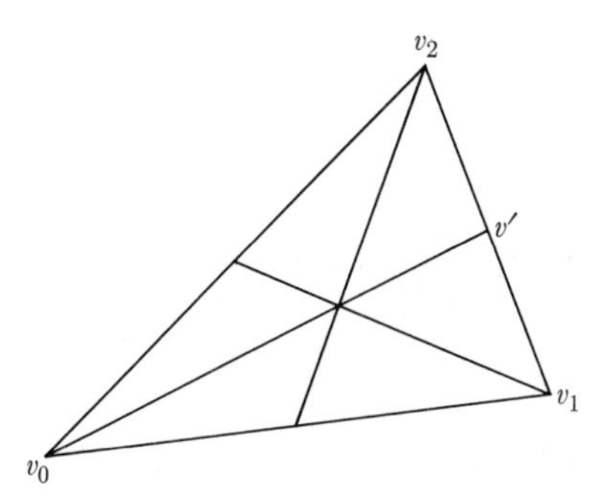

FIGURE 8–7

the polytope $|K|$, the carrier $|\mathring{\text{S}}\text{t}(v^{(n)})|$ is an open set, and the collection of all such open sets constitutes a finite open covering of $|K|$ (see Section 5–4).

THEOREM 8–6. *Let K be a finite geometric simplicial complex, and let $n > k \geqq 0$. If there is a sequence of vertices $v^{(i)}$, $i = k, k+1, \ldots, n$, with $v^{(i)}$ a vertex of $K^{(i)}$ and $v^{(k)}$ barycentrically related to $v^{(n)}$, then $\mathring{\text{S}}\text{t}(v^{(n)})$ is contained in $\mathring{\text{S}}\text{t}(v^{(k)})$.*

Proof: If $n = k + 1$, let s^q be an open simplex of $K^{(n)}$ having $v^{(n)}$ as a vertex. The vertices of s^q are barycenters of simplexes of $K^{(k)}$. Let $\dot{\sigma}_i$, $i = 0, 1, \ldots, q$, denote the vertices of s^q. As ordered, the simplex σ_q in $K^{(k)}$ is the carrier of s^q and, as open sets, σ_q contains s^q. By assumption, $v^{(n)}$ is barycentrically related to $v^{(k)}$, a vertex of that simplex of $K^{(k)}$ of which $v^{(n)}$ is the barycenter. Thus $v^{(k)}$ is a vertex of σ_q, and for any open simplex s^q in $K^{(n)}$ having $v^{(n)}$ as a vertex, we know that s^q lies in $\mathring{\text{S}}\text{t}(v^{(k)})$. This establishes the theorem for $n = k + 1$, and a finite induction completes the proof. □

For the remainder of this section, we adopt the following conventions. Let K be a finite geometric simplicial complex with polytope carrier $|K|$, and

(1) let v_i^v, $i = i, 2, \ldots, i(n)$, be the vertices of the nth barycentric subdivision $K^{(n)}$ of K,

(2) let $\mathfrak{U}_n = \{|\mathring{\text{S}}\text{t}(v_i^n)|\}$, $i = 1, \ldots, i(n)$, be the covering of $|K|$ by the open stars of vertices of $K^{(n)}$ (we denote the collection $\{\mathfrak{U}_n\}$ by Σ' and will show that Σ' is a cofinal family of coverings of $|K|$), and

(3) let f_n be the simplicial mapping of the nerve of $\mathfrak{U}_n$ onto $K^{(n)}$ defined by $f_n(|\mathring{\text{S}}\text{t}(v_i^n)|) = v_i^n$, $i = 1, 2, \ldots, i(n)$.

It is easily shown that f_n is a one-to-one simplicial mapping of $\mathfrak{U}_n$ onto $K^{(n)}$ and that f_n^{-1} is also simplicial. Therefore $\mathfrak{U}_n$ and $K^{(n)}$ are isomorphic, and we have the following result.

LEMMA 8–7. $K^{(n)}$ is a geometric realization of the complex $\mathfrak{U}_n$ for each n.

Also, we can state the next lemma, which is merely a rewording of the remark following Theorem 6–23.

LEMMA 8–8. The simplicial mapping f_n induces an isomorphism of the groups $Z_p(\mathfrak{U}_n, G)$ onto $Z_p(K^{(n)}, G)$ for each dimension p and any coefficient group G. Hence f_n induces an isomorphism of $H_p(\mathfrak{U}_n, G)$ onto $H_p(K^{(n)}, G)$.

LEMMA 8–9. The collection Σ' is a cofinal family of open coverings of $|K|$.

Proof: From Section 5–4, as was remarked above, we know that each set $|\mathring{\mathrm{S}}\mathrm{t}(v_i^n)|$ is open and that $\mathfrak{U}_n$ is a covering of $|K|$. Now let $\mathfrak{U}$ be any finite open covering of the compact metric space $|K|$. Let ϵ be the Lebesgue number of $\mathfrak{U}$ (see Theorem 1–32). In view of Theorem 5–20, there is an integer N such that the mesh of $K^{(n)}$ is less than $\epsilon/2$ if $n > N$. Thus, since each simplex has diameter $< \epsilon/2$, each open star has diameter $< \epsilon$. It follows that for $n > N$, $\mathfrak{U}_n$ is a refinement of $\mathfrak{U}$. This proves that Σ' is a cofinal family (actually a cofinal sequence) of coverings. □

By virtue of Theorem 8–4, we may now state that the Čech homology group $H_p(|K|, G)$ is isomorphic to the pseudo Čech group $H_p(|K|, \Sigma', G)$, defined on the cofinal sequence $\Sigma' = \{\mathfrak{U}_n\}$, for each dimension p. Our goal in this section will be attained by showing that this pseudo Čech group is isomorphic to the simplicial homology group $H_p(K, G)$.

THEOREM 8–10. If K is a finite geometric simplicial complex with polytope carrier $|K|$, then for each dimension p the simplicial homology group $H_p(K, G)$ is isomorphic to the pseudo Čech group $H_p(|K|, \Sigma', G)$.

Proof: Considering $K^{(n)}$ as the first barycentric subdivision of $K^{(n-1)}$, the construction used in proving Theorem 6–23 yields a chain-mapping of $C_p(K^{(n-1)}, G)$ into $C_p(K^{(n)}, G)$. This chain-mapping was denoted by u in the proof of Theorem 6–23, but it is convenient to use the symbol $u_{n-1,n}$ here. For $n > k + 1$, we iterate this mapping to obtain

$$u_{k,n} = u_{n-1,n} \cdot u_{n-2,n-1} \cdots u_{k,k+1}.$$

Also in analogy to the simplicial mapping u' of Theorem 6–23, we define a simplicial mapping $w_{n,k}$ of $K^{(n)}$ into $K^{(k)}$, $n > k$, by setting $w_{n,k}(v_i^n)$ equal to any vertex v_j^k of that lowest-dimensional simplex in $K^{(k)}$ that contains v_i^n as a point. We also denote by $w_{n,k}$ the induced chain-mapping. Again citing Theorem 6–23, we see that

(i) $u_{k,n} \cdot w_{n,k}$ is chain-homotopic to the identity on $K^{(n)}$ and

(ii) $w_{n,k} \cdot u_{k,n}$ is the identity on $K^{(k)}$.

By Theorem 6–20, this proves that the induced homomorphisms ${}_*u_{k,n}$ and ${}_*w_{n,k}$ on the homology groups are actually isomorphisms onto.

In view of Theorem 8–6, we know that $\mathring{\mathrm{S}}\mathrm{t}(v_i^n)$ lies in $\mathring{\mathrm{S}}\mathrm{t}[w_{n,k}(v_i^n)]$ since v_i^n is barycentrically related to $w_{n,k}(v_i^n)$, by definition. Our notation will be simplified if we denote K by $K^{(0)}$ and the identity mapping of K onto itself by $u_{0,0}$. Then $u_{0,n}$ is a chain-mapping of K into $K^{(n)}$, etc.

We next define a transformation r on the cycles of K as follows. Let z_p be an element of the simplicial cycle group $Z_p(K, G)$. Define the transformation r of $Z_p(K, G)$ into $Z_p(|K|, \Sigma', G)$ by setting

$$r(z_p) = \{{}_*f_n^{-1}[u_{0,n}(z_p)]\}.$$

Analyzing this, we see that $u_{0,n}(z_p)$ is a cycle on $K^{(n)}$ which is carried over by ${}_*f_n^{-1}$ to a cycle on the isomorphic complex $\mathfrak{U}_n$. On homology cosets, r induces a transformation r_* defined by

$$r_*([z_p]) = [r(z_p)].$$

We will show that r_* is an isomorphism onto.

It is inherent in the definition that there is a coordinate of $r(z_p)$ on every element $\mathfrak{U}_n$ of Σ'. We need only establish the requisite projection property to prove that $r(z_p)$ is a pseudo Čech cycle. To this end, assume that $\mathfrak{U}_n > \mathfrak{U}_k$, implying that $n > k$. Define the simplicial mapping

$$\pi_{n,k} = f_k^{-1} w_{n,k} f_n.$$

Note that $f_n{:}\mathfrak{U}_n \to K^{(n)}$, $w_{n,k}{:}K^{(n)} \to K^{(k)}$, and $f_k^{-1}{:}K^{(k)} \to \mathfrak{U}_k$. Since each of its factors is simplicial, so is $\pi_{n,k}$. We prove that $\pi_{n,k}$ is a projection of $\mathfrak{U}_n$ into $\mathfrak{U}_k$. Consider a "vertex" $|\mathring{\mathrm{S}}\mathrm{t}(v_i^n)|$ of $\mathfrak{U}_n$. By our definition, $f_n(|\mathring{\mathrm{S}}\mathrm{t}(v_i^n)|) = v_i^n$. Then $w_{n,k}(v_i^n)$ is barycentrically related to v_i^n, and by Theorem 8–6, $\mathring{\mathrm{S}}\mathrm{t}(v_i^n)$ lies in $\mathring{\mathrm{S}}\mathrm{t}[w_{n,k}(v_i^n)]$. Then we have $f_k^{-1}[w_{n,k}(v_i^n)] = \mathring{\mathrm{S}}\mathrm{t}[w_{n,k}(v_i^n)]$, and $\mathring{\mathrm{S}}\mathrm{t}(v_i^n)$ lies in $f_k^{-1}w_{n,k}f_n[\mathring{\mathrm{S}}\mathrm{t}(v_i^n)]$. Hence $\pi_{n,k}$ is a projection. As usual, we also let $\pi_{n,k}$ denote the induced chain-mapping of $\mathfrak{U}_n$ into $\mathfrak{U}_k$.

Now, for $n > k$, we wish to show that

$$\pi_{n,k}({}_*f_n^{-1}[u_{0,n}(z_p)]) \sim {}_*f_k^{-1}[u_{0,k}(z_p)].$$

Writing out all the factors of $\pi_{n,k}$, we have

$${}_*f_k^{-1}{}_*w_{n,k*}f_{n*}f_n^{-1}{}_*u_{0,n}(z_p) = {}_*f_k^{-1}w_{n,k}u_{0,n}(z_p)$$

By construction,

$${}_*f_k^{-1}w_{n,k}u_{0,n}(z_p) \sim {}_*f_k^{-1} \cdot w_{n,k} \cdot u_{k,n} \cdot u_{0,k}(z_p)$$

and, since $w_{n,k} \cdot u_{k,n}$ is the identity on homology cosets,

$$_*f_k^{-1} w_{n,k} u_{k,n} u_{0,k}(z_p) \sim {_*f_k^{-1}} u_{0,k}(z_p).$$

This proves that $r(z_p)$ is a pseudo Čech cycle. Since both $_*f_n^{-1}$ and $_*u_{0,n}$ are homomorphisms, it follows that r_* is also a homomorphism:

$$r_* : H_p(K, G) \to H_p(|K|, \Sigma', G).$$

To prove that r_* is an isomorphism onto, we first show that r_* *is single-valued.* If $z_p^1 \sim z_p^2$ on K, then for each $\mathfrak{U}_n$, we have

$$_*f_n^{-1}[u_{0,n}(z_p^1)] \sim {_*f_n^{-1}}[u_{0,n}(z_p^2)]$$

on $\mathfrak{U}_n$ because both $_*f_n^{-1}$ and $_*u_{0,n}$ are isomorphisms on homology groups. Thus r_* is well-defined.

Next we show that r_* is one-to-one. To do this, assume that $r(z_p^1) \sim r(z_p^2)$; that is, $r_*[z_p^1] = r_*[z_p^2]$. This homology relation holds only if the corresponding coordinates in each covering are homologous. This applies to $\mathfrak{U}_0$, too! Thus

$$_*f_0^{-1} u_{0,0}(z_p^1) \sim {_*f_0^{-1}} u_{0,0}(z_p^2).$$

Since $u_{0,0}$ is the identity, we have

$$_*f_0^{-1}(z_p^1) \sim {_*f_0^{-1}}(z_p^2),$$

and applying the isomorphism $_*f_0$ to both sides of this homology gives the desired result, $z_p^1 \sim z_p^2$.

Finally we show that r_* *is onto.* Let $\{z_p(\mathfrak{U}_n)\}$ be any pseudo Čech cycle of $|K|$. We find a cycle z_p of K such that $r(z_p)$ is homologous to $\{z_p(\mathfrak{U}_n)\}$. To do so, take the coordinate of $\{z_p(\mathfrak{U}_n)\}$ on $\mathfrak{U}_0$ and move it over onto $K^{(0)} = K$ by means of the mapping f_0. That is, let $z_p = f_0[z_p(\mathfrak{U}_0)]$. We must show that, with this choice of z_p, $r(z_p) \sim \{z_p(\mathfrak{U}_n)\}$. This entails proving that

$$f_n^{-1}[u_{0,n}(z_p)] \sim z_p(\mathfrak{U}_n) \qquad \text{for each } n.$$

Projecting the coordinate $z_p(\mathfrak{U}_n)$ into $\mathfrak{U}_0$, we have

$$\pi_{n,0}[z_p(\mathfrak{U}_n)] = f_0^{-1} w_{n,0} f_n[z_p(\mathfrak{U}_n)].$$

Since $\{z_p(\mathfrak{U}_n)\}$ is a pseudo Čech cycle, we must have that

$$\pi_{n,0}[z_p(\mathfrak{U}_n)] \sim z_p(\mathfrak{U}_0) = f_0^{-1}(z_p),$$

by our choice of z_p. Applying $f_n^{-1} u_{0,n} f_0$ to both sides of this homology, we obtain

$$f_n^{-1} u_{0,n} f_0 f_0^{-1} w_{n,0} f_n[z_p(\mathfrak{U}_n)] \sim f_n^{-1} u_{0,n} f_0 f_0^{-1}(z_p).$$

Since ${}_*f_0$, ${}_*f_0^{-1}$, ${}_*f_n^{-1}$, ${}_*f_n$, and ${}_*u_{0,n*}w_{n,0}$ are all identity isomorphisms on homology classes, it follows that

$$z_p(\mathcal{U}_n) \sim f_n^{-1}u_{0,n}(z_p).$$

This proves that r_* is onto. □

THEOREM 8–11. The simplicial homology groups of a triangulation K of a finite polytope $|K|$ are topological invariants of the polytope and do not depend upon the particular triangulation K.

Proof: Combine Theorems 8–10 and 8–4 with the fact that the Čech homology groups are topologically invariant. □

This last result is the justification for our use of simplicial theory. In spite of the noninvariant machinery of oriented complexes, etc., used in constructing simplicial homology groups, we actually obtain an invariant of the underlying polytope. We might note also that the argument above suffices to establish the existence of Čech homology groups in the case of finite polytopes, with no restriction upon the coefficient group G.

8–3 Čech homology theory (continued). The reader will have discovered an inverse limit system hidden behind the development of Čech theory as given in Section 8–1. We pause to formulate the Čech theory in terms of limit systems, which is a more modern viewpoint. Our brief exposition is intended largely to indicate a direction for further study.

As was remarked in Section 8–1, the collection $\Sigma(X)$ of all finite open coverings of a space X is a directed set under the partial ordering of refinement. For each element $\mathcal{U}$ of $\Sigma(X)$, we may define the groups $H_p(\mathcal{U}, G)$ and $H^p(\mathcal{U}, G)$, the pth homology and cohomology groups of the simplicial complex $\mathcal{U}$ over an abelian group G. And whenever $\mathcal{V} > \mathcal{U}$ in $\Sigma(X)$, we have the projection-induced homomorphisms

$${}_*\pi_{\mathcal{U}\mathcal{V}}: H_p(\mathcal{V}, G) \to H_p(\mathcal{U}, G)$$

and

$${}^*\pi_{\mathcal{U}\mathcal{V}}: H^p(\mathcal{U}, G) \to H^p(\mathcal{V}, G).$$

These satisfy the condition that if $\mathcal{W} > \mathcal{V} > \mathcal{U}$ in $\Sigma(X)$, then

$${}_*\pi_{\mathcal{U}\mathcal{V}*}\pi_{\mathcal{V}\mathcal{W}} = {}_*\pi_{\mathcal{U}\mathcal{W}}$$

and

$${}^*\pi_{\mathcal{V}\mathcal{W}}{}^*\pi_{\mathcal{U}\mathcal{V}} = {}^*\pi_{\mathcal{U}\mathcal{W}}.$$

If we let $H_p(\Sigma)$ denote the collection $\{H_p(\Sigma, G)\}$ and ${}_*\pi$ denote the collection $\{{}_*\pi_{\mathcal{U}\mathcal{V}}\}$, the pair $[H_p(\Sigma), {}_*\pi]$ is an inverse limit system over the directed set $\Sigma(X)$, the pth *Čech homology system* of X with coefficients in

G. Similarly, if $H^p(\Sigma) = \{H^p(\mathfrak{U}, G)\}$ and $^*\pi = \{^*\pi_{\mathfrak{U}\mathfrak{V}}\}$, then the pair $[H^p(\Sigma), {}^*\pi]$ is a direct limit system over $\Sigma(X)$, the pth *Čech cohomology system* of X with coefficients in G. It becomes simply a matter of checking the definitions to see that the pth Čech homology group $H_p(X, G)$ is precisely the inverse limit group of the Čech homology system $[H_p(\Sigma), {}_*\pi]$. And we now *define* the pth Čech cohomology group $H^p(X, G)$ to be the direct limit group of the Čech cohomology system $[H^p(\Sigma), {}^*\pi]$.

The existence of these Čech groups clearly depends upon the theory of inverse and direct limit groups. We will simply state the conditions under which they can be shown to be defined. The Čech homology groups $H_p(X, G)$ can be shown to exist for any space X and any module G over a ring, and then $H_p(X, G)$ will be a module over the same ring. In addition, if X is compact, we may take G to be a compact abelian topological group, and then $H_p(X, G)$ will be the same. On the other hand, while the cohomology groups $H^p(X, G)$ are also defined for any space and any module G over a ring, they are not meaningfully defined for compact topological groups.

As was stated explicitly in Lemma 2–93, elements of a direct limit group such as $H^p(X, G)$ are easier to construct than are elements of an inverse limit group such as $H_p(X, G)$. In an inverse limit group, any particular coordinate of a given element controls only those coordinates which precede it in the ordering. Hence, to construct a Čech cycle, one must find coordinates on every covering of (at least) a cofinal family of coverings. In general, this can be a troublesome task. On the other hand, if one finds a cocycle on any covering of the space X, then he has a Čech cocycle! For, speaking intuitively, all refinements of any given covering $\mathfrak{U}$ constitute a cofinal family and, since a cocycle $z^p(\mathfrak{U})$ determines the cohomology class of $z^p(\mathfrak{V})$ for each $\mathfrak{V} > \mathfrak{U}$, it follows that $z^p(\mathfrak{U})$ determines an element of $H^p(X, G)$. This fact, together with the algebraic duality theorems mentioned in Section 7–16, has simplified many arguments in homology theory. For an important example of this procedure, the reader is referred to Chapters VII and VIII of Wilder [42].

The technique of limit systems also may be applied to define the relative Čech groups. First, let A be a closed subset of a space X, and let $\mathfrak{U}$ be an element of $\Sigma(X)$. A *simplex* $\langle U_0 \cdots U_p \rangle$ of $\mathfrak{U}$ is *on* A if and only if the intersection $\bigcap_{i=0}^p U_i$ meets A.

LEMMA 8–12. The collection of all simplexes of $\mathfrak{U}$ on A is a closed subcomplex $\mathfrak{U}_A$ of $\mathfrak{U}$.

The proof is left as an exercise.

In view of Lemma 8–12, we may define the relative simplicial groups $H_p(\mathfrak{U}/\mathfrak{U}_A, G)$ and $H^p(\mathfrak{U}/\mathfrak{U}_A, G)$ over a coefficient group G. Another easy exercise will provide a proof of the following result.

LEMMA 8-13. If $\mathcal{V} > \mathcal{U}$ in $\Sigma(X)$, and if $\pi_{\mathcal{U}\mathcal{V}}$ is a projection of $\mathcal{V}$ into $\mathcal{U}$, then $\pi_{\mathcal{U}\mathcal{V}}$ projects $\mathcal{V}_A$ into $\mathcal{U}_A$.

According to the definitions in Section 7-3, Lemma 8-13 states that each projection $\pi_{\mathcal{U}\mathcal{V}}$ is a simplicial mapping of the pair $(\mathcal{V}, \mathcal{V}_A)$ into the pair $(\mathcal{U}, \mathcal{U}_A)$. We know, too, that any two projections $\pi_{\mathcal{U}\mathcal{V}}$ and $\pi'_{\mathcal{U}\mathcal{V}}$ of $\mathcal{V}$ into $\mathcal{U}$ are chain-homotopic. It follows that the induced homomorphisms

$$ {}_*\pi_{\mathcal{U}\mathcal{V}}:H_p\left(\frac{\mathcal{V}}{\mathcal{V}_A}, G\right) \to H_p\left(\frac{\mathcal{U}}{\mathcal{U}_A}, G\right) $$

and

$$ {}^*\pi_{\mathcal{U}\mathcal{V}}:H^p\left(\frac{\mathcal{U}}{\mathcal{U}_A}, G\right) \to H^p\left(\frac{\mathcal{V}}{\mathcal{V}_A}, G\right) $$

depend only upon the order relation between $\mathcal{U}$ and $\mathcal{V}$ and not upon the particular projection $\pi_{\mathcal{U}\mathcal{V}}$. Thus the collection $H_p(\Sigma/\Sigma_A) = \{H_p(\mathcal{U}/\mathcal{U}_A, G)$ and the collection ${}_*\pi = \{{}_*\pi_{\mathcal{U}\mathcal{V}}\}$ together constitute an inverse limit system over $\Sigma(X)$, while the pair $[H^p(\Sigma/\Sigma_A), {}^*\pi]$, defined analogously, is a direct limit system over $\Sigma(X)$. The pth *relative Čech homology group of X mod A*, $H_p(X/A, G)$, is the inverse limit group of the system $[H_p(\Sigma/\Sigma_A), {}_*\pi]$, and the pth *relative Čech cohomology group of X mod A*, $H^p(X/A, G)$, is the direct limit group of the system $[H^p(\Sigma/\Sigma_A), {}^*\pi]$.

By combining these definitions with the concept of a mapping of one inverse (direct) limit system into another, it is possible to construct the Čech homology (cohomology) sequence of the pair (X, A). For complete details of this construction, the reader is referred to Eilenberg and Steenrod [7], Chapters VIII and XI. The necessity for restricting the coefficient group G also arises here. For the Čech homology sequence to be an exact sequence, the space X must be compact, and the group G must be either a compact abelian topological group or a vector space over a field. This restriction may be lifted, however, if (X, A) is a triangulated pair, in which case the exactness theorem can be established with any coefficient group.

By means of similar techniques, one may also prove the excision theorem for relative Čech homology groups. Indeed, in the reference made above, a complete verification of the Eilenberg-Steenrod axioms (see Section 7-7) is given. We will merely exemplify this procedure with the single instance presented in the next section.

8-4 Induced homomorphisms. We have mentioned frequently that a continuous mapping induces homomorphisms on homology groups. This fact will be verified in this section, thereby proving the homology analogue of the corresponding situation in homotopy theory (see Theorem 4-28).

Let $f:X \to Y$ be a continuous mapping of X into Y, where both X and Y are compact Hausdorff spaces. Then the inverse transformation f^{-1}

carries each open covering $\mathfrak{U}$ in $\Sigma(Y)$ onto an open covering $f^{-1}(\mathfrak{U})$ in $\Sigma(X)$. Since all intersections of elements in $f^{-1}(\mathfrak{U})$ are preserved by f, we have the following result.

LEMMA 8–14. If $f:X \to Y$ is continuous, and if $\mathfrak{U}$ is an element of $\Sigma(Y)$, then $f^{-1}(\mathfrak{U})$ is isomorphic to a subcomplex of $\mathfrak{U}$ under an injection $f_{\mathfrak{U}}$.

Proof: If $\langle U_0 \cdots U_p \rangle$ is a simplex of $\mathfrak{U}$, then $\bigcap_{i=0}^{p} U_i$ is not empty, and neither is $\bigcap_{i=0}^{p} f^{-1}(U_i)$. This implies that $\langle f^{-1}(U_0) \cdots f^{-1}(U_p) \rangle$ is a simplex of $f^{-1}(\mathfrak{U})$. Let $f_{\mathfrak{U}}$ be defined by setting

$$f_{\mathfrak{U}}(f^{-1}(U_i)) = U_i$$

for each nonempty set $f^{-1}(U_i)$. Clearly $f_{\mathfrak{U}}$ is a one-to-one simplicial mapping of $f^{-1}(\mathfrak{U})$ into $\mathfrak{U}$. Since f was not assumed to be onto, there may be open sets in $\mathfrak{U}$ which have empty inverse images in X. Hence $f^{-1}(\mathfrak{U})$ is isomorphic only to a subcomplex of $\mathfrak{U}$. □

LEMMA 8–15. If $f:X \to Y$ is continuous, and if $\mathfrak{V} > \mathfrak{U}$ in $\Sigma(Y)$, then $f^{-1}(\mathfrak{V}) > f^{-1}(\mathfrak{U})$ in $\Sigma(X)$. Furthermore, if $\pi_{\mathfrak{U}\mathfrak{V}}:\mathfrak{V} \to \mathfrak{U}$ is a projection of $\mathfrak{V}$ into $\mathfrak{U}$, then $\pi_{\mathfrak{U}\mathfrak{V}}$ carries $f^{-1}(\mathfrak{V})$ into $f^{-1}(\mathfrak{U})$ (these being considered as subcomplexes of $\mathfrak{V}$ and $\mathfrak{U}$ after Lemma 8–14). If $\pi'_{\mathfrak{U}\mathfrak{V}}$ is the mapping of $f^{-1}(\mathfrak{V})$ into $f^{-1}(\mathfrak{U})$ in $\Sigma(X)$ defined by $\pi_{\mathfrak{U}\mathfrak{V}}$, then $\pi'_{\mathfrak{U}\mathfrak{V}}$ is a projection, and we have commutativity in the diagram

$$\begin{array}{ccc} f^{-1}(\mathfrak{V}) & \xrightarrow{\pi'} & f^{-1}(\mathfrak{U}) \\ {\scriptstyle f_{\mathfrak{V}}}\downarrow & & \downarrow{\scriptstyle f_{\mathfrak{U}}} \\ \mathfrak{V} & \xrightarrow[\pi]{} & \mathfrak{U} \end{array}$$

Proof: The first two parts of the lemma are obvious. Since π' is defined by $\pi'(f^{-1}(V)) = f^{-1}(\pi(V))$ for each V in $\mathfrak{V}$, and since $f_{\mathfrak{U}}$ and $f_{\mathfrak{V}}$ are identity injections, it follows that π' is a projection and has the desired commutative property. □

THEOREM 8–16. Let X and Y be compact Hausdorff spaces, and let $f:X \to Y$ be continuous. Let $f^{-1}:\Sigma(Y) \to \Sigma(X)$ be the associated mapping of coverings, and for each element $\mathfrak{U}$ in $\Sigma(Y)$, let $f_{\mathfrak{U}}$ be the injection of $f^{-1}(\mathfrak{U})$ into $\mathfrak{U}$. Then the induced homomorphisms

$${}_*f_{\mathfrak{U}}:H_p(f^{-1}(\mathfrak{U}), G) \to H_p(\mathfrak{U}, G),$$

together with f^{-1}, constitute a transformation Φ of the pth Čech homology system of X into that of Y.

The proof is merely a matter of checking the definition (in Section 2–14) of such a transformation of inverse limit systems, the necessary commutative relations being given in Lemma 8–15. Again looking at the definitions in Section 2–14, we see that the transformation Φ on the homology systems induces a homomorphism on the inverse limit groups

$$f_*:H_p(X, G) \to H_p(Y, G).$$

This is the *homomorphism induced by the continuous mapping* f. That it satisfies the axioms of Eilenberg and Steenrod is proved in Chapter IX of their book [7].

8–5 Singular homology theory. As is the case in most of the sections in this chapter, the present section merely introduces an important topic whose scope does not admit of a complete study in an introductory course in topology. For a more complete exposition on the subject of singular homology theory, see Eilenberg [75] or Eilenberg and Steenrod [7].

Let X be a topological space. With X we will associate a complex $S(X)$, the total singular complex of X, as follows. Consider a geometric simplex $s^p = \langle v_0 \cdots v_p \rangle$ with ordered vertices in some Euclidean space E^n and a continuous mapping $f:s^p \to X$ of s^p into X. The pair (s^p, f) will represent a singular simplex in X.

We define an equivalence relation between such pairs by setting

$$(s_1^p, f_1) \simeq (s_2^p, f_2)$$

if the (unique) affine transformation ψ determined by $\psi(v_i) = w_i$ (the vertices of s_2^p), $i = 0, 1, \ldots, p$, where ψ is considered only on the simplex s_1^p, satisfies the following criterion:

$$f_1 = f_2\psi.$$

It is easily shown that this is indeed an equivalence relation on pairs (s^p, f), and we define a *singular simplex* on X to be an equivalence class $\sigma^p = [(s^p, f)]$ of pairs (s^p, f) under this relation.

The *total singular complex* $S(X)$ of the space X consists of the singular simplexes on X with the necessary functions defined as follows:

(i) $\dim \sigma^p = \dim [(s^p, f)] = p$, and

(ii) if $s^p = \langle v_0 \cdots v_p \rangle$ and if $t^q = \langle v_{i_0} \cdots v_{i_q} \rangle$ is a face of s^p, the vertices of t^q taken in the same order as they appear in s^p, then we define the incidence relation

$$[(s^p, f)] > [(t^q, f|t^q)],$$

where $f|t^q$ is f restricted to t^q as usual. This may be rephrased in terms of incidence numbers by saying that if $s_i^{p-1} = \langle v_0 \cdots \hat{v}_i \cdots v_p \rangle$, then

$$[[(s^p, f)], [(s_i^{p-1}, f|s_i^{p-1})]] = (-1)^i$$

and otherwise $$[\sigma^p, \sigma^{p-1}] = 0.$$

The *integral singular chains* are taken to be formal finite linear combinations of singular simplexes with integral coefficients, $c_p = \sum^f \eta_i \sigma_i^p$, each η_i being an integer, and each σ_i^p an element of $S(X)$. Using the natural componentwise addition, the chains form a group $C_p(X)$, the pth *singular chain group* of X.

The boundary operator ∂ is defined on elementary chains by the formula

$$\partial[(s^p, f)] = \sum_{i=0}^{p} (-1)^i [(s_i^{p-1}, f|s_i^{p-1})],$$

where $s_i^{p-1} = \langle v_0 \cdots \hat{v}_i \cdots v_p \rangle$ as usual. It is easy to prove that $\partial\partial = 0$ for elementary chains, and the same property holds when the above definition is extended linearly to obtain the boundary of an arbitrary singular chain. The necessary algebraic requirements for a homology theory are now present, and we may define the pth *singular homology group* $\mathcal{H}_p(X)$ with integral coefficients as $\mathcal{H}_p(X) = H_p(S(X))$.

Given two spaces X and Y and a continuous mapping $f:X \to Y$, there is an induced transformation $f_\#: S(X) \to S(Y)$ such that $f_\#\partial = \partial f_\#$. This transformation is defined as follows. Given a singular simplex $[(s^p, \varphi)]$ in $S(X)$, we set

$$f_\#[(s^p, \varphi)] = [(s^p, f\varphi)].$$

One need only check that $(f\varphi)|s_i^{p-1} = f(\varphi|s_i^{p-1})$ to prove that $f_\#$ commutes with ∂. But this is obvious. It follows that $f_\#$ is a chain-mapping of $S(X)$ into $S(Y)$ (see Section 6–11) and hence induces homomorphisms

$$f_* : \mathcal{H}_p(X) \to \mathcal{H}_p(Y).$$

It is easily shown that *if f is a homomorphism, then f_* is an isomorphism onto* and that *if $f:X \to Y$ and $g:Y \to Z$, then $(gf)_* = g_*f_*$*. The reader may wish to prove these statements as exercises. We prove only the following theorem as an illustration of the methods used in singular homology theory.

THEOREM 8–17. If f and g are homotopic mappings of X into Y, then the induced homomorphisms f_* and g_* of $\mathcal{H}_p(X)$ into $\mathcal{H}_p(Y)$ coincide.

Proof: This involves setting up a chain-homotopy between the chain-mappings $f_\#$ and $g_\#$. Having such a chain-homotopy, Theorem 6–19 will apply to prove this theorem. We must define the operator $\mathfrak{D}:C_p(X) \to C_{p+1}(Y)$ such that

$$\partial\mathfrak{D} = f_\# - g_\# - \mathfrak{D}\partial,$$

and this requires some preparation.

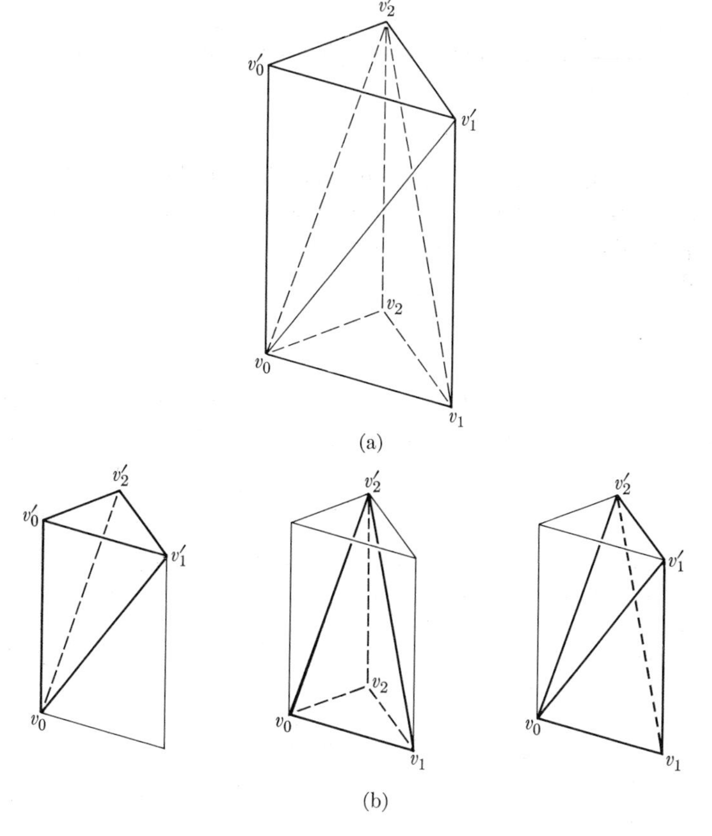

FIG. 8-8. (a) $P = s^2 \times I^1$. (b) Three 3-simplexes in $s^2 \times I^1$.

Given a geometric simplex $s^p = \langle v_0 \cdots v_p \rangle$, we construct the prism $P = s^p \times I^1$ and decompose P into $(p + 1)$-simplexes of the form

$$t_i^{p+1} = \langle v_0 \cdots v_i v_i' \cdots v_p' \rangle,$$

where the v_j are vertices on the bottom face of P, $s^p \times 0$, and the v_j' are vertices on the top face of P, $s^p \times 1$. This subdivision of P is illustrated for a 2-simplex in Fig. 8-8.

Now consider the chain on P,

$$d(s^p) = \sum_{i=0}^{p} (-1)^i t_i^{p+1}.$$

It is an easy exercise to show that

$$\partial d(s^p) = {'s^p} - s^p - \sum_{i=0}^{p} (-1)^i \, d(s_i^{p-1}),$$

where $s_i^{p-1} = \langle v_0 \cdots \hat{v}_i \cdots v_p \rangle$ again. This operator d can be extended linearly to chains c_p of geometric simplexes to yield an operator, still called d, satisfying

$$\partial d(c^p) = {}'c_p - c_p - d(\partial c_p).$$

Now suppose that F is a continuous mapping of the prism P into Y. Then any ordered simplex t^q in P gives rise to a singular simplex $\tau^q = [(t^q, F|t^q)]$ in $S(Y)$, and this can be extended linearly to chains on P. Now define

$$c_F(s^p) = \sum_{i=0}^{p} (-1)^i [(t_i^{p+1}, F|t_i^{p+1})].$$

This is a singular chain in $S(Y)$. In the same way, we define $c_F(s_i^{p-1})$ for each face s_i^{p-1} of s^p and extend linearly to chains, thereby obtaining the relation

$$\partial c_F(s^p) = {}'s^p - s^p - c_F(\partial s^p).$$

Returning to the proof of the theorem, consider the two mappings f and g and the assumed homotopy $h: X \times I^1 \to Y$ such that $h(x, 0) = f(x)$ and $h(x, 1) = g(x)$. To construct the desired chain-deformation $\mathfrak{D}$, let $\sigma^p = [(s^p, \varphi)]$ be a singular simplex in $S(X)$, where $s^p = \langle v_0 \cdots v_p \rangle$. Construct the prism $P = s^p \times I^1$. A mapping $F: P \to Y$ is defined by

$$F(x, t) = h(\varphi(x), t)$$

for each point (x, t) in P. It is evident that, in the notation introduced above, the chain $c_F(s^p)$ has the property

$$\partial c_F(s^p) = g_\#({}'s^p) - f_\#(s^p) - c_F(\partial s^p).$$

Setting $\mathfrak{D}(\sigma^p) = c_F(s^p)$, we have the desired chain-deformation. □

We do not develop relative singular theory and its consequences here. Of course, the first results of this theory are those taken as the Eilenberg-Steenrod axioms (see Section 7–8). For reasonable spaces, for instance compact metric spaces locally connected in all dimensions (definition later), the singular theory and the Čech theory coincide. For other spaces the two theories do not agree. An example of this last statement is afforded by the topologist's closed sine curve pictured in Fig. 8–2, which carries a nonbounding Čech 1-cycle but does not carry a nonbounding singular 1-cycle.

Before leaving singular homology theory, we may mention a variation called *singular cubic homology theory*. In this theory, we define a *singular n-cube* in a space X to be a mapping $f: I^n \to X$. Such a singular n-cube is called *degenerate* if f does not depend upon all its coordinates. For instance, if $f(x, x_2, \ldots, x_n) = f(y, x_2, \ldots, x_n)$, where $x_2, \ldots, x_n$ are fixed, we say that f is degenerate along its first coordinate. Let $Q_n(X)$ be the free abelian group generated by the set of all singular n-cubes in X, and let $D_n(X)$ be the free abelian group generated by the set of all degenerate singular n-cubes in X. Then the *singular cubic chain groups* of X are defined as the quotient groups

$$C_n(X) = \frac{Q_n(X)}{D_n(X)}.$$

The boundary operator ∂ is defined as follows. Let f be a singular n-cube in X. For each integer $i = 1, 2, \ldots, n$, there are two singular $(n - 1)$-cubes f_i^0 and f_i^1 given by

$$f_i^0(x_1, \ldots, x_{i-1}, x_{x+1}, \ldots, x_n) = f(x_1, \ldots, x_{i-1}, 0, x_{i+1}, \ldots, x_n)$$

and

$$f_i^1(x_1, \ldots, x_{i-1}, x_{i+1}, \ldots, x_n) = f(x_1, \ldots, x_{i-1}, 1, x_{i+1}, \ldots, x_n).$$

We define $\partial: Q_n(X) \to Q_{n-1}(X)$ by setting

$$\partial f = \sum_{i=1}^{n} (-1)^i (f_i^0 - f_i^1)$$

for elementary chains f and extending this linearly to arbitrary chains. It is not difficult to show that ∂ maps $D_n(X)$ into $D_{n-1}(X)$, and hence ∂ induces an operator, still called ∂, of $C_n(X)$ into $C_{n-1}(X)$. The basic property $\partial\partial = 0$ is also easy to prove, and again we have the algebraic requirements for a homology theory. It is this homology theory which is most convenient when discussing relations between the homology groups and the homotopy groups of a space (see below).

EXERCISE 8-1. Prove that the singular simplicial homology groups and the singular cubic homology groups of a space X are isomorphic.

In Section 4-7 we defined the higher homotopy groups $\pi_n(X, x)$ by considering certain mappings of the n-cube I^n into X. The singular cubic homology groups $H_n(X)$ are also defined by considering mappings of I^n into X. It is natural to ask how the two groups $\pi_n(X, x)$ and $H_n(X)$ compare, particularly if we free the homotopy group of its dependence upon the base point x by taking the space X to be arcwise-connected. We will quote two results that provide a partial answer to this question.

THEOREM 8-18. Let X be an arcwise-connected space. Then there is a natural homomorphism $h: \pi_n(X) \to H_n(X)$, $n \geq 1$. Furthermore, if

$f:X \to Y$ is a continuous mapping of X into an arcwise-connected space Y, then the induced homomorphisms on the homotopy and homology groups provide commutativity in the following diagram:

$$\begin{array}{ccc} \pi_n(X) & \xrightarrow{f_*} & \pi_n(Y) \\ h\downarrow & & \downarrow h \\ H_n(X) & \xrightarrow[f_*]{} & H_n(Y) \end{array}$$

The image of an element of $\pi_n(X)$ in $H_n(X)$ under the homomorphism h is called a *spherical homology class.*

One of the first results relating homotopy and homology groups is the following theorem due to Hurewicz [85].

THEOREM 8–19. If each homotopy group $\pi_n(X)$, $1 \leqq p < n$, is trivial, then the homomorphism $h:\pi_n(X) \to H_n(X)$ is an isomorphism onto. If $n = 1$, then the homomorphism $h:\pi_1(X) \to H_1(X)$ is onto, and its kernel is precisely the commutator subgroup of $\pi_1(X)$.

The reader who is interested in exploring this line of inquiry is again referred to Hilton [13] or Hu [14(*a*)].

8–6 Vietoris homology theory. The Vietoris homology theory was the first of the Čech-type homology theories to appear. It was introduced by Vietoris [129] in 1927 and in this form applies only to metric spaces. While this theory has been used in many research papers, it has not been discussed so extensively as has the more general Čech theory. Again, for the sake of brevity, we consider only compact spaces in this presentation. We may refer the reader to Begle [53] for generalizations.

Let M be a compact metric space, and let ϵ be a positive number. We construct the simplicial complex $K_\epsilon = \{\mathcal{V}, \Sigma\}$, where the vertices in $\mathcal{V}$ are the points of M and where a finite subcollection of vertices $p_0, p_1, \ldots, p_n$ forms an n-simplex in Σ if and only if the diameter of the set $\bigcup_{i=0}^n p_i$ $[= \max d(p_i, p_j)]$ is less than ϵ. It is easy to prove that for each $\epsilon > 0$, K_ϵ is a simplicial complex (see Exercise 5–4). Therefore, for each $\epsilon > 0$ and each integer $n \geqq 0$, we may construct the simplicial homology group $H_n(K_\epsilon)$ of K_ϵ with integral coefficients.

Given $\epsilon_1 > \epsilon_2 > 0$, it is evident that each simplex of K_{ϵ_2} is also a simplex of K_{ϵ_1} and hence that there is an identity injection $j_{\epsilon_1\epsilon_2}$ of K_{ϵ_2} into K_{ϵ_1}. This injection then induces a homomorphism $_*j_{\epsilon_1\epsilon_2}$ of $H_n(K_{\epsilon_2})$ into $H_n(K_{\epsilon_1})$. Furthermore, if $\epsilon_1 > \epsilon_2 > \epsilon_3 > 0$, then the induced homomorphisms satisfy the relation

$$_*j_{\epsilon_1\epsilon_2}{}_*j_{\epsilon_2\epsilon_3} = {}_*j_{\epsilon_1\epsilon_3}.$$

Since the positive real numbers constitute a directed set, the collection

$\{H_n(K_\epsilon)\}$ together with the injection-induced homomorphisms $\{_*j_{\epsilon\delta}\}$ form an inverse limit system of groups and homomorphisms. The inverse limit group of this system is the nth *Vietoris homology group* $V_n(M)$.

Clearly the complexes K_ϵ are much too large for convenient manipulation (they can certainly have a nondenumerable number of simplexes and infinite dimension). The usual technique in using Vietoris theory involves discussing the existence or, more often, the nonexistence of certain essential (nonbounding) cycles. In this way, one studies the connectivity properties of the space M without becoming involved with the complexes K_ϵ. It is known that the Vietoris groups, the singular groups, and the Čech groups coincide if the underlying space is sufficiently well-behaved. For instance, all these coincide with the simplicial homology groups on a finite polytope.

We close this section by stating the result of Vietoris [129], for which he invented this theory. Let M and N be compact metric spaces, and let $f:M \to N$ be continuous. If for each point y in N, the inverse set $f^{-1}(y)$ has trivial Vietoris homology groups $V_p(f_p^{-1}(y))$ for all dimensions $p \leqq n$, then f is an *n-monotone mapping*. We use augmented homology groups in dimension zero so that 0-*monotone* agrees with *monotone* as defined in Section 3–7.

THEOREM 8–20. Let M and N be compact metric spaces, and let f be an n-monotone mapping of M onto N. Then the Vietoris homology groups $V_p(M)$ and $V_p(N)$ are isomorphic for each dimension $p \leqq n$.

Incidentally, this theorem has been generalized by Begle [53] to compact Hausdorff spaces.

8–7 Homology local connectedness. The higher-dimensional connectivity property of a space X that is reflected in the vanishing of the Čech homology group $H_p(X, G)$ may be localized by the standard procedure (see Section 3–1). In doing so, we obtain a natural generalization of point-set local connectedness (Section 3–1). This is in direct analogy to Section 4–9, in which homotopy local connectedness is introduced as a generalization of local arcwise connectedness.

Let X be a locally compact Hausdorff space and $\Sigma(X)$ be the family of all finite open coverings of X. If P is a subset of X, and if $\mathfrak{U}$ is an element of $\Sigma(X)$, then $\mathfrak{U} \wedge P$ denotes the subcomplex of $\mathfrak{U}$ consisting of all simplexes of $\mathfrak{U}$ that meet P; that is, $\langle U_0 \cdots U_p \rangle$ is a simplex of $\mathfrak{U} \wedge P$ if and only if $(\bigcap_{i=0}^{p} U_i) \cap P$ is not empty. Then a chain $c_p(\mathfrak{U})$ is said to be on P if and only if $c_p(\mathfrak{U})$ is on $\mathfrak{U} \wedge P$. A Čech cycle $z_p = \{z_p(\mathfrak{U})\}$ is on P if and only if, for each $\mathfrak{U}$ in $\Sigma(X)$, the coordinate $z_p(\mathfrak{U})$ is on P. Similarly, z_p bounds P if and only if, for each $\mathfrak{U}$ in $\Sigma(X)$, the coordinate $z_p(\mathfrak{U}) = \partial c_{p+1}(\mathfrak{U})$, where $c_{p+1}(\mathfrak{U})$ is a chain on P.

The locally compact Hausdorff space X is *locally connected in dimension* n, in the sense of Čech homology, *at a point* x (abbreviated "n-lc at x") if, given any open set P containing x, there exists an open set Q containing x and contained in P such that every Čech n-cycle on Q bounds on P. (We make a few remarks about this "two-set" definition later.) Then X is n-lc if it is n-lc at every point, and X is lc^n if it is p-lc for each $p \leqq n$. For reasons explained in Section 8–3, the Čech cycles here are taken to have coefficients in a field (or in a vector space over a field).

Let us show that 0-lc in the sense of this definition corresponds with local connectedness.

THEOREM 8–21. The locally compact Hausdorff space X is 0-lc if and only if X is locally connected.

Proof: We use augmented homology. Assume that X is locally connected, and let x be a point of an open set P in X. By definition, there exists an open connected set Q containing x and lying in P. Let z_0 be an augmented Čech 0-cycle on Q. If $\mathfrak{U}$ is any covering of X, let $z_0(\mathfrak{U}) = \sum_{i=1}^k a_i\sigma_i^0$, $a_i \neq 0$, be the coordinate of z_0 on $\mathfrak{U}$. By definition, $\sum_{i=1}^k a_i = 0$. Each σ_i^0 is an element of $\mathfrak{U}$, of course, and by using the simple chain theorem (Theorem 3–4), we obtain a sequence of elements of $\mathfrak{U}$ starting with σ_1^0 and ending at σ_j^0, $j = 2, 3, \ldots, k$. This sequence need not be a simple chain of sets, but we can associate with it a 1-chain c_1^j such that $\partial c_1^j = \sigma_j^0 - \sigma_1^0$. Then

$$\partial \sum_{j=2}^{k} a_j c_1^j = \sum_{j=2}^{k} a_j \sigma_j^0 - \sum_{j=2}^{k} a_j \sigma_1^0.$$

But $a_1 = -\sum_{j=2}^k a_j$, so we have

$$\partial \sum_{j=2}^{k} a_j c_1^j = \sum_{i=1}^{k} a_i \sigma_i^0 = z_0(\mathfrak{U}).$$

Since each element of $\mathfrak{U}$ used in $z_0(\mathfrak{U})$ is on Q, it follows that the chain $\sum_{j=2}^k a_j c_1^j$ also lies on Q and hence on P. This proves that X is 0-lc.

To prove the converse, suppose X is 0-lc but not locally connected. Then there must be a point x in X and an open set P containing x such that every open set Q in P containing x meets at least two components of P. We choose open sets R, Q, Q', and P' such that (1) x lies in R, R lies in Q, Q is closure-contained in Q', Q' is closure-contained in P', and P' is closure-contained in P and (2) every augmented Čech 0-cycle on R bounds on Q. There are two points x_1 and x_2 in R and a decomposition $P = P_1 \cup P_2$, where P_1 and P_2 are separated, with x_1 in P_1 and x_2 in P_2. Let $\mathfrak{U}$ be the covering of X consisting of the open sets $X - \overline{P}'$, $(P - \overline{Q}) \cap P_1$, $(P - \overline{Q}) \cap P_2$, $Q' \cap P_1$, and $Q' \cap P_2$. By definition, the coordinate on $\mathfrak{U}$ of a nontrivial 0-cycle on $x_1 \cup x_2$ must have the form $a(U_1 - U_2)$, where $U_1 = Q' \cap P_1$ and $U_2 = Q' \cap P_2$. But $U_1 - U_2$ is not homologous to zero on Q. This contradiction of the 0-lc assumption completes the proof. □

This last result shows that the n-lc property of a space is a valid generalization of local connectedness. It follows that one should expect the n-lc property to be exploited in much the same way as is local connectedness. Understandably, the latter is a very difficult task. We will merely refer the reader to Chapter VII of Wilder's *Topology of Manifolds* [42], which will bring him up to the point of reading the current literature.

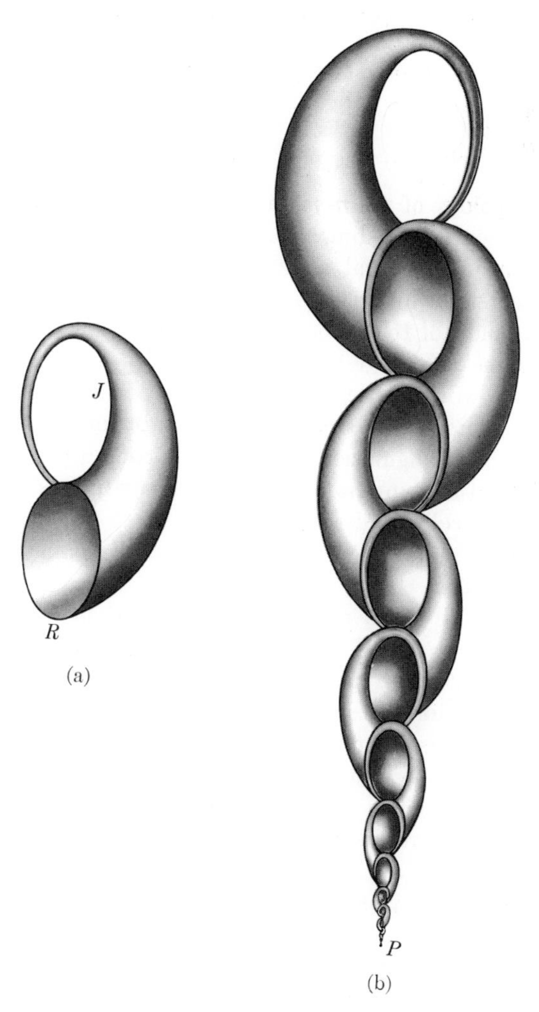

FIGURE 8-9

Local connectedness and the n-lc property are actually defined differently in other than just the dimension. A space is locally connected if and only if it has a basis of open connected sets. A corresponding formulation for higher dimensions would claim the existence of a basis of open sets on each of which every Čech n-cycle bounds. This is true for $n = 0$, of course, but it cannot be proved even for 1-lc spaces. The continuum in Fig. 8–9(b) is constructed of a sequence of finite cones $C_1, C_2, \ldots$, each C_i having its vertex identified with a point on its rim as in Fig. 8–9(a). Each C_i has two simple closed curves singled out, its rim R_i and the curve J_i shown in Fig. 8–9(a). To construct the continuum, we identify R_i with J_{i+1} for each i and have the sets C_i converge to a point p. Then $C = p \cup \bigcup C_i$ is the desired continuum. Any open set U containing the point p must contain a first curve R_j. The 1-cycle on R_j can bound only on C_j and C_j cannot be in U, or else $J_i = R_{j-1}$ would also be in U. Thus C is 1-lc at p, but no open set U containing p has the property that every Čech 1-cycle on U bounds on U.

Again we refer the reader to Section 8, Chapter VI, of Wilder [42] for a brief discussion of these relatively unexplored matters.

8–8 Some topology of the n-sphere. In this final section, we study the simplest of the compact n-dimensional manifolds, the n-sphere. Our purpose is to introduce several results, in particular the Jordan-Brouwer separation theorem, which the reader will see in generalized form if he proceeds to a study of the theory of manifolds. (For comparable separation theorems, see Section 6–17.)

We follow the work of Alexander [45] closely in using the strongly geometric mod 2 homology theory and by introducing a cell subdivision of S^n in place of the triangulations we have considered previously. This results in a substantial computational advantage, which is evidenced by the fact that our first cell subdivision of S^n has $2n + 2$ cells, two of each dimension $i = 0, 1, \ldots, n$, whereas the minimum triangulation of S^n has $2(2^n - 1)$ simplexes.

Let S^n denote the set of points $(x_1, \ldots, x_{n+1})$ in E^{n+1} satisfying the equation $\sum_{i=1}^{n+1} x_i^2 = 1$. We say that S^n is in *standard position.* Let $P_1, \ldots, P_n$ be distinct hyperplanes through the origin in E^{n+1}. (For purposes of illustration, we may take P_i to be the hyperplane with equation $x_{i+1} = 0$.) It is clear that P_n intersects S^n in (a set isometric to) S^{n-1}. Using well-known properties of real numbers, it is easy to prove that P_n separates S^n into two topological n-cells s_1^n and s_2^n. Next, the hyperplane P_{n-1} intersects S^{n-1} in S^{n-2} and separates S^{n-1} into two $(n-1)$-cells s_1^{n-1} and s_2^{n-1}. (We ignore the separation of S^n by P_{n-1}.) In general, then, the hyperplane P_i intersects the sphere S^i in a sphere S^{i-1} and separates S^i into two i-cells s_1^i and s_2^i; P_1 intersects S^1 in two 0-cells s_1^0 and s_2^0. We consider the cells s_1^i and s_2^i to be relatively open sets in S^i. This gives a cell subdivision of S^n, which we will denote by K_0, the first of a sequence of such subdivisions.

Note that each i-cell in K_0 has as its point-set boundary the collection of all j-cells, $j < i$. In particular, the two cells s_1^{i-1} and s_2^{i-1} are called the boundary cells of the i-cell. Moreover, each i-cell, $i < n$, is a boundary cell of exactly two $(i + 1)$-cells. One more fact is useful later. Each i-cell in K_0 is a convex subset of S^i in the metric of S^i; that is, given any two points of the i-cell, each great circle in S^i through the two points has an arc joining the two points that lies entirely in the i-cell.

The homology theory modulo 2 of the cell complex K_0 is very simple. Recalling that Z_2 denotes the group of integers mod 2, we define the chain groups $C_p(K_0, Z_2)$ as usual, and it is easily seen that each such chain group is isomorphic to the direct sum $Z_2 \oplus Z_2$. Given an elementary i-chain, we define its boundary by setting

$$\partial(0 \cdot s_j^i) = 0 \cdot s_1^{i-1} + 0 \cdot s_2^{i-1} \qquad (j = 1, 2; i > 0),$$

$$\partial(1 \cdot s_j^i) = 1 \cdot s_1^{i-1} + 1 \cdot s_2^{i-1} \qquad (j = 1, 2; i > 0),$$

and

$$\partial(c_0) = 0 \qquad \text{(nonaugmented theory).}$$

We verify that $\partial\partial = 0$ by noting that

$$\begin{aligned}\partial(1 \cdot s_1^{i-1} + 1 \cdot s_2^{i-1}) &= \partial(1 \cdot s_1^{i-1}) + \partial(1 \cdot s_2^{i-1}) \\ &= (1 + 1) \cdot s_1^{i-2} + (1 + 1) \cdot s_2^{i-2} = 0.\end{aligned}$$

Following the familiar procedure, we construct the mod 2 homology groups $H_p(K_0, Z_2) = Z_p(K_0, Z_2) - B_p(K_0, Z_2)$. Since there are no $(n + 1)$-cells in K_0, $H_n(K_0, Z_2) = Z_n(K_0, Z_2)$, and since every 0-chain is a 0-cycle, $Z_0(K_0, Z_2) = C_0(K_0, Z_2) = Z_2 \oplus Z_2$. There are but four p-chains to be checked in each dimension. These are $0 \cdot s_1^p + 0 \cdot s_2^p$, $1 \cdot s_1^p + 1 \cdot s_2^p$, $0 \cdot s_1^p + 1 \cdot s_2^p$, and $1 \cdot s_1^p + 0 \cdot s_2^p$. One easily shows that the first two are cycles while the last two are not (for $p > 0$). Hence we know that $Z_p(K_0, Z_2)$, $p > 0$, is isomorphic to Z_2. But also each such p-cycle, $p < n$, is the boundary of an elementary $(p + 1)$-chain, and so $B_p(K_0, Z_2)$, $p < n$, is also isomorphic to Z_2. It follows that

$$\begin{aligned}H_n(K_0, Z_2) &= Z_2, \\ H_p(K_0, Z_2) &= 0, \qquad (0 < p < n). \\ H_0(K_0, Z_2) &= Z_2,\end{aligned}$$

Next we construct a sequence of subdivisions $K_1, K_2, \ldots$ of S^n such that K_i is a refinement of K_{i-1} and such that the maximum of the diameters of cells in K_i approaches zero as i increases indefinitely. The subdivisions are constructed with the aid of further hyperplanes through the origin as follows. Given a hyperplane P, it will intersect each 1-cell of K_0 in a point, each 2-cell in a 1-cell, etc. In particular, we will want to subdivide an n-cell of K_0 into two smaller n-cells. This must be done quite carefully.

The hyperplane P intersects one of the two n-cells of K_0 in an $(n - 1)$-cell which itself has two boundary $(n - 2)$-cells, and each of these has two boundary $(b - 2)$-cells, etc. Ignoring what might happen in the n-cell of K_0, we could form a new subdivision including all these new k-cells at once. For technical reasons, however, it is convenient to do the subdividing more slowly. We will insist that, before we introduce a new 1-cell, we already will have introduced the two 0-cells which will be its boundary cells, that before we introduce a new 2-cell we already will have the 1-cells forming its boundary, etc. Furthermore, we will introduce just one new cell at a time, which cell may or may not be a boundary cell of a newly added cell. (It will be a boundary cell of two old cells, of course.)

In summary, suppose that the subdivision K_i, $i \geqq 0$, has been defined and that we intersect a $(k + 1)$-cell s^{k+1} of K_i with a hyperplane through the origin in E^{n+1}. This intersection is a k-cell. If in K_i there are two $(k - 1)$-cells forming the boundary of this k-cell, then K_{i+1} will be the collection of cells of K_i with the new k-cell added and the $(k + 1)$-cell s^{k+1} subdivided into two $(k + 1)$-cells. The additional requirement that, given any positive number ϵ, there is an integer N such that, whenever $i > N$, every cell of K_i has diameter less than ϵ may be achieved by construction. In any case we select some sequence $K_0, K_1, K_2, \ldots$, which will remain fixed for the remainder of the discussion.

Figure 8–10 illustrates the top hemisphere of S^2 as it would appear under some choice of subdivisions. The curved lines are arcs of great circles, and where stages are omitted we may assume that subdivision of the lower hemisphere is taking place unseen by us.

At the seventeenth stage in Fig. 8–10, the top hemisphere has been subdivided into five 2-cells, thirteen 1-cells, and nine 0-cells. Two of these 2-cells have five boundary 1-cells, one 2-cell has four boundary 1-cells, and two have three boundary 1-cells. But each of the 1-cells has exactly two boundary 0-cells and can have no more and no less.

Given a particular cell complex K_i, we form the mod 2 chain groups $C_p(K_i)$ (we will understand that the coefficient group is Z_2 hereafter). Any p-chain mod 2, say c_p, may be identified with the union of those p-cells s^p for which $c_p(s^p) = 1$. Adding to these p-cells the minimum number of cells of the lower dimensions necessary to form a subcomplex, we obtain the *subcomplex* $L(c_p)$ *associated with the chain* c_p. We note that such a subcomplex is uniformly p-dimensional, so not every subcomplex of K_i is the associated complex of some p-chain.

Given any subcomplex L of K_i, we may form its mod 2 homology group

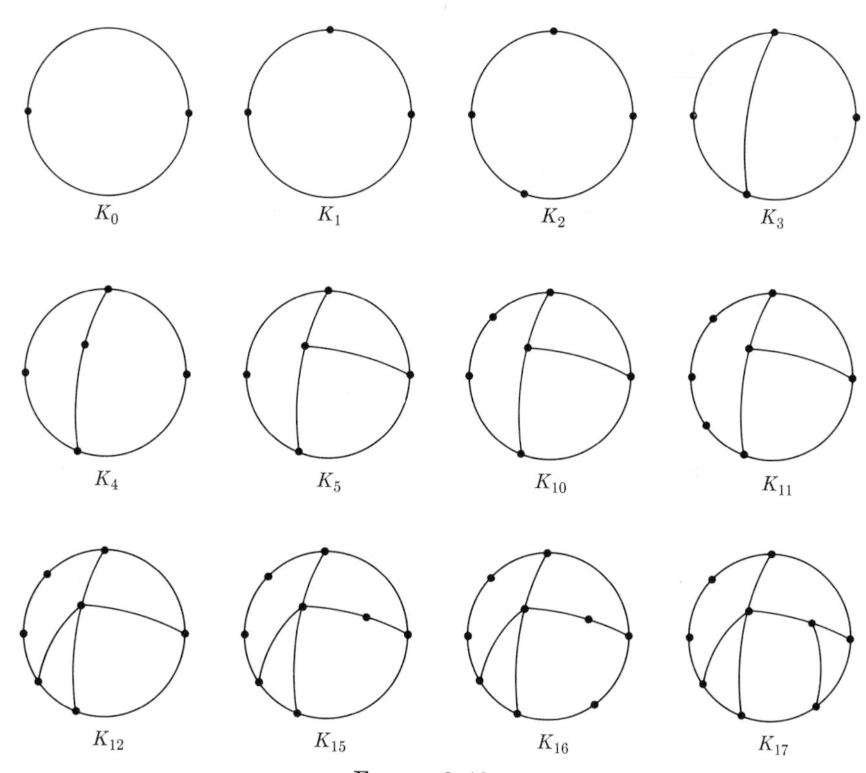

FIGURE 8–10

$H_p(L)$. We recall that each such homology group is isomorphic to a direct sum of cyclic groups of order 2. The number of generators of $H_p(L)$ is the mod 2 Betti number of L, which we may denote by $r_p(L)$. Now in passing from K_i to K_{i+1}, it may be that one of the cells s^p of L is that cell subdivided in the process. If so, then L no longer exists as a subcomplex of K_{i+1}, but there is the obvious subcomplex L' in K_{i+1} consisting of all the cells of L except s^p together with the $(p-1)$-cell that divides s^p and the two new p-cells into which s^p has been divided. It is obvious that there are more chains in $C_p(L')$ than in $C_p(L)$, and hence more p-cycles and more boundaries are possible. The important fact, of course, is that we do not increase the number of homology classes in this process. The following result is thus the analogue of Theorem 6–24.

THEOREM 8–22. If L is a subcomplex of K_i, and if L' is its subdivision in K_{i+1}, then $r_p(L) = r_p(L')$ for all p.

Proof: Suppose that as we pass from K_i to K_{i+1}, the p-cell s^p is subdivided by a new $(p-1)$-cell s^{p-1} into two new p-cells, s_1^p and s_2^p. No change will be made in the chain groups $C_j(L)$, $j \neq p, p-1$, and we cannot have altered the boundaries except perhaps in dimensions $p-2$, $p-1$, p, and $p+1$. Hence the only mod 2 Betti numbers that can possibly have been altered are $r_j(L)$, $j = p-2, p-1, p, p+1$. We investigate each of these.

Suppose that z_{p+1} is a cycle on L. Since the chain group $C_{p+1}(L)$ is unaltered by the subdivision, z_{p+1} is a chain on L'. We show that z_{p+1} is also a cycle on L'. This is easy because if $\partial z_{p+1} = 0$ on L, then each p-cell s^p in L is a face of an even number of $(p+1)$-cells in z_{p+1} (it occurs an even number of times in ∂z_{p+1}). Then each of the p-cells s_1^p and s_2^p is a face of each of the same cells of z_{p+1} and hence occurs an even number of times in ∂z_{p+1} as the boundary is taken in L'. Therefore $\partial z_{p+1} = 0$ in L', too. Conversely, if z_{p+1} is a cycle on L', a similar argument proves z_{p+1} to be a cycle on L. It follows that $Z_{p+1}(L)$ and $Z_{p+1}(L')$ are isomorphic. Then, since the chain groups $C_{p+2}(L)$ and $C_{p+2}(L')$ are the same, we have $B_{p+1}(L)$ isomorphic to $B_{p+1}(L')$ and hence $H_{p+1}(L)$ isomorphic to $H_{p+1}(L')$.

Next, if z_p is a cycle on L, and if the p-cell s^p is not in the subcomplex associated with z_p, then z_p is also a cycle on L'. If s^p is in the subcomplex associated with z_p, then we may write $z_p = 1 \cdot s^p + z_p^1$. Then the chain $\gamma_p = 1 \cdot s_1^p + 1 \cdot s_2^p + z_p^1$ is a cycle on L' for $\partial(1 \cdot s_1^p + 1 \cdot s_2^p) = \partial(1 \cdot s^p)$, the $(p-1)$-cell s^{p-1} occurring twice. Furthermore, γ_p bounds if and only if z_p bounds. Also, no cycle on L' can contain s_1^p without containing s_2^p and conversely because s^{p-1} is a face of just these two p-cells in K_{i+1}. This gives the desired isomorphisms between $Z_p(L)$ and $Z_p(L')$ and between $B_p(L)$ and $B_p(L')$.

A cycle z_{p-1} of L is still a cycle of L' since no new $(p-2)$-cells are added in passing from K_i to K_{i+1}, and z_{p-1} bounds on L' if and only if it bounds on L. However, there may be new cycles on L' of the form $\gamma_{p-1} = 1 \cdot s^{p-1} + \gamma_{p-1}^1$. Consider the cycle $z_{p-1} = \gamma_{p-1} + \partial s_1^p$. This cycle does *not* contain s^{p-1}, because s^{p-1} is in ∂s_1^p and in γ_{p-1}. Then $z_{p-1} - \gamma_{p-1} = \partial s_1^p$, showing that γ_{p-1} is homologous to z_{p-1} on L'. Thus every new cycle is homologous to an old cycle, and this shows that no new independent cycles are introduced in dimension $p-1$ by subdivision.

Finally, the mod 2 Betti number r_{p-2} can be changed only by having a cycle z_{p-2} bound on L' while not bounding on L. This is conceivable because there are extra $(p-1)$-chains on L', those involving s^{p-1}. Suppose however that $z_{p-2} = \partial(1 \cdot s^{p-1} + c_{p-1})$, c_{p-1} being a chain on L. If we add to this the chain $\partial(\partial s_1^p) = 0$, we obtain

$$\partial(1 \cdot s^{p-1} + \partial s_1^p + c_{p-1}) = z_{p-2}.$$

Hence z_{p-2} also bounds in L if it bounds in L'. $\square$

COROLLARY 8–23. For all $i \geqq 0$ and $0 < p < n$,

$$H_n(K_i, Z_2) = Z_2 = H_0(K_i, Z_2)$$

and

$$H_p(K_i, Z_2) = 0.$$

Our next task is to define homology groups for an open subset of S^n. To do this, let D denote an open subset of S^n, and for each integer $i \geqq 0$, let L_i denote the subcomplex of K_i consisting of all cells s^p of K_i whose closures lie in D. Note that L_{i+1} either is L_i (if the new cells added in subdivision are not in D), or is a subdivision of L_i, or is L_i plus cells added in K_{i+1}. Figure 8–11 illustrates these three possibilities in two dimensions.

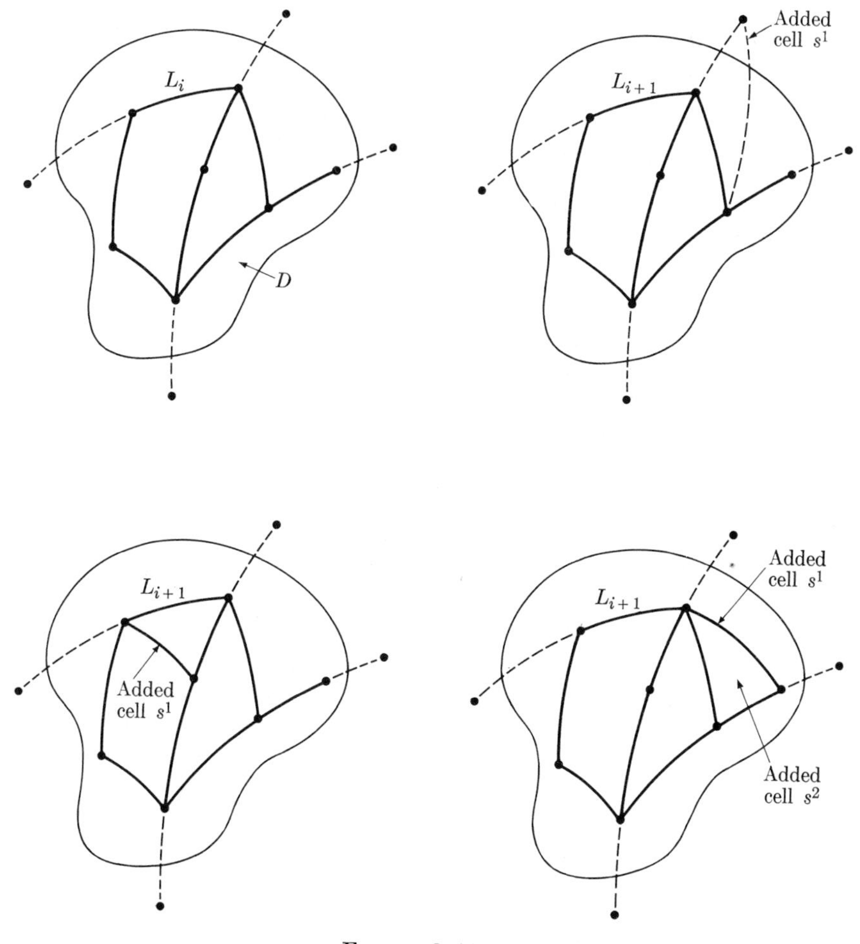

FIGURE 8–11

Each subcomplex L_i has its own chains, cycles, etc. Any cycle on L_i gives rise to a *cycle on* D as follows. If z_p^i is a cycle on L_i, then in every L_{i+j}, there is a cycle z_p^{i+j} obtained from z_p^i by subdivision. We define z_p^{i+j} to be equivalent to z_p^i. The collection of all such equivalent cycles $\{z_p^i, z_p^{i+1}, \ldots\}$ is said to be a cycle on D and is denoted by $[z_p^i]$.

Addition of two cycle $[z_p^i]$ and $[z_p^k]$ with, say, $k > i$, is defined by forming the subdivision of z_p^i in L_k and adding this to z_p^k. Then that each L_i is a complex implies that the collection of all p-cycles on D constitutes a group $Z_p(D)$ under this addition.

A cycle $[z_p^i]$ of D is a *bounding cycle* of D if there is some $k \geqq i$ such that the subdivision of z_p^i in L_k bounds in L_k. Again we have a group $B_p(D)$ of all bounding p-cycles, and $B_p(D)$ is a subgroup of $Z_p(D)$. The homology group $H_p(D) = Z_p(D) - B_p(D)$ is defined as usual. The pth *Betti number of the open set* D, $r_p(D)$, is the number of generators of $H_p(D)$ and may be infinite.

Since $H_p(D)$ is constructed by means of an infinite process, it is not always easy (or even possible) to compute $H_p(D)$ for a given set. Some useful results will be obtained, however.

THEOREM 8–24. If D is an open subset of S^n, then the number of components of D is $r_0(D)$.

Proof: Let the components of D be $C_0, C_1, C_2, \ldots$ (possibly infinite in number), the total number being $N + 1$ (or ∞). For each $k = 0, 1, \ldots, N$, let s_k^0 be a vertex of some K_i in C_k. Then the 0-cycles $z_0^k = 1 \cdot s_0^0 + 1 \cdot s_k^0$ are nonbounding. Also the cycle $z^0 = 1 \cdot s_0^0$ is nonbounding. Furthermore, no nonzero linear combination of these cycles z_0^k bounds. For in such a combination $z_0 = z_0^{k_1} + z_0^{k_2} + \cdots + z_0^{k_r}$, there is at most one vertex in each component C_k. If $z_0 = \partial c_1$ in some L_i, and if s_k^0 is in z_0, then s_k is the boundary of the chain formed by all the 1-cells of c_1 lying in C_k. But no 1-chain has an odd number of vertices in its boundary. This proves that $r_0(D) \geqq N + 1$.

Next, let $z_0 = \sum s_i^0$ be a 0-cycle. For each k, let z_0^k be the chain of all 0-cells of z_0 lying in C_k. There is a polygonal arc from the previously chosen vertex s_k^0 in C_k to each 0-cell $s_{k_j}^0$ in z_0^k; the union of these arcs is a connected compact subset T_k of D such that T_k contains s_k^0 and all the 0-cells of z_0^k. Since there is at most a finite number of 0-cells in z_0, we need consider only a finite number of sets T_k. The distance $d(T_k, S^n - D)$ is a positive number, say, η_k. Let K_j be a subdivision of S^n of mesh less than the minimum of the numbers $\eta_k/2$ and such that all vertices s_k^0 and all those vertices in z_0 are in K_j. Then in each C_i, the union of the cells of K_j that intersect the set T_k is connected, and these cells plus all their boundary cells form a connected subcomplex L_j^k containing the vertex s_k^0 and all the vertices $s_{k_j}^0$ in C_k. Each cycle $1 \cdot s_k^0 + 1 \cdot s_{k_j}^0$ then bounds in L_j^k; let c_1^{kj} be a 1-chain in L_j^k having $1 \cdot s_k^0 + 1 \cdot s_{k_j}^0$ for its boundary. Then we have

$$z_0 + \sum \partial c_i^{kj} \sim z_0.$$

In short, every 0-cycle z_0 is homologous to a sum of the cycles z_0^k previously chosen (that is, $z_0^0 = 1 \cdot s_0^0$ and $z_0^k = 1 \cdot s_0^0 + 1 \cdot s_k^0$). Therefore $r_0(D) \leqq N + 1$. □

The dimension $n - 1$ also interests us.

THEOREM 8–25. If z_{n-1} is a cycle on some K_i, then there are exactly two chains c_n^1 and c_n^2 in K_i such that $\partial c_n^1 = z_{n-1} = \partial c_n^2$. Furthermore, the carriers $|c_n^1|$ and $|c_n^2|$ intersect in the carrier $|z_{n-1}|$.

Proof: Since $r_{n-1}(K_i) = 0$, we know that z_{n-1} bounds a chain c_n^1 of K_i. Also we know that there is a fundamental n-cycle z_n on K_i. We let $c_n^2 = z_n + c_n^1$. Then $\partial c_n^2 = \partial z_n^2 + \partial c_n^1 = 0 + \partial c_n^1 = z_{n-1}$, so there are at least two n-chains bounded by z_{n-1}.

Suppose that there is a cell s^k in $|c_n^1| \cap |c_n^2| - z_{n-1}$. Choose a point p in the (relative) interior of s^k. Let U be a spherical neighborhood of p whose closure $\overline{U}$ does not meet $|z_{n-1}|$. The boundary of U is an $(n - 1)$-sphere S^{n-1}, which we may take so as not to contain a vertex of K_i. The intersection of S^{n-1} with a cell of K_i is then a cell of S^{n-1} of one lower dimension. Also there are $(n - 1)$-cells in this intersection which come from n-cells of both c_n^1 and c_n^2. Let ${}_*c_{n-1}^1$ and ${}_*c_{n-1}^2$ be the corresponding $(n - 1)$-chains on S^{n-1}. Since $\partial c_n^1 = z_{n-1}$ and $S^{n-1} \cap |z_{n-1}|$ is empty, we have that $\partial {}_*c_{n-1}^1 = 0 = \partial {}_*c_{n-1}^2$. Therefore ${}_*c_{n-1}^1$ and ${}_*c_{n-1}^2$ are nonintersecting $(n - 1)$-cycles on S^{n-1}. But this is impossible since $r_{n-1}(S^{n-1}) = 1$.

Finally, if there were a third chain c_n^3 with $\partial c_n^3 = z_{n-1}$, then $c_n^1 + c_n^3$ and $c_n^2 + c_n^3$ would be independent n-cycles of K_i, contradicting the fact that $r_n(K_j) = 1$. □

COROLLARY 8–26. If x is any point of S^n, and if z_{n-1} is a cycle of the open set $S^n - x$, then z_{n-1} bounds in $S^n - x$. Thus $r_{n-1}(S^n - x) = 0$.

Proof: In every subdivision K_i of S^n, one of the two chains c_n^1 and c_n^2 on which z_{n-1} bounds can always be taken to lie in $S^n - x$. □

THEOREM 8–27. Let x and y be distinct points of S^n. Then

$$r_{n-1}(S^n - x - y) = 1.$$

Proof: Let $d(x, y) = \eta$, and choose K_i such that all cells of K_i are of diameter less than $\eta/2$. Let H be the subcomplex of K_i composed of all n-cells whose closure contains x and of their faces. Then there is an n-chain c_n whose carrier is $|H|$. Let $z_{n-1} = \partial c_n$. In K_i, the cycle z_{n-1} also bounds a chain k_n not intersecting c_n. But the carrier $|c_n|$ contains x, and $|k_n|$ contains y. Therefore no chain that does not have a carrier containing either x or y can have z_{n-1} for its boundary. Hence

$$r_{n-1}(S^n - x - y) \geqq 1.$$

Then suppose that there were two cycles z_{n-1}^1 and z_{n-1}^2 in $S^n - x - y$. Then there are chains c_n^1 and c_n^2 such that both $|c_n^1|$ and $|c_n^2|$ contain x; and $\partial c_n^1 = z_{n-1}^1$ and $\partial c_n^2 = z_{n-1}^2$. We may choose i so large that the subcomplex H defined above does not intersect $|z_{n-1}^1| \cup |z_{n-1}^2|$. Letting $z_{n-1} = \partial c_n$ as in the previous paragraph, we have that $z_{n-1} + z_{n-1}^1 = \partial(c_n + c_n^1)$ and $z_{n-1} + z_{n-1}^2 = \partial(c_n + c_n^2)$. Now the point x is not in the carriers $|c_n + c_n^1|$ and $|c_n + c_n^2|$, so adding the chains $z_{n-1} + z_{n-1}^1$ and $z_{n-1} + z_{n-1}^2$, we obtain $z_{n-1}^1 + z_{n-1}^2 = \partial(c_n^1 + c_n^2)$, and neither x nor y is in the carrier $|c_n^1 + c_n^2|$. Thus z_{n-1}^1 and z_{n-1}^2 are homologous in $S^n - x - y$. Hence

$$r_{n-1}(S^n - x - y) = 1. \square$$

We come now to one of the most important results in this section and, indeed, in the topology of the n-sphere.

THEOREM 8–28 (Alexander addition). Let A and B be closed subsets of S^n, and let z_r, $r < n - 1$, be a cycle of $S^n - (A \cup B)$. Suppose there are chains c_{r+1}^1 of $S^n - A$ and c_{r+1}^2 of $S^n - B$ such that $\partial c_{r+1}^1 = z_r = \partial c_{r+1}^2$, and that there exists a chain k_{r+2} of $S^n - (A \cap B)$ such that $\partial k_{r+2} = c_{r+1}^1 + c_{r+2}^2$. Then z_r bounds in $S^n - (A \cup B)$. For the case $r = n - 1$, if $A \cap B$ is not empty, and if the chains c_{r+1}^1 and c_{r+1}^2 exist as before, then again z_r bounds in $S^n - (A \cup B)$.

Proof: Since k_{r+2} lies in $S^n - (A \cap B)$, no cell in its carrier $|k_{r+2}|$ intersects $A \cap B$. Hence the sets $A' = A \cap |k_{r+2}|$ and $B' = B \cap |k_{r+2}|$ are disjoint closed subsets of S^n, and $d(A', B') = \eta > 0$. Select a subdivision K_i with mesh less than $\eta/2$ and such that K_i contains a refinement of all the chains mentioned in the hypotheses. Let $k_{r+2}(A)$ be the chain of all $(r + 2)$-cells meeting A'. Then the carrier $|k_{r+2} + k_{r+2}(A)|$ lies in $S^n - A$. Let $\gamma_{r+1} = \partial k_{r+2}(A)$. We may write the equation

$$\partial(k_{r+2} + k_{r+2}(A)) = (c_{r+1}^2 + \gamma_{r+1}) + c_{r+1}^1. \tag{1}$$

Since $\partial\partial = 0$, our hypotheses imply

$$\partial(c_{r+1}^2 + \gamma_{r+1}) + \partial c_{r+1}^1 = 0; \qquad \partial c_{r+1}^1 = z_r. \tag{2}$$

That is,

$$\partial(c_{r+1}^2 + \gamma_{r+1}) = z_r. \tag{3}$$

Now $|k_{r+2} + k_{r+2}(A)|$ is in $S^n - A$. Hence by (1), $c_{r+1}^2 + \gamma_{r+1}$ is a chain of $S^n - A$. By our choice of K_i, $|\gamma_{r+1}|$ is in $S^n - B$, and by hypothesis, $|c_{r+1}^2|$ is in $S^n - B$. Hence $|c_{r+1}^2 + \gamma_{r-1}|$ is in $S^n - (A \cup B)$, and from (3), z_r bounds in $S^n - (A \cup B)$.

In case $r = n - 1$, either $c_n^1 = c_n^2$ or $|c_n^1| \cup |c_n^2| = S^n$ by our preceding results. In the second case, either $|c_n^1|$ or $|c_n^2|$ meets $A \cup B$, contradicting the hypotheses. And if $c_n^1 = c_n^2$, then c_n^1 is in the set $S^n - (A \cup B)$. □

Our first application of the Alexander addition theorem is in a proof of the fact that the n-sphere S^n has the *Phragmen-Brouwer properties*, which are listed following the proof of the next result.

THEOREM 8–29. If A and B are disjoint closed subsets of S^n, $n > 1$, and if neither A nor B separates the point x from the point y in S^n, then $A \cup B$ does not separate x from y in S^n.

Proof: Let A and B be closed disjoint sets in S^n, neither of which separates the point x from the point y. Let η be a positive number so small that no point of $A \cup B$ is within a distance η of $x \cup y$. Choose a subdivision K_i of mesh $<\eta/2$. Let s_x^0 and s_y^0 be vertices of the cells of K_i containing x and y, respectively. Clearly, neither A nor B separates s_x^0 from s_y^0 in S^n. Since $1 \cdot s_x^0 + 1 \cdot s_y^0$ is a cycle in one component of $S^n - A$, it bounds a 1-chain c_1^1 in $S^n - A$ and, similarly, it bounds a chain c_1^2 in $S^n - B$. Then $c_1^1 + c_1^2$ is a cycle, and since $n > 1$, $c_1^1 + c_1^2$ bounds a chain k_2 and, trivially, k_2 lies in $S^n - (A \cap B)$. By Theorem 8–28, $1 \cdot s_x^0 + 1 \cdot s_y^0$ bounds in $S^n - (A \cup B)$, thus implying that s_x^0 and s_y^0 lie in one component of $S^n - (A \cup B)$. It follows that x and y also lie in one component of $S^n - (A \cup B)$. □

The Pragmen-Brouwer properties are special connectivity properties, all of which hold for the n-sphere. They are defined as follows for any space S.

Property 1. If A and B are disjoint closed subsets of S, and if x and y are points of S such that neither A nor B separates x from y in S, then $A \cup B$ does not separate x from y in S.

Note that Theorem 8–29 says that S^n, $n > 1$, has Property 1.

Property 2 (Phragmen-Brouwer). If neither of the disjoint closed subsets A and B of S separates S, then $A \cup B$ does not separate S.

Property 3 (Brouwer). If M is a closed and connected subset of S, and if C is a component of $S - M$, then the boundary of C is closed and connected.

Property 4 (unicoherence). If A and B are closed connected subsets of S, and if $S = A \cup B$, then $A \cap B$ is connected.

Property 5. If F is a closed subset of S, and if C_1 and C_2 are disjoint components of $S - F$ having the same boundary B, then B is closed and connected.

Property 6. If A and B are disjoint closed subsets of S, if a is a point in A, and if b is a point in B, then there exists a closed connected subset C of $S - (A \cup B)$ which separates a from b.

The following sequence of theorems (8–30 through 8–35) may be proved as exercises by the interested reader. The proofs may also be found in Chapter II of Wilder [42].

THEOREM 8–30. If the space S is connected and locally connected, then Properties 1 and 2 are equivalent.

THEOREM 8–31. If the space S is connected and locally connected, then Properties 1 and 3 are equivalent.

THEOREM 8–32. If the space S is connected and locally connected, then Properties 1, 4, and 5 are equivalent.

THEOREM 8–33. If the space S is connected and locally connected, then Property 6 implies Property 4.

THEOREM 8–34. If the normal space S is connected and locally connected, then Property 3 implies Property 6.

THEOREM 8–35. If a metric space is connected and locally connected and has one of the Properties 1, 2, . . . , 6, then it has all of the other properties.

We conclude from Theorem 8–29 that the n-sphere S^n, $n > 1$, has all the Phragmen-Brouwer properties listed above.

THEOREM 8–36. Let c^k be a homeomorph of the closed k-cell I^k imbedded in S^n. The $r_p(S^n - c^k) = 0$ for all $p > 0$.

Proof: We give an inductive argument. If $k = 0$, then c^0 is a point x, and for $p = n - 1$ we have $r_{n-1}(S^n - x) = 0$ by Corollary 8–26. Suppose that $p < n - 1$. There is no loss of generality in assuming that the point x lies in some open n-cell of every subdivision K_i of S^n. Let z_p be a cycle of $S^n - x$ in some K_i. Then z_p bounds in K_i. But if $z_p = \partial c_{p+1}$, then every cell in the carrier $|c_{p+1}|$ is of dimension $\leqq n - 1$, and $|c_{p+1}|$ does not contain the n-cell of K_i that contains x. Thus c_{p+1} is a chain of $S^n - x$.

Now suppose that the theorem is true for every topological closed cell of dimension less than k. We may decompose the cell c^k into two n-cells $c_1^k \cup c_2^k$, with $c_1^k \cap c_2^k = c^{k-1}$, a topological closed $(k-1)$-cell. If z_p is a nonbounding cycle of $S^n - c^k$, then either z_p fails to bound in $S^n - c_1^k$ or z_p fails to bound in $S^n - c_2^k$. For if this were not true, then $z_p = \partial c_{p+1}^1$ in $S^n - c_1^k$ and $z_p = \partial c_{p+1}^2$ in $S^n - c_2^k$. Then $c_{p+1}^1 + c_{p+1}^2$ is a cycle of $S^n - (c_1^k \cap c_2^k) = S_n - c^{k-1}$. By our induction hypothesis, $c_{p+1}^1 +$

$c^2_{p+1} = \partial k_{p+2}$ in $S^n - c^{k-1}$. Thus the conditions of Theorem 8-28 are satisfied, and z_p bounds in $S^n - (c^k_1 \cup c^k_2) = S^n - c^k$, a contradiction.

Repetition of this argument establishes the existence of a decreasing sequence $c^k_1, c^k_2, \ldots$ of closed k-cells whose intersection $\cap c^k_j$ is a point x and such that for no j is z_p a bounding cycle in $S^n - c^k_j$. However z_p does bound in $S^n - x$, as we showed above. Thus there is a chain c_{p+1} in $S^n - x$ such that $z_p = \partial c_{p+1}$. Now there is an open set U containing x and not meeting the closed carrier $|c_{p+1}|$, and there is a j sufficiently large so that c^k_j lies in U. Since $c^k_j \cap |c_{p+1}|$ is empty, it follows that z_p bounds in $S^n - \gamma^k_j$, a contradiction. Therefore z_p bounds in $S^n - c^k$. $\square$

If M is a closed subset of S^n, and if z_p is a cycle of $S^n - M$ which does not bound in $S^n - M$, then z_p is said to *link* M in S^n. Note that if M is not empty, then p cannot equal n because no n-chain on $S^n - M$ is a cycle.

THEOREM 8-37. Let Σ^k be a topological k-sphere in S^n. Then

$$r_p(\Sigma^k) = r_{n-p-1}(S^n - \Sigma^k) = 0 \quad (1)$$

and $(p \neq k)$.

$$r_k(\Sigma^k) = r_{n-k-1}(S^n - \Sigma^k) = 1 \quad (2)$$

Proof: We apply an induction on the dimension k. For $k = 0$, Σ^k is a pair of points. In Theorem 8-27, we proved that $r_{n-1}(S^n - \Sigma^0) = 1$. For $p > 0$, it requires only a simple modification of the proof of Theorem 8-27 to prove that $r_{n-p-1}(S^n - \Sigma^0) = 0$.

Now suppose that the theorem is true for all dimensions less than k. Let Σ^k be a k-sphere in S^n, and let $\Sigma^k = A \cup B$, where A and B are closed k-cells and $A \cap B$ is a $(k-1)$-sphere Σ^{k-1}. Let z_p be a cycle of $S^n - \Sigma^k$, $p \neq n - k - 1$. By Theorem 8-36, there is a chain c^1_{p+1} in $S^n - A$ and a chain c^2_{p+1} in $S^n - B$, such that $\partial c^1_{p+1} = z_p = \partial c^2_{p+1}$. Then $c^1_{p+1} + c^2_{p+1}$ is a $(p+1)$-cycle of $S^n - (A \cap B) = S^n - \Sigma^{k-1}$. By the induction hypothesis, $c^1_{p+1} + c^2_{p+1}$ does not link Σ^{k-1} since $p + 1 \neq n - (k-1) - 1$. Thus there exists a chain k_{p+2} in $S^n - \Sigma^{k-1}$, with $\partial k_{p+2} = c^1_{p+1} + c^2_{p+1}$. Then by the Alexander addition theorem, z_p bounds in $S^n - \Sigma^k$, which proves (1).

Also by our induction hypothesis, there is a cycle z_{n-k} that links Σ^{k-1}. Then the intersection $A' = |z_{n-k}| \cap A$ cannot be empty, for if it were, then z_{n-k} would not bound in $S^n - A$, contradicting Theorem 8-28. And in turn, this implies that z_{n-k} would bound in $S^n - \Sigma^{k-1}$, which contains $S^n - A$. Similarly, $B' = |z_{n-k}| \cap B$ is not empty. Since z_{n-k} is in $S^n - \Sigma^{k-1} = S^n - (A \cap B)$, A' and B' are disjoint. Let $d(A', B') = \eta > 0$, and take K_i to be a subdivision of S^n with mesh less than η. Let γ_{n-k} be the chain of all $(n-k)$-cells of z_{n-k} having at least one face in B'. Then we have

$$\partial\gamma_{n-k} = z_{n-k-1} \qquad \text{in } S^n - A \tag{a}$$

and

$$\partial(z_{n-k} + \gamma_{n-k}) = z_{n-k-1} \qquad \text{in } S^n - B. \tag{b}$$

Now if there were a chain c_{n-k} in $S^n - \Sigma^k$ with $\partial c_{n-k} = z_{n-k-1}$, we could apply Theorem 8–28 to obtain

$$\gamma_{n-k} + c_{n-k} \qquad \text{is a cycle bounding in } S^n - A \tag{c}$$

and

$$(z_{n-k} + \gamma_{n-k}) + c_{n-k} \qquad \text{is a cycle bounding in } S^n - B. \tag{d}$$

Hence each would bound in $S^n - \Sigma^{k-1}$. But adding (c) and (d), we would have z_{n-k} bounding in $S^n - \Sigma^{k-1}$, a contradiction. We must conclude that z_{n-k-1} links Σ^k and hence that $r_{n-k-1}(S^n - \Sigma^k) \geqq 1$.

Suppose then that some other cycle γ_{n-k-1} also links Σ^k. We show that $z_{n-k-1} + \gamma_{n-k-1}$ bounds in $S^n - \Sigma^k$, which will show that $r_{n-k-1}(S^n - \Sigma^k) \leqq 1$. By Theorem 8–28 again, there are chains c^1_{n-k} in $S^n - A$ and c^2_{n-k} in $S^n - B$, with $\partial c^1_{n-k} = \gamma_{n-k-1} = \partial c^2_{n-k}$. Using Theorem 8–29, we see that the cycle $c^1_{n-k} + c^2_{n-k}$ links Σ^{k-1}. By the induction hypothesis, we have

$$c^1_{n-k} + c^2_{n-k} \sim z_{n-k} \qquad \text{in } S^n - \Sigma^{k-1}. \tag{e}$$

Then

$$\partial(\gamma_{n-k} + c^1_{n-k}) = z_{n-k-1} + \gamma_{n-k-1} \qquad \text{in } S^n - A \tag{f}$$

and

$$\partial[\gamma_{n-k} + z_{n-k} + c^2_{n-k}] = z_{n-k-1} + \gamma_{n-k-1} \qquad \text{in } S^n - B. \tag{g}$$

From (e) we obtain

$$\partial c_{n-k+1} = z_{n-k} + c^1_{n-k} + c^2_{n-k} \qquad \text{in } S^n - \Sigma^{k-1}. \tag{h}$$

Thus (f), (g), and (h) imply that

$$z_{n-k-1} + \gamma_{n-k-1} \sim 0 \qquad \text{in } S^n - \Sigma^k, \tag{i}$$

by the Alexander addition theorem. □

The above theorem is a special case of the Alexander duality theorem, which asserts that, for any closed polyhedron $|K|$ imbedded in S^n, $r_p(K) = r_{n-k-1}(S^n - |K|)$. Indeed, if we use the Čech homology theory, we can replace the polyhedron by any closed set. (See Borel [66].) It is also worth noting in this connection that the sphere Σ^k in the above theorem may be wildly imbedded as described in Section 4-6.

As a consequence of Theorem 8-37, we can prove the famous separation theorem of Jordan and Brouwer.

THEOREM 8-38 (Jordan-Brouwer separation). If Σ^{n-1} is a topological $(n - 1)$-sphere imbedded in S^n, then it separates S^n into exactly two components, of which Σ^{n-1} is the common boundary.

Proof: It follows from Eq. (1) of Theorem 8-37 that Σ^{n-1} separates S^n into two components, say A and B. Let x be a point of Σ^{n-1}, and let s^{n-1} be an open $(n - 1)$-cell in Σ^{n-1} containing x and having diameter less than ϵ. Then $X = \Sigma^{n-1} - s^{n-1}$ is a closed $(n - 1)$-cell in S^n, so $S^n - X$ is connected. In $S^n - X$, there is an arc L from a point of A to a point of B. Since L intersects Σ^{n-1}, it must do so in the cell s^{n-1}. Thus s^{n-1} contains a limit point of both A and B. Since ϵ is arbitrary, it follows that the point x is a limit point of both A and B. □

REFERENCES

Books

1. Alexandroff, P., *Combinatorial Topology,* vol. 1. Rochester, N.Y.: Graylock, 1956.

2. —, and Hopf, H., *Topologie.* Berlin: Springer-Verlag, 1935. (Ann Arbor: Edwards Brothers, 1945.)

3. Birkhoff, G., *Lattice Theory.* New York: A.M.S. Colloquium Publication No. 25 (1948).

4. Birkhoff, G. D., *Dynamical Systems.* New York: A.M.S. Colloquium Publication No. 9 (1927).

5. Bourbaki, N., *Topologie Générale.* Paris: Actualites Sci. Ind. 858 (1940), 916 (1942), 1029 (1947), 1045 (1948), 1084 (1949).

6. Chevalley, C., *Theory of Lie Groups.* Princeton: Princeton University Press, 1951.

7. Eilenberg, S., and Steenrod, N., *Foundations of Algebraic Topology.* Princeton: Princeton University Press, 1952.

8. Fraenkel, A., *Abstract Set Theory.* Amsterdam: North-Holland, 1953.

8(a). — and Bar-Hillel, Y., *Foundations of Set Theory,* Amsterdam: North-Holland, 1953.

9. Hall, D. W., and Spencer, G. L., *Elementary Topology.* New York: Wiley, 1955.

10. Halmos, P. R., *Finite Dimensional Vector Spaces.* Princeton: Van Nostrand, 1958.

11. —, *Introduction to Hilbert Space.* New York: Chelsea, 1951.

12. Hausdorff, F., *Mengenlehre.* Berlin: de Gruyter, 1927, 1935.

13. Hilton, P. J., *An Introduction to Homotopy Theory.* Cambridge: Cambridge University Press, 1953.

14. Hodge, W. V. D., *Harmonic Integrals,* 2nd ed. New York: Cambridge University Press, 1952.

14(a). Hu, S. T., *Homotopy Theory,* New York: Academic Press, 1959.

15. Hurewicz, W., and Wallman, H., *Dimension Theory.* Princeton: Princeton University Press, 1941.

16. Kaplan, W., *Advanced Calculus.* Reading, Mass.: Addison-Wesley, 1952.

17. Kelley, J. L., *General Topology.* New York: Van Nostrand, 1955.

18. Kuratowski, K., *Topologie,* vols. 1 and 2, 2nd ed. Warsaw: 1948.

19. Lefschetz, S., *Topology.* New York: A.M.S. Colloquium Publication No. 12 (1930).

20. —, *Algebraic Topology.* New York: A.M.S. Colloquium Publication No. 27 (1942).

21. —, *Topics in Topology.* Princeton: Princeton University Press, 1942.

22. —, *Introduction to Topology.* Princeton: Princeton University Press, 1949.

23. Moore, R. L., *Foundations of Point Set Theory.* New York: A.M.S. Colloquium Publication No. 13 (1932).

24. Newman, M. H. A., *Elements of the Topology of Plane Sets of Points.* Cambridge: Cambridge University Press, 1939.

25. Pólya, G., *How to Solve It.* Princeton: Princeton University Press, 1945.

26. Pontrjagin, L. *Topological Groups.* Princeton: Princeton University Press, 1939.

27. —, *Combinatorial Topology.* Rochester, N.Y.: Graylock, 1952.

28. Rado, T., *Length and Area.* New York: A.M.S. Colloquium Publication No. 30 (1948).

29. —, and Reichelderfer, P. V., *Continuous Transformations in Analysis.* Berlin: Springer, 1955.

30. Reidemeister, K. *Knotentheorie.* Berlin: Springer, 1932.

31. —, *Topologie der Polyeder.* Leipzig: Akademischer Verlag, 1938.

32. Schoenflies, A., *Die Entwicklung der Lehre von den Punktmannigfaltigkeiten,* II. Leipzig: Teubner, 1908.

33. Seifert, H., and Threlfall, W., *Lehrbuch der Topologie.* Leipzig: Teubner, 1934.

34. Sierpinski, W., *Introduction to General Topology.* Toronto: University of Toronto Press, 1934.

35. Steenrod, N., *The Topology of Fibre Bundles.* Princeton: Princeton University Press, 1951.

36. Thrall, R. M., and Tornheim, L., *Vector Spaces and Matrices.* New York: Wiley, 1957.

37. Vaidynathaswamy, R., *Treatise on Set Topology,* Part 1. Madras: Indian Mathematical Society, 1947.

38. Veblen, O., *Analysis Situs,* 2nd ed. New York: A.M.S. Colloquium Publication No. 5 (1931).

39. Wallace, A. H., *An Introduction to Algebraic Topology.* New York: Pergamon, 1957.

40. Whyburn, G. T., *Analytic Topology.* New York: A.M.S. Colloquium Publication No. 28 (1942).

41. —, *Topological Analysis.* Princeton: Princeton University Press, 1958.

42. Wilder, R. L., *Topology of Manifolds.* New York: A.M.S. Colloquium Publication No. 32 (1949).

43. —, *Introduction to the Foundations of Mathematics.* New York: Wiley, 1952.

44. —, and Ayres, W. L. (editors), *Lectures in Topology.* Ann Arbor: University of Michigan Press, 1941.

Papers, etc.

45. Alexander, J. W., "A proof and extension of the Jordan-Brouwer separation theorem," *Trans. A.M.S.* **23,** 333–349 (1922).

46. —, "An example of a simply connected surface bounding a region which is not simply connected," *Proc. Nat. Acad. Sci.* **10,** 8–10 (1924).

47. Alexandroff, P. S., "Über die Metrization der im kleinen kompakten topologischen Raume," *Math. Ann.* **92,** 294–301 (1924).

48. —, "Untersuchungen über Gestalt und Lage abgeschlossener Mengen beliebiger Dimension," *Ann. Math.* (*2*) **30,** 101–187 (1928).

49. ANDERSON, R. D., "On monotone interior mappings in the plane," *Trans. A.M.S.* **73,** 211–222 (1952).

50. ARTIN, E., and FOX, R. H., "Some wild cells and spheres in three-dimensional space," *Ann. Math.* **49,** 979–990 (1948).

51. BAIRE, R., "Sur les fonctions de variables réeles," *Ann. di Mat.* **3** (1899).

52. BANACH, S., "Über die Bairesche Kategorie gewissen Funktionenmengen," *Studia Mathematich* **3,** 174–179 (1931).

53. BEGLE, E. G., "The Vietoris mapping theorem for bicompact spaces," *Ann. Math. (2)* **51,** 534–543 (1950).

54. BERNSTEIN, F., "Zur Theorie der trigonometrischen Reihe," *Leipzig Bericht* **60,** 329 (1908).

55. BING, R. H., "A homogeneous indecomposable plane continuum," *Duke Math. J.* **15,** 729–742 (1948).

56. —, "A convex metric for a locally connected continuum," *Bull. A.M.S.* **55,** 812–819 (1949).

57. —, "Metrization of topological spaces," *Can. J. Math.* **3,** 175–186 (1951).

58. —, "Snake-like continua," *Duke Math. J.* **8,** 653–663 (1951).

59. —, "Concerning hereditarily indecomposable continua," *Pacific J. Math.* **1,** 43–51 (1951).

60. —, "Higher-dimensional hereditarily indecomposable continua," *Trans. A.M.S.* **71,** 267–273 (1951).

61. —, "A connected countable Hausdorff space," *Proc. A.M.S.* **4,** 474 (1953).

62. —, "Locally tame sets are tame," *Ann. Math.* **59,** 145–158 (1954).

63. BIRKHOFF, G. D., "On the combination of topologies," *Fund. Math.* **26,** 156–166 (1936).

64. —, and KELLOGG, O. D., "Invariant points in function spaces," *Trans. A.M.S.* **23,** 96–115 (1922).

65. BLANKINSHIP, W. A., and FOX, R. H., "Remarks on certain pathological open subsets of 3-space and their fundamental groups," *Proc. A.M.S.* **1,** 618–624 (1950).

66. BOREL, A., "The Poincaré duality in generalized manifolds, "*Mich. Math. J.* **4,** 227–240 (1957).

67. BORSUK, K., "Sur les retractes," *Fund. Math.* **17,** 152–170 (1931).

68. —, "Über Schnitte der n-dimensionalen Euclidischen Räume," *Math. Ann.* **106,** 239–248 (1932).

69. —, "Über die Abbildungen der metrischen kompakten Räume auf die Kreislinie," *Fund. Math.* **19,** 220–242 (1932).

70. BROUWER, L. E. J., "Beweis der Invarianz der Dimensionahl," *Math. Ann.* **70,** 161–165 (1911).

71. —, "Über den natürlichen Dimensionsbegriff," *J. für Math.* **142,** 146–152 (1913).

72(a). BROWN, MORTON, "A proof of the generalized Schoenflies Theorem," *Bull. A.M.S.* **66,** 74–76 (1960).

72. ČECH, E., "Theorie générale de l'homologie dans une espace quelconque," *Fund. Math.* **19,** 149–183 (1932).

73. DIEUDONNE, J., "Une généralization des espaces compacts," *J. Math. Pures Appl.* **23,** 65–76 (1944).

74. DOWKER, C. H., "Mapping theorems for noncompact spaces," *Am. J. Math.* **69,** 200–242 (1947).

75. EILENBERG, S., "Singular homology theory," *Ann. Math.* **45,** 407–447 (1944).

76. ERDOS, P., "The dimension of the rational points in Hilbert space," *Ann. Math.* **41,** 734 (1940).

77. FLORES, G., "Über n-dimensionale Komplexe die im R_{2n+1} absolute selbstverschlungen sind," *Ergebnisse eines mathematischen Kolloquium* **6,** 4–7 (1934).

78. FOX, R. H., "On topologies for function spaces," *Bull. A.M.S.* **51,** 429–432 (1945).

79. —, "Recent developments of knot theory of Princeton," *Proc. International Congress of Math.*, vol. 2, Cambridge, 453–457 (1950).

80. FREUDENTHAL, H. "Über die Klassen der Sphären-abbildungen I," *Composito Math.* **5,** 299–314 (1937).

81. HALL, D. W., and PUCKETT, W. T., "Conditions for the continuity of arc-preserving transformations," *Bull. A.M.S.* **47,** 468–475 (1941).

82. HARROLD, O. G., "Euclidean domains with uniformly abelian local fundamental groups, II," *Duke Math. J.* **17,** 269–272 (1950).

83. HOPF, H., "Über die Abbildungen der dreidimensionalen Sphäre auf die Kugelfläche," *Math. Ann.* **104,** 637–665 (1931).

84. HUREWICZ, W., "Über oberhalf-stetige Zerlegungen von Punktmengen in Kontinua," *Fund. Math.* **15,** 57–60 (1930).

85. —, "Beiträge zur Topologie der Deformationen. I. Höherdimensionale Homotopiegruppen," *Proc. Akad. Wetenschappen, Amsterdam* **38,** 112–119 (1935); "II. Homotopie und Homologiegruppen," *ibid.* 521–528; "III. Klassen und Homologietypen von Abbildungen," *ibid.* **39,** 117–126 (1936); "IV. Asphärische Räume," *ibid.* 215–224.

86. —, "On duality theorems," *Bull. A.M.S.* **47,** 562 (1941).

87. JONES, F. B., "A theorem concerning locally peripherally separable spaces," *Bull. A.M.S.* **41,** 437–439 (1935).

87(a). —, "Connected and disconnected plane sets and the functional equation $f(x) + f(y) = f(x + y)$," *Bull. A.M.S.* **48,** 115–120 (1942).

88. —, "Concerning normal and completely normal spaces," *Bull. A.M.S.* **43,** 671–677 (1937).

89. KELLEY, J. L., "The Tychonoff product theorem implies the axiom of choice," *Fund. Math.* **37,** 75–76 (1950).

90. KLEE, V. L., "Some topological properties of convex sets," *Trans. A.M.S.* **78,** 30–45 (1955).

91. KNASTER, B., "Un continu dont tout sous-continu est indecomposable," *Fund. Math.* **3,** 247 (1922).

92. —, and KURATOWSKI, C., "A connected and connected im kleinen point set which contains no perfect set," *Bull. A.M.S.* **33,** 106–109 (1927).

93. KURATOWSKI, C., "Sur les coupures irreducibles du plan," *Fund. Math.* **6,** 130–145 (1924).

94. —, "Sur le problème des courbes gauches en Topologie," *Fund. Math.* **15,** 271–283 (1930).

95. LEFSCHETZ, S., "The residual set of a complex on a manifold and related questions," *Proc. Nat. Acad. Sci.* **13,** 614–622 and 805–807 (1927).

96. —, "Chain-deformations in topology," *Duke Math. J.* **1,** 1–18 (1935).

97. LERAY, J., and SCHAUDER, J., "Topologie et equations functionelles," *Annales Scientifiques de l'École Normale Superieure* **51,** 45–78 (1934).

98. LERAY, J., "La théorie des points fixés et ses applications en analyse," *Proc. International Congress of Math.*, vol. 2, Cambridge, 202–208 (1950).

99. LUBANSKI, M., "An example of an absolute neighborhood retract, which is the common boundary of three regions in the 3-dimensional Euclidean space," *Fund. Math.* **40,** 29–38 (1953).

100. MAZUR, B., "On imbedding of spheres," *Bull. A.M.S.* **165,** 59–65 (1959).

101. MAZURKIEWICZ, S., "Sur les continus homogenes," *Fund. Math.* **5,** 137–146 (1924).

102. MENGER, K., "Über umfassendste n-dimensionale Mengen," *Proc. Akad. Wetenschappen, Amsterdam* **29,** 1125–1128 (1929).

103. MOISE, E. E., "An indecomposable plane continuum which is homeomorphic to each of its nondegenerate subcontinua," *Trans. A.M.S.* **63,** 581–594 (1948).

104. —, "Affine structures in 3-manifolds. VII. Invariance of the knot-type; local tame imbedding," *Ann. Math. (2)* **59,** 159–170 (1954).

105. MOORE, R. L., "An extension of the theorem that no countable point set is perfect," *Proc. Nat. Acad. Sci.* **10,** 168–170 (1924).

106. —, "Concerning upper semi-continuous collections of continua," *Trans. A.M.S.* **27,** 416–428 (1925).

107. —, "A connected and regular point set which contains no arc," *Bull. A.M.S.* **32,** 331–332 (1926).

108. —, "Concerning triods in the plane and the junction points of plane continua," *Proc. Nat. Acad. Sci.* **14,** 85–88 (1928).

109. NÖBELING, G., "Über eine n-dimensionale Universalmenge in R_{2n+1}," *Math. Ann.* **104,** 71–80 (1930).

110. NAGUMO, M., "A theory of degree of mappings based on infinitesimal analysis," *Am. J. Math.* **73,** 485–496 (1951).

111. —, "A note on the theory of degree of mappings in Euclidean spaces," *Osaka Math. J.* **4,** 1–9 (1952).

112. OXTOBY, J. C., and ULAM, S. M., "On the equivalence of any set of first category to a set of measure zero," *Fund. Math.* **31,** 201–206 (1938).

113. POINCARÉ, H., "Analysis situs," *J. de l'École Poly., Paris (2)* **1,** 1–123 (1895).

114. PONTRJAGIN, L., "Sur une hypothèse fondamentale de la théorie de la dimension," *Compt. rend.* **190,** 1105 (1930).

115. —, "Über den algebraischen Inhalt topologischer Dualitätssätze," *Math. Ann.* **105,** 165–205 (1931).

116. —, "The theory of topological commutative groups," *Ann. Math. (2)* **35,** 361–388 (1934).

117. ROBERTS, J. H., "Concerning atriodic continua," *Monatsh. Math. und Phys.* **37,** 223–230 (1930).

118. —, "Concerning collections of continua not all bounded," *Am. J. Math.* **52,** 551–562 (1930).

119. Seifert, H., and Threlfall, W., "Old and new results on knots," *Can. J. Math.* **2,** 1–15 (1950).

120. Serre, J-P., "Homologie singulière des espaces fibrés. Applications," *Ann. Math.* (*2*) **54,** 425–505 (1951).

121. Sierpinski, W., "Sur les espaces métriques localement séparables," *Fund. Math.* **21,** 107–113 (1933).

122. Smirnov, Yu. M., "On metrization of topological spaces," *Uspekhi Matem. Nauk* **6,** 100–111 (1951).

123. —, "A necessary and sufficient condition for metrizability of a topological space," *Doklady Akad. Nauk SSSR* **77,** 197–200 (1951).

124. Stone, A. H., "Paracompactness and product spaces," *Bull. A.M.S.* **54,** 977–982 (1948).

125. —, "Metrizability of unions of spaces," *Proc. A.M.S.* **10,** 361–365 (1959).

125(a). Treybig, B., "Concerning local separability in locally periphally separable spaces," *Proc. A.M.S.* **10,** 957–958 (1959).

126. Tychonoff, A., "Über einen Funktionenraum," *Math. Ann.* **111,** 762–766 (1935).

127. Urysohn, P., "Über metrization der kompakten topologischen Räume," *Math. Ann.* **92,** 275–293 (1924).

128. —, "Über die Mächtigkeit zusammenhängender Mengen," *Math. Ann.* **94,** 262–295 (1925).

129. Vietoris, L., "Über den höheren Zusammenhang kompakten Räume und eine Klasse von zusammenhangstreuen Abbilddungen," *Math. Ann.* **97,** 454–472 (1927).

130. Wallace, A. D., "The structure of topological semi-groups," *Bull. A.M.S.* **61,** 95–112 (1955).

131. Whitney, H., "On products in a complex," *Ann. Math.* **39,** 397–432 (1938).

132. Wilder, R. L., "The sphere in topology," *A.M.S. Semicentennial Publication*, vol. 2, New York, 136–184 (1938).

133. Yoneyama, K., "Theory of continuous sets of points," *Tohoku Math. J.* **11–12,** 43–158 (1917).

134. Young, G. S., "The introduction of local connectivity by change of topology," *Am. J. Math.* **68,** 479–494 (1946).

INDEX